Banditry, Rebellion and Social Protest in Africa

Banditry, Rebellion and Social Protest in Africa

Edited by DONALD CRUMMEY

James Currey
LONDON

Heinemann
PORTSMOUTH N.H.

James Currey Ltd
54b Thornhill Square, Islington, London N1 1BE, England

Heinemann Educational Books Inc.
70 Court Street, Portsmouth, New Hampshire 03801, USA

British Library Cataloguing in Publication Data

Banditry, rebellion and social protest in Africa.
  1. Social conflict — Africa — History
  I. Crummey, Donald
  303.6'096   HN 773

  ISBN 0-85255-004-9
  ISBN 0-85255-005-7 pbk

Library of Congress Cataloguing in Publication Data

Banditry, rebellion, and social protest in Africa.
  Includes bibliographical references and index.
  1. Social conflict — Africa — History — Addresses, essays, lectures.
  2. Brigands and robbers — Africa — History — Addresses, essays, lectures.
  3. Dissenters — Africa — History — Addresses, essays, lectures.
  4. Radicalism — Africa — History — Addresses, essays, lectures.
  5. Africa — Colonialization — Addresses, essays, lectures.
  I. Crummey, Donald.
  HN773.B36 1986   303.6'096   85-27352

  ISBN 0-435-08011-3

Typeset in 10/11pt Baskerville by Pen to Print, Colne
Printed in Great Britain

CONTENTS

SECTION IV: Rebellion

LIST OF MAPS

FOREWORD

This collection originated in a symposium on Rebellion and Social Protest in Africa which was held at the University of Illinois in Urbana-Champaign in April 1982. Not all the papers presented at the symposium have been published here. Some of the symposium papers which are published here have been revised extensively, while Terence Ranger's is completely different. Some of the papers which are published here, those by Robert Gordon, David Prochaska, and Cynthia Brantley, were not presented at the original symposium. Funding for the symposium came from a grant to the African Studies Program at Illinois from the US Department of Education under Title VI of the Higher Education Act. The program has subsequently supported this publication in other ways as well, in part with additional Title VI support. I am happy to acknowledge that support.

During the preparation of this book I have been conscious of the loss of four people, each in their own way associated with it. During the original symposium many of its participants first heard of the death of Thomas Hodgkin, whose inspiration informs much of what follows. A few months later Ruth First, a courageous practitioner and perceptive analyst of protest and resistance, died in the course of struggle. In April 1983 we lost Zekaria Abdulahi, a participant in the seminar which preceded the Urbana symposium, and an examplar of a whole generation of Ethiopian youth whose lives have been taken over by a consuming search for justice and a new social order. Within that generation Zekaria stood out in his commitment to reason and scholarship as inherent to the search. Finally, in September of that same year, one of our contributors, Richard Caulk, died. It was he who began the study of Ethiopian banditry. His loss, like that of the others, was grievous. I dedicate the book to their memory.

Donald Crummey
Urbana

NOTES ON CONTRIBUTORS

Ralph Austen is a Professor of African history at the University of Chicago. He has written extensively on African economic history, colonial history, and the Islamic slave trade, based partially on field research in Cameroon and Tanzania. His essay in this volume represents a transition in the focus of his work from political economy to culture.

Cynthia Brantley is an Associate Professor of history at the University of California, Davis. She received her PhD from UCLA in 1973. She has done field work in Kenya on a number of occasions. In addition to her monograph, *Giriama and Colonial Resistance in Kenya* (Berkeley, 1981) she has published articles in a number of scholarly journals.

Stephen Bunker teaches sociology at The Johns Hopkins University. He has done field research on peasants and rural development programmes in Uganda, Brazil, Guatemala, and Peru. In addition to monographs on Uganda and the Amazon, published in 1985 and forthcoming, he has contributed articles to journals in a variety of academic fields.

Richard Caulk died in 1983. At the time he was a member of the faculty of Rutgers University in Camden, New Jersey. He had received his PhD from the School of Oriental and African Studies in 1966, following which he taught in Addis Ababa for eleven years. He was the author of numerous articles on nineteenth and twentieth century Ethiopian social, political and diplomatic history.

Donald Crummey is Professor of African history and director of the African Studies Program at the University of Illinois at Urbana-Champaign. After receiving his PhD from the School of Oriental and African Studies in 1967 he taught at Addis Ababa University for six years. He has published two books and a number of articles dealing with the social, political and religious history of Christian Ethiopia.

Alison des Forges specializes in the study of states of the interlacustrine region. Her latest research in Rwanda focused on relationships between indigenous religions and the state in the eighteenth and nineteenth centuries. She has taught most recently at the Institutes of African Studies at Peking University and at the Chinese Academy of Social Sciences.

Timothy Fernyhough is a doctoral candidate in African history at the University of Illinois at Urbana-Champaign. He holds BA and MA degrees from London and Sussex universities respectively. He has lived and travelled extensively in Africa,

Latin America and Europe. His current research focuses on the interaction of land, state and society in southern Ethiopia since the nineteenth century.

Bill Freund is the author of *Capital and Labour in the Nigerian Tin Mines* (London, 1981) and *The Making of Contemporary Africa* (Bloomington, Indiana, 1984) as well as numerous shorter pieces on African history and politics. He has taught at Ahmadu Bello University in Nigeria and the University of Dar es Salaam in Tanzania. He is presently Professor of Economic History at the University of Natal in Durban, South Africa.

Robert Gordon teaches anthropology at the University of Vermont. He holds degrees from Stellenbosch University and the University of Illinois at Urbana-Champaign. He has done research in Namibia and Papua New Guinea, and is the author of *Mines, Migrants and Masters* (Johannesburg: 1977) and (with M. Meggitt) *Law and Order in the New Guinea Highlands* (Hanover, New Hampshire: 1985).

Ray Kea is Professor of African history at Carleton and St Olaf colleges in Northfield, Minnesota. He is a graduate of London University and the author of *Settlements, Trade and Polities in the Seventeenth Century Gold Coast* (Baltimore, 1982). He has done research in Ghana.

Shula Marks is Director of the Institute of Commonwealth Studies and Professor of Commonwealth history in the University of London. A past editor of the *Journal of African History*, and a member of the editorial board of the *Journal of Southern African Studies*, she has written widely on South African history.

David Prochaska is an Assistant Professor at the University of Illinois at Urbana-Champaign. He received his PhD from the University of California at Berkeley in 1981. His research interests centre on the social history of colonialism and the colonial sociology of knowledge. He has published several articles, and is writing a book on the social history of an Algerian town during the period of French settler colonialism.

Terence Ranger taught at the University College of Rhodesia and Nyasaland between 1956 and 1963 when he was deported. His early research, published in two books, dealt with African protest movements. Research in Zimbabwe in 1980 led to *Peasant Consciousness and Guerrilla War in Zimbabwe. A Comparative Study* (London, 1985). He is now researching the modern agrarian history of southern Matabeleland and is Professor of Modern History at the University of Manchester.

Allen Roberts, PhD (Chicago), is a social anthropologist with twenty years' experience in Africa. His research and writing deal with political economy, systems of thought, African art and applied social science, this last with emphasis upon renewable-energy initiatives in Africa. He teaches at Albion College, Michigan, and is engaged in research at the Center for Afroamerican and African Studies, the University of Michigan (Ann Arbor).

Leroy Vail has taught at the universities of Malawi and Zambia and has conducted research into the linguistics and history of southeast Africa. With Landeg White he has written *Capitalism and Colonialism in Mozambique* (London,

1980). He is presently an Associate Professor of African history at Harvard University.

Julia Wells holds a PhD in African History and Education from Columbia University — Teachers College. On several occasions she has conducted research in South Africa on black women's resistance. While in South Africa, she developed a textbook on women in Africa as part of an alternative curriculum for local black high school students, and she now administers a programme of study in the United States for black South Africans.

Landeg White has taught at the universities of the West Indies, Malawi, Zambia and Sierra Leone. He has published several books on literáture, both of the West Indies and Africa, and is, with Leroy Vail, the author of *Capitalism and Colonialism in Mozambique* (London, 1980). He is currently Director of the Centre of Southern African Studies at the University of York, England.

Larry Yarak is an Associate Professor of history at Texas A & M University. He received his PhD from Northwestern University for a dissertation on 'Asante and the Dutch: A Case Study in the History of Asante Administration, 1744–1872'. He has done research in Ghana, the Netherlands and the United Kingdom; and taught at George Williams College in Downers Grove, Illinois.

Introduction[1]: 'The great beast'

DONALD CRUMMEY

Writing of the 1960 Tiv rebellion in Nigeria, B. A. Nkemdirim dismisses many of the rebel actions as 'disorder' and 'sheer banditry', perpetrated by 'criminal hordes'. He quotes approvingly H. L. Nieburg's general comment that 'Beneath all the norms of legal, and institutional behaviour in natural societies lies the great beast, the people's capacity for outraged, uncontrolled, bitter and bloody violence.'[2] This book is about the great beast. However, its thrust differs sharply from that of Nkemdirim and Nieburg holding that, while the people indeed have a capacity for outrage and may resort to bitter and bloody violence, judgments about 'disorder' and 'uncontrol' betray more about the class prejudices of those who make them than they do about the people themselves. Underlying this book are different assumptions which require a critical examination of the meaning of banditry and criminality. The contributors to this volume believe that both the roots of popular violence and the shapes and direction which it takes will yield to critical examination. We also believe that popular violence needs to be seen in the light of other kinds of popular protest and politics. It especially needs to be seen in the context of state structures, for popular violence arises in situations where states claim a monopoly on the legitimate use of violence and direct it against the people. Most popular violence is a response to state or ruling-class violence. Indeed, there is another beast altogether.

Allen Roberts, in his chapter on the Tabwa in this collection, explores the ways in which this Zairian people remembers the period of its incorporation into the colonial state.[3] He recounts the conflicts and forms of political violence which arose in response to the demands of changing times and as a result of the ambitions of the Tabwa chiefs. The central metaphor and symbol whereby the Tabwa understand these processes is that of the 'beast': they believe their chiefs to have become lions, voracious, devouring creatures, who under cover of darkness overthrew the very legal system and political order which they daily maintained. The situation with which Roberts deals compels one to think of the incipient state, and of a situation of incipient class relations. The Tabwa, in their wrestling with the meaning and nature of the incipient state, produced an image to rival that of Nieburg's 'great beast': an equally arbitrary, oppressive and dangerous projection. This collection identifies more strongly with the Tabwa, and finds their image to reflect more accurately the experience of Africa's people at the hands of their various rulers. So this book is also about another beast: the state.

Popular history can resolve the rivalry between the conflicting beastly images. Popular history lies at the intersection of social and political history. Its subject is the people and their public activity. Popular history refuses to take rulers, African or otherwise, on their own terms, but tries to see them from the perspective of the

1

ruled. It defines the people as the producers of social wealth: tillers, herders, artisans, miners, factory workers, and transporters. These producers work within social structures of varying degrees of inequality and exploitation. Social history examines these structures and calls particular attention to the views and actions of those marginal to the male- and class-dominated institutions of traditional historiography: women, workers, peasants and criminals. In their responses to the social structures in which they found themselves, and in their efforts to bend or remake those structures along lines more congenial to themselves, 'marginal' groups engaged in diverse kinds of political activities. Historians of Africa have paid too little attention to these actions and have thereby ignored a vital factor in the history of the continent. In the following pages, the contributors to this book try to suggest some of the ways we can make up for this neglect.

Developments both in Africa and in African studies during the 1970s called into question some of the framework of the first generation of African studies. That generation closely identified with African nationalism and took for one of its seminal texts Thomas Hodgkin's *Nationalism in Colonial Africa*, first published in 1956. Conceptually that framework claimed for the new regimes of Africa legitimacy resting on their having led successfully a popular struggle to replace alien rule with indigenous rule. It also assumed that such regimes held the key to the continent's political and economic progress. Yet in the late 1960s developments began to challenge that framework. First, direct challenges to the legitimacy of postcolonial regimes came in the forms of revolts, dissident religious movements and coups.[5] Second, the level of political violence in Africa rose rapidly, following revolts in Rwanda (1959–63) and Biafra (1966). Finally, in quite different ways the passing of the Portuguese African empire and Haile Sellassie in Ethiopia brought revolutionary regimes to Africa, regimes which derived inspiration from new sources and exercised power through new institutions.

Meanwhile, frustrations in the development process also called for new concepts and themes of interpretation. Accordingly underdevelopment studies and class analysis have flourished. In this way the African studies field has opened itself to new, comparative perspectives. Such perspectives have come from the economic history of Latin America, from attempts to understand world history and economic development as a single process, from peasant studies in Latin America and Asia, and from the social history of Europe in the early modern period.

During the past ten years the chronological limits of earlier studies of resistance and popular politics have also emerged more sharply. Those studies were bounded too closely by European colonial rule. They excluded protest, resistance, and rebellion in precolonial Africa.[6] Secondly, they encouraged the identification of these themes rather too closely with the period of colonial rule as if Africans had had no previous experience of alien rule or oppression and no earlier traditions of protest and resistance. This identification in turn encouraged the assumption that European rule itself was the prime cause or sole focus of protest and resistance during the colonial period. Finally, it encouraged the interpretation of protest and resistance in the era of renewed independence as colonial hangovers. The combined effect exaggerated the importance of the colonial period, and rendered more difficult our understanding of colonial developments.

The way forward lies in combining the concerns of the studies of African resistance with the concerns and methodologies of radical social history. Resistance studies insist on the importance of politics and emphasise the element of political struggle, 'now hidden, now open', that is so important a feature of life in all social formations where rulers stand apart from their subjects. On the other hand, radical social history allows us to deal with the initiatives taken by the dominated and with the many ways in which they have organised their lives in areas which lay beyond their dominators' grasp. The combination of these two approaches will lead us to a fresh appreciation of violence; a subject to which we have not done full justice.[7] Neither the role of repressive state violence, nor that of popular violence has received adequate treatment from historians of Africa, although we must reject the contention that Africa is a continent intrinsically more violent than others, as A. A. Bozeman argues.[8] The real challenge is to see violence within its social setting, to appreciate its roots in social conflict, and to understand how and why the people turn to it.

The first two sections of this book look at the popular implications for Africa of a theme developed by historians of eighteenth-century England: criminality and its meaning for the social order within which it operates. Crime is inherently a form of protest, since it violates the law. Crime may be accompanied by many forms of consciousness. Where laws are clearly directed by one class against another, as is the case with laws inhibiting poaching or restricting access to forests or common lands, they engender a class-conscious defiance. Banditry is simply a form of criminality common to agrarian societies. The later sections of the book deal with more conventional cases of protest and resistance: boycotts and the like. Here we find an uneasy relationship between violence, authority, and the law, a relationship which collapses with the outbreak of rebellion, the form of protest and defiance with which we close our book.

Criminality

Engels claimed that theft is 'the most primitive form of protest'.[9] In a situation of radical inequality, with wealth on one side and poverty on the other, theft offers a direct path of redress and redistribution. Engels also argues that poverty and exploitation degrade, and that degradation brings moral decline and a rise in violence which afflicts the relations of workers with each other. In this he shared a growing assumption that poverty is a cause of crime. We find the same argument in Marx. In a famous passage he excoriates the process of primary accumulation in the development of capitalist society as criminal in character, and then argues that the immiseration so produced caused a further rise in crime as conventionally understood.[10] Elsewhere, in an article dealing with Germany in the 1840s, he developed a different line of argument. He saw the contemporary struggle over forest rights and the theft of wood as an aspect of class conflict, between the rural German working class and the industrialising bourgeoisie.[11] By continuing to exploit the resources of the forest in defiance of recent laws the workers resisted proletarianisation. Here Marx's key contribution of direct pertinence to Africa was his highlighting of the role that property law plays in the primary accumulation of capital, and of how resistance to such law, viewed as criminal by the authorities, marks a stage in the development of class society.

In the development of English class society the enclosure of common lands, the introduction of absolute private tenures, and the imposition of new kinds of taxes, combined to ruin the peasants and create a landed class of capitalists. In a valuable collection, historians of eighteenth-century England have shown how criminality reflects this process and reveals the nature of popular resistance to it.

Douglas Hay has shown how criminal laws of unbridled ferocity and an elaborate ritual of criminal justice maintained the class structure. He also investigates poaching by a rural population which viewed game laws as an abridgement of traditional rights and a threat to its standard of living.[12] Other essays in the same collection deal with riots, smuggling, plunder, and extortion, using these themes to throw light on popular attitudes and behavior. Many other works could be cited in this vein, not least E. P. Thompson's *Whigs and Hunters*.[13]

In spite of its rich potential, few of the insights and little of the methodology of this field have been applied to Africa, even in cases where Marx's original observations seem directly pertinent, such as the processes of primary accumulation associated with colonial rule and most particularly in the European colonies of white settlement. Stephen A. Lucas and J. M. Wasikhongo have tried to use crime to gain insights into the processes of rural and urban development, but their results are pretty thin, even if Wasikhongo's data on armed robbery in Nairobi suggest interesting possibilities.[14] Wasikhongo notes a link between crime and political process, but for him, as for the police officials whom he interviewed, it is a purely historical one, limited to the colonial era. Then 'an act of armed robbery involving violence would easily be classified as political terrorism and vise [*sic*] versa'. But, he continues, 'after independence most of the political activity either declined or totally stopped'.[15] Michael L. Stone's paper on poaching in Kenya focuses on the British administration, missing the element of incipient class conflict; an element which Jack Goody hints at in his study of the Green Revolution in northern Ghana, but fails to develop seriously.[16] One study which does make suggestive links between crime and political process is Lutakome A. Kayiira's dissertation on *kondoism* in independent Uganda. Kayiira develops a detailed picture of this form of gang robbery with violence, showing how its incidence fluctuated according to the political atmosphere. Unfortunately, he makes few additional cross-links to Ugandan social development, leaving the class dimension unacknowledged.[17]

The most ambitious, and disappointing, work on crime and society in Africa is the collection entitled *Crime and Punishment in South Africa*.[18] Apart from several pieces dealing with the Coloured community, the work contains few insights into the social conditions of South Africa's subject populations, and even fewer into their attitudes and actions. But South Africa is also the setting for the one truly exemplary piece which uses criminal data to explore the social milieu and adaptations of Africans, Charles van Onselen's: '"The regiment of the hills": South Africa's lumpenproletarian army 1890–1920'.[19] Van Onselen's paper is rich in information and insight. It argues that the criminal gang which forms the focus of the study possessed a consciousness of wider concerns, and is an example of resistance and revolt. Some of the papers below try to follow van Onselen's lead. But none of them deals with the urban setting which is so prominent in his study, and a large field awaits further work, cities like Salisbury, Luanda, Kinshasa (Leopoldville), Lagos, Ibadan, Accra, Kumasi, Dakar, Nairobi, and Addis Ababa offering extensive opportunities.

We offer three new studies of crime as a dimension of resistance and social protest. Larry Yarak uses Dutch records to look at the social context and meaning of crime on the Gold Coast between 1815 and 1830.[20] He discusses some twenty cases, mostly of theft and murder, and argues that his material affords insights into the class structure of Elmina under Dutch dominance. Theft allows valuable inferences about inequality in Elmina, but tells little about resistance. Rather different are his murder cases. Here he finds 'an unmistakable social dimension' in the evidence that murder expressed 'the tensions inherent in Elmina's class and ethnic divisions'. Particularly poignant and vivid are the details which Yarak offers on the anomic life of the slave Donkobo. Donkobo resisted the growing of export crops and plotted to murder his mistress. Sentenced to hard labour, he escaped and lived as a solitary eccentric until executed in 1829 for child murder.

William Freund takes as his text Engels' passage on crime as the earliest form of workers' rebellion against capitalism.[21] He qualifies Engels by arguing for the continuing significance of theft and by showing how it worked closely with more vocal forms of protest. He focuses on the Birom of Nigeria's Jos Plateau and their engagement with tin mining. Central to his analysis is the capitalist notion of property. On the one hand he argues that theft was a way in which Nigerian entrepreneurs tried to share in mine profits. On the other hand he argues that much 'theft' arose from a basic Birom rejection of capitalist notions of property, in the name of at least 'the germ of a populist "moral economy"'. In the former instance the 'criminal' character of the action seems less open to dispute, but in the latter case the charge of criminality is directly at issue. Freund demonstrates the coincidence of periods of tin theft and periods of heightened political consciousness. The underlying link is the Birom view 'that wealth has unjustly been taken from the people and appropriated by outsiders'.

Criminality is a concept upon which Allen Roberts does not directly call in his chapter on Tabwa resistance,[22] and yet it is pertinent to this theme, and further serves as a bridge to later sections of this collection. Like Freund, Roberts investigates a conflict of points of view, although his setting is markedly earlier and belongs much more to the era of the establishment of colonial rule than to Freund's developed colonial economy. Property is not at issue; authority and power are. Murder and terror are the instruments that Roberts discusses. He tries to plumb the Tabwa rationale for a series of attacks on missionaries, and mission-related chiefs, by lion-men between the 1890s and the 1930s. He notes the deeply ambiguous attitude which the Tabwa have towards law and order, an ambiguity rooted in their cosmology and manifested, in this case, in the symbol of the lion. As already remarked this seems a clear case of incipient criminality in the context of an incipient state. Moreover, the contradictions which Roberts finds in Tabwa attitudes towards law and order are probably much more widespread, and are certainly confirmed by my own work on the Amhara. By focusing on detailed cases from 1886 to 1896 Roberts also argues that while resistance to colonial intrusion was part of what was going on, the Tabwa 'lion' attacks arose primarily from local chiefly politics. In his emphasis on the local context he anticipates much recent work on rebellion. Finally, Roberts suggests that Tabwa lion-men may be viewed as social bandits. In so doing, he carries us ahead to Section II.

Banditry

Banditry is a form of criminality very widespread in agrarian class societies. It seems to have flourished from China and India through Turkey and the Mediterranean on to the Americas. It has inspired works of art and spawned novels, films, poems, and tales.[23] Eric Hobsbawm has caught the richness of the bandit and of his meaning for protest and resistance.[24] Hobsbawm pins his discussion on the Robin Hood legend and then considers more modern figures who act out similar scripts. He argues that we must distinguish two forms of banditry. The more common is simply criminal and venal. However, some bandits rise above preying on the industrious and productive and direct their activities to redressing the wrongs of the peasants, upon whom they depend for protection and support. Drawn by a dialectic of social demand and interdependence they find themselves acting as protectors, redistributors, and avengers. These are 'social bandits' and they reach their most glorious careers in moments of great social upheaval when they may become leaders of rebellions and even revolutions. The Mexican bandit, Pancho Villa stands at the centre of this apotheosis.

Hobsbawm has stimulated a lively debate. One line of development has explored the context of social banditry and has sharpened or refocused Hobsbawm's emphasis on peasants undergoing transformation by agrarian capitalism.[25] Another line, scrutinising more closely both the concept of the social bandit and the heroic claims made on its behalf, has concluded that the concept is mythical.[26] Yet even this line would leave us with a fascinating myth whose recurrence over centuries and on all continents still requires explanation.

Africanists have done little with bandits. Edmond Keller was first into the field with an article on Mau Mau, which he interpreted in the light of social banditry.[27] This interpretation has been effectively challenged.[28] Yet anyone familiar with the texture of Mau Mau, or the ethos of Bildad Kaggia's Nairobi, will readily grasp the rich potentiality which Mau Mau offers for exploring the links between criminality and rebellion.[29] Allen Isaacman returned to the theme, with an article which details the careers of two figures, Mapondera and Dambukashamba, active along the frontier between Portuguese and Rhodesian territory at the turn of the century.[30] He shows the origins of resistance in 1894 thanks to the pressures created by colonial conquest, and how Mapondera and Dambukashamba exploited the rugged terrain and a poorly defined border to harass the Portuguese. What he does not show is why the concept of social bandit is preferable to the concept of primary resistance in the understanding of their cases. Mapondera, who emerges as much the larger of the two heroes in Isaacman's account, was a traditional chief, who, 'well before . . . the imposition of colonialism . . . had gained an unparalleled reputation as the guardian of the "traditional" order'.[31] Only in the eyes of his would-be subjugators, who had yet to establish their own legitimacy, was there anything criminal in his actions.

Finally, Gervase Clarence-Smith has an account of banditry in the Huila highlands of southern Angola in the years from 1860 to about 1910.[32] Clarence-Smith calls this 'social banditry', following Hobsbawm, and with reference to the social dislocations which followed white settlement in the area and which gave rise to it. Struggles for land and cattle between Boers and Portuguese on the one hand, and the Nyaneka on the other, drove some Nyaneka peasants to the hills where soldiers and slaves joined them. These bands often recruited their leaders from chiefly families. Clarence-Smith distinguishes between social banditry and common banditry, and as an example of common banditry he discusses the case of Oorlog, who moved back and forth between police service and banditry with an ease and frequency which may amaze those less familiar with bandits elsewhere.

In spite of the problems in much of the literature, we should pursue the study of banditry. Contemporary African governments are quick to make pejorative use of the term. In October 1982, the Associated Press reported that the Ugandan government had launched a major offensive against unspecified guerrillas; noted that the government 'refers to guerrillas as "bandits"'; and cited high officials as describing the object of the offensive as stamping out 'banditry and thuggery'.[33] The Zimbabwe government similarly explained its security problems. A *New York Times* story of August 1982 reported that 1000 members of the ZAPU guerrilla forces had 'defected and taken to the bush as bandits or dissidents'; epithets which recurred almost daily in press reports.[34] Finally, neighbouring Mozambique also dismissed the actions of the opposition Mozambique National Resistance as 'banditry'.[35]

The pertinence of banditry to Africa has not escaped Hobsbawm. In the first edition of *Bandits* he drew attention to the activities of Ghanaian smugglers. In the second edition he vividly recounts the careers of the Masāzgi brothers of Eritrea, in his view social bandits if ever there were any. Hobsbawm could not have known of Mushala, headlined in the *New York Times* as '"Robin Hood" of the Zambians'.[36] Mushala perfectly exemplifies the social-bandit type. He began his career as a game ranger, but rebelled when his desire for promotion was blocked. He exploited remote and borderland territories, gathering a band of about two dozen. About 1975 he and his gang 'revealed themselves, depending on the sources, as plunderers and highway robbers, or as more benevolent figures taking from the rich what was to be given to the poor'. Poor people protected Mushala and attributed such magical powers to him as the ability to make himself invisible and to turn bullets into water. His death in an ambush late in 1982 coincided with a crime wave on the distant Copper Belt. This crime wave had been marked by bold daylight acts of robbery which had humiliated the Zambian police; so the officials made an example of the dead Mushala. Television gave prominence to his corpse, while it was publicly suggested that his body be frozen and taken round the country to prove his death.

We offer six new studies, inspired directly by Hobsbawm's discussion of banditry. Five of them are grouped to form the section on banditry, while the sixth, Terence Ranger's study of the Zimbabwe war of liberation, is included in the section on rebellion. These studies do not exhaust the influence of Hobsbawm's seminal study on this collection for a number of our other chapters also betray its influence. We have already noted that Roberts suggests that his Tabwa lion-men may have acted as social bandits. Richard Caulk, in his chapter on the 1894 rebellion against Italian rule in Eritrea, points out that Bāhta Hagos, the rebellion's leader, had begun his career as a bandit, and that his son later followed him in this path. Alison Des Forges finds bandits to have played an important role in the Rwandan revolt of 1912. Finally, Shula Marks sees one of the key figures in the Bambatha Revolt in Natal in 1906 as revealing 'many of the continuities between banditry, social banditry and resistance'.[37]

Our contributors found plenty of ordinary bandits, but with the exception of Ranger, few social ones. Ralph Austen offers a survey of the Africanist literature inspired by Hobsbawm's primitive rebels.[38] He shares my dissatisfaction with previous Africanist forays into social banditry and is very sceptical that the idea merits much application in Africa. Austen organises his discussion around the concept of 'heroic criminality'. He finds five types in Africa: the self-helping frontiersman; the populist redistributor; the professional underworld; the *picaro*; and the urban guerrilla. His first two categories reflect Hobsbawm's bandits and social bandits, and, while admitting that examples of both may be found, claims them to be of limited applicability. Austen also believes that African banditry is largely confined to Southern Africa where he sees it as a function of the frontier of white dominance. Austen's chapter is not limited to banditry. In his review of the professional underworld he stresses the importance of the state. Austen believes that picaresque figures of deep moral ambiguity are keys to an understanding of twentieth-century Africa. Lastly he points out that urban guerrillas have been a force of only minor importance in Africa, limited to Algeria, South Africa, and Ethiopia. He concludes with the suggestion that magic and witchcraft remain key terms for expressing and understanding social deviance in Africa, in this way returning to territory explored by Allen Roberts. He also affirms that his review vindicates 'Hobsbawm's fundamental project of understanding criminal deviance as an expression of positive alternatives to dominant social values, although often in archaic terms'. Based as it is on a review of the literature, Austen's chapter is necessarily limited to previous studies, and affected by the problem of negative evidence. He may be a little too pessimistic about the applicability of both bandit and social-bandit concepts to Africa; but his corresponding strength is that he is the only contributor to this collection to grapple theoretically with the challenge of segmentation and its implications for our understanding of the dynamics of social and political change in Africa.

Ray Kea's chapter on Gold Coast banditry offers a nuanced and detailed survey of brigandage and crimes against property during the seventeenth and eighteenth centuries.[39] He relates changing patterns of crime to changing forms of political economy, and to articulation with the world economy. Drawing on a rich variety of sources he establishes a vivid and colourful picture of endemic banditry and highway robbery in Akan country, emanating from both peasant and ruling-class milieus. He suggests a wider applicability of his bandit theme within the West African region, and in an earlier version of the paper, issued a challenge which still stands. Kea argued that it is historians, not bandits, who are lacking in West Africa. He believes that the literature on the precapitalist societies of West Africa ignores outlawry and social marginality because of the ideologically grounded refusal of its authors to consider these societies 'as peasant societies in which exploitative social relations existed or as societies capable of experiencing structural crises and tensions engendered by class conflicts'.

Ethiopianists are much inclined to agree with Kea, not only because they deal with what is manifestly just such a society as he describes, but also because they have seen only too clearly the effect of ideological restrictions on scholarly work. Ethiopian studies were conceived and born in the era of high imperialism, reached adolescence under Fascism, matured under an odd combination of an indigenous monarchy and a paternalistic liberalism, and now face the buffeting of institutionalised Marxism–Leninism. Ethiopianists also share with Kea a comparative wealth of historical material. In this they do differ markedly from scholars working on many other parts of the continent. Here we publish two studies of Ethiopian banditry: one by myself, the other by Tim Fernyhough.[40] My chapter deals with the nineteenth century, Fernyhough's with the twentieth; mine deals with the dialectical relationship between banditry and revolt, Fernyhough's with details of banditry itself. Both chapters agree that banditry was very widespread in highland Christian Ethiopia, of considerable social significance, and was dominated by the ruling classes. Thus they tend to reject the notion of social banditry, although Fernyhough, influenced by the Latin American literature, argues that banditry was an institution permitting social mobility, and my chapter documents the existence of the social–bandit myth with respect to the nineteenth-century emperor Tēwodros. The Ethiopian chapters cover the same geographical region, generalising about the entire area inhabited by Amharic and Tegreññā-speakers, but focusing for much of their detail on the province of Bēgamder and the territories around Lake Tānā. Together they, too, challenge the view that banditry is a phenomenon of little importance in Africa.

We conclude our section on banditry with Robert Gordon's study of the Bushmen and their relations with the state in twentieth-century Namibia.[41] He begins by arguing that the very name 'Bushman' meant, in origin, bandit; thereby dispelling romantic notions of Bushman isolation, and firmly drawing them into the history of the white-dominated Cape. He then turns to look at the interaction between the German administration and the Bushmen in the years before and into the First World War. At this time white settlers appropriated Bushman lands and water holes, and the Bushmen responded with cattle rustling and other acts of violence and harassment. At first the administration and settlers responded to Bushman resistance with a practice little short of genocide. The South African administration which followed the Germans was little different. By the Second World War appropriation of Bushman lands was complete, and the settlers and administration had joined hands to appropriate their labour. Administrative policy pursued this goal down to the present. Meanwhile other processes were at work which were to culminate in the 1980s in one of the classic bandit transitions, a process Gordon describes as praetorianisation. Their hunting skills still highly refined, the Bushmen are employed in pursuit of SWAPO guerrillas, and richly rewarded by Namibia's South African occupying force.

The argument of Sections I and II adds up to a formidable case in favour of further studies exploring what is conventionally viewed as criminality or other forms of social marginality, and suggests that such studies will bear rich fruit for deepening our understanding of protest and resistance. However, Section III equally suggests that there is much still to be gained by more-conventional and traditional treatments of these themes.

Protest and resistance

Resistance studies in Africa have a long lineage, but they experienced the arrival of a vigorous new age set in the 1960s in work dealing with East and Central Africa. Terence Ranger staked out much of the terrain with his *Revolt in Southern Rhodesia* and a widely discussed article in the *Journal of African History* which argued for a fresh assessment of movements of resistance to the establishment of colonial rule.[42] Ranger helped to make current the term 'primary' resistance, a term which gives European colonialism a determining place in the history of resistance in Africa. Ranger rehabilitated armed resistance as an important dimension of the African response to European colonial intrusion in the late nineteenth century, arguing that the developing literature on imperialism downplayed this factor. Then he tried to get people to look at this resistance in a new way. He disagreed diametrically with those who had argued that armed resistance was motivated by reactionary nostalgia and generally brought bad consequences to those who waged it. On the contrary, he averred, it was often progressive and forward-looking and brought tangible gains. Finally, he tried to demonstrate direct and concrete links between 'primary' resistance and the mass nationalism of the 1950s and 1960s. He succeeded in changing the terms of the debate.

Michael Crowder developed similar themes for West Africa, focusing rather more narrowly at first on cases of resistance to the establishment of European rule, but subsequently turning to studies of protest against an established colonial order.[43] Rather more than Ranger, Crowder uncritically professed the African nationalist beliefs underlying most early studies of resistance. He claimed that: 'European colonial rule did not win the permanent acceptance of any known African group'[44] but provided no justification for the claim. Moreover, his comments tended to juxtapose rather stark and crude alternatives of noble resistance and wicked collaboration.

Resistance studies have flourished as the books by Cynthia Brantley and E. I. Steinhart, to name only two, indicate.[45] Brantley and Steinhart register several distinct advances. Brantley extends the ‘range of resistance studies. Chronologically she encompasses both the nineteenth-century background and the maturing colonial order, since she places the climax of Giriama resistance in their rebellion of 1914. Thematically, she emphasises the economic impact of colonial rule as a factor leading to violent resistance. Steinhart's study ranges across a number of distinct societies in western Uganda and takes on a comparative dimension lacking in previous analyses. Besides, he is concerned with a continuum of responses to the establishment of colonial rule, ranging from military rejection to peaceful acceptance. Finally, he takes as a key point of departure for his analysis the internal divisions of African society into classes and strata.[46]

Barbara and Allen Isaacman sound similar notes in a general review of resistance and collaboration in Southern and Central Africa.[47] They join Steinhart in stressing the variety of African responses to the advent of colonial rule, drawing particular attention to the importance of collaboration. 'Privilege and exploitation,' they argue, 'were not the exclusive prerogative of the dominating racial minority.'[48] They, too, argue that internal African social differentiation into strata and classes was a major source of the variety of African responses to colonial rule. And with Brantley they extend the idea of resistance to cover African responses to the capitalist demands of an established colonial order. Such an extension I would push even further to embrace the economic demands of post-colonial African governments, in this way loosening the too close association of the notion of resistance with the period of European political domination of the continent. All periods of African history offer us examples of internal resistance: resistance to conquest; and popular resistance to unjust exactions by rulers.[49] We now have one substantial study of internal African resistance which, through its precolonial focus, marks a major conceptual breakthrough: Iris Berger's *Religion and Resistance. East African Kingdoms in the Precolonial Period.*[50] As we will see when we come to look at rebellion, Berger's concern with religion is particularly fruitful.[51]

Protest studies, so far as we can distinguish them from resistance studies, imply somewhat different contexts. Protest entails a higher degree of vocalisation. By contrast, resistance may appear mute, and stealth may be one of its essential features. Protest assumes some common social and political order linking protesters and those to whom the protesters appeal. Protesters tend to direct their energy to the redressing of grievances arising from within that order. Protest may take many forms: strikes, boycotts, campaigns of defiance, riots and 'disorders'. Recent work suggests no shortage of incidents. Three authors, all dealing with what is today Ghana provide a sketch of both the prospects for, and potential limitations of, protest studies. Stanley Shaloff looks at two sets of riots, one in 1931 and the other in 1932. He sets the riots in the context of the origins of nationalist politics, and explores their impact on attitudes and rivalries amongst the colonial African elite.[52] Unfortunately he tells us too little about the popular dimension to these riots, the 'people' acting here as a largely undifferentiated whole. Rather more penetrating and perceptive is Roger G. Thomas's study of popular protest at an increase in chiefly power in the north-eastern part of the colony during the First World War.[53] Equally interesting is the rich circumstantial detail which Jeff Crisp provides about the activities of miners at the Tarkwa goldfields. Crisp discusses a wide range of actions which gain coherence when seen from the perspective of the workers themselves, and directly challenges the notion that workers have little or no national political consciousness. He also offers a mature assessment of the limits of local protest.[54] Crisp goes on to make the obvious point, all too little recognised in practice, that strikes are incidents which should reveal major aspects of working-class formation and consciousness. While Southern Africanists have made headway with such questions, other regions of the continent are less well-served.[55]

We offer below four new contributions to the study of protest and resistance. Leroy Vail and Landeg White provide a fascinating look at protest and resistance in colonial Mozambique.[56] They draw on a source far-too-neglected by historians: popular songs. Some of these songs they collected themselves; others have been available to scholarship for over thirty years. Some of their songs are pithy and earthy and produced by peasants themselves; others are the elegant and refined productions of court performers. Yet both kinds yield complex and rewarding insights into how several different peoples of Mozambique perceived and responded to the colonial order. Vail and White bring out strongly the bitter hostility which Portuguese rule invoked; but equally stress the limits to popular consciousness as revealed in the songs. In central Mozambique, in the Zambesi Valley, they find a focus on either the immediate conditions of work or on the household effects of migrant labour. Amongst the Chopi they find a growing ethnic pride and a fairly willing embracing of migrant labour, although not of all of its conditions. Their probing of popular consciousness anticipates a growing theme amongst historians of African rebellions, as we will see below.

David Prochaska explores the relations between French colonisers and Muslim Algerians in the latter decades of the nineteenth century.[57] He sets the context in describing the precolonial economy of the forests and plains of the Annaba region. Then he recounts the massive expropriation of Algerian lands, particularly forest lands, by the French. Finally, he discusses the dramatic outbreaks of forest fires which followed the expropriations. Blessed with an abundance of materials – contemporary newspaper accounts, administrative reports, the observations of consular officials, petitions and court depositions, he is able to plumb both the incidents themselves and contemporary attitudes towards them. But the richness of his sources leads him to no simple conclusions. The settlers, against whom many of the fires were set, found many reasons to exaggerate their magnitude and malign the Algerians who probably started them. The Algerians, who probably started many of them, did not start them all, and had many reasons to deny their complicity. We are left on a note of ambiguity which reflects accident, and the uncertain relationship between protest and criminality. This ambiguity persists into contemporary Algeria.

Julia Wells looks at the resistance by black women in 1913 to the imposition of passes in the Orange Free State.[58] Her chapter deals with an urban environment, and with a fairly highly developed national market which drew on both male and female labour. Wells analyses a pattern of peaceful protest and defiance, inspired in part by contemporary British suffragettes; and examines the circumstances which made peaceful means successful.

Together with Cynthia Brantley, whose chapter on the Giriama is included in the final section of this book, Wells raises the extremely important question of the participation of women in movements of protest and resistance; and the related, but distinct, question of what insights a feminist perspective might offer on the themes of this collection. As Austen has remarked, one aspect of the continuing interest in banditry is 'a romantic view of masculine violence'. Only a careful attention to the role of women in the movements under review, together with an equal sensitivity to feminist reinterpretations, will allow us eventually to see how many of our other concerns arise from masculine bias, whether romantic or not. The literature on women in Africa is expanding rapidly and may in time produce substantial shifts in our perception of oppression in Africa and how people coped with it. However, as Margaret Stroebel observes, the current accounts of resistance and nationalism contain 'little mention of women's roles', roles which remain 'largely hidden'.[59] Yet we have a few glimpses of what we might expect. Two of our studies of rebellions, those by Des Forges and Brantley, emphasise the role of women as inspirers and organisers; and Des Forges is by no means alone in showing how their roles as diviners helped women to do this.[60] Ranger, in his chapter below, touches on the importance of young women as intermediaries between the Zimbabwe guerrillas and the civilian population. Judith van Allen rescued the Igbo 'Women's War' from earlier accounts of the 1929 'Aba Riots', uncovering effective political action by Igbo women and an incapacity on the part of their British rulers to understand what was happening openly in front of them.[61] Wells' chapter is only one of a growing number of studies of the role of women in resistance to racist and class oppression in South Africa.[62]

Finally several major African artists have taken up the theme of women and their potential for leadership in times of political crisis. Sembene Ousmane's *God's Bits of Wood* reaches its climax and resolution when the women affected by the 1947–8 Dakar–Niger railway workers' strike take matters into their own hands and march in protest to Dakar.[63] Sembene returned to the leadership potential of women with *Ceddo*, a film set in nineteenth-century Senegal. In this account the heroine's act of individual rebellion inspires her people to free themselves. Ngugi wa Thiong'o's most controversial novel, *Petals of Blood*, features as a central character, Wanja, a former bar girl who constitutes the book's dramatic and moral focus.[64] She too rebels against degradation and oppression. Romanticised as these accounts are, they indicate a sensitivity to the potential of women as actors in the political arena, a sensitivity lacking among most African historians.

Stephen Bunker returns the discussion to a rural milieu dominated by men. Bunker deals with peasants, cultivators significantly engaged with the production of commodities for the international market.[65] He takes the now unfashionable view that underdevelopment and dependency notwithstanding, African peasants enjoy autonomy in their relations with the state, whether colonial or nationalist. The economic basis for this autonomy is that the state's dependence on the revenues derived from taxing agricultural production exceeds the peasants' dependence on their cash earnings. The key to the difference lies in the peasants' 'exit option'. Peasants can cease to produce, relying instead on subsistence. The state cannot forgo revenue. Bunker then explores the political space opened by this option and shows how peasant demands and priorities shaped the careers of their leaders. He shows how violence directed against property, and the threat of such violence, further defined the options available to politicians. Like Prochaska, he deals with fire as a weapon of protest.

Bunker shows that protest was as alive in rural settings as in urban ones. His case also reminds us that protest may be as closely related to 'prosperity' as it is to deprivation. In its wider relations the protest analysed by Bunker articulated with nationalist politics in colonial Uganda. In none of these respects did it approach the scale or intensity of a classical rebellion.

Rebellion

Unlike banditry and criminality, rebellion as a form of resistance in Africa has been generously treated. The importance of rebellion has grown over the past fifteen years as regimes have emerged in the former Portuguese territories and in Zimbabwe through guerrilla struggle. Rebellion has also marked Chad, Namibia, Eritrea, and Western Sahara. Analysis of rebellion has grown in sophistication, and in its grasp of the complexity of forces involved. Without losing sight of the profound influence of European colonialism, analysis is starting to move beyond the confident anti-colonial explanations of earlier work. As we will see, one important breakthrough has come through an appreciation of the vital role played by such African religious practices as magic and divination in mobilising people for revolt and in sustaining them in combat. Fifteen years ago magic was a subject commonly passed over in political discussions. Yet the study of rebellion in Africa remains almost entirely concentrated in the last hundred years, bounded by the framework of colonial rule. African historians have yet to do justice to the level of social unrest and conflict in the preceding centuries.

Three areas leap to mind as fertile ground for the study of precolonial revolts: the western Sudan, Ethiopia, and North Africa. Walter Rodney suggested a broader framework for the interpretation of *jihads* in the Sudanic zone of West Africa, which no one has yet developed.[66] Basil Davidson has raised the question of the popular factor in the downfall of Songhay; and the troubles of Kanem in the fourteenth and fifteenth centuries would bear serious attention in this light.[67] Christian Ethiopia in the early seventeenth century was wracked by social unrest and peasant revolt, which scholarship continues to try to confine within a procrustean mould of religious controversy.[68] F. Lawson has provided a subtle and convincing account of three provincial revolts in Egypt in the early 1820s, which, together with other accounts, suggests that North Africa has many more examples to offer.[69] And finally, while we have little work to guide us, the history of precolonial African states in the forest areas of West Africa, in the Central African savanna and around the East African lakes, all ought to be looked at from this angle.

While there have been few studies of precolonial African rebellions, those rebellions which fit the colonial framework have continued to receive attention. The following crude typology exists: 'early' rebellions accompany the establishment of colonial rule. 'Post-pacification' revolts follow. Torpor marks the interwar years; but the late 1940s and the 1950s see some massive 'nationalist' revolts. Finally, in the 1960s we find 'post-colonial' revolts. The typology is useful, but exaggerates the peaceful nature of the interwar years, and overplays the immediate importance of European colonial rule as a causal factor.

Ranger's path-breaking study of the 1896–7 Ndebele and Shona revolt remains the leading study of 'early' rebellions, and became the point of departure for a sub-historiography.[70] In 1977 Julian Cobbing successfully challenged two of Ranger's major arguments. He showed that Ranger was mistaken in attributing either a central organisational role or a progressive ideology to certain religious mediums, and anachronistic in perceiving the rising as possessing a larger and more forward-looking consciousness than had previously characterised African responses to European intrusions in the area.[71] Underlying Cobbing's criticism is the charge that Ranger's original study was shaped too much by the African nationalist concerns of the 1960s. Two years later Cobbing was seconded by D. N. Beach, who returned to the attack. Beach demonstrated that Ranger had exaggerated the extent of Shona organisation and spontaneity in 1896 and had read a forward-looking ideology into the record where none was to be found. Like Cobbing, Beach finds the leaders and ideologues of the rising to have been thoroughly 'traditionalist' in outlook.[72]

Although much of Ranger's original study remains untouched by these criticisms, they are damaging. In this respect, *Revolt in Southern Rhodesia* has stood up less well than its counterpart in the preceding historiographical generation, Shepperson and Price's study of John Chilembwe, and his revolt in Malawi in 1915.[73] Extensive archival findings by later scholars have confirmed virtually every point of importance to Shepperson and Price, although Vail and White rightly take them to task for their very inadequate view of Chilembwe's relations with the local peasants, and their poor analysis of peasant behaviour during the revolt.[74] Another revision has emphasised the influence of millenarianism on the rising; but on the debit side has been some dubious psychohistory.[75]

John Chilembwe and his revolt carry us over into the turbulent early years of the twentieth century. They saw some major upheavals: Herero and Nama revolts in South West Africa in 1904; Maji Maji in Tanganyika in 1905; Bambatha in Zululand in 1906; Rwanda in 1912.[76] In his study of Maji Maji, John Iliffe struck a note which we will hear again and again in looking at rebellion when he successfully demonstrated the importance of religious leadership and the emergence of a transcendent ideology, in this case arising from earlier, anti-witchcraft movements, which temporarily created a new and broader African unity.[77] However, the close proximity in time of all these revolts raises intricate questions about causation, since their contexts were rather different. Maji Maji and Bambatha provide a useful contrast.

The origins of Maji Maji lie first in the German conquest of East Africa in the 1880s and 1890s, and more immediately in German administrative attempts to impose cotton growing on African cultivators. The origins of Bambatha date back to the 1830s and the Boer intrusion into Zululand. A prolongued process of land alienation ensued, accompanied by administrative subordination and labour extraction, a process which intensified with the final military subjugation of the Zulu kingdom in 1879.[78] Whereas Maji Maji raises the question of resistance to becoming peasants within a capitalist nexus, Bambatha raises the question of proletarianisation within the context of the combined pressure of settler colonialism and international mining capital. Where Maji Maji drew its inspiration and organisation from a tradition of witchcraft eradication movements, Bambatha raised the question of popular consciousness during a period of revolutionary transformation.

The First World War was a watershed in the colonial history of Africa. It ushered in a brief generation of European supremacy, and of mass African adjustment to the social and economic forces which that supremacy promoted. The war itself saw considerable social unrest in Africa, as Osuntokun has shown for West Africa.[79] In contrast, even in those parts of Africa outside the formal empires, the interwar years witnessed a reduction in the scale and intensity of social protest,[80] although in 1921 South Africa experienced the largest bloody confrontation between the government and its African subjects in all the years from the Bambatha rising to the late 1950s, and three years later saw a major insurrection by white mine workers.[81] However, the 1921 'Bullhoek massacre', in which some two-hundred members of a millenial sect were killed by the South African police, is more a case of state-initiated violence than of insurrection, since Enoch Mgijima's Israelites looked upon themselves as having withdrawn from the world, and took no steps directly to challenge their rulers.

Ethiopia between the wars was on an even more independent trajectory as factions of the country's nobility struggled amongst themselves to reconstitute the imperial institutions compromised by the death of Emperor Menilek II in 1913. Feudal wars and provincial revolts occurred, and the Italian invasion of 1935 brought the period to an end. However, with the exception of the occupation period which followed the invasion until 1941, we have little information about the popular dimension to these revolts.[82]

More typical of the interwar years in colonial Africa was the situation in the Kenyan district of Nandi in 1923. Here unrest built to a point just short of militant action. Discontents over increases in taxation, pressures for labour, restraints on movement, and the continuing effects of large-scale land alienation, channelled themselves through the traditional religious leader, the *orkoiyot*, and focused on a religious ceremony, which was to be the occasion for united resistance. The British anticipated Nandi resistance and forestalled it by arresting five key leaders, banning the ceremony and introducing reforms.[83] Nonetheless, even at the height of this era of 'improvement and differentiation', some colonial subjects resorted to insurrection.[84] Mahdism in the Sudan retained a tradition of violent defiance which declined in frequency as the Condominium wore on, but which still produced one sizeable revolt in Darfur in 1921.[85] In 1928 a rebellion broke out amongst the Gbaya of the present-day Cameroon–Central African Republic borderland, the largest insurrection in either Cameroon or French Equatorial Africa between the wars.[86] Deeply oppressed by a colonialism marked by forced labour and a ready use of violence, and by private concessionaires, the Gbaya responded to the call of Karnou, a religious prophet, who couched his message in the idiom of Gbaya traditional culture and institutions. The rebellion spread to a number of neighbouring peoples and burned on into the 1930s. In 1934 Burundi was torn by a major rebellion.[87] Once again religion, in the form of a prophetess, proved a major force. J.-P. Chrétien interprets this rebellion in the light of popular uprisings in medieval and early modern Europe. He convincingly shows that a determining role was played by colonial forces, particularly economic forces in the shape of the Great Depression. He provides a subtle account of how the Barundi peasants responded to these forces with a total rejection of colonial rule in the name of an idealised past. He leaves open the question of links to modern mass. nationalism.[88]

Not until the late 1940s do we find a popular rebellion which embodied a nationalist consciousness, and had links to a secular, nationalist political organisation; but even this rebellion, in Madagascar in 1947, and its kindred uprisings of the 1950s, was deeply imbued with attitudes, values and aims derived from the past, whether imagined or remembered. The 1940s actually opened with a major revolt in Tegrē against the restored regime of Haile Sellassie.[89] This rebellion, reminiscent of the earlier risings in Rwanda and Burundi, was known as Wayānē, and consumed the eastern half of the province for much of 1942. Wayānē had a strong popular dimension, and like some of the examples discussed by Hobsbawm drew on the resources of banditry, but as in many other feudal risings, the gentry and nobility assumed leadership roles, and directed the movement towards the goals of provincial autonomy and the restoration of a local dynasty. However, the Madagascar revolt of 1947 was characteristic of an attenuated colonial order.[90] This massive revolt burned intensely for only three months, yet left some 60 000 dead. J. Tronchon writes powerfully of the revolt, locating its roots firmly in the development of Malgache nationalism. He argues that the rebels were engaged in a war of national liberation. While the movement was overwhelmingly peasant in membership, it also drew on urban elements. Tronchon stands alone as a serious scholar of the Malgache rebellion. By contrast, Mau Mau has received a high degree of attention, and given rise to a large literature.[91]

Mau Mau was a peasant insurrection with some radical *petit-bourgeois* leaders and links to an urban trade-union movement. The colonial administration acting under pressure from the settler community triggered Mau Mau, but a foundation for the rising existed in the violent actions of defiance which the radical wing of the nationalist movement had already embarked upon, in both rural and urban areas. Peasant rebellion in Kenya drew considerable urban support, but the objectives of the forest fighters focused on access to land and a more equitable distribution of the colony's resources for the poorer African segments of society. Only from the standpoint of that settler community were these objectives revolutionary. Mau Mau confirmed the emerging African nationalist consensus by not challenging the capitalist nexus of property relations in Kenya. Mau Mau attitudes to Christianity were ambivalent, and a strong current of Kikuyu religious concepts and practices, not least of them oathing, inspired the movement and bound it together. So far as organisation is concerned, current scholarship on Mau Mau stresses a high degree of spontaneity and a weak overall structure. In this, as in a number of other respects, Mau Mau contrasts with nationalist insurgency in Cameroon.

In Cameroon in December 1956 the Union des Populations du Cameroun (UPC) launched a rebellion against French colonial rule.[92] The UPC under the leadership of Ruben Um Nyobe was a nationalist party, which had developed through conflict with the French. Its insurrection was crushed by the combined forces of French colonialism and conservative Africans. As an unequivocally nationalist rebellion, one in which the initiative was taken by the leadership of a political party, the Cameroon rising was unique in either French or British tropical Africa in the 1950s, and had to await the 1961 União das Poblações do Norte de Angola revolt in northern Angola before it was replicated. Thereafter the insurrections sparked by the PAIGG in Guinea and by Frelimo in Mozambique successfully developed this tradition, which was picked up yet again by the Zimbabweans at the end of the 1960s. There is irony here. African nationalist movements have looked back to the violent upheavals of the early colonial period for legitimation. Yet many of those same movements have, at best, an equivocal attitude towards violence in their own era. In Cameroon the nationalist regime of the Union Nationale Camerounaise led by Ahmadou Ahidjo emerged through the suppression of the popular UPC rebellion. KANU, the triumphant party of African nationalism in Kenya, has avoided too close an association with the memory of the Mau Mau struggles. And in some cases, as in Zaire, nationalist regimes have themselves provoked major rebellions.[93]

Zaire has had a turbulent history. The process of independence in that country was traumatic for the whole continent. Nor did independence bring social harmony. So grievous was the failure of the postcolonial elite, and so blatant its efforts at collective advancement, that in 1964 massive uprisings broke out in the name of a second independence. Not even Mau Mau has received quite the level of attention which the Zaire rebellions of 1964 have stimulated. The outstanding scholar of the rebellions is Benoit Verhaegen, whose two volumes are a monument to diligence and imagination.[94] Verhaegen's work covers all aspects of the rebellions: socio-economic background; political context; leadership; ideology; and military affairs. As in Mau Mau, we find a peasant movement heavily influenced by the *petit bourgeoisie*. This *petit bourgeoisie* was something of an intelligentsia, but its 'modern' tendencies in no way lessened the importance of indigenous religious beliefs and practices so far as the peasants were concerned. Medicine men accompanied the fighting units, providing inspiration, ideological purity, and invulnerability to bullets. A valuable aspect of Verhaegen's account is the sheer volume of original documents which he published, particularly those emanating from rebel hands. In this respect his work is virtually unique.

Verhaegen's study of the Zaire rebellions of 1964 dramatically shifts the analytical framework away from that of anticolonial resistance to that of peasant rebellion.[95] This allows analysis to draw on the work of such non-Africanists as Eric Wolf and James Scott.[96] A wider comparative framework is thereby gained, as well as a much greater sensitivity to internal differentiation within peasant society. Peasant studies emphasise the impact of movements within the world economy on peasant societies and reinforce the importance of indigenous religious traditions in inspiring, sustaining and limiting the insurrectionary potential of rural populations in Africa. Finally, the comparative study of peasant rebellion highlights the dramatic juxtaposition of revolution and reaction in peasant-based movements.

These themes dominate available interpretations of recent revolts in Africa, or seem applicable to them. Gebru Tareke has provided sensitive and convincing accounts of two provincial rebellions in Ethiopia in the later 1960s, in addition to his account of the Wayānē in Tegrē to which I have already referred.[97] His cases refer to Gojjām, an Amharā area where grievances focused on taxes, and to multi-ethnic Bālē, where land alienation and the rise nearby of an independent Somali Republic ignited popular discontents. J.-L. Amselle has produced an equally sophisticated analysis of a 1968 rebellion in Ouolossebougou, Mali.[98] Amselle points out how Malian peasants transferred their opposition from the colonial state to its nationalist successor. He argues convincingly that resistance to the state is a fundamental aspect of peasant consciousness. He also shows how peasants manipulate the past to serve their present ends. By contrast, in their account of the Agbekoya rebellion in western Nigeria, C. E. F. Beer and Gavin Williams stress the limitations of peasant action.[99] Agbekoya was a movement of small Yoruba farmers infuriated by a rise in taxes at a time of declining cocoa prices and growing alienation from the nationalist political elite of Nigeria's Western State. The Agbekoya rebels attacked palaces and district council offices, invaded Ibadan, set up road blocks, ambushed police, and engaged them in pitched battles. Following concessions from state officials, the organisation declined and its leaders were co-opted. Just as striking as these examples of stimulating and fruitful analysis are the cases which await development: Eritrea in the 1960s and 1970s;[100] the Burundi rebellion of 1972;[101] the Biafran revolt; and the Tiv rebellion of 1960.[102]

Richard Caulk and Alison Des Forges reflect the concerns of recent peasant studies in their chapters below.[103] Both explore their local class contexts, of northern Rwanda and of northern Ethiopia, and place their rebellions firmly within them, showing how inherited class relations created the tinder which colonial interventions set alight. They further show how those class relations constrained the insurrections, although in different ways. Caulk's central figure is Bāhta Hagos, a ruling nobleman, who in 1894 defiantly opposed Italian threats to peasant land rights in Eritrea. Bāhta's revolt failed, Caulk argues, because the peasants did not trust Bāhta, with good reason. His own exercise of power had oppressed them. Des Forges argues that the discontents of cultivators in Rwanda had been created by the Rwanda nobles through land alienation and degradation of the cultivators' status. She points out that the Rwanda rebels focused their discontents primarily on their own monarch, rather than on his German 'protectors', thereby blurring the anti-colonial focus of this revolt. Both Caulk and Des Forges argue that earlier experiences of resistance and rebellion moulded popular actions, and indicate that banditry played a considerable role in their stories. In the Eritrean case, banditry had been a major path to office for aspirant gentry, such as Bāhta, or a source of support for officeholders dislodged from their offices. In Rwanda the Abatwa ethnic group had adopted a relationship to the state and to the cultivators of the north which was essentially one of banditry. The Abatwa under their leader Basebya played a major role in the course of the 1912 rising. Religion was also important. In Eritrea a colonial power closely identified with Catholicism threatened the Orthodox culture of its subject peasants, but the potential which this antagonism held for intensifying opposition and consolidating peasant support was blunted by the fact that the rebellious Bāhta was himself a Catholic. By contrast, the key force in Rwanda in precipitating the move towards rebellion proved to be the militant anticolonialism of the Nyabingi cult and its leader Ndungutse.[104]

Cynthia Brantley's chapter below develops some distinctive themes.[105] She locates the resistance of the Giriama of coastal Kenya in an economic context. She argues that the Giriama, although long in contact with coastal and imperial forces which had sought to dominate and exploit them, had avoided a confrontation with those forces, well into the twentieth century, albeit at the cost of the erosion of many of their institutions. The colonial state eventually forced a confrontation in 1914 through applying pressure in the form of taxes and orders to headmen in the effort to obtain Giriama labour for coastal plantations. The Giriama rose in violent opposition. What particularly distinguished the rising is that it was led by a woman, Mekatalili. Normally men played the leading political and military roles among the Giriama. Brantley argues that the erosion of their principal institutions in the immediately preceding decades had created a vacuum in Giriama affairs, and that Mekatalili stepped into this vacuum, mobilising women, young men, and fading elders against the British and their collaborators. Mekatalili's leadership, although drawing on religion, was essentially secular and political and rested on her charisma and oratorical powers. Part of its efficacy undoubtedly came from her special ability to mobilise women whose solidarity spread to the men, their husbands and sons.

Brantley's paper impels the discussion forward into the context of the developed colonial state and its capitalist economy. However, we should be careful lest we lose sight of the challenge of Mekatalili *en route*. It would be foolish to think her entirely unique, the only woman to play a role of secular leadership in a movement of rebellion or violent protest. But there is a deeper challenge, raised by the deeper meaning of the Giriama rising; and that is the role played by ordinary women as participants in such movements. It may not be so central as that played by the followers of Mekatalili, but the case for research to determine such roles seems a strong one indeed.

Although Shula Marks's chapter below deals with events belonging to much the same period as those dealt with by Des Forges and Caulk, the themes she raises belong as much with those raised by Ranger.[106] Far from being the subjects of a newly established colonial state, the inhabitants of Zululand and Natal at the beginning of the twentieth century faced the results of two generations of settler pressure lately reinforced by the demands of mining capital. Pressure on the land, evictions of 'squatters' from ancestral lands, and administrative demands to pay increased taxes and to participate in non-remunerative wage labour, all these combined to create deep popular unrest and dissatisfaction. Marks explores the class dimensions of this unrest on the basis of evidence derived from many sources, not least the rumours reported by government-paid informers. She pays attention to the attitudes and actions of the Zulu Christians and considers what influence their nascent class position had on them. She also explores the interaction between criminality, protest and popular consciousness, highlighting the case of Cakijana, the major assistant to Bambatha, the revolt's leader. The developed colonial context of the Natal rising of 1906 allows Mark's discussion to form a useful bridge onward to Terence Ranger's chapter.

Ranger's chapter, in its turn, with its emphasis on banditry and rebellion, returns the collection to its starting point, bringing us back to an exploration of the meaning of resistance and its relationship to criminality.[107] He sketches the colonial history of Rhodesia with the social-bandit type in mind. He finds few examples, but suggests that poaching in the south-eastern part of the country would bear closer examination in this respect, and singles out one Chitokwa who eluded the law until 1964. Ranger's main story concerns the relations between guerrillas and peasants in the 1970s and the complex question of the grass-roots legitimacy of guerrilla struggle. He argues that acceptance by the peasantry and leadership by a 'liberation movement' are the two necessary conditions for guerrilla legitimacy. Pure bandits have neither; while armed bands enjoying one or other, but not both, of these kinds of acceptance, operate on ambiguous terrain. Here he finds some potential social bandits.

The bulk of his account establishes the legitimacy of the struggle of the ZANLA warriors under the leadership of Robert Mugabe's ZANU in the eastern parts of Zimbabwe in the mid-1970s. However, he also explores the tensions between guerrillas and peasants which intensified between 1977 and 1979, and points out that those tensions threatened the guerrillas' legitimacy, guerrilla warfare shading frequently into banditry. Moreover, he discusses the social violence which has persisted in Matabeleland since independence in 1980, and the disbanding of the guerrilla armies of national liberation. He points out that ZAPU, the nationalist party whose ZIPRA guerrillas had led the struggle in that area, disavows all connection with armed bands there, in this way denying them one of the two essential criteria for guerrilla legitimacy which he identifies. He also mentions the activities of South African-supported elements. He concludes that Matabeleland today is afflicted by large-scale banditry. However, a close reading of the Zimbabwean press and of official pronouncements leads him to conclude that social banditry is a significant phenomenon in Matabeleland. At least some of the armed bands at work there clearly have legitimacy in the eyes of the peasants and receive active support from them. At least some of the armed bands articulate peasant demands and operate on behalf of peasants dissatisfied at either the slow rate of change, or its direction in independent Zimbabwe.

Ranger's chapter moves the discussion beyond the terrain covered by our other studies of peasant rebellion. The Zimbabwe peasants did not rebel on their own. From the 1940s to the early 1970s their resistance took many shapes, some violent, but none had the intensity, scale or durability of a rebellion. The armed uprising was launched from outside Rhodesia and did not reach significant proportions until the successful guerrilla struggle of Frelimo in Mozambique had opened the entire eastern border. Yet even contemporary guerrilla warfare, fully developed into a struggle for national liberation, does not mean an end to many of our thematic concerns. Take the case of religion. Ranger's work establishes yet again the enduring importance of African religious institutions.[108] He points out how the guerrillas made effective use of spirit mediums in winning the confidence of the peasants, and in guaranteeing their own psychological and material security. He also shows that this relationship was reciprocal: that the mediums acted as moral checks on guerrilla activities. Finally, he points out that the war encouraged a dramatic increase in the incidence of, and concern about witchcraft.

Guerrilla warfare launched by a nationalist political party brings us into new territory, only briefly touched on before in the case of the UPC insurrection in Cameroun or the Zaire uprisings of 1964. Like conventional warfare, guerrilla warfare entails an army, training, discipline, chains of command, and therefore a considerable separation of military and civilian affairs, although these are normally integrated again at the political level in this most political of all forms of warfare. Guerrilla warfare has increased in importance in Africa over the past ten to fifteen years, and its importance is likely to continue to grow.[109] Consider the situations in Angola, Namibia, the Horn of Africa, Western Sahara and Chad; all with active guerrillas at the time of writing: and Zaire, daily offering prospects of guerrilla reinstatement.

Finally, the subjects of guerrilla warfare and of the Horn of Africa raise the question of revolution. Revolution is a common interest linking students of the Horn with those of Southern Africa. The former are trying to puzzle out the aftermath of Africa's first classic revolution, while the latter grapple with the incipient stages of its next.[110] In the intervening territories the regimes which succeeded Portuguese rule in Guinea-Bissau, Angola, and Mozambique seek a revolutionary transformation based on the logic of anti-imperialist struggle.[111]

We present no unequivocal cases of revolution. Yet, many of the movements, incidents, and figures we discuss had revolutionary objectives or implications. Freund, in his chapter on the Nigerian tin mines, writes of a 'populist moral economy', and demonstrates it at work at a time of nationalist political agitation. He could have demonstrated such a moral economy as one of the underpinnings of peasant support for nationalist movements, as Ranger discovered. We can discern a peasant moral economy underlying rebellions like Mau Mau, and at this point we can begin to glimpse its revolutionary dimension. Even if the forest fighters themselves never fully articulated a revolutionary programme, such was the logic of their struggle. Only in a Kenya free of capitalist patterns of property in land could landlessness and severe socio-economic deprivation cease to exist. A populist moral economy contains many contradictions. Populism is a response to capitalism against which it reacts, not in the name of a dialectically opposed system, but in the name of moral principles. Hence its goal is a capitalism reformed through the application of those principles. Yet capitalism is inherently an amoral system, and therefore one incapable of reform along these lines. Thus the populist moral economy gives rise to movements which contain many contradictions, contradictions which can only be resolved at the level of revolution and transformation.

The collection emphasises the contradictions. Peasants, seeking redress from grievances, turn for leadership to the class which helped create the grievances. Peasants, oppressed by the conditions of their attachment to the world economy, try to intensify that attachment. Impelled into the future, people turn to the past. Downtrodden by material circumstances, people seek religious release. Such are the contradictions of peasant movements. To do justice to the efforts to transcend these contradictions and to describe their revolutionary resolution would require another collection.

In the meantime, I would reaffirm the basic argument of this book. At this simplest level the argument claims that any attempt to account for social, political, or economic change which ignores the role of the producers of society's wealth is empirically inadequate and theoretically unsatisfying. Change and conflict are inseparable, and the people play active roles in these processes. In so doing, they alarm their rulers and give rise to such fantastic projections as 'the great beast'. From the popular angle the same beast disappears, and in its place there emerges another beast, the agent of oppression, evident in many parts of Africa over extended periods of time. European rule was not the first experience many Africans had either of alien rule or of oppression, nor did these experiences cease with the colonial withdrawal. Popular responses to alien rule and to oppression have been varied and complex, frequently marked by direct opposition and protest.

Much of the literature on African history and politics fails adequately to deal with oppression or, more generally, with popular views and actions. Instead the literature remains resolutely elitist, both in its focus on elites and ruling classes, and in its assumptions of their intrinsic importance. The first shortcoming is within reach of redemption, provided that students view elites within frameworks which recognise how the masses constrain, influence, and mould them. The second shortcoming is obscurantist.

Our contributors suggest new ways of getting at popular views and actions, through a refinement of resistance studies and the taking up of themes developed by radical social history. Engels has shown that the study of criminality may uncover many forms of protest and rebellion and, at its basic level, reveal the pathology of particular socioeconomic formations. Criminality, property relations, and class rule are intimately interconnected, and the study of criminal forms of behaviour has much to tell us about the strains and stresses of social change in Africa and of popular responses to that change. Banditry offers a rich tapestry for study, one on which myth and reality are tightly interwoven. I believe we still have a great deal to learn about bandits in Africa, and expect, as studies of African banditry develop, to read of fresh examples of the popular tendency to make heroes of the defiant, however grossly and harshly their defiance may actually bear on the humble and productive. But the people resist in many other ways than by crime. They use force, boycotts, protests, and defiance; and in most cases they show a capacity to understand their problems at least as well as their rulers and to express that understanding. They draw on many resources and institutions, from the village well or washhouse to the mosque, shrine, and church. On rare occasions the people rebel, but frequency is no measure of rebellion's significance. Rebellions are cataclysmic events toward which people find themselves driven, often reluctantly. The study of rebellion usually reveals deep fissures and tensions within the social fabric invisible in normal times. The record of rebellion in Africa reminds us of two things: that the people will resort to arms in the struggle for a more just ordering of society; and that they will not do so unless provoked to an extreme.

NOTES

1 Research for this introduction was supported by a grant from the Research Board of the Graduate College, University of Illinois at Urbana-Champaign. I am particularly indebted to Tim Fernyhough whose contribution went far beyond that of a conventional research assistant. Many friends and colleagues read and commented on earlier versions. Particularly helpful were: D. Cordell, A. Des Forges, W. Freund, B. Jewsiewicki, and C. C. Stewart.

2 Harold L. Nieburg, *Political Violence: The Behavioural Process* (New York: St Martin's Press, 1969), p. 104; quoted by B. A. Nkemdirim, 'The Tiv rebellion in Nigeria 1960: a study of political violence', *Mawazo*, vol. 4, no. 3 (1975), pp. 35–52; see particularly p. 42.

3 Chapter 3, below.

4 Thomas Hodgkin, *Nationalism in Colonial Africa* (London: Muller, 1956).

5 Some of these developments are reflected in Robert I. Rotberg and Ali A. Mazrui (eds), *Protest and Power in Black Africa* (New York: Oxford University Press, 1970). But the assumptions underlying that collection were resolutely nationalist.

6 But see here Walter Rodney, 'Jihad and social revolution in Futa Djalon in the eighteenth century', *Journal of the Historical Society of Nigeria*, vol. 4, no. 2 (1968), pp. 269–84.

7 For a generally balanced view, which rather idealises the precolonial situation, see B. Bernardi, 'La violenza nel continente africano', *Problemi di Ulisse*, no. 86 (Firenze, 1978), pp. 50–7. See also Victor T. LeVine's early schematic survey: 'The course of political violence' in William H. Lewis (ed.), *French Speaking Africa. The Search for Identity* (New York: Walker, 1965), pp. 58–79; and the recent, and thoughtful, R. Anifowose, *Violence and Politics in Nigeria. The Tiv and Yoruba Experience* (New York, 1982). John McCracken makes useful suggestions from a promising angle in 'Coercion and control in Nyasaland: a study of a colonial police force' (unpublished paper, History Department, Chancellor College, University of Malawi, 1983). Frantz Fanon appears to be the forgotten figure of the 1980s: *The Wretched of the Earth* (Harmondsworth: Penguin, 1967).

8 A. A. Bozeman, *Conflict in Africa. Concepts and Realities* (Princeton: Princeton University Press, 1976). See also B. Nkemdirim's rather more sober 'Reflections on political conflict, rebellion, and revolution in Africa', *Journal of Modern African Studies*, vol. 15, no. 1 (1977), pp. 75–90.

9 Frederick Engels, *The Condition of the Working Class in England. From Personal Observation and Authentic Sources* (Introduction by Eric Hobsbawm) (London: Panther Books, 1969), p. 240; also pp. 159–62. The text differs from that in the stimulating collection edited by David F. Greenberg, *Crime and Capitalism: Readings in Marxist Criminology* (Palo Alto: Mayfield, 1981), pp. 48–50.

10 K. Marx, *Capital* (London, Vol. 1, 1967), as cited in Greenberg (ed.), *Crime and Capitalism*, pp. 45–8.

11 P. Linebaugh, 'Karl Marx, the theft of wood, and working class composition', in

Greenberg (ed.), *Crime and Capitalism*, pp. 79–97.

12 Chapter 1, Douglas Hay, 'Property, Authority and the Criminal Law', and 'Poaching and the Game Laws on Cannock Chase', Chs. 1 and 5 in Douglas Hay, Peter Linebaugh, John G. Rule, E. P. Thompson and Cal Winslow, *Albion's Fatal Tree. Crime and Society in Eighteenth Century England* (London: Allen Lane, 1975).

13 E. P. Thompson, *Whigs and Hunters. The Origin of the Black Act* (London: Allen Lane, 1975). See also the special summer issue of the *Journal of Social History*, VIII (1975); and the classic, Louis Chevalier, *Classes laborieuses et classes dangereuses à Paris pendant la première moitié du XIX siècle* (Paris: Plon, 1953); with application to the Third World by Torcuato S. di Tella, 'The Dangerous Classes in Early Nineteenth Century Mexico', *Journal of Latin American Studies*, vol. 5, no. 1 (1973), pp. 79–105.

14 Stephen A. Lucas, 'Social deviance and crime in selected rural communities of Tanzania', *Cahiers d'études africaines*, vol. 16, no. 3 (1976), pp. 499–518; J. M. Wasikhongo, 'Armed robbery and the development process in Africa: trends in Nairobi, Mombasa, and Abidjan, and criminal processes in Nairobi' (unpublished PhD dissertation, University of Wisconsin-Madison, 1979).

15 Wasikhongo, 'Armed robbery', pp. 92–3.

16 Michael L. Stone, 'Organised poaching in Kitui District: a failure in district authority, 1900 to 1960', *International Journal of African Historical Studies*, vol. 5, no. 3 (1972), pp. 436–52; and Jack Goody, 'Rice-burning and the Green Revolution in northern Ghana', *Journal of Development Studies*, vol. 16, no. 2 (1980), pp. 136–55.

17 Lutakome A. Kayiira, 'Violence in kondoism: the rise and nature of violent crime in Uganda' (unpublished PhD dissertation, State University of New York at Albany, 1977). See also Ali A. Mazrui, *Soldiers and Kinsmen in Uganda. The Making of a Military Ethnocracy* (Beverly Hills: Sage, 1975), pp. 89–90, 128–9; and the fertile vistas hinted at in Nwokocha Nkpa's 'Armed robbery in post-civil war Nigeria: the role of the victim', *Victimology: an International Journal*, vol. 1, no. 1 (1976), pp. 71–83.

18 James Midgley, Jan H. Steyn and Roland Graser (eds), *Crime and Punishment in South Africa* (Johannesburg: McGraw Hill, 1975).

19 *Past and Present*, no. 80 (1978), pp. 91–121; revised and reprinted as 'The regiment of the hills – *Umkosi Wezintaba*: The Witwatersrand's lumpenproletarian army, 1890–1920', in Charles van Onselen, *Studies in the Social and Economic History of the Witwatersrand 1886–1914*, vol. II, *New Nineveh* (Harlow: Longman, 1982), pp. 171–201.

20 Chapter 1, below.

21 Chapter 2, below.

22 Chapter 3, below.

23 See for example the novels: Yashar Kemal, *Memed My Hawk* (New York: Pantheon, 1961) about Turkey; Sahle Sellassie, *Warrior King* (London: Heinemann, 1974) about Ethiopia; and João Guimarães Rosa, *The Devil to Pay In the Backlands* (New York: Knapf, 1963) about Brazil. The Jamaican film starring Jimmy Cliffe, *The Harder They Come*, is a powerful Third World adaptation of a well-developed Hollywood theme.

24 E. J. Hobsbawm, *Primitive Rebels. Studies in Archaic Forms of Social Movement in the 19th and 20th Centuries* (Manchester: Manchester University Press, 1959); and idem, *Bandits* (Harmondsworth: Penguin, 1972; 2nd edn., 1981).

25 Pat O'Malley, 'Social bandits, modern capitalism and the traditional peasantry. A critique of Hobsbawm', *Journal of Peasant Studies*, vol. 6, no. 4 (1979), pp. 489–501; idem, 'Class conflict, land and social banditry. Bush-ranging in nineteenth century Australia', *Social Problems*, vol. 26, no. 3 (1979), pp. 271–83; Billy Jaynes Chandler, *The Bandit King: Lampiao of Brazil* (College Station: Texas A & M University Press, 1978); John McQuilton, *The Kelly Outbreak, 1878–1880: The Geographical Dimension of Social Banditry* (Carlton: Melbourne University Press, 1979); and P. Underwood, *Disorder and Progress. Bandits, Police and Mexican Development* (Lincoln, Nebraska, 1981).

26 Hobsbawm's most trenchant critic has been Anton Blok, 'The peasant and the brigand: social banditry reconsidered', *Comparative Studies in Society and History*, vol. 14, no. 4 (1972), pp. 494–505. See also L. Lewin, 'The oligarchical limitations of social

banditry in Brazil: the case of the "good" thief Antonio Silvino', *Past and Present*, no. 82 (1979), pp. 116–46.

27   E. J. Keller, 'A twentieth century model: the Mau Mau transformation from social banditry to social rebellion', *Kenya Historical Review*, vol. 1, no. 2 (1973), pp. 189–205.

28   D. A. Maughan Brown, 'Social banditry: Hobsbawm's model and "Mau Mau",' *African Studies*, vol. 39, no. 1 (1980), pp. 77–97.

29   The best introduction to Mau Mau is Donald L. Barnett and Karari Njama, *Mau Mau from Within. Autobiography and Analysis of Kenya's Peasant Revolt* (London: MacGibbon & Kee, 1966). See also B. Kaggia, *Roots of Freedom* (Nairobi: East African Publishing House, 1975), and Frank Furedi, 'The African crowd in Nairobi: popular movements and elite politics', *Journal of African History*, vol. 14, no. 2 (1973), pp. 275–90. Scott Myerly has explored some of these issues in his unpublished paper: 'Crime, justice and rebellion: the duality of Kikuyu crime in pre-Emergency Kenya, 1945–1952' (Urbana, 1981). I am indebted to Myerly for stimulating comments on criminality in general.

30   Allen Isaacman, 'Social banditry in Zimbabwe (Rhodesia) and Mozambique, 1894–1907: an expression of early peasant protest', *Journal of Southern African Studies*, vol. 4, no. 1 (1977), pp. 1–30.

31   ibid., p. 10.

32   W. G. Clarence-Smith, *Slaves, Peasants and Capitalists in Southern Angola 1840–1926* (Cambridge: Cambridge University Press, 1979), pp. 82–8.

33   *Champaign-Urbana News Gazette*, 7 October 1982.

34   *Toronto Globe and Mail*, 9 August 1982. See also Chapter 17, below by T. O. Ranger for a much fuller development of this theme.

35   *Africa News*, 13 September 1982.

36   Hobsbawm, *Bandits* (1972 edn), p. 14; idem (1981 edn), *New York Times*, 2 January 1983.

37   Chapters 13, 14, 16, below. J. P. Chrétien reinforces Des Forges study in Chapter 14, below with his own finding of the social-bandit myth in connection with a 1934 revolt in neighboring Burundi: J.-P. Chrétien, 'Une révolte au Burundi en 1934', *Annales, E.S.C.*, vol. 25, no. 6 (1970), p. 1700.

38   Chapter 4, below.

39   Chapter 5, below.

40   Chapters 6 and 7, below.

41   Chapter 8, below.

42   T. O. Ranger, *Revolt in Southern Rhodesia 1896–7. A study in African Resistance* (London: Heinemann, 1967); *idem*, 'Connexions between "Primary resistance" movements and modern mass nationalism in East and Central Africa', *Journal of African History*, vol. 9, nos. 3 & 4 (1968), pp. 437–53, 631–41. Similar ideas underpin his *The African Voice in Southern Rhodesia 1898–1930* (London/Nairobi: Heinemann, 1970).

43   Michael Crowder (ed.), *West African Resistance. The Military Response to Colonial Occupation* (London: Hutchinson, 1971); and his introduction with A. Asiwaju to the special issue of *Tarikh*, vol. 5, no. 3 (1977), entitled 'Protest against Colonial Rule in West Africa'.

44   Crowder in *Tarikh*, vol. 5, no. 3, p. 2.

45   Cynthia Brantley, *The Giriama and Colonial Resistance in Kenya, 1800–1920* (Berkeley: University of California Press, 1981); and E. I. Steinhart, *Conflict and Collaboration. The Kingdoms of Western Uganda 1890–1907* (Princeton: Princeton University Press, 1977).

46   Gerry Caplan showed more sensitivity to internal stratification and its implications for the nature of the response of African states to the advent of colonial forces than did most students of the era of the Scramble: Gerald L. Caplan, *The Elites of Barotseland, 1879–1969; A Political History of Zambia's Western Province* (Berkeley: University of California Press, 1970).

47   Allen and Barbara Isaacman, 'Resistance and collaboration in Southern and Central

Africa, *c.* 1850–1920', *International Journal of African Historical Studies*, vol. 10, no. 1 (1977), pp. 31–62.

48 ibid., p. 61.

49 Issues touched on, but not developed, by Yves Person in his treatment of the great anti-Samori revolt of 1888, in Crowder (eds), *West African Resistance*, pp. 131–2 and his 'Samori and Resistance to the French', in Rotberg and Mazrui (eds), *Protest and Power*, pp. 98–9.

50 I. Berger, *Religion and Resistance. East African Kingdoms in the Pre-colonial Period* (Tervuren, Belgium: Musée Royale de l'Afrique Centrale, 1981, Annales, Série In.8°, Sciences Humaines, No. 105).

51 For an account of religion and resistance in a colónial context see Bonnie Keller, 'Millenarianism and resistance: the Xhosa cattle killing', *Journal of Asian and African Studies*, vol. 13, nos. 1–2 (1978), pp. 95–111. Islam has received generous attention as a vehicle for resistance. See Hodgkin, *Nationalism*; also J. Goody, 'Reform, renewal and resistance: a Mahdi in northern Ghana' and the other papers in the section 'Colonial Misfortune and Religious Response', in Christopher Allen and R. W. Johnson (eds), *African Perspectives. Papers in the History, Politics and Economics of Africa* (Cambridge: Cambridge University Press, 1970). A contemporary edge to the theme of Islam, protest, and violence derives from the fate of the Maitatsine sect, tens of thousands of whose adherents have died in 'disturbances' in Kano in December 1980; in Kano, Kaduna, and Maiduguri in October 1982; and in Yola, Maiduguri and Kaduna in February and March 1984. See *Africa News* of 26 January 1981, 29 November 1982, and 12 March 1984; also *The Economist*, 6 November 1982; *The New York Times*, 16 November 1982; and the Associated Press report carried in the *Champaign-Urbana News Gazette* of 22 April 1984. The 1984 symposium of the African Studies Program of the University of Illinois on 'Popular Islam in 20th Century Africa' had two valuable papers on the Maitatsine phenomenon: Paul Lubeck, 'Islamic protest under semi-industrial capital: 'Yan Tatsini explained', forthcoming in a special issue of *Africa* devoted to the symposium; and Allen Christelow, 'Religious protest and dissent in northern Nigeria: from Mahdism to Quranic integralism'.

52 S. Shaloff, 'The income tax, indirect rule, and the Depression: the Gold Coast riots of 1931', *Cahiers d'études africaines*, vol. 14, no. 2 (1974), pp. 359–75; and *idem* 'The Cape Coast asafo company riot of 1932', *International Journal of African Historical Studies*, vol. 7, no. 4 (1974), pp. 591–607.

53 Roger G. Thomas, 'The 1916 Bongo "riots" and their background; aspects of colonial administration and African response in eastern upper Ghana' *Journal of African History*, vol. 24, no. 1 (1983), pp. 57–75. L. Plotnicov's note on riots in Jos in 1945 draws attention to a Nigerian incident which deserves closer analysis: 'An early Nigerian civil disturbance: the 1943 Hausa–Ibo riot in Jos', *Journal of Modern African Studies*, vol. 9, no. 2 (1971), pp. 297–305.

54 Jeff Crisp, 'Union atrophy and worker revolt: labour protest at Tarkwa Goldfields, Ghana, 1968–1969', *Canadian Journal of African Studies*, vol. 13, nos. 1–2 (1979), pp. 265–93.

55 Wale Oyemakinde, 'The Nigerian General Strike of 1945', *Journal of the Historical Society of Nigeria*, vol. 7, no. 4 (1975), pp. 693–710.

56 Chapter 9, below.

57 Chapter 10, below.

58 Chapter 11, below.

59 See Margaret Stroebel's review article: 'African Women', *Signs: Journal of Women in Culture and Society*, vol. 8, no. 1 (1982), p. 124; and one of the few works which she fails to mention: Edna G. Bay (ed.), *Women and Work in Africa* (Boulder: Westview Press, 1982). Bay's collection arose from the sixth annual symposium of the African Studies Program of the University of Illinois. Two articles which start to explore the role of women in African nationalism are: LaRay Denzer, 'Towards a history of West African women's participation in nationalist politics: the early phase, 1935–1950', *Africana*

Research Bulletin, vol. 6, no. 4 (1976), pp. 65–85; and Cheryl Johnson, 'Grass roots organizing: women in anticolonial activity in southwestern Nigeria', *African Studies Review*, vol. 25, nos. 2–3 (1982), pp. 137–57.

60   Ranger, *Revolt in Southern Rhodesia*; Chrétien, 'Une révolte au Burundi'. But see below, material at note 71, for a revision of Ranger's account.

61   Judith van Allen, '"Aba Riots" or Igbo "Women's War"? Ideology, stratification and the invisibility of women', in Nancy J. Hafkin and Edna G. Bay (eds), *Women in Africa. Studies in Social and Economic Change* (Stanford: Stanford University Press, 1976), pp. 59–85.

62   One of the earliest accounts was: N. van Vuuren, *Women Against Apartheid: the Fight for Freedom in South Africa, 1920–1975* (Palo Alto, 1979). But now see also: Richard E. Lapchick and Stephanie Urdang, *Oppression and Resistance. The Struggle of Women in Southern Africa* (Westport, Conn., and London: Greenwood Press, 1982). Urdang has also written on women in Guinea-Bissau: *Fighting two Colonialisms. Women in Guinea-Bissau* (New York and London: Monthly Review Press, 1979).

63   Sembene Ousmane, *God's Bits of Wood* (New York: Anchor, 1970).

64   Ngugi wa Thiong'o, *Petals of Blood* (London: Heinemann, 1977). The similarities in theme are not merely coincidence. *Petals* contains a direct reference to *God's Bits*, p. 338. In lighter, less solemn vein, but equally alive to the political concerns and activities of women, is Wole Soyinka's *Ake. The Years of Childhood* (London: Rex Collings, 1981), pp. 180–222.

65   Chapter 12, below.

66   Rodney, 'Jihad and Social Revolution'. In contrast, see Philip D. Curtin, 'Jihad in West Africa: early phases and interrelationships in Mauritania and Senegal', *Journal of African History*, vol. 12, no. 1, pp. 11–24. See also T. Büttner, 'Der Charakter der Erhebung des Uthman dan Fodio and die Grundung des Sokotoreiches zu Beginn des 19. Jahrhundert in Nordnigeria', *Zeitschrift fur Geschichtswissenschaft*, vol. 22, no. 1 (1974), pp. 47–63, for a preliminary class analysis of the origins and consequences of the *jihad* of Usman dan Fodio.

67   Basil Davidson, *The People's Cause. A History of Guerrillas in Africa* (Harlow: Longman, 1981), Ch. 3; John Hunwick, 'Songhay, Bornu and Hausaland in the sixteenth century', in J. F. Ajayi and Michael Crowder (eds), *A History of West Africa*, Vol. 1 (London: Longman, 1972), pp. 177–81.

68   Two useful works, exceptions to my strictures, are: Merid W. Aregay, 'Southern Ethiopia and the Christian kingdom 1508–1708 with special reference to the Galla migrations and their consequences' (unpublished PhD thesis, University of London, 1971); and A. Bartnicki and J. Mantel-Niecko, 'The role and significance of the religious conflicts and people's movements in the political life of Ethiopia in the seventeenth and eighteenth centuries', *Rassegna di studi etiopici*, vol. 24 (1969–70, pp. 5–39).

69   F. Lawson, 'Rural revolt and provincial society in Egypt, 1820–1824', *International Journal of Middle East Studies*, vol. 13, no. 2 (1981), pp. 131–53.

70   Ranger, *Revolt in Southern Rhodesia* stands head and shoulders above the two studies of the East African anti-German rising of 1888–90: Robert D. Jackson, 'Resistance to the German invasion of the Tanganyikan coast, 1888–1891', in Rotberg and Mazrui (eds), *Protest and Power*, pp. 37–79; and G. Akinola, 'The East African coastal rising, 1888–1890', *Journal of the Historical Society of Nigeria*, vol. 7, no. 4 (1975), pp. 609–30.

71   Julian Cobbing, 'The absent priesthood: another look at the Rhodesian risings of 1896–1897', *Journal of African History*, vol. 18, no. 1 (1977), pp. 61–84. Although the Shona religious mediums probably played a more peripheral role in the Shona-Ndebele revolt of 1896–7 than Ranger allowed, religion was nonetheless a centrally contributing factor in many instances of violent resistance to European colonialism in Africa. See, for instance, Elizabeth Hopkins, 'The Nyabingi cult of southwest Uganda', in Rotberg and Mazrui (eds), *Protest and Power*, pp. 258–336; Anne King, 'The Yakan cult and Lugbara response to colonial rule', *Azania*, vol. 5 (1970),

pp. 1–25; and many of the citations and articles below.

72 D. N. Beach, '"Chimurenga": the Shona rising of 1896–97', *Journal of African History*, vol. 20, no. 3 (1979), pp. 395–420.

73 George Shepperson and Thomas Price, *Independent African. John Chilembwe and the Origins, Setting and Significance of the Nyasaland Rising of 1915* (Edinburgh: Edinburgh University Press, 1958).

74 Chapter 9, below.

75 Ian and Jane Linden, 'John Chilembwe and the New Jerusalem', *Journal of African History*, vol. 12, no. 4 (1971), pp. 629–51; R. Rotberg, 'Psychological stress and the question of identity: Chilembwe's revolt reconsidered', in Rotberg and Mazrui (eds), *Protest and Power*, pp. 337–73. See also the document edited by Rotberg: George Simeon Mwase, *Strike a Blow and Die. A Narrative of Race Relations in Colonial Africa* (Cambridge, Mass.: Harvard University Press, 1967).

76 Jon M. Bridgman, *The Revolt of the Hereros* (Berkeley and Los Angeles: University of California Press, 1981); J.-P. Chrétien, 'Forces traditionelles et pression coloniale au Rwanda allemand', *Revue française d'Histoire d'Outre-Mer*, vol. 59, no. 217 (1972), pp. 40–61. For Rwanda see also Chapter 14 by Des Forges, below. John Tosh rightly protests at African historiography's fixation with large-scale risings, pointing out that a great deal of resistance to colonialism was small in scale: 'Small-scale resistance in Uganda: the Lango "rising" at Adwar in in 1919', *Azania*, vol. 9 (1970), pp. 51–64.

77 John Iliffe, 'The organization of the Maji Maji rebellion', *Journal of African History*, vol. 8, no. 3 (1967), pp. 495–512.

78 Shula Marks, 'The Zulu disturbances in Natal', in Rotberg and Mazrui (eds), *Protest and Power*, pp. 213–57; idem, *Reluctant Rebellion. The 1906–8 Disturbances in Natal* (Oxford, 1970). See also Jeff Guy, *The Destruction of the Zulu Kingdom* (London: Longman, 1979). The distinctive periodisation of South African history, compared with that of much of the rest of the continent, is well exemplified by Newton-King and Malherbe's papers on the Khoikhoi rebellion of 1799–1803, the context of which was essentially colonial: S. Newton-King and V. C. Malherbe, *The Khoikhoi Rebellion in the Eastern Cape (1799–1803)* (Centre for African Studies, University of Cape Town, 1981). See also I. B. Sutton, 'The digger's revolt in Griqualand West, 1875', *International Journal of African Historical Studies*, vol. 12, no. 1 (1979), pp. 40–61.

79 J. Osuntokun, 'West African armed revolts during the First World War', *Tarikh*, vol. 5, no. 3 (1977), pp. 6–17; and Thomas, 'The 1916 Bongo "riots"'.

80 Tosh, 'The Lango "rising"' contains a caution here. Although the Lango incidents which he discusses occurred after the war in 1919, thematically they still belong to the era of 'pacification', the initial establishment of colonial rule.

81 R. Edgar, 'The prophet motive: Enoch Mgijima, the Israelites, and the background to the Bullhoek massacre', *International Journal of African Historical Studies*, vol. 15, no. 3, (1982), pp. 401–22; N. Herd, *1922. The Revolt on the Rand* (Johannesburg: Blue Crane Books, 1966).

82 See the chapter by R. A. Caulk in A. D. Roberts (ed.), *The Cambridge History of Africa*, Vol. 7, *c. 1905 to c. 1940* (forthcoming).

83 Diana Ellis, 'The Nandi protest of 1923 in the context of African resistance to colonial rule in Kenya', *Journal of African History*, vol. 17, no. 4 (1977), pp. 555–75. L. J. Greenstein's debunking account of the events is less satisfactory: 'The Nandi "uprising" of 1923', *Pan-African Journal*, vol. 9, no. 4 (1976), pp. 397–406.

84 The phrase is John Iliffe's. See John Iliffe, 'The age of improvement and differentiation (1907–45)', in A. Temu and I. Kimambo (eds), *A History of Tanzania* (Nairobi: East African Publishing House, 1969), pp. 123–60.

85 Hassan Ahmed Ibrahim, 'Mahdist risings against the Condominium government in the Sudan, 1900–1927', *International Journal of African Historical Studies*, vol. 12, no. 3 (1979), pp. 440–71.

86 T. M. Bah, 'Karnou et l'insurrection des Gbaya (La situation au Cameroun 1928–1930)', *Afrika Zamani*, vol. 3 (Douala, 1974), pp. 105–61.

87   Chrétien, 'Une révolte au Burundi'.
88   Similar themes to those of the Gbaya and Burundi revolts are raised by the Pende revolt of 1931, the largest revolt in the Congo against Belgian rule. An intensely harsh colonial exploitation, exacerbated by the Great Depression, and administrative responses to it, led to insurrection sparked off by the message of a prophet of ancestral religion, called Muluba, and the unpopular actions of a colonial-appointed chief: Sikitele-Giza a Sombula, 'Les causes principales de la revolte pende en 1931', *Likundoli*, Serie C, 1, 2 (Lubumbashi, 1976), pp. 181–200; and *idem*, 'Les racines de la revolte pende de 1931', *Etudes d'Histoire africaine*, vol. 5 (1973), pp. 99–153.
89   Patrick Gilkes, *The Dying Lion. Feudalism and Modernization in Ethiopia* (London: Friedmann, 1975), pp. 187–91. Gebru Tareke has greatly advanced the discussion of this revolt: 'Peasant resistance in Ethiopia: the case of *Weyane*', *Journal of African History*, vol. 25, no. 1 (1984), pp. 77–92.
90   J. Tronchon, *L'insurrection malgache de 1947. Essai d'interprétation historique* (Paris: Maspero, 1974). For a brief account in English, see Raymond K. Kent, *From Madagascar to the Malagasy Republic* (New York: Praeger, 1962), Ch. 6.
91   In addition to Barnett and Njama, *Mau Mau from Within*, and Kaggia, *Roots of Freedom*, see also: C. Rosberg and J. Nottingham, *The Myth of Mau Mau, Nationalism in Kenya* (New York, 1966); Robert Buijtenhuijs, *Le mouvement 'Mau Mau'. Une révolte paysanne et anti-coloniale en Afrique noire* (Paris/The Hague: Mouton, 1971); *idem*, *Essays on Mau Mau. Contributions to Mau Mau Historiography* (Leiden: African Studies Centre, 1982); and J. Newsinger, 'Revolt and repression in Kenya: the "Mau Mau" rebellion, 1952–1960', *Science and Society*, vol. 45, no. 2 (1981), pp. 159–85. Claude Welch provides a useful thematic summary in a comparative context which also includes the 1964 Kwilu rising in Zaire: *Anatomy of Rebellion* (Albany, New York, 1980).
92   Richard A. Joseph, 'Ruben um Nyobe and the "Kamerun" rebellion', *African Affairs*, vol. 74, no. 293 (1974), pp. 428–48; *idem*, *Radical Nationalism in Cameroun. Social Origins of the UPC Rebellion* (Oxford: Clarendon Press, 1977). Joseph's account supercedes Willard Johnson, 'The Union des Populations du Cameroun in rebellion: the integrative backlash of insurgency' in Rotberg and Mazrui (eds), *Protest and Power*, pp. 671–92.
93   See, for example, Robert Buijtenhuijs, *Le Frolinat et les révoltes populaires du Tchad, 1965–1976* (The Hague: Mouton, 1978).
94   Benoît Verhaegen, *Rébellion au Congo* (Kinshasa/Brussels: Institut de recherches économiques et sociales, 2 vols, 1966, 1969). For a brief summary, see his 'Les rébellions populaires au Congo en 1964', *Cahiers d'études africaines*, vol. 7, no. 26 (1967), pp. 345–59. See also Renée C. Fox, Willy de Craemer and Jean-Marie Ribeaucourt, '"The second independence": a case study of the Kwilu rebellion in the Congo', *Comparative Studies in Society and History*, vol. 8, no. 1 (1965–6), pp. 78–109; T. Turner, 'Peasant rebellion and its suppression in Sankuru, Zaire', *Pan-African Journal*, vol. 7, no. 3 (1974), pp. 193–215; Turner's introduction with Warren Weinstein to this special issue of the *Pan-African Journal*, 'Introduction to peasant rebellion and ethnic conflict in Africa', pp. 185–92; and Welch, *Anatomy of Rebellion*. As a feat of individual scholarship, John Marcum's study of the Angolan revolution bears some comparison with Verhaegen's work: *The Angolan Revolution* (Cambridge, Mass: M.I.T. Press, 2 vols, 1969, 1978).
95   For intensity of peasant violence even Zaire could not fully match the Rwanda rising of 1959. See R. Lemarchand, 'The coup in Rwanda', in Rotberg and Mazrui (eds), *Protest and Power*, pp. 877–923; *idem*, *Rwanda and Burundi* (New York: Praeger, 1970).
96   Eric Wolf, *Peasant Wars of the Twentieth Century* (New York: Harper & Row, 1969); James Scott, *The Moral Economy of the Peasant. Rebellion and Subsistence in Southeast Asia* (New Haven and London: Yale University Press, 1976).
97   Gebru Tareke, 'Rural Protest in Ethiopia, 1941–1970: a study of three rebellions' (unpublished PhD dissertation, Syracuse University, 1977). See also Gilkes, *Dying Lion*, pp. 182–6, 214–19; John Markakis, *Ethiopia. Anatomy of a Traditional Polity*

(Oxford: Clarendon Press, 1974), pp. 376–87; and Allan Hoben, *Land Tenure among the Amhara of Ethiopia. The Dynamics of Cognatic Descent* (Chicago: University of Chicago Press, 1973), pp. 219–26. Shallow and unsatisfactory is Gildas Nicolas, 'Peasant rebellions in the socio-political context of today's Ethiopia', *Pan-African Journal*, vol. 7, no. 3 (1974), pp. 235–62.

98 J.-L. Amselle, 'La conscience paysanne: la révolte de Ouolossebougou (Juin 1968, Mali)', *Canadian Journal of African Studies*, vol. 12, no. 3 (1978), pp. 339–55.

99 C. E. F. Beer and Gavin Williams, 'The Politics of the Ibadan Peasantry', in Gavin Williams (ed.), *Nigeria. Economy and Society* (London: Rex Collings, 1976), and C. Beer, *The Politics of Peasant Groups in Western Nigeria* (Ibadan, 1976). See also the rather less satisfactory Wasihi A. Sambo, 'The Agbekoya uprising: a study in political and economic conflict', *Journal of African Studies*, vol. 3, no. 2 (1976), pp. 246–66: and Tunde Adeniran, 'The dynamics of peasant revolt. A conceptual analysis of the *Agbekoya Parapo* uprising in the Western State of Nigeria', *Journal of Black Studies*, vol. 4, no. 4 (1974), pp. 363–75.

100 Richard Sherman, *Eritrea. The Unfinished Revolution* (New York: Praeger, 1980); Bereket Habte Selassie, *Conflict and Intervention in the Horn of Africa* (New York and London: Monthly Review Press, 1980).

101 See, for example, W. Weinstein, 'Conflict and confrontation in Central Africa: the revolt in Burundi, 1972', *Africa Today*, vol. 29, no. 4 (1972), pp. 17–37.

102 Nkemdirim, 'Tiv rebellion'. Anifowose, *Violence and Politics*, marks a considerable advance over Nkemdirim, and also provides a lucid account of the later events in the Western Province; but he shows no sensitivity to class as a factor in these uprisings, and betrays little interest in the dialects of peasant politics.

103 Chapters 13 and 14, below.

104 In his earlier account of the rebellion Chrétien, 'Forces traditionelles', also emphasised the role of Ndungutse. See also Hopkins, 'The Nyabingi cult'.

105 Chapter 15, below.

106 See Chapter 16, below.

107 See Chapter 17, below.

108 See T. O. Ranger, *Peasant Consciousness and Guerrilla Warfare in Zimbabwe: A Comparative Study* (London: James Currey, 1985), a rare account of the details of recent revolutionary transformation in Africa based on varied and solid documentation.

109 Consider the great difference between the ground covered by Kenneth Grundy, *Guerrilla Struggle in Africa. An Analysis and Preview* (New York: Grossman, 1971), and Davidson, *The People's Cause.*

110 Fred Halliday and Maxine Molyneux, *The Ethiopian Revolution* (London: NLB, 1981); M. and D. Ottaway, *Ethiopia. Empire in Revolution* (New York, 1978); Theda Skocpol, *States and Social Revolutions. A Comparative Analysis of France, Russia and China* (Cambridge: Cambridge University Press, 1979), pp. 287, 292, 350 note 10. Even the blue-ribbon establishment Study Commission on US Policy toward South Africa in its report, *South Africa: Time Running Out* (Berkeley, 1981), is inclined to diagnose the South African situation as pre-revolutionary.

111 Basil Davidson has been an eloquent interpreter of these developments. See particularly his *The People's Cause.*

SECTION 1
Criminality

1 Murder and theft in early nineteenth-century Elmina

LARRY W. YARAK

Dutch officers stationed at the Gold Coast town of Elmina in the early nineteenth century recognised the normative basis of Elmina's social and political order.[1] As Prosecutor (*Fiscaal*) J. Cremer wrote:

> Although composed of human beings, [indigenous] society here [at Elmina] cannot, unfortunately, be said to consist of rational creatures. Nevertheless, just as in civilised countries, this society does distinguish virtue from vice, innocence from misdeed, justice from injustice, and above all, order from disorder or rebellion.[2]

Elmina's laws (or 'customs' as the Dutch preferred to term them) served, to use the words of Eugene Genovese, as 'a principal vehicle for the hegemony of the ruling class'.[3] The primary purpose of the present paper is to investigate the hegemonic function of Elmina's laws, as revealed in the acts of criminal adjudication recorded in the town in the early nineteenth century. Following Genovese, it will be assumed that 'the law cannot be viewed as something passive and reflective, but must be viewed as an active, partially autonomous force, which mediated among the several classes and compelled the rulers to bend to the demands of the ruled.'[4]

In the case of Elmina, we are fortunate to possess detailed records of the investigation and adjudication of a number of cases of criminal acts in which the social dimension is discernible. In addition, these documents provide fascinating and often vivid glimpses into otherwise obscure conditions of life enjoyed or endured by members of Elmina's diverse class and ethnic groupings. At a more analytical level, however, these records document significant aspects of fundamental social tension within Elmina's class structure, and the mediating function of Elmina law and legal practice.

The documents upon which this paper is based are to be found in the records of the Dutch coastal administration's Council of Senior Officials.[5] The Council was chaired by the Dutch Governor (variously called 'Governor', 'Governor-General', or 'Commander') and met quarterly in the main Dutch fort at Elmina. I have examined the Council's records for the years between 1815 and 1830. In 1815 the Dutch home government took actions to renew its involvement in the affairs of its 'possessions' on the Gold Coast in the wake of the Napoleonic upheavals, dispatching both funds and a number of new officials to the coast. As a result documentation of administrative actions improved dramatically. In 1830, the documentation deteriorated once again, in part as a result of the need for increased administrative secrecy as the Dutch commenced an ill-fated revival of their overseas slave trade under the guise of 'recruitment' for their East Indies army.[6]

In the period under review, the Council reconstituted itself into a judicial body on more than twenty occasions, hearing testimony with regard to crimes alleged to have been committed by various residents of Elmina town, which the main Dutch fort adjoined. Ten of these cases involved theft; eight concerned murder or conspiracy to commit murder; three involved battery; one concerned a case of slander that resulted in the suicide of the wronged party; and a final case concerned sedition.

Elmina's communities

Before going into the cases themselves, it is necessary to describe briefly the general circumstances which led to them being brought before the Dutch Council. These circumstances arose from the peculiar set of relationships that governed the affairs of the four semi-autonomous communities (as I will call them) that comprised early nineteenth-century Elmina.[7] Each of these communities was organised into more-or-less independent hierarchies of authority; at the same time it is important to note that there was a considerable degree of intermarriage between prominent members of all four communities. Such informal linkages may indeed have been ritualised, though Dutch documentation of this matter is disappointing.[8] The largest of the communities was that of the citizens of Elmina town itself, or *Edena* as it is called in Akan. The town's free citizenry (*Edenafo*), together with slaves and pawns, was said to number some 12 000–14 000, most of whose livelihood came from fishing. Next in size was the Asante community at Elmina, which numbered roughly 1000 at its peak during 1824–30. It was headed by a resident who was appointed by the Asantehene.[9]

A third community was made up of the so-called *vrijburgers* or 'free citizens' of Elmina and their employees, retainers and slaves, whose numbers have not as yet been ascertained for this period. The *vrijburgers* themselves were a small, generally literate group of persons of mixed African and European parentage. What appears to have distinguished them from the larger group of mulatto residents, the so-called *tapoeyers*, was a matter of law: the *vrijburger* was said to have recourse to Dutch law, rather than Edena 'customs'; however, this matter would bear further investigation.[10] Finally, there were the Dutch officials themselves, most of whom resided in the two forts at Elmina, while some chose to live in the town. Their numbers varied during 1815–30 from a low of perhaps five to a high of about twenty. A remarkable amount of authority was concentrated in the hands of the governor. Among the Dutch community must also be reckoned an as yet undetermined but sizeable number of employees, servants and slaves. Dutch officials estimated the total population of the town to be about 15 000.

Elmina town's political hierarchy consisted of a king (*Edenahen*) and his advisors or elders; the 'senior captain' of the seven Asafo companies (*ekuw* or 'quarters' as the Dutch called them); the 'captains' of each of the companies; and, finally, a council of prominent merchants (the *Besonfo* council). Under these luminaries resorted the mass of Elmina citizens and slaves, grouped into ten named and numbered Asafo companies, of which however only numbers one through seven were accorded full citizenship rights. It was at the broader company level that some formal effort to integrate Elmina's four communities is in evidence. The slaves, common soldiers and artisans owned or employed by the Dutch, for example, lived in Elmina town occupying their own 'quarter', and so were at times regarded as virtually on a par with the seven 'true' Asafo companies – today their descendants are fully accorded that status as company number eight.

Similarly, the *vrijburgers'* less well-to-do brethren, the *tapoeyers*, constituted the tenth company. (The ninth company was composed of the slaves and retainers of one wealthy merchant, Kwamena Konua, whose ancestors had been powerful slave-traders. Most of Kwamena Konua's people did not reside in Elmina town, but in two villages located a short distance away.) Thus for the overwhelming majority of Elmina's inhabitants, disputes over debts, inheritance, petty theft, adultery, slander and the like were either settled before less formal 'family' councils or, in the last resort, before the Edenahen and his councillors.

By mutual assent of the Dutch and the Edenahen, cases involving serious offences – grand theft or murder – were brought before the Dutch governor and his council for settlement.[11] Further, any crime against the person, property or dignity of a white official or visiting merchant was similarly reserved for Dutch adjudication. By definition, a *vrijburger* was entitled to bring his or her case into the Dutch fort for a hearing; the less exalted *tapoeyers*, on the other hand, appear not to have had this right. Finally, criminal acts involving residents of neighbouring towns or polities were settled either directly by the Dutch or by their intercession. Thus, the right of the Dutch to hear certain criminal cases was recognised as legitimate by the other communities at Elmina, and would seem to have derived from shared mercantile interests. Seldom did the need arise for the Dutch to attempt to compel compliance by a show of force.[12]

The Asante community comprised something of a special case. Its position *vis-à-vis* the other communities, and that of its head, the Asante resident, changed markedly during the period under review. From 1815 to 1826 the Edena authorities – and, to a certain extent, the Dutch as well – saw themselves as under the 'protection' of the Asantehene. Consequently, the Asante community in Elmina – a changing group of itinerant traders, resident mercantile agents, and the political staff of the resident – enjoyed a privileged position. Thus resident Kwadwo Akyampon (at Elmina 1822–32) was taken in as a co-opted member of the Edenahen's council. Accordingly, during 1815–26, disputes between Asantes and Elminas appear to have been settled before that body, with the active participation of an Asante official. After the major defeat of the Asante army at Katamanso in 1826, however, the status of the Asante community declined significantly, as the ability of the Asantehene to act as the town's 'protector' was brought into question. The resident's judicial powers were consequently reduced to that of overseeing the internal affairs of the Asante community at Elmina. After 1826 serious disputes between Asantes and Elminas tended to be brought into the Dutch fort for mediation.[13]

Thus the records of the twenty or so cases heard by the Dutch Council during 1815–30 encompass by no means all incidents of the adjudication of criminal acts in Elmina, nor can they be said to contain a proportionally representative sample. Most of the recorded cases involved either white officials or *vrijburgers* as the injured parties. However, it should be borne in mind that each of the three permanent communities resident at Elmina – the Edenafo, the *vrijburgers* and the Dutch – were similarly stratified: prominent members of each drew upon the labour of a class of slaves. Moreover, all four communities possessed similar notions of individual rights in private property, at least with regard to the commodities of trade; and it was trade, after all, that brought the various communities into contact with one another at Elmina. Thus, it seems reasonable to conclude that fundamental social tensions evident in one Elmina community were probably indicative of those found in the others.

Cases of theft

All of the cases of theft heard by the Dutch Council involved the purloining of items valued in the trade at Elmina: cloth, guns, gold dust, etc. Fully half the cases involved acts allegedly committed by slaves. Most of the remaining incidents of theft were attributed to *tapoeyers* or *vrijburgers*. Only one involved a 'free citizen' of Edena; but even in this case, his accomplices in an alleged 'cattle-stealing' ring were slaves. Thus the Dutch were broadly aware of motivations for theft beyond the mere 'desire for gold'; 'poverty and misery' they accepted as a rationale, though not an excuse. And in the case of slaves, mistreatment by one's master held a similar status.[14]

Therefore, in cases of theft by slaves, the Council frequently inquired if destitution or mistreatment lay behind the act. Surprisingly – or perhaps not so surprisingly when one considers that accused slaves were questioned by, or in the presence of their masters – no slave criminal was able to demonstrate to the satisfaction of the Council that poverty or mistreatment was the prime motivation for his act. Consequently, in these cases it is difficult for the historian to distinguish 'social crime' from 'crime without qualification', a situation not unlike that noted for eighteenth-century England.[15] Accordingly, in the following examination of some of the more interesting cases of theft, attention will be focused on the aspects of daily life at Elmina revealed in the sources, and on differing forms of punishment accorded to various thieves on the basis of their social standing.

One slave who explicitly cited poverty as the reason for his illegal action was Kwasi Anan, a slave owned by the Dutch government who served as a soldier in the fort garrisons.[16] Accused of having stolen two pieces of cloth (value: less than one ounce of gold), Kwasi Anan asserted that he had been 'driven by hunger, because he had been forced to pay for his father's funeral with the subsistence pay he received' from his Dutch masters. However, the Council remained unmoved by this explanation: why had he not then asked for an advance on his pay, or ever complained of hunger to the Dutch? Kwasi Anan is said to have given no reply to this question.

Anan's crime had been committed in the town. The pieces of trading cloth had been removed from the bedroom of a white official, J. Oosthout, who maintained a warehouse and lodgings outside the fort. Oosthout was in fact asleep in the house when Kwasi Anan broke in and seized the goods. Kwasi Anan managed to flee and escape identification as Oosthout awoke detecting the break-in; but suspicion immediately fell on the 'slave-soldier' since it was quickly noted that he was absent without leave from his post on the fort's night guard.

Kwasi Anan subsequently admitted the theft and directed the Dutch to where he had hidden the cloth. He was found guilty by the Council and sentenced to two years of 'forced labour on the public roads', six days in the week, while chained by the leg to a block of wood. But first he was to be brought before the Dutch garrison and declared unworthy to be a soldier, in other words, disgraced publicly before his fellow soldiers. At the end of his prison term, he was to be ejected 'as a scoundrel' from the fort. However, during the period of his detention Kwasi Anan was to continue to receive half of his subsistence pay.

A similarly severe punishment was meted out to two Asantes at the end of 1827.[17] Yaw Gyawa and Kwabena Nyaako, social standing unknown, were found guilty of having stolen a chest full of personal effects belonging to a mulatto woman named Akua Martha. On the morning that she discovered it missing, Akua Martha informed the Elmina town authorities of the theft of the chest, which contained 'clothes, coral beads, jewelry, gold, etc.' The Edenahen told her to keep the incident a secret while his people made an investigation. It took three months for the guilty parties to be found. It seems that Yaw Gyawa and Kwabena Nyaako had immediately removed the items from the chest and taken most of them to an Elmina village in the interior for sale. The proceeds they used for their daily subsistence. They were discovered only when one of them attempted to sell a remaining cloth in Elmina town three months after the theft.

The accused Asantes confessed to the crime before the Edenahen and his council, but the latter saw fit to turn the pair over to the Dutch governor for sentencing. Notwithstanding that the Asantes had clearly been motivated by the severe deprivation that generally characterised life in the Asante community during 1826–30,[18] the Dutch governor sentenced the pair to three years of forced labour and one hundred lashes each. On the day of sentencing, the Asante resident, his retainers and the Edena authorities were called into the fort to hear the Council's decision. Subsequently Yaw Gyawa and Kwabena Nyaako were whipped in 'a public place', presumably in town. Three weeks later the Dutch surgeon reported that the men had recovered sufficiently from their wounds to begin their period of forced labour.

A very different kind of punishment was reserved for *vrijburger* thieves, even though their acts were clearly not motivated by want. The case of *vrijburger* Alexander Pieter was heard by the Dutch council in April 1827.[19] In the previous month, Captain Veyssey of the English schooner, *Pott*, reported to the Dutch governor that while his ship lay anchored off the Dutch fort at Axim some three ounces, four and one-half *engels* of gold had been stolen from his cabin. Veyssey strongly suspected that Pieter had done the deed since he was the only man whom the captain had left alone in his cabin for a brief period. Questioned about the accusation, Pieter at first denied it. Later, however, a witness from Axim town stated that through an open window on board he had seen Pieter reach into the captain's gold box and remove a handful. Thus confronted, the *vrijburger* admitted to the crime, but maintained that he had taken less than one ounce, 'only what he could squeeze between his thumb and forefinger'.

As the Council discussed an appropriate punishment, the governor pointed out that acts like Pieter's adversely affected the confidence of coastal merchants in the honesty of Dutch traders, and that:

> if such acts are not suppressed and punished severely and as a deterrent to others, the door would be thrust wide open to more and perhaps more serious crime, whereby the good name of all would be blemished, and dishonour done to our national reputation.

Consequently, Pieter was condemned to stand in front of the main entrance to the fort for two hours, bound to a pole with a sign hung above his head on which his name and his crime were to be clearly written. He was subsequently imprisoned for a minimum of three months (no mention of forced labour) or until such time as he paid a 'restitution' of three times the value of the gold he had stolen. This was determined to be the amount claimed by the English ship captain.

By early August, Pieter had paid off his fine and was released. However he was told that he should never again be allowed to enter the Dutch fort for any reason, unless it were for another period of incarceration. This was a serious blow for any merchant at Elmina. Sometime later, however, Pieter acquired a high position in the mulatto community, serving as its representative to the Edena government. This seems to have been the motivation for the Edenahen's request to the Dutch governor on 30 April 1830 that Pieter be allowed once again to enter the Dutch fort, since his behaviour had been good ever since his release from detention. The Council complied with the request.[20]

A final case of theft provides further confirmation that punishment for theft was meted out differentially according to the social standing of the thief.[21] Additionally it hints at the tensions implicit in the master–slave relationship when the slave was held not far from his homeland. Finally, it gives a remarkably detailed look into the *modus operandi* of an individual contemplating a major theft and flight from Elmina.

On 22 August 1826 Kwasi Ako, his wife Akosua and a youth named Okum were seized near the Fante town of Anomabo, about thirty kilometres east of Elmina, in possession of forty-seven ounces of stolen gold. All three individuals were slaves belonging to Kuman Badu, one of the wealthiest merchants in Elmina town, and a member of the *Besonfo* council of the Edena government. The three were detained in the English fort at Anomabo, then extradited to Elmina at the request of Kuman Badu and the Dutch governor. Questioned by the Dutch prosecutor and the Council on several occasions, Kwasi Ako's story was as follows.

Born at Akrodo, a Fante village near Apam (about one-hundred kilometres east of Elmina), he had been purchased by Kuman Badu and brought to Elmina during the governorship of H. W. Daendels (1816–18). He rose to become a trusted slave of his master, having responsibility for his master's keys. He thus had access to Kuman Badu's gold dust, which was contained in a box which lay at the foot of the merchant's bed. Kwasi Ako removed some of the gold one evening when his master was out of the house and then waited until three o'clock in the morning to make his escape. Asked if he had been mistreated by his master or denied food, Kwasi Ako is said to have replied, no, 'that he was only punished when he did wrong'. Why then had he committed the crime? The 'devil' made him do it, was his recorded reply.

Questioning by the Council then turned to the woman Akosua. She claimed that she knew nothing of the gold theft until after she and Kwasi Ako had left Elmina. She had run away with Kwasi Ako only because he had forced her to do so with a loaded gun, though she allowed that he had urged her on previous occasions to join him in his planned flight. She stated that she had not been with Kwasi Ako long and so had no children by him. On one occasion he had beaten her, but that matter had been settled. She insisted that she had left her master only because she had been forced to by Kwasi Ako. She then revealed that as part of the preparation for his act, Kwasi Ako had consulted the 'fetish priest' Akyene, who was also a slave but belonged to the *tapoeyer* Christina Lydings. Akyene lived in a village owned by Lydings in the hinterland of Elmina.

Under further questioning, Kwasi Ako admitted that he had consulted Akyene, apparently as a means of ensuring the success of his contemplated crime. His payment for Akyene's services was to be made subsequent to the successful completion of the theft. In the matter of the gun, Kwasi Ako stated that when he was seized near Anomabo he indeed was carrying his gun, an English 'Tower' musket, and he admitted that it was loaded. Asked why he brought his gun he replied, 'because it is customary that if someone goes on a trip he carries a weapon, either loaded or unloaded as he wishes.' He denied that he had forced Akosua to flee with him. The Dutch Council acknowledged that it was customary on the coast for the traveller to carry a gun, but that it was not at all normal for it to be loaded. Hence they ruled that Akosua had indeed been forced to accompany Kwasi Ako in his flight from Elmina.

Interrogation of Okum revealed that he was about 15 years old, had been born at 'Jumba' (an as yet unidentified coastal village located west of the Pra River) and had been brought to Elmina as a slave of Kuman Badu in February 1823. Okum claimed that he had been 'seduced' by Kwasi Ako into running away and that he had been offered Kwasi Ako's gun, once they had arrived safely in the latter's homeland. Like Akosua, Okum asserted that he had had no knowledge of the gold theft until after the three had left Elmina. *En route* to Apam their first step had been at Cape Coast where they had rested at the house of one Abenaba Komfo, a 'half-sister' of Kwasi Ako.

The Dutch Council accepted Okum's explanation of his flight from his master at Elmina. Both the threat to Akosua and the 'seduction' of Okum were seen therefore as exacerbating Kwasi Ako's already serious crime of theft. Consequently, while two of the three members of the Council pressed for a prison term of ten years, the Dutch governor argued for the death sentence. His primary concern, so he stated, was the need to strike a blow against local belief in the power of 'fetish priests'; the execution of Kwasi Ako would demonstrate unequivocally the failure of the efforts of Akyene. In the end, however, the governor allowed his opinion to be overruled by the majority sentiments of the Council.

Cases of murder, attempted murder, or conspiracy to commit murder

In contrast to the instances of theft, where 'social crime' was difficult to distinguish from 'crime without qualification', virtually all cases that involved homicide or efforts to commit homicide had an unmistakable social dimension. The implication of the evidence is that the tensions inherent in Elmina's class and ethnic divisions were more likely to find expression in murder than in social banditry. It appears that in a society such as Elmina's where wealth found its highest expression in the ownership or overlordship of people, the supreme act of rebellion was the taking of human life. It is telling to observe that the Dutch, while acknowledging potential social motivations (such as poverty) for theft at Elmina, emphasised personal or emotional reasons in the cases of homicide; as Prosecutor Cremer put it:

> only cannibals or savages shed blood out of instinct or custom; but those who possess the concept that murder and manslaughter are punishable crimes can therefore only be motivated to commit such acts by such pernicious passions as hatred and revenge.[22]

Nevertheless, the Dutch were sufficiently concerned about the incidents of suspected, attempted, or successful murder to investigate and record in remarkable detail the circumstances surrounding such acts.

In the evening of 16 March 1827, Governor J. C. van der Breggen Paauw, seated at his work table, was brought a fresh cup of tea.[23] It was still hot when he took his first tentative drink, so he swallowed only a small amount; but he was immediately

aware that something was amiss, for the tea tasted extremely salty. He called in his
valet who also tasted it and declared it unfit for consumption. Van der Breggen
Paauw was convinced that an attempt had been made to poison him.

The governor's suspicion fell upon one Kwao Afraba, a government servant
whose primary duty was to serve as a member of the team of men who pulled the
governor's coach through town. On 19 March, van der Breggen Paauw personally
conducted the initial interrogation of Kwao Afraba. Kwao Afraba stated that he
was currently in the employ of the governor, having served white men for fifteen
years. At first he denied that he was even in the fort on the evening of the incident.
When witnesses were produced who testified to his presence near the fort's kitchen,
Kwao Afraba changed his story and declared that on that evening he had indeed
passed through the kitchen. He had just entered the fort from town, having
purchased there his dinner, consisting of some *kenke* (fermented maize bread) and
salt. As he passed the pot that held water boiling for the governor's tea, he lifted
the lid to see what was cooking. Accidentally, so he claimed, his handful of salt fell
into the pot. Van der Breggen Paauw then asked Kwao Afraba to re-enact the
incident; according to the governor, he was unable to do so.

Kwao Afraba nonetheless stuck to his story. He admitted having erred in not
informing the governor's kitchen staff of the accident. He further admitted that
other members of his 'family' had once been found guilty of attempting to poison
the Dutch commander of the fort at Shama. Asked if he had any reason to
complain of the governor, he is reported to have replied, the governor 'has treated
him very well, paid him fairly, fed him at times, and gave him a present at New
Year. He has never been beaten and has not the slightest reason to complain'.
Finally, asked where he had obtained the salt, Kwao Afraba stated that he had got
it from his 'aunt' Amma in town.

The next day van der Breggen Paauw called Amma and her (female) slave,
Okumpaa, into the fort. Both denied that Kwao Afraba had come to them for salt
on the 16th; neither had even seen him for months, for there had been an
unspecified dispute between Amma and her nephew. Amma stated that neither
she nor Okumpaa had been at home that evening, as they had gone out to attend a
wake that lasted through the night. Did she keep salt in a pot on the stoop of her
house as Kwao Afraba claimed? Her recorded answer draws an interesting social
portrait:

> She is a poor woman, and lives from one day to the next . . . each day she buys
> enough salt and other things to eat for that day which is then consumed on the
> same day. The next day she again buys what she needs for the day, because she
> is too poor to buy a lot at once.

Despite her professed poverty, Amma did own a slave, whose testimony confirmed
that of her mistress. Kwao Afraba's and Amma's testimony attest to the
remarkable development of market relations in Elmina town society in the early
nineteenth century.

On 22 June 1827 the Dutch Council met to decide the case. Kwao Afraba was
brought in and, under questioning, consistently maintained his innocence. His
initial confession to the crime, he declared, had been given out of fear of being
beaten. How often had he been beaten in the past? Now he answered that he had in
fact been beaten on one occasion, when he was given twenty-four strokes for
showing up tardily at his post of coach-puller. Had he not confessed to the crime
before six of his fellow servants, and before 'the fetish', when he was first

apprehended? This Kwao Afraba denied. But the Council reasoned that his initial confession, made for fear of the powers of 'the fetish', was more believable than his present denials, since these were motivated merely by fear of punishment. Therefore the members voted for a guilty verdict. But when van der Breggen Paauw demanded as sentence the immediate execution of Kwao Afraba, contrary to explicit standing instructions from the Ministry of Colonies, the other members of Council baulked at his decision.

Instead the Council decided to allow the prosecutor to make a full investigation. On 4 July 1827 the prosecutor submitted his report, which recommended that the charge be dropped and Kwao Afraba be released for lack of evidence. Incensed, van der Breggen Paauw ordered that the accused be held until the Minister's opinion could be heard in the affair. On 26 July 1828 Kwao Afraba was released from custody on instructions from Holland, and told never to enter the fort again.[24] Whatever the truth regarding Kwao Afraba's alleged actions, the case demonstrates at the least the lingering fears of the white officials in Elmina, dependent, as van der Breggen Paauw put it, 'for nearly everything' on the residents of the town.

A much less equivocal case of conspiracy to commit murder was detected in mid-1821.[25] The object of the conspiracy was the *vrijburger* Carolina Huydecoper, widow of Matthys Ruhle, who had also been a *vrijburger*. The conspirators were Kokoroko Panin, Ata, Kokoroko Kakraba, and Donkobo, all slaves belonging to Jacob Ruhle, Sr of Amsterdam, Matthy's father. All but Ata were hired out by Ruhle to Huydecoper, and had been put to work on a plot of land the latter owned in the hinterland of Elmina.

In a memorandum to the Dutch Council, Carolina Huydecoper explained the circumstances of the affair. Kokoroko Panin, Donkobo and Kokoroko Kakraba were originally slaves belonging to the estate of Carel Ruhle, a wealthy *vrijburger* who had died in December 1818. Carel's brother Jacob Ruhle, Sr took over ownership of the men in liquidation of Carel's debts to him. They were placed under the supervision of Carolina Huydecoper, Jacob's daughter-in-law, in May 1820. She paid a fixed sum to Jacob for the right to employ them on the land she owned near Diapem, a small hamlet inhabited by these and other of her slaves.

She was initially impressed with the work done by Kokoroko Panin and Donkobo, and so promised to buy two females to give to them as wives. This she did, providing them with three youths as well. Kokoroko Panin she raised to the status of foreman (*bomba* in the Dutch coastal argot) over the other slaves. The entire labour force was put to work cultivating foodstuffs – yams and plaintain – at Diapem. But larger events soon intervened which led to a change in the relations between Huydecoper and her slaves, and in the character of the work done at Diapem.

Towards the end of 1820 Anglo–Asante hostility increased, culminating in the February 1821 British expedition against Pantsil, a Fante-born sub-chief of the Asantehene who resided at Dunkwa, about forty kilometres from Elmina.[26] The result was a substantial decrease in the level of Asante trade at Elmina, as the trading routes to the interior became insecure. Thus deprived of her normal source of income, Carolina Huydecoper decided to attempt to engage her labour force in the cultivation of crops for overseas export: cotton and coffee. However, she immediately encountered labour problems with her slaves at Diapem. As she reported, young crop plantings were neglected, weeds were allowed to grow between the cotton plants, insufficient water was given to the coffee trees. Huydecoper then acquired more labourers (pawns) and hired a free *tapoeyer*,

Abraham Andriessen, to supervise the plantings and to 'spur on' the work-force.

Andriessen soon reported that Kokoroko Panin and his wife were the source of the trouble. Huydecoper called Kokoroko Panin to Elmina and warned that if his work did not improve she would take away the wife she had given him. Soon thereafter she followed through with this threat; when the wives of Kokoroko Panin and Konkobo came to Elmina bringing firewood from Diapem, Huydecoper told them to remain in town until further orders. Kokoroko Panin came to Elmina the next day to find out what had happened; Huydecoper said that she felt that his wife was causing the problems. But she then released the women with the warning that if there were any further problems, she never again would allow his wife to return to Diapem.

These events took place sometime during the dry season of 1820–21. When the rains came in April–May 1821 the time arrived for the planting of food crops. Huydecoper instructed the slaves that they could now have three days in the week to work their own plots at Diapem, for which she provided the necessary tools, but the remainder of their time had to be spent on her cotton and coffee plantations. She subsequently paid a visit to Diapem and found her plantation in a 'state of anarchy', and concluded that her slaves 'did no work other than for themselves'. Taken suddenly ill however, Huydecoper returned to Elmina but received constant reports from Diapem that Kokoroko Panin and his wife still refused to work; they even threatened to run away to the neighbouring Akan kingdom of Wasa, where Kokoroko Panin had been born.

On 18 June, Huydecoper called all her slaves at Diapem to Elmina and asked them what the problem was. Kokoroko Panin asserted that the main culprit was Donkobo. Having heard many other complaints against Donkobo, she ordered him put in irons. At this point Donkobo revealed that Kokoroko Panin had been plotting for some time to poison his mistress, Carolina Huydecoper.

It emerged from subsequent investigations by the Dutch prosecutor and Council that Kokoroko Panin had conceived of the idea himself. In his defence, he claimed at first that she had mistreated him, but then said, 'actually she did not mistreat me, but always expressed dissatisfaction with me, and also she took my wife from me for two days and for this reason I sought revenge.' Kokoroko Panin had approached another of the Ruhle slaves, Ata, a 'fetish priest', who, it was known, had knowledge of the preparation of poison. Ata instructed Kokoroko Panin to extract bile from a crocodile for the preparation of the potion. The plan was to smear the poison onto Huydecoper's eating table in the hope that her bread might absorb it.

Kokoroko Panin had apparently been unable to obtain the needed bile, though Huydecoper testified that he had gone so far as to ask her for a gun, powder and shot, in order, he had said, to kill a crocodile that was plaguing the residents of Diapem. Ata and Kokoroko Panin eventually confessed their guilt; but the Dutch were especially concerned to discover who among the slaves had had knowledge of the plot but had failed to disclose it to the authorities. Kokoroko Kakraba stated that he had learned of the plot from Donkobo, and used this knowledge to extort better treatment and a promise of money from Kokoroko Panin. Donkobo said he had overheard Ata and Kokoroko Panin plotting in the latter's hut, but claimed he had not had the opportunity to inform Huydecoper until he was called to Elmina. As it later emerged, Donkobo was in any case embittered because his wife had also been withheld from him by Huydecoper.

The Dutch Council decided that Ata and Kokoroko Panin were guilty and liable to banishment from the coast, and so informed the Minister of Colonies. The Minister concurred and on 2 March 1824 the two were put on board a Dutch warship and sent to exile in Java.[27] Kokoroko Kakraba and Donkobo were sentenced to lifelong detention and forced labour on the public roads of Elmina. Meanwhile, Kokoroko Kakraba had died on 30 March 1822, leaving only Donkobo with the prospect of long-term incarceration in the Dutch fort.[28]

Beyond the richness of detail on the lives of those involved, this case suggests that norms existed in Elmina society regulating the treatment of slaves and limiting the appropriation of their labour. The general context of the slaves' grievances seems to have been that of Huydecoper's demand for increased work discipline and her reduction in the time allotted to the tending of the slaves' own plots. The specific incident that set off the plot seems to have been Huydecoper's detention of two of the slaves' wives. The slaves' resistance to their mistress's demands took the form of a plot to murder her. Though the conspiracy was clearly unsuccessful, it would be interesting to know if the incident had nevertheless stimulated Caroline Huydecoper to change the way that she dealt with her slaves; on this the documents are silent.

Within about a year of Kokoroko Kakraba's death, Donkobo seized an opportunity that presented itself when his prison overseer fell asleep; he broke his chains with a stone and made a successful escape.[29] He remained in the vicinity of Elmina, however, and for some five years lived as a kind of outcast near Diapem.[30] It is clear that his presence there was tolerated by his former fellow-slaves, as well as the free inhabitants of the Elmina village of Amoanda, which lay in the vicinity. While no apparent effort was made to inform the Dutch that their former prisoner was still in the Elmina district, it should also be noted that Donkobo's haunts were at the boundaries of Elmina, near the villages of the ever hostile Fante, in a no-man's-land that Elminas avoided if possible. By report, Donkobo subsisted by hunting in the surrounding forest, using a cudgel and machete, and by purloining cultivated foodstuffs from farms located near Diapem and Amoanda. When these were insufficient he occasionally begged for food in Amoanda.

Donkobo maintained a solitary hut not far from Diapem and directly adjoining a plot of land cultivated by one Ahemma and her sons Kwaku Sakyi, aged about 18, and Kwaku Asamankoma, aged 12, all slaves of Jacob Ruhle, Sr of Amsterdam, but supervised by his son Abraham at Elmina. On several occasions Kwaku Sakyi had caught Donkobo attempting to steal food from their farm. On 15 September 1828, the two brothers went to the farm to harvest yams, even though an army of Fantes, Wasas and Denkyeras was gathering in the vicinity to attack Elmina. The boys failed to return home that afternoon, and Ahemma went out to look for them. Twenty steps from Donkobo's hut she encountered the body of her elder son, his throad slit, and his genitals hacked off; his gun lay across his chest and his cloth still contained a small amount of gold dust.

The following day Abraham Ruhle sent thirty armed men to recover the body and investigate the scene. They were unable to locate either the other boy's body or to find any trace of the murderer. Three days later Donkobo appeared in Amoanda carrying a knife, a machete and a blood-stained cudgel. As he asked for some food, saying he had not eaten for three days, he was seized by the inhabitants of Amoanda and taken to the Edenahen, who transferred him to the Dutch fort for detention.

Subsequent investigations by the Dutch revealed an interesting train of events that led up to the murders. Kwaku Sakyi and one Kofi Abi, an artisan-slave also owned by the Ruhles, had both had their food-plots plagued repeatedly by theft. According to Kofi Abi's testimony, he and Kwaku Sakyi had caught Donkobo in the act of plundering their farms on two occasions; to prevent a recurrence, they told him that he ought to come to live with the other slaves at Diapem. Apparently the two felt that Donkobo's actions could best be controlled by forcing him back into the social community of slaves at Diapem. Caught a third time on their farms, they reportedly said to him: 'We have warned you; and in order to compel you to come live with us, we are going to destroy your hut.' As Donkobo fled the scene, Kwaku Sakyi and Kofi Abi pulled down Donkobo's hut and set fire to the remains. A short time later the murders occurred.

In the initial interrogation by the prosecutor, Donkobo was said to have confessed to the crime. But brought subsequently before the Council, and the Elmina public that was allowed to witness the trial, Donkobo conveyed an impression of total alienation from the proceedings. He refused to answer any questions, 'remaining seated without emotion, acting as if he were asleep'. Under insistent questioning from the governor, Donkobo finally began to respond. Denying any wrongdoing, his answers were largely limited to the words, 'they burned my hut'. Then he asserted that 'Nyankopan [God] had made them pay'. At the end of the day's session Donkobo is said to have related to the Council the story of his life, which the Dutch unfortunately failed to record, except to note his insistent claim that once his wife had been taken from him. Carolina Huydecoper's action more than seven years earlier apparently still loomed large in Donkobo's mind.

The next day's session failed to elicit any further response from Donkobo, as various witnesses were brought before him. Asked how he had managed years earlier to escape from detention, he is said to have replied, 'Nyankopan is my father'. On the next day, 25 February 1829, the Dutch called in a 'fetish priest' to question Donkobo, but he too was unable to elicit any further information. Finally, after hearing testimony that ruled out the possibility that the Fantes might have been responsible for the crime, or that Kwaku Asamankoma might have killed his elder brother and then fled, the Dutch Council returned a verdict of guilty. Since Donkobo was a repeat offender, punishment was set at death by hanging. On 26 February 1829 Donkobo was publicly executed at Elmina.

The trajectory of Donkobo from favoured slave, to co-conspirator in a projected murder, to anomic thief and self-imposed social exile, had seemingly all the makings of the classic social bandit. But instead Donkobo engaged in the senseless harassment of his fellow slaves at Diapem – it was reported that at times he stole more than he himself could consume, throwing the excess into the paths where Elmina women passed – and finally committed one of the ultimate anti-social acts, child murder and mutilation.

Conclusion

The available documentation regarding crime in pre-capitalist Elmina provides little evidence of social banditry, at least as that concept has customarily been defined. That this should be so is not surprising, given the structure of material life in the coastal town. Eric Hobsbawm has noted that social banditry is most likely to emerge 'at two moments of historical evolution: that at which primitive and communally organized society gives way to class-and-state society, and that at which the traditional rural peasant society gives way to the modern society.'[31] Neither situation could be said to have obtained in early nineteenth-century Elmina, a society with a highly developed market economy in which slave labour played a significant role in production.

Nevertheless there was an unmistakably social dimension to most of the incidents of crime and their adjudication at Elmina. Examination of the records of 'criminal' activity reveals both the social tension that existed between classes – in particular between master and slave – as well as the way in which the law recognised the legitimate demands of the subordinate classes for adherence to normative treatment by their social betters. It may indeed be argued that the resistance of Elmina's lower classes to any deterioration in their rights and privileges was at least partly responsible for the failure of a slave-based export agriculture to emerge here, as entrepreneurs such as Carolina Huydecoper no doubt would have liked.[32] The available evidence conveys the impression of a well-knit, hegemonic class structure at Elmina in which differences of wealth, status and privilege were legitimated through a general adherence by the town's dominant classes to broadly accepted norms of appropriate treatment of slaves and commoners. These norms were in turn set and enforced at least in part through occasional threats or actual outbursts of violent resistance on the part of members of Elmina's subordinate classes.

NOTES

1 The research on which this paper is based was carried out with financial support from the Asante Collective Biography Project (National Institute for the Humanities funded), Social Science Research Council, and Fulbright-Hays (Department of Education), to whom I am grateful. I would also like to thank Donald Crummey, Allen Isaacman, David Roediger and Ivor Wilks for reading and commenting on earlier drafts of this paper. Of course, any deficiencies that remain are my own.

2 Algemeen Rijksarchief, The Hague, Archief van de Nederlandsche Bezittingen ter Kuste van Guinea (hereafter NBKG), inventory number 804: Minutes of Council dd Elmina, 28 June 1827, Prosecutor's Report dd 28 June 1827.

3 E. Genovese, *Roll, Jordan, Roll: The World the Slaves Made* (New York: Pantheon, 1974), p. 26.

4 ibid.

5 NBKG 796x to 805.

6 For a brief introduction to the 'recruitment' effort, see Albert van Dantzig, 'The Dutch military recruitment agency in Kumasi', *Ghana Notes and Queries*, no. 8 (1966), pp. 21–4; and René Baesjou (ed.), *An Asante Embassy on the Gold Coast* (Leiden: Afrika-Studiecentrum, 1979), pp. 23–4.

7 The following discussion of the structure of Elmina's communities is based generally on several years of archival research in eighteenth- and nineteenth-century Dutch documents, the records of the Secretary for Native Affairs in the National Archives of Ghana, and on field research conducted in Elmina during 1978–9. See also, H. Feinberg, 'A history of Elmina, Ghana' (unpublished PhD thesis, Boston University, 1969) and Baesjou, *Asante Embassy*, 'Introduction'.

8 For example, an American ship's captain provided the following vivid description of

Governor F. Last and his coastal family: 'The Governor is a short, jovial fellow, five feet high, his wife a black shiney negro weighing about 200.lb [sic] by whom he has had several children. He also had two other negro wives by whom he had children, but the first one only presided at his table;' see Norman R. Bennett and George E. Brooks (eds), *New England Merchants in Africa* (Brookline: Boston University Press, 1965), pp. 118–19. Needless to say, one would search Last's official journal in vain for such specific and enlightening information.

9   For a reconstruction of the history of the Asante residency at Elmina, see chapters 1, 3, and 4 of my dissertation, 'Asante and the Dutch: a case study in the history of Asante administration, 1744–1873' (unpublished PhD thesis, Northwestern University, 1983).

10   Dutch ideas concerning the *vrijburger* were probably derived from their somewhat earlier colonial experience in the East Indies. See the discussion on the status of mixed-descent offspring in Indonesia and the Cape in G. Frederickson, *White Supremacy: A Comparative Study in American and South African History* (New York: Oxford University Press, 1981), pp. 96–7.

11   Relations between the Edena authorities and the Dutch were regulated by a 'contract', which was subject to occasional revision. For a specimen, see J. T. Furley (tr.), *H. W. Daendels Journal and Correspondence* (Legon: Institute of African Studies, 1964), pp. 196–208; note especially article 14.

12   The Dutch were naturally reluctant to resort to force, since their only viable threat was to destroy a portion or all of the town with cannon. It was of course from the town and its trade links with the interior that the Dutch derived their material sustenance and mercantile income.

13   Indeed, a majority of the twenty cases heard by the Dutch Council date from 1826–30.

14   See Prosecutor Cremer's comments in NBKG 804: Minutes of Council dd 4 July 1827, Report of the *Fiscaal*.

15   See Douglas Hay, Peter Linebaugh, John G. Rule, E. P. Thompson and Cal Winslow, *Albion's Fatal Tree. Crime and Society in Eighteenth-Century England* (London: Allen Lane, 1975), p. 14.

16   Information regarding this case is drawn from NBGK 796x: Minutes of Council dd 13 December 1816. I wish to acknowledge the assistance of Mr D. Owusu-Ansah in the rendering of Dutch transliterations of local names into modern Akan.

17   Information regarding this case is drawn from NBKG 805: Minutes of Council dd 7 December 1829; also NBKG 358: Elmina [Governor's] Journal, entries for 1, 8 and 27 December 1829.

18   See Yarak, 'Asante and the Dutch', p. 40.

19   Information regarding this case is drawn from NBKG 804: Minutes of Council dd 2 April and 2 August 1829.

20   NBKG 805: Minutes of Council dd 30 April 1830.

21   Information regarding this case is drawn from NBKG 804: Minutes of Council dd 26 and 30 August, and 6 September 1826.

22   NBKG 804: Minutes of Council dd 4 July 1827, Report of the *Fiscaal*.

23   Information regarding this case is drawn from NBKG 804: Minutes of Council dd 19 and 22 June, and 26 July 1827.

24   NBKG 358: Elmina [Governor's] Journal, entry for 26 July 1828.

25   Information regarding this case is drawn from NBKG 800: Minutes of Council dd 6 and 23 July 1821.

26   On the wider context of events, see Pre-Code Sheet 27 (Opentry), *Asante Seminar '76*, no. 4 (1976), and Career Sheet 39 (Owusu Dome), *Asantesem: The Bulletin of the Asante Collective Biography Project*, no. 10 (1979).

27   NBKG 802: Minutes of Council dd 2 March 1824.

28   NBKG 801: Minutes of Council dd 13 April 1822.

29   Information regarding Donkobo's second case is drawn from NBKG 806: Minutes of Council dd 23, 24, 25 and 26 February 1829.

30   Donkobo's apparent reluctance to leave the Elmina district entirely probably derived

from his being so far from his homeland; his name suggests that he was born in the northern area of present-day Ghana.

31   E. J. Hobsbawm, 'Social banditry', in Henry A. Landsberger (ed.), *Rural Protest: Peasant Movements and Social Change* (London: Macmillan, 1974), p. 149.

32   Of numerous attempts to establish a form of plantation agriculture in commodities such as cotton, coffee, and tobacco by Dutch and *vrijburgers*, all failed. In a final report on the Dutch possessions on the Gold Coast after their cession to the British, the Dutch Minister of Colonies noted to the Dutch parliament that the failure of export agriculture to take root was due primarily to 'the unwillingness of the negroes to perform disciplined work'. See *Handelingen der Staten-General*, 1873–74: Bijlagen II 156–2, p. 2: Missive of the Ministers of Colonies and Foreign Affairs dd The Hague, 30 January 1874.

2 Theft and social protest among the tin miners of northern Nigeria[1]

WILLIAM FREUND

... The revolt of the workers began soon after the first industrial development, and has passed through several phases ... The earliest, crudest, and least fruitful form of this rebellion was that of crime. The working-man lived in poverty and want, and saw that others were better off than he. It was not clear to his mind why he, who did more for society than the rich idler, should be the one to suffer under these conditions. Want conquered his inherited respect for the sacredness of property, and he stole. We have seen how crime increased with the extension of manufacture; how the yearly number of arrests bore a constant relation to the number of bales of cotton annually consumed.

The workers soon realized that crime did not help matters. The criminal could protest against the existing order of society only singly as one individual; the whole might of society was brought to bear upon each criminal, and crushed him with its immense superiority.[2]

Engels' view of property theft among English proletarians in the early phases of the Industrial Revolution quoted above proposes an historical progression in the forms taken by class conflict within the capitalist mode of production.[3] Theft and other individualised types of resistance to the iron rule of private property dominate early forms of proletarian class consciousness, but give way subsequently to collective organisation and more directly challenging industrial and political struggle. Engels' analysis is strongly and imaginatively seconded in the work of E. J. Hobsbawm, who describes as 'primitive rebels' the bandits, *mafiosi*, and other bold men who rob from the rich to help the poor or who appear to protect communities against the predatory ravages of an emergent capitalism; the same men bow ultimately to collective and rational forms of protest that strive to meet capital head on. In the light of trade unionism and socialism, 'primitive rebels' must appear as inchoate, anti-social, and gangsterlike.

Other historians of social protest and class conflict have found that the historical progression suggested by Engels has not manifested itself so neatly in practice. Moreover, there have been many occasions on which industrial-age working-class organisation has been comparatively or totally ineffective in taking up a general class interest, and it is precisely at the points when and where working-class organisation has failed that a study of theft may prove to be of most interest. Theft, as the following discussion will suggest, can play a major role alongside the more conventionally developed forms of working-class protest; indeed, it can provide an effective vehicle of protest when and where conventional forms fail. Far from being crude or fruitless, theft can be an effective economic response by proletarians to the totalising surplus demands of corporate capital.

One cannot study such responses without investigating, at the same time, the material and social conditions of the labour process and broader social forces. For instance, the idea of theft makes no sense apart from its relationship to the general notion of property. Theft may be seen as part of the ordinary process of capitalist competition and accumulation, just as swindling and fraud were recognised by Marx as a logical consequence of the appropriation of surplus value through exchange. In this broader theoretical framework the distinction between theft and *re*-appropriation may be less important than their similarities.

Such would be the case for northern Nigeria, where a significant proportion of thefts took the form of an appropriation of mining profits by petty entrepreneurs. When, in the nationalist era, these entrepreneurs were themselves Nigerians, rivalry over appropriation evolved into a crude struggle by national capital to secure a share of the take and to control a mass clientele through the disposition of patronage. At the same time, social and communal aspects of theft had developed by the 1940s, reflecting a view of property that clashed with the assumptions enshrined in the colonial legal system and that, as a consequence, implictly allied itself with contemporary political and social protest. If the persistent community appropriation of tin ore failed to achieve the ideological coherence of so much European resistance to capitalist infringement of 'traditional' popular rights (as demonstrated in the work of E. P. Thompson and George Rudé), it nonetheless contained the germ of a populist 'moral economy' that conflicted with the political economy of the mining companies and the state.

Theft has always been an important theme in the history of mining men, wherever minerals are scarce or some individuals are accorded exclusive rights to mine. The appeal of theft has been all the greater where more precious minerals are mined, particularly gold and diamonds. Tin is not a gem and has been used primarily as an alloy for industrial purposes, but it is relatively scarce and very valuable as an industrial metal. In Cornwall, England, the illegal mining and sale of tin became a significant feature of the industry after the Middle Ages.[4] In those parts of the world which the West has underdeveloped (the source of most tin ore since the late nineteenth century), tin theft has plagued the arrangements of mining capital.[5]

In the diamond and gold mines of South Africa, a very thorough, dehumanising, and expensive system of controls was initiated before 1900 to prevent poor African workers from bringing home a share of their toil. These controls were intimately linked to a broader strategy of extreme labour coercion.[6] However, this link was lacking in northern Nigeria where labour control was left very largely in the hands of African middlemen, and labour supply was left successfully to the indirect pressure of market forces. The quantity of tin mined through alluvial and open-cast techniques in Nigeria required a huge labour force but did not create profits large enough to justify mounting a vast security system. As a result, the organisation of production in northern Nigerian tin mines provided the historical conditions for theft, and its attractions for workers have been commensurately great.

Like many of the important capitalist mining operations developed elsewhere in colonial Africa, Nigerian tin mining had its origins in a precolonial and precapitalist setting. Tin was traded and used over a wide area, but was produced essentially for a regional market. At least in the nineteenth century, little tin appears to have passed beyond the confines of the central savanna of West Africa. In tin-mining centres, labour remained fairly closely integrated into the

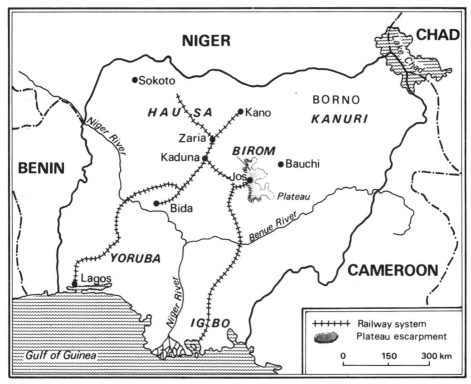

Map 1: *Nigeria, showing the tin-mining areas*

production of extended households, including retainers and slaves. Such house-holds engaged in activities other than tin mining, and, in particular, produced most of their own food requirements. By the nineteenth century, most of what became northern Nigeria had experienced the development of class society, but a class given over entirely to wage labour had only begun to emerge. The lack of free labour ultimately limited the expansion of commodity production and accumu-lation of capital, and held tin production at a small scale. Surplus value from tin mining, smelting and smithing seems to have been appropriated by the smelter owners and merchants to a great degree. Some was extracted by the state in the form of taxes.[7]

Smelting occurred close to the point of production while most manufacture of tin metal into useful articles through smithing took place in urban centres, notably Kano, Bauchi, and Bida. Tin was used largely for ornamenting plates, horsegear, weaponry, and jewelry to be sold in urban markets. Most mining was performed by Hausa-speaking workers near the northern foot of the Jos Plateau, from which trading routes stretched to the commercial entrepôts of the Sokoto Caliphate. The riverbeds grew richer in tin as one ascended the escarpment to the high plateau, but here the writ of the Caliphate did not run.

Hausa-speaking peoples on the plains, where most of the precolonial tin mining took place, had been Islamised and transformed into a peasantry with only vestiges of clan organisation. On the other hand, the Plateau peoples had successfully prevented their incorporation into the Muslim emirates under Sokoto.

Plateau inhabitants comprised a multitude of small communities, speaking a variety of languages. One of the largest language groups was Birom, and most of the richest tin-bearing ground lay in northern and central Birom country.[8] The Plateau indigenes very probably mined and smelted some tin for their own purposes, and they also traded tin in small quantities; but on the whole they closed off the exploitation of the best sources of tin on the Plateau.

British merchants had become aware of local tin mining on the Plateau by the 1880s, when Niger Company factors began to purchase straws of tin metal at their trading posts on the Benue River, but they did not proceed to investigate the commercial possibilities until the conquest of northern Nigeria in the early 1900s. In 1902, when the British occupied Bauchi – the emirate in which they expected to find the greatest source of tin – they quickly made inquiries and prepared for the development of a European mining industry. Within the next few years, expeditions conquered the Plateau, often using extremely brutal methods to clear the way for miners. The British held that throughout northern Nigeria all land was the property of the Crown by conquest. As a result, mining companies were able to gain control of property through prospecting licenses and leases by direct arrangement with the colonial regime. At the same time, the companies completely disregarded the rights of Nigerians as owners of the land. While mining firms had to provide some compensation to farmers whose fields they had destroyed, the amount was not fixed by law and was often settled in the form of a small bribe to the local 'chief', himself usually dependent on the British for the legitimacy of his office. The farmer or chief had no right at all to forbid mining activity.

The British takeover demolished the existing tin industry. After 1902, imported tin began to undersell and replace the local product. Nigerian smelters were forced to shut down and until 1961 all Nigerian tin was exported to Britain as ore to be smelted on Merseyside. In addition, most of the labour force went to work for the British companies, where they provided an essential core of skills in prospecting and washing tin ore. For over a decade after the infusion of British mining capital, all major new discoveries were actually made by Nigerian workmen. Late in 1910, the British Treasury approved a loan for a light railway to extend from the main line being built through northern Nigeria. The feeder line reached the foot of the Plateau in 1912, and was extended up to the new mining centres of Jos and Bukuru in 1915. It proved indispensable for exporting large quantities of ore for which there was not sufficient portage labour.

The decisive requirement for establishing British capital in Nigerian tin mining was an enormous expansion of the labour force. Until hydro-electric power became available in the 1920s, and even thereafter, the possibilities of substituting machinery for men were very limited. Supplying labour to the mining sites was indirectly aided by the British policy of demanding taxation in cash. While taxation was not new in northern Nigeria, a portion of it had always been collected in kind, and the sums fluctuated to take into account the effect of frequent drought, flood, or insects upon the harvest. But once the British began to demand cash exclusively, natural crises made tax payments very difficult. Demands for 'political labour' to build railways and roads were also very onerous at times, and peasant households sometimes collapsed under the strain.[9] As a result, permanent individual workers and a much larger stream of migrant, dry-season labour came to the Plateau to work tin. Migrant labourers tended to come from parts of northern Nigeria that did not produce cash crops, but some trekked over the border from French territory. The largest number were Hausa- and Fulfuldespeakers, entering from Bauchi, Sokoto, and Zaria provinces, but there were also many Kanuri from Borno in the north-east, people from the hill country of eastern Bauchi and the Borno-Kano borderlands.

Mining managers, however, found these workers insufficiently reliable and too expensive. For these reasons, they sought to obtain a bigger proportion of their labour force from the Birom-speaking people who lived atop the major minefields, as well as from other native Plateau inhabitants. The colonial administration looked upon this reorientation in labour recruitment with some favour since it hoped to enforce its authority through the regular collection of taxes in cash from the 'Plateau pagans'.

The precapitalist Plateau economy, it should be added, had been delicately poised. In general, the quality of the soil on the Plateau was poor. Only a carefully balanced husbandry that created necessary labour for households throughout the year made viable the rather large population densities to be found in many districts. Plateau peoples produced little surplus that could be marketed or appropriated, and much of it was in the form of ironware goods, not foodstuffs. Even limited, dry-season labour had a bad effect on agriculture. When possible, Birom communities migrated short distances to get away from the mining activities or tried to earn cash through crop sales. However, armed tax-gathering patrols and steady, if less direct, pressure through the agency of the colonial Native Authority system increasingly propelled workmen toward the mines.

Initially, pressure on the Birom was a fairly straightforward case of corvée labour extracted by a colonial state in circumstances where a non-capitalist mode of production prevailed. However as time passed, the agricultural system of the Birom became acutely strained.[10] With the introduction of British-protected markets around 1906–7, the Birom's need for cash increased, notably for the purchase of cotton clothing. Extended families tended to break up into smaller units; the population continued to increase, while Fulani herdsmen from the plains brought large herds of cattle up to the Plateau to graze, thus destroying its grass cover. Mining operations alienated much of the best land alongside streams and destroyed the brush cover which the Plateau inhabitants used for firewood. As a result, the manure made available by the Fulani herds had to be used as fuel rather than fertilizer. The impressive system of local terracing was abandoned, erosion spread, and fallow land became even scarcer. By the 1940s, land previously left fallow for three out of four, or five out of six years, was used almost continually. The results were, of course, poor crops and further soil exhaustion.

Under these pressures, the Plateau people provided the speedy and obedient workmen sought by management. Moreover, there was no need to provide them with housing or foodstuffs; they were intimidated by the conditions of work and eager to return to their households. As early as 1918, the Birom were working in large numbers for several weeks a year and accepting consistently lower wages than other workers.

The Great Depression marked a notable watershed in the wages and conditions of labour in northern Nigeria. Earlier, labour could command something of a premium. When mine labour was first in high demand from 1909–10, the price of labour throughout the territory rose from 6d to 9d per day, an increase of 50 per cent. Attempts to lower wages were successfully resisted by withdrawals of labour from the minefields. During the Depression, prices of commodities and wages fell dramatically, but the regime continued to collect taxes at high levels; far more wage labour and cash-crop production were required for people to meet tax demands. As a result, while wages fell by the end of 1931 to one-third of their level in the mid-1920s, mine managers could not employ all the hands seeking work. As the Depression continued, estimates for Plateau workers suggest that eight weeks, rather than four, were beginning to be the norm for dry-season labour. Furthermore, the proportion of people who had to work for a wage to secure enough to eat was increasing.

Conditions deteriorated rapidly in the 1940s. Under the pressure of wartime demand, tin mining expanded over even more cultivable land. As the quality of tin in the ground decreased, more and more workers were needed. But while the companies had to employ a larger labour force than before the war at a time of mass inflation they were unprepared to raise wages substantially. After the Second World War, Birom villagers often worked seven months a year in the mines, and a miner's wage was more than half used to provide a meager diet for himself.

Direct control over labour was not, however, very vigorous. In the early years of the domination of capitalist mining, managers realised that any large-scale hiring of European supervisors would eat away at profits. The ratio of white to African in the mines remained extremely low.[11] Gangs of workmen were supervised and often recruited by headmen, who formed part of a system of labour contracting that prevailed throughout the Nigerian labour market.

Mine managers did not know who worked for them or lived in their camps. Payment and tax collection were in the hands of the headmen, apart from the categories of skilled labour that grew up when heavy machinery began to be introduced in the 1920s. As headmen were often creditors to their workers, indebtedness proved decisive in subordinating labour even though it did not contribute to rigorous enforcement of production goals. In contrast to Southern African mine compounds, the Plateau camps contained all manner of people, including traders, Islamic teachers, craftsmen and farmers. Mine managers did not generally seek to, and were not able to, control squatting.

References to tin theft go back to the earliest days of the capitalist takeover of the industry. A letter of 1913 listed three main forms of theft: 1) highway robbery; 2) theft by interpreters and headmen; and 3) tributers working on the property of a company other than the one with which tributing arrangements had been made.[12] In his report of 1913, the Government Inspector of Mines, Langslow Cock, considered that 'this is the commencement of a very large question that is becoming acute.'[13] Official discussions of theft placed the brunt of responsibility on management and its lack of effective supervision. But without direct supervision, it was impossible to determine where tin ore came from or to confine workers to the spots where they could mine legally. Early tin theft could be very substantial; a manager reported tin stolen worth £5–£10 000 early in 1912.[14] However, mining companies were unprepared to lay out large sums to provide security staff. During the pre-First World War period, most capital attracted to Nigerian tin mining was speculative and looked to stock-exchange dealings rather than production for profits. The amount of tin which could be extracted through simple techniques did not warrant expenditure on hundreds of well-paid European supervisors. Rather, the export of tin ore was made illegal by any other than an authorised lease-holder, thereby assuring that the profit from every pound of tin would accrue to some European entrepreneur.[15] After 1913, Nigerians were forced to close down their smelters, in part because European miners accused them of accepting stolen tin.[16]

Because of the lease-holder's monopoly, fences of stolen tin were primarily European before the Second World War, while thieves were usually 'Hausas', mine company employees from the plains. In 1916, the Nigerian government created a mining 'right' which allowed European petty capitalists to operate as 'private miners' on their own behalf for the first time. Private miners flourished in the immediate post-First World War tin boom and in renewed prosperous times for the tin trade in the mid-1920s and mid-1930s. The government suspected that they were the principal buyers of stolen tin, as they operated on very low costs, used little machinery, and generally paid tributers better prices for tin than the sizeable companies.[17]

Tributing, the system whereby miners actually sold their ore to lease-holders at an agreed rate, made theft enormously easier. Mining managers found tributing a very effective way of controlling labour cheaply and were unwilling to replace it by more expensive and direct means of supervision. They were equally reluctant to see greater administrative authority in the mining camps, which they admitted to be veritable thieves' dens full of 'casuals of ill-repute'.[18]

During the interwar years, tin theft was associated exclusively with plains area 'Hausas' in the mine camps. The local Plateau population had neither the commercial contacts nor the expertise in mining to take an interest. They were far more exercised by the British ban on iron smelting which struck at the very heart of their economy. In the 1920s, iron tools were stolen with more alacrity than tin ore. Hausa workers were reported to be stealing iron picks in order to sell them to the local Birom. In 1932, an Ngas man from farther south on the Plateau was electrocuted while attempting to steal iron wire, and, for a time thereafter, the companies agreed to hire no Ngas workers.[18]

In 1937, however, there appears in the records the case of Hammod, a farmer of the village Fayoral near Ropp in the southern end of the minefields, who died in an accident while mining illegally with tools acquired from a mining camp. He was selling tin as tribute to a private miner by means of the miner's interpreter. The interpreter got a share of the take and was presumed to be the real initiator of the poaching. Another aspect in the story of Hammod was more interesting: an entire village of Birom was regularly involved in illegal mining and would 'put on Hausa clothes' in order to make tin sales.

European fences at this date were apparently unwilling to buy directly from any but Hausa middlemen.[20] The story of Hammod is the first reference I have located to what would become the principal form of tin stealing after the Second World War. So long as theft lay in the hands of Hausa or Europeans, it represented a direct appropriation by petty entrepreneurs. When taken up by entire Birom communities it acquired a qualitatively different and distinctive social aspect.

During and after the war, tin theft increased precipitously, and continued to rise into recent years.[21] Moreover, officials readily recognised that Birom villagers were largely responsible for the increase in theft.[22] A 1952 committee listed seventeen communities where good evidence existed that much of the population engaged in illicit tin mining.[23] A more recent observer suggests that perhaps 80 per cent of Birom villagers in the most heavily mined section of the Plateau are regularly involved.[24] The techniques of illegal tin mining are graphically described in the essay of a sociology student at Ahmadu Bello University, to which the following pages are very indebted. Villagers call theft *lambar gudu*, or running from the license in Hausa, lambar being a corruption of the word 'number', and the numbered license referring to the police.[25] To this day, the Birom go out particularly at night, when they can hope to avoid detection, and during the dry season, when they are able to make use of tunnels and when money is especially

Table 1: Incidence of tin theft

Date	Amount	Value in pounds sterling	Percentage of production	Source
1944		60–100 000	5	NAK, JOSPROF III, S.112, v.1, B. E. Frayling, Chief Inspector of Mines to Chief Secretary, Lagos, 6 November 1945.
1945			10	NAK, JOSPROF III, S.112, v.1, Manager, Ex-lands to Chief Secretary, Lagos, 13 April 1945.
1950	205 tons	250 000		NAK, JOSPROF III, S.112, v.1, Resident, Plateau (Rex Niven) to SNP, 6 December 1950.
1952	400 tons (ATMN)			NAK, JOSPROF III, S.112, v.1.
1971		1 000 000	33 per cent for seven biggest producers	P. Harrigan, 'Nigeria's Tin Mines – Progress, Prospects and Problems'. *West Africa* no. 2897, 18 December 1972.

short. They also work tin in the daylight, especially on small streams, but this is more dangerous. Usually a party of half a dozen will go out at a time, typically twice in an evening. One will serve as a lookout while others will dig or wash tin. It was crucial for Birom to acquire the techniques of tin miners in order to be able to pan efficiently. A large number of the thieves are mine employees, able to judge where a paddock is rich in tin.[26] The danger from landfalls is considerable, and a notable proportion of serious accidents in the minefields are caused by illegal mining.

Villagers return with their ore concentrate to a rendezvous arranged with middlemen, typically in the backyards of drinking houses early in the morning before any police might be likely to arrive. The middlemen are agents of private miners who served as the fences. These miners have the vehicles to transport ore and, most importantly, the legal right to sell tin. Since the smelter first opened on the Plateau in 1961, all tin has been sold to it, with the government taking its royalty share at the time of each transaction. One middleman interviewed for the essay aspired to become such a private miner himself. A brother of his had recently dismissed him because he was doing more and more business on the side. While the middlemen claimed that profits might be acquired easily, villagers have apparently become more sophisticated in the 1970s: they are ceasing to use middlemen and are selling directly to private miners whom they approach themselves.[27]

Another form of illicit tin mining has involved private miners organising large parties directly. A document from 1952 records the apprehension of 250 illegal miners working government land around the Kwall Reservoir, which provides the minefields with hydroelectric power, with the apparent knowledge of local officials. They were in the employ of a Greek national who had recently expatriated £100 000 to France. The government ordered him deported.[28]

Police protection for the mine companies has never been adequate. Despite receiving extra subventions from the companies and being organised into a special minefields force, police have been strongly subject to the temptation of accepting bribes, given their poor salaries and the sizeable sums offered. They are often little more than nightwatchmen who share the responsibility of patrolling a vast area. When Birom men began to enter Native Authority employment during the 1950s, the mining companies lost all hopes for the possibility of protection. Punishment for illegal mining, moreover, has tended to remain mild and only a minor deterrent for villagers.[29]

Through the 1950s, it was Europeans who played the main role as fences for stolen tin, and no doubt they were the ones who generally profited from it. However, an increasing number of Nigerians, mainly Southerners but also Birom, began to acquire a mining right. In the last decade, these groups have come to play a pivotal role. The most successful Birom private miner, Dan Boyi Zang, comes from Gyel, one of the most congested mining disricts. Amalgamated Tin Mines of Nigeria (ATMN), the London Tin Corporation affiliate which has controlled nearly half of Nigerian tin production since its formation in 1939, found Zang the person most responsible for theft from their properties. According to Zang, ATMN arranged to have his house searched, resulting in a frame-up in 1955. In 1959, employees of his were found guilty of illegal mining and he was questioned by ATMN; accusations were renewed in 1963. At the beginning of 1964, one hundred Gyel villagers were found mining ATMN property illegally, and the police suspected that Zang instigated the thefts and received the stolen property. The ATMN area manager claimed that Zang threatened to have him deported from Nigeria through his political contacts if ATMN persisted in investigating his activities.[30]

Other private miners emulated Zang's success. By the early 1970s, there were Birom private miners in Du, Shen and other villages.[31] It was their presence which enabled villagers to avoid middlemen entirely. They functioned as the big men of their communities, available for the provision of jobs and services. Village cohesion remains strong enough and the alternative possibilities for inhabitants weak enough so that the private miners are generally viewed today as 'brothers', not as exploiters.[32]

The evidence presented in two Birom students' essays and in government documents indicates that villagers do not and never have seen tin theft as a violation of any moral code which makes sense to them. Obtaining tin by theft is merely a modest compensation for the expropriation of land and difficulties faced by Birom farmers; it expresses the belief that wealth has been appropriated unjustly by outsiders.[33] One of the students wrote:

> . . . a man living in a village like Shen will go for 'illegal mining' around the mining paddock at Timtim because he feels that this is part of his ancestral land and therefore quite legal under all rights of ownership to the land.[34]

The other disliked even applying the word 'stealing' to such activity.[35] Proceeds from illicit mining go toward the fulfilment of the cash needs of the villagers: purchase of food and clothing, taxation, school fees, etc. It may be more equally and less corruptly distributed than the compensation money mine companies pay for damage to land – money which is often pocketed by 'chiefs'.[36]

Before the Second World War, there were only very sporadic incidents that one could consider as resistance to the reign of mining capital on the Jos Plateau. In the 1940s, however, with the collapse of Birom agricultural production on the one hand, and the intense pressures of inflation on the other, the situation changed rapidly. In 1947, a political movement, the Birom Political Union (BPU) emerged, spearheaded by a corps of ex-servicemen. The Union operated in two distinct ways. It crystallised a wave of peasant protest by fighting against the inadequacies of land compensation and reclamation programmes, against the extension of mining activity (actually halting operations from time to time), and against British plans to remove a section of the Birom from the Plateau and into a new settlement located in an unhealthy area. At the same time, the BPU responded to the reorganisation of Nigerian politics along ethnic lines by attempting to secure effective Birom control of local government on the Plateau and to enable the Birom to function competitively as a corporate unit.

The late 1940s witnessed the growth of the trade union movement as well. Unionism spread to the Plateau with the formation of the Nigerian African Mineworkers Union (NAMU) from company unions in 1947, but remained strongest among artisans and clerical workers from the south, of whom the upper tiers formed a labour aristocracy with its own distinctive set of demands. This upper tier, in fact, constituted the only part of the Nigerian labour force which could actually live and support a family from mine earnings. Overall, trade union organisation remained weak and dependent upon the intervention of the colonial regime for the legitimacy and success of its strike actions. In 1954 the union movement was split by the organisation of the Northern Mineworkers' Union (NMU), a group hostile to southerners, opposed on principle to strikes, and controlled by labour contractors closely linked to the Northern Peoples' Congress (NPC), the dominant political party in the Northern Region of Nigeria.[37] Despite its weaknesses, NAMU was able to lead strikes which effectively crippled production, and, by embracing demands for better pay of the unskilled majority, it at times gained substantial support from the mass of northern miners, even though the Northern Mineworkers' Union was hostile to its efforts.[38]

Far from disappearing in the shadow of these more sophisticated and complex means of protest, however, tin theft grew alongside the militancy of the BPU and the unions during the late 1940s and early 1950s. By disrupting the social hegemony of mine management, strikes effectively encouraged theft to expand. Official records, in fact, link strikes and thefts in time and mood.[39]

From the mid-1950s, however, the intensity of protest on both peasant and worker fronts tended to decline. Tin prices dropped during this period, while production cutbacks and lay-offs sapped union strength. At the same time, the NMU had its hand strengthened by the growing efficacy of the patronage that the regional government in Kaduna could supply. As a result, wage levels in the area rose under direct pressure from the government, thus increasing union dependency and cementing the bonds through which labour contractors controlled the mass of the workforce. In 1961 and again in 1964, the NMU was able to limit significantly the effects of NAMU-called strikes, while NAMU was encouraged to look inward toward an orientation at once more ethnic and more narrowly concerned with the aspirations of a labour aristocracy. This reorientation tended to exclude Birom workers, who were among the least skilled and most marginal of the workforce, and who, when they entered unions at all, tended to join the NMU. In 1958, Birom workmen formed a Middlebelt Mine Workers' Union, but it was far smaller and weaker than the other two.

The force of Birom peasant protest had weakened somewhat earlier. BPU militants were able to use ethnic politics to transform themselves into an elite eligible for government jobs and patronage in Jos and in Kaduna. Ambitious young Birom turned in ever greater numbers to the missions and the schools, and sought ways to make a living that would sever their ties of dependence on a deteriorating agriculture. While Nigerian nationalist politics emerged in the 1940s with mass support stemming from radical protest over a broad range of issues, during the 1950s it developed in ways which stifled such protest or channelled it away from class-conscious politics, notably by enhancing ethnic tensions and rivalries.

For the mass of Birom on the minefields, however, this decline in militancy was coupled with an extension of illicit mining. For one thing, the Nigerian state began to evolve as a vehicle for indigenous capital accumulators to improve their situation *vis-à-vis* foreign capital. Tin theft became more and more prevalent as the political order began to break down several years ʼafter Britain granted independence to Nigeria in 1960. On the Plateau, tensions ran high between the ruling NPC and the rival and initially more radical Nigerian Elements' Progressive Union, which had the support of most Northern Muslims living in the trading centers. Birom dissatisfaction with the NPC, despite the incorporation of part of the Birom elite within it, was also strong. The NMU campaigned with an increasingly chauvinistic and shrill voice against the southern, particularly Igbo, skilled workers and the NAMU.

Beginning in 1963, there was a veritable explosion of illegal mining on the central minefields.[40] By the beginning of 1965, armed illegal miners were working in the daytime and resisting police intervention. In an incident at Zawan that year, policemen had their revolvers taken away from them by the people mining. The climate of tension only increased after the military coup in January 1966, when the NPC Premier of the Northern Region and the NPC Prime Minister of the Federal government were assassinated. Between April and June 1966, the number of armed raids connected with tin theft rose monthly from five to seven to ten. Then came a second coup, in which General Gowon, himself an Ngas from the Plateau, took power, followed by the massacres of Igbo during September–October, the murder of NAMU stalwarts, the abolition of regional governments, and the drift into civil war.[41] During the Civil War and thereafter, the atmosphere of violence declined in the minefields, but theft did not fade away. It attained greater significance than ever in the 1970s and remains a characteristic feature of minefield life to the present.

The relationship of theft to social protest, when placed in the general context of the resistance of labour to the prerogatives of capital, is thus rather more complex than the pure progression which Engels once proposed. This, at least, is what the case of the tin miners in colonial and post-colonial Nigeria suggests. Theft in the Nigerian tin mines has a serious economic rationale for producers, for it is directly related to the structure of production and exchange. While tributers and, more particularly, illegal miners may not extract ore as efficiently as do foreign firms, hand labour continues to compete effectively with heavy machinery in the recovery of tin ore. To the extent that tin mining can proceed without the use of heavy capital investment – an extent still considerable today – the mining capitalist must appear nakedly as a predator and parasite who uses his export facilities to obtain a monopoly over the mineral. Were substantial portions of the minefields simply expropriated from the companies and given over to the local inhabitants to exploit as they liked, it is unclear whether anyone but the companies would lose as a result. If the share of the surplus taken by private miners and the state would be limited, such an expropriation would likely be directly beneficial to mine workers.

Tin stealing cannot be considered entirely as an individual or isolated form of protest. It has clearly evolved as a social practice, requiring not only the co-operation of groups acting together but a substantial degree of community cohesion as well. Theft in the 1940s became entangled in a more general pattern of social protest in northern Nigeria, and so it must be examined in the same material context against which the more explicit protest was articulated. Theft continues to form the basis of an alternative economy for the Birom (and indeed for much of the Nigerian labour force generally) and, as such, reflects the resistance of Nigerian countrymen and -women with rights in land-subsistence production and without total recourse to the market. But it also reflects the continued importance of petty commodity production and exchange.

The essential role of private miners in the Nigerian economy suggests that illicit tin mining has historically served as a means of capital accumulation for petty, and in their own way, parasitic entrepreneurs. And, in this respect, much of theft's social import derives from the limited success of other forms of protest in shattering the framework of clientelist social relations and ethnic politics, which places labour at the mercy of contractors. When the private miners are then paired with the national state, in its rentier role *vis-à-vis* the oil multinationals, a clearer and more focused image emerges of the lumpen-capitalism which dominates Africa's richest and most populous country. The strands of protest and adaptation by workers are seamless on such a loom.[42]

Nor is the appropriation of tools, finished products, or money receipts by workers an aberration in more industrialised or unionised societies. Here, too, capitalism and its attendant property laws encounter substantial 'illicit' resistance and adaptation. Engels's and Hobsbawm's view of private property theft as an archaic form of protest may be borne out in a very long-term or theoretically abstract socialist perspective, but its continued strength and viability requires a more thorough assessment of how workers react to capital and how petty accumulators react to the imposition of monopoly.

NOTES

1 I am grateful for the criticism of Peter Linebaugh, Allen Howard and Jean-Christophe Agnew, and for that offered at seminars where the paper was presented at the Centre for the Study of Social History, University of Warwick and the Department of History, University of Dar es Salaam. This chapter was originally published in *Radical History Review*, no. 26 (October 1982), pp. 68–86.

2 Frederick Engels, *The Condition of the Working Class in England* (London: Panther Books, 1969), p. 240.

3 I have chosen to use the word 'theft' throughout. Peter Linebaugh has suggested that 'direct appropriation' is more accurate and less pejorative. It is, however, cumbersome, and limited. Modern connotations of this ancient Anglo-Saxon word 'theft' are those given by generations of capitalist development; excepting this caveat, it seems an adequate term for my purposes here.

4 G. Lewis, *The Stannaries* (Truro: D. Bradford Barton, repr. 1965), p. 155; J. G. Rule, 'Labouring men in Cornwall c. 1740–1870; a study in social history' (unpublished PhD, University of Warwick, 1971), pp. 59–61.

5   *Tin International*, December 1969. For Bolivia, see *Tin International*, September 1964; Charles P. Geddes, *Patino, The Tin King* (London: Hale, 1972), pp. 86–7.

6   Martin Legassick, 'Gold, agriculture and secondary industry in South Africa 1885–1970', in Robin Palmer and Neil Parsons, *The Roots of Rural Poverty in Central and Southern Africa* (London: Heinemann, 1977), p. 178; Charles van Onselen, *Chibaro: African Mine Labour in Southern Rhodesia 1900–1933* (London: Pluto Press, 1976), p. 131.

7   The following pages summarise arguments developed at greater length and more rigorously in my book *Capital and Labour in the Nigerian Tin Mines* (London: Longman, 1981).

8   The ethnography of the Plateau region is spotty at best. For the Birom, see Tanya Baker, 'The social organization of the Birom' (unpublished PhD thesis, University of London, 1954). Baker selected for her study the section of the Birom least affected by mining.

9   For an understanding of strong and weak peasant households, and the rhythm of peasant economy observed in (somewhat artificial) isolation, see, for Russia, A. V. Chayanov, *The Theory of Peasant Economy* (Homewood, Illinois: Irwin, 1966); for northern Nigeria, Polly Hill, *Rural Hausa: A Village and a Setting* (Cambridge: Cambridge University Press, 1972).

10 The southernmost Birom speakers were much less disrupted by the mines than several smaller neighbouring Plateau 'tribes'. For an example see Stanley Diamond, 'The Anaguta of Nigeria: suburban primitives', in Julian Steward (ed.), *Contemporary Change in Traditional Societies*, Vol. 1 (Urbana: University of Illinois Press, 1967).

11 White-black worker ratios were 1:89 in 1915, 1:80 in 1920, 1:104 in 1925, 1:178 in 1935. Nigeria, Mines Department: *Annual Reports*.

12 National Archives, Kaduna (henceforth NAK), SNP 11N, To Government Inspector of Mines, 6 November 1913, 1415m/1913, Relative to Tin Stealing.

13 NAK, SNO 11N, Cock to Secretariat, Zungeru, November 12 1913, 1415m/1913.

14 Rhodes House Oxford (RH), Niger Company Papers, MS. Afr. s. 89, W. R. Humbold to District Superintendent of Police, Naraguta.

15 NAK, SNP 11N, Ag. Government Inspector of Mines R. G. Williams to Secretariat, Zungeru, 415m/1913; RH, MS. Afr. s. 89, Niger Company Papers, H. W. Laws to Secretary, Niger Company, 1 July 1909; RH, MS. Afr. s. 90, Scarborough draft letter, 7 January 1912.

16 NAK, SNP 17/3, Prohibition of Tin Smelting in Lirue-n-Dalma by Natives, 29185.

17 NAK, JOSPROF I/1, Secretariat, Northern Provinces to Resident Plateau Province, 22 April 1927, Prosecutions for Mining Offences, 726; NAK, JOSPROF III, Manager, Jos Tin Areas Co. to Resident, Plateau Province 19 October 1938, S.112, v.1, Tin Stealing; NAK, JOSPROF III, Reports on Mine Operators by Chief Inspector of Mines 1945–51, 272. Rob Turrell reports striking similarities among the illegal diamond buyers of nineteenth-century Kimberley, 'The crisis of illegal diamond buying and the social history of Kimberley 1883–1885', (Seminar paper, Queen Elizabeth House, Oxford, 1 November 1978). Indeed, fencing illegally obtained diamonds may lie at the heart of some of the greatest South African mining fortunes. For gold mining in Rhodesia, see van Onselen, *Chibaro*, p. 241.

18 NAK, SNP 17/2, Meeting of the Lt.-Governor, II Northern Provinces with Local Council, Nigerian Chamber of Mines, 3 November 1926, 16053, Prevalence of Theft by Pagans on the Plateau.

19 *idem*; 340/1914, Third Quarterly Report, 1914, Bauchi Province; 3/1933, Annual Report, Plateau Province, 1932; NAK, JOSPROF I/1, 29/1933, Annual Report; Jos Division, 1932.

20 NAK, SNP 17K, Case of Hammod (d. 13 April 1937), 2109; v.10, Mining Accidents and Inquests.

21 See Table 1.

22 NAK, JOSPROF III, Rex Niven, Resident, Plateau Province to Secretariat, Kaduna, 6 December 1950, S. 112 v.1; Memorandum no. 2, Nigerian Chamber of

Mines, 3 March 1966; NAK, JOSPROF IV, PRO/S. 16, Tin Stealing and Illegal Mining.

23 The villages, some of them Birom settlements, and other mine camps or Hausa villages, were: Gona, Bassa, Mai Idan Toro, Barrakin Auka, Swifts camp, Naraguta Karama camp, Sho, Du, Zawan, Gyel, Kul, Genawarri Bisa. Between Jos Tin Areas property and Tollemache's house, Gindiri, Tilden Fulani, Mai Juju (Bauchi Province) and Rishin (Bauchi Province); NAK, JOSPROF III, Committee on Tin Theft, 14 January 1952, S.112 v.1.

24 Aishatu Oumar-Shittien, 'New economic factor: social impact of mining on the Birom' (Part III essay, Department of Sociology, Ahmadu Bello University, 1973), p. 63.

25 ibid., p. 73.

26 ibid., pp. 68, 72–3.

27 ibid., pp. 62–79.

28 NAK, JOSPROF III, Enclosure in Resident, Plateau Province to Civil Secretary, Kaduna, 5 December 1952, S.112 v.1. A similar tale is recorded for the same area five years later in NAK, JOSPROF IV, Report on Tin Theft, 12 February 1957, RO/S.16.

29 NAK, JOSPROF III, Resident, Plateau Province to Secretariat, Northern Provinces, June 1950; Minutes of the Meeting of the Sub-committee on tin theft, Nigerian Chamber of Mines, the Provincial Resident, the District Officer, Jos, the Chief Inspector of Mines and the Superintendent of Police, 21 June 1950; Sub-committee meeting minutes, 29 September 1950; Rex Niven, Resident, Plateau Province to Secretariat, Northern Provinces, 6 December 1950; Meeting of the Committee on tin theft, 1 March 1951; all in S.112 v.1. NAK, JOSPROF IV, Special Meeting on Tin Theft Organized by the Provincial Secretary in Kaduna, 16 May 1966, RO/S.16; Peter Harrigan, 'Nigeria's tin mines – progress, prospects and problems', *West Africa*, no. 2897, 18 December 1972.

30 See the documents in NAK, JOSPROF IV, RO/S.16, including the Security Officer's Report to ATMN, 1963 and Zang to Provincial Secretary, 21 July 1966.

31 Paul Chunum Logams, 'Birom integration into Nigeria'. (Part III essay, Department of Government, Ahmadu Bello University, 1973), p. 13.

32 Oumar-Shittien, 'New economic factor', p. 75.

33 NAK, JOSPROF IV, Resident, Bauchi Province to District Officer, Jos, 7 February 1957; Special Meeting on Tin Theft Organised by the Provincial Secretary in Kaduna, 16 May 1966; both RO/S.16; Abbas Dabo Sambo, District Officer, Jos to Ag. Provincial Secretary, Jos, June 1965; Mining Other than Tin, RO/S.44.

34 Oumar-Shittien, 'New economic factor, p. 104.

35 Lomgas, 'Birom integration', p. 66.

36 Oumar-Shittien, 'New economic factor', p. 76.

37 In developing my analysis of the trade union movement on the Plateau, here briefly encapsulated, I have made use of: Wogu Ananaba, *The Trade Union Movement in Nigeria* (London: Hurst, 1969); Robin Cohen, *Labour and Politics in Nigeria 1945–1971* (London: Heinemann, 1974); B. J. Dudley, *Parties and Politics in Northern Nigeria* (London: Cass, 1968); and Mohammed Jibrin, 'Conflict and cooperation: trade union activity on the Plateau tin field (1942–73)', Part III essay. Department of Government, Ahmadu Bello University, 1974.

38 This was notably true in the 1956 strike, the culmination of more than a decade of union organisation.

39 NAK, JOSPROF III, Local Council, Nigerian Chamber of Mines, Annual Address, 8 February 1957; MIN 1/v.1, Minutes of the Nigerian Chamber of Mines 1941–1960. For a more recently made correlation, see the *Nigerian Standard*, 24 January 1976.

40 NAK, JOSPROF IV, Memorandum number 2, 3 March 1966, Nigerian Chamber of Mines, RO/S.16. They pointed particularly to the villages of Zaramaganda, Du, Gyel, and Zawan. This was also a period of improving tin prices.

41 NAK, JOSPROF IV, Musa Tanko, Ag. Provincial Secretary to Premier's Secretary, 17 February 1965 and Annual Report, Nigerian Chamber of Mines, 1966, in RO/S.16;

Abbas Dabo Sambo, District Officer, Jos, to Ag. Provincial Secretary, Jos, June, 1965, RO/S.44. Also see Jibrin, 'Conflict and cooperation'.

42 See my 'Oil boom and crisis in contemporary Nigeria'. *Review of African Political Economy*, no. 13 (1979), pp. 91–100. Among other important assessments there is Yusufu Bala Usman, *For the Liberation of Nigeria* (London, 1979).

3 'Like a roaring lion': Tabwa terrorism in the late nineteenth century[1]

ALLEN F. ROBERTS

> Awake! be on the alert! Your enemy the devil, like a roaring lion, prowls round looking for someone to devour. *New English Bible*, 1 Peter 5:8.

Introduction

In the mid-1890s, 'lions' killed scores of Tabwa and others settled about the newly founded White Fathers' mission at Baudouinville (Moba), Congo Free State (now Zaire). Tabwa told the priests these were theriomorphic men, not beasts, and implored them to proclaim the banning of lion-men. Their response was to deride the 'superstitions' of these 'heathen', and to seek permission from the civilian authority to prohibit panicked flight from mission environs. This granted, the missionaries used force to maintain the presence of the population they wished to proselytise. Lions were hunted with little success. Unscathed by the fire within, brazen 'lions' broke into homes, and more and more people fell their victims when venturing into the bush. Finally, lions *were* killed, and the human carnage *did* abate, but only temporarily. 'Lion'-related deaths would resurge several times, then apparently cease in the late 1930s, when fifty Tabwa died and the colonial administration recognised the agents to be human. Several men were apprehended, tried and sentenced to hang.

To study these events, I have combined consultation of diaries and letters of early European visitors and settlers, colonial court records and publications discovered in Zaire, Belgium and Rome; with conversations and informants' exegeses collected over four years of anthropological fieldwork among Lakeside Tabwa. I was struck first by the singularly, even sublimely obtuse interpretation of the killings presented by missionaries who were recording them in their diaries. The events described seem so un-lionlike, such as when a man who heard scratching at his door thrust a brand outside and was severely clawed on the arm by a 'lion'. The victims were those living in proximity to the missions at Baudouinville, and were often intimately involved in mission affairs.[2] The priests *chose* to disbelieve what the would-be victims were telling them because of their sincere conviction that such statements illustrated child-like lack of reasoning of 'their savages', and because of self-interest during the very last years of slave-taking by Swahili, Nyamwezi and ambitious locals, when their own lives were threatened.

The missionaries would not admit that they were being rebuffed by Tabwa. The point is worth stressing. In a recent essay, Harold Schneider has examined lion-man beliefs among Turu and suggests that 'Belief in lion men . . . appears to be a relatively modern phenomenon', largely because a report by a magistrate and an ethnography by resident White Fathers both written early in this century make no mention of Turu lion-men, whereas over one-hundred deaths in 1946–8 were attributed either to man-eating lions or lion-men by the colonial government. He then attributed the 'rise' of lion-men to an effort by men to control their women who have become insubordinate to them in a colonial context that allowed women greater freedom than did precolonial independence.[3] Schneider is correct to assume that lion-man beliefs, like any others, may be changed to suit new circumstances; and his hypothesis concerning increased conflict between Turu men and women is intriguing. The lion-man strategy might be adapted to such a circumstances, as it has been to many others elsewhere. Yet Schneider fails to mention if there were deaths by 'lions' earlier in the century which might have been misunderstood and misrepresented by expatriate writers, and assumes that reporting by European observers is 'objective'. The hypothesis to be developed below is that Tabwa lion-men did operate at a time when missionaries dismissed their existence as fantasy. Furthermore, in some cases, lion-men were acting to resist directly the Fathers and, more specifically, the growing prominence of Africans loyal to the mission. Such an assertion is consistent with suggestions made in the synoptic works of Birger Lindskog and P. Joset on theriomorphic societies elsewhere in Africa.[4]

Authors such as J. Maes, referring to the *anioto* leopard-man incidents of north-eastern Congo, find that 'the whole movement was a conservative reaction against . . . collaborators' with the colonisers.[5] Others hold with Emil Torday[6] that 'leopard and similar societies . . . were formed to defend the population against the greed of kings and chiefs who attempted to sell their own subjects to the slavers', and that the energies of these were redirected toward 'the expulsion of the White Man from Africa' soon thereafter. These might be 'xenophobic or nationalistic movements among the population' then, or even as Albert Schweitzer intimated, 'native movements' manipulated by 'communist organizers'.

One might wonder if, in Eric Hobsbawm's terms, lion-men acting in the 1890s were 'social bandits': 'heroes . . . even leaders of liberation, and in any case . . . men to be admired, helped and supported' by an agrarian peasantry 'oppressed and exploited by someone else'.[8] In the cases to be discussed below, some Tabwa must have lent support to lion-men operating from their midst while others, the victims, might denounce but apparently would not take direct action against them. As Ralph Austen suggests in his contribution to this collection, 'social bandit' is a romantic term inappropriate to what we know of the African historical and cultural context, yet, to use his phrase, Tabwa lion-men were 'purveyors of violence' and, in some sense, 'heroic criminals'.[9] Their violation of 'law' (the current definition of 'crime') could however be for a variety of reasons, some decidedly less 'heroic' than others, as that word is generally understood.

A further generality may be mentioned. Hobsbawm asserts that banditry tended to become epidemic in times of pauperisation and economic crisis.[10] One may consider circumstances obtaining for Tabwa in the late 1880s and early 1890s. Besides great politico-economic upheaval wrought by slavers and by those who opposed them, Tabwa were wracked by epidemics of smallpox and influenza, a rinderpest epizootic, severe drought, locust plagues and famine. Other unusual environmental phenomena seemed to underscore the menace of radical change. Brilliant Sungrazer comets in 1880, 1882 and 1887 were felt by Tabwa to portend

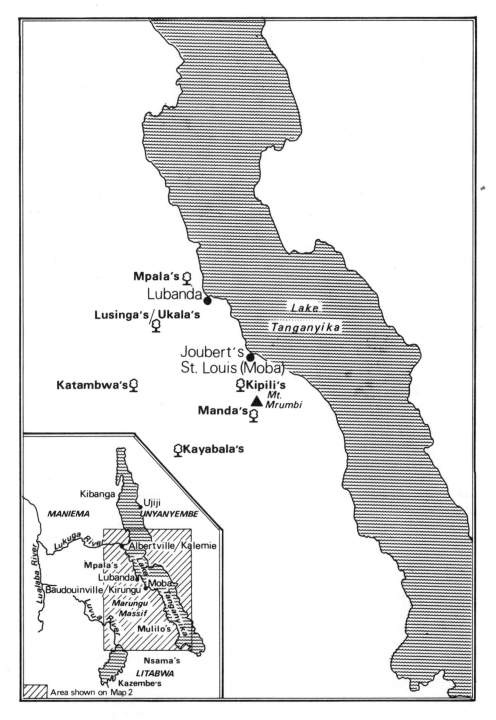

Map 2: *Sites of Tabwa lion-man incidents in South-Eastern Zaire*

the deaths of chiefs and (in afterthought) the advent of Europeans. With the sudden bursting of a barrier upstream, the Lukuga River again became the sole outlet of Lake Tanganyika, the level of which sank some 15 m between 1880 and 1890; waters retreated as much as a kilometre from their former reaches, changing lake-shore geography and stranding some villages in new swamplands. Most important, however, were the social changes occasioned by the establishment of Catholic missions among Tabwa. As I have described elsewhere, communal celebration of Tabwa religion and corporate politico-economic activities were outlawed as missionaries introduced a colonial society based on initiative and personal salvation.[11] It will be suggested here that lion-men attacks were one strategy of resistance to some of these social changes.

Incidents of inordinate Tabwa death by 'lions' reported in a variety of sources can be charted. The years cluster: (1) 1886–7; (2) 1894–6; (3) 1904, 1909–12; (4) 1918–22; (5) 1928–31; (6) 1939–42. One can speculate that these outbreaks are a reaction to stressful events and conditions: (1) and (2) as above, the advent of Europeans and attendant disruption; (3) the (temporary) demise of the important Tabwa chief Manda, first implementation of taxation and other far-reaching policies by a new Belgian colonial administration; (4) post-war malaise augmented by famine and the influenza pandemic of 1918 which killed hundreds of Tabwa; (5) effects of the Great Depression when the economy of the Belgian Congo came to a standstill; (6) shockwaves from the newly implemented Belgian version of indirect rule, exacerbated by war mobilisation. Yet such a correlation is overly simplistic, since during other periods of severe strain (e.g., the years of the First World War when all Tabwa men were conscripted to the Home Guard or as porters, or were forced to grow extra crops to feed Anglo-Belgian forces),[12] no deaths are attributed to lions in sources consulted. Ultimately, since we lack life histories of, and focused interviews with lion-men themselves, which might reveal their motivations, such a correlation between moments of extreme social stress and lion-man incidents remains 'superbly untestable', as Mary Douglas finds the somewhat congruous relation between disorder and an increase of witchcraft accusations,[13] and for some of the same reasons.

Reporting in missionary diaries, travelogues or other nineteenth-century sources is neither consistent nor disinterested. 'Disorder' is often only a matter of perspective. Furthermore, as both Lindskog and Joset note, cases of leopard or lion-men attacks had particular explanations, including revenge, vendetta, ritual or 'gangsterism'. Close scrutiny of the Tabwa data, such as it is, suggests that all these are possible explanations. However, these do not obviate a general correlation between incidents and their historical contexts (lion-men attacks as a product of disorder); rather, particular case studies allow an understanding of the process or dynamics of the incidents. One can suggest, then, that there is an explanation of the instrumentality of lion-man murder, a kind of functionalist 'bottom line' to all the various incidents.

My hypothesis is that killing by lion-men represents for Tabwa a strategy to oppose the fundamental principle of political power, namely, the recruitment of people to a residence group over which one can influence and impose decisions. As Tabwa say, *sultani ni watu*, 'a chief is people'. Terror is centrifugal. Other killers are suspected within the community: sorcerers (*ndozi* in Kitabwa). Murder by, and then of, sorcerers is *not* random, and though disruptive, is historically comprehensible to particular actors, a chain of events with a beginning and an end. Lion-men, however, kill without measure and their victims may have no direct relation to the object of their wrath. Flight is the only rational reaction. If 'a chief is people', a chief whose people have been killed or dispersed is no longer a chief, or at least only nominally so until he counters his adversary. Attacks by lion-men are a political strategy of terrorism.[14]

Such an assertion follows exegesis by present-day Tabwa. For instance, when he was younger, my informant Kanengo[15] was a gun-hunter who occasionally pursued lions. He told me that lions will not attack men without provocation, but will lie in the grass and let one pass. Stories of lions breaking down doors to kill people are lies, Kanengo added: these are men. 'Lions have shame (*haya*), man does not!'

Space constraint precludes a discussion here of man-eating lions according to Western naturalists. My amateur review of the literature suggests that human predation by lions has not received systematic analysis and critical review. Lions may eat humans, as George Schaller has documented, yet such an event seems 'very exceptional'.[16] Celebrated cases like that reported by J. H. Patterson in *The Man-eaters of Tsavo* are still cited in popular accounts (which attract buyers with their titillating exotica) as 'proof' of the great cats' treachery, even as such authors insidiously describe the African and (in the case of the Tsavo murders) East Indian victims as 'half-mad with fright', 'terrorized and ridden by darkest superstition . . . utterly apathetic and quite unwilling to give the authorities any help at all' as they 'huddled together whimpering while they listened to the . . . cats fighting over and eating friends and relatives'.[17] A proper study would consider the cases themselves *and* the way they have been reported. At issue in the latter regard is not whether lions occasionally devour humans, but that people *like* to believe they do so with significant frequency. To paraphrase a line from William Arens' study of 'cannibalism' among Africans and others, whether or not lions eat people 'is interesting but moot. But if the idea they do is commonly accepted without adequate documentation, then the reason for this state of affairs is an even more intriguing problem'.[18]

Visanguka are not lions at all, but men whose wiles (*mayele*), according to Tabwa, included the ability to make themselves invisible. They would slash the throats of victims with hooks or claws of iron, either fixed onto a sort of glove or as a hand-held implement that . 'looks like a garden rake'. *Visanguka* would disguise themselves with genet skins or, as one contended, those of the sort of wild cat that brazenly takes chickens from one's yard. Wooden replicas of lion footprints were fashioned for their feet.[19] Kanengo said that *visanguka* were ones so attired, who might murder persons of the *clan* of one who refused to repay a debt, cuckolded, or who committed other grievous insult. Consensus among others, however, was that *visanguka* might be summoned for reasons like these, but would then take victims *hovyo*, that is, indiscriminately and without measure, sometimes (as in a case at Chief Kayabala's in 1939), not touching anyone related to the intended victim. Lion-men would cut up and scatter the remains of their victims in the bush but never consume the meat themselves.

The term *kisanguka* (*visanguka*, pl.), 'lion-man', is from the root verb **sanga*, 'to find, meet on a path, or be on the verge of' in Tabwa and Bemba.[20] The derivative verb **sanguka* is 'to become, to change oneself into', 'to be unfaithful to (as a wife leaving her husband or a man leaving a chief for another)', 'to resurrect'.[21] Change is implicit to **sanga* (discovery, encounter, potential), realised in **sanguka*, and personified in *kisanguka*, a noun shifted from the class of sentient beings (usually prefixed by *mu/ba*) to the amplicative, abstracted or objectified *ki/vi*.[22] It has been argued that 'the assignment of words to particular noun classes in Bantu languages is not at all fortuitous and is a definite reflection of the feelings of the speakers about particular objects'.[23] One may speculate that *visanguka* change not so much in form (since that is a fiction recognised by all but the most naive), as they do in essence: they remove themselves from the most elementary responsibilities defining human community.

A Tabwa chief is 'father of his people', leader and fount of wisdom. Analogy is drawn between a chief and the moon, both of which afford guidance and 'light'.[24] Tabwa as freely speak of a chief's 'unenlightened' self-interest, (and the word is not foreign to their logic). In effect, Tabwa chiefs are deemed the greatest sorcerers of the land. Like the moon, they have a dark phase. Like lions, although usually 'benign', they are ruthless in attack.

Tabwa chiefs do not dance naked at crossroads and otherwise mock conventions, nor do they use medicines to afflict or assassinate enemies within their communities and then consume their flesh, as do ordinary sorcerers. They *do* condone this by others, since illness and suffering are a part of life, and sorcery is an important cause of such misfortune, according to Tabwa explanation. From the front porch of his/her house (traditionally at the 'head' of the village's central avenue perpendicular to other dwellings), the chief watches all that transpires, even those monstrous activities others cannot see, but the dire effects of which they feel to the quick in misfortune, illness and death. As I was told, 'If a sorcerer wants to kill someone, he tells the chief . . . If the chief agrees, he kills him. Yes, he dies, he dies; and who does this? Isn't it the chief?'[25]

So a bargain is struck. Yet just as the chief's vision is greater than any particular individual's affairs, since he is justice and sees *all* that transpires; so he must not consume his share of the human meat brought to him by the sorcerers he has allowed to kill within his community, in the course of particular conflicts. Instead, he is thought to sell the meat to other sorcerers, and the wealth so acquired will be recirculated in the networks of dependency that constitute his political base. Such ghoulish prestations join others, notably those offered by hunters in the lands identified with the chief, and by farmers from fields tilled at the chief's indulgence.

The moon in the heavens, lions in the bush, and the Tabwa chief at the head of his community are metaphorically linked, each better understood by reference to the other. That lions and lion-men should be associated with Tabwa chiefs is the enacting of these metaphors, their personification in the realm of political action. Were space to permit, more could be said of the cosmological meaning, general attributes and activities of Tabwa lion-men; yet the above will permit a discussion of 'lion'-related deaths of the 1880s and 1890s.[26]

The historical setting

In order to understand the means and meanings of the deaths by 'lions' which occurred in the 1890s, it is necessary to sketch the history of the inhabitants of the Marungu Massif, and that of the White Fathers' mission that would be founded at Kirungu/'Baudouinville' in 1893. Throughout the middle and later decades of the nineteenth century, the Marungu Massif received successive influxes of population. Tabwa to the south were invaded by 'Matuta Kyalo', an epithet for the Ngoni who 'destroyed the country'. Those living upon the plateaux of the Massif felt the effects as refugees moved northward. Kazembe's Lunda empire on the

Luapula (in what is now northern Zambia) attracted caravan commerce, and several Tabwa chiefs (most notably Nsama) shared the politico-economic stimulation and sought to create empires of their own.[27] Again, as a consequence of the fierce competition for control of lands and scarce resources such as salt pans, some were forced or encouraged to move northward into the Marungu, among these the ancestor of Manda, a chief of the Zimba (or 'Leopard') Clan who will be discussed below. Nyamwezi or 'Yeke' ivory hunters entered the area south-west of Lake Tanganyika at about this same time and established strongholds in the Massif and along the Luvua River to its west, some as satellites of, a few in overt opposition to Msiri, the most famous and powerful of their number.[28] Coastal slave-traders and their WaNgwana henchmen came, starting in mid-century, making pacts with ambitious Tabwa chiefs to act in a local context to mutual profit. As traffic in slaves grew and other territories of conquest like the Maniema became depleted or otherwise problematic to further exploitation, Tabwa quickly became both predators and prey.

Two Tabwa chiefs deserve mention in this regard: Lusinga of the Sanga (or 'Bushpig') clan, and Katele of the 'Leopard'. Lusinga was among those of his clan who settled for a time to the west of Tabwa land, to participate in (or resist, depending upon the narrator's perspective) the Luba sphere of influence outward from its own centre at the lakes of the Lualaba.[29] In the 1870s, Lusinga moved eastward again, to settle near salt springs in the mountains west of Mpala. He 'descended like an avalanche upon the more peaceably disposed inhabitants near the lake, and swept off the entire population of thirty or forty thriving villages, turning the country into a perfect desert', wrote Joseph Thomson[30] on his visit to the area in December 1879. Lusinga visited UNyanyembe, east of Lake Tanganyika where he is said to have acquired muskets which he was the first to introduce to the Marungu and adjoining lands.

The notoriety of this 'sanguinary potentate'[31] grew through the decade, as he laid waste to lands ever farther from his mountain fastness. He attacked to the north-east to establish a base at Cape Tembwe, where Lake Tanganyika is more easily crossed than elsewhere.[32] In his own area, Lusinga was menaced by Ukala, a Nyamwezi. Lusinga requested the assistance of warriors of the Leopard Clan to combat Ukala, and together they defeated and beheaded him. Lusinga granted certain lands to those who had helped him.

Such an account may be an example of the genre known as *milandu*: stories told in the course of litigation when points known to all are manipulated to present the best case against one's adversaries. Whether or not this was the manner in which the Leopard Clan obtained particular lands is disputed, but the conflict with intrusive Ukala was real enough. Lusinga's relations with other Leopard Clan members were less cordial. To his south, Lusinga assaulted Leopard Clan chief Manda, killing him and razing villages along the path.[33] This incident is described differently in a document to be discussed shortly.

Emile Storms, Fourth Expedition leader of the International African Association (AIA), established an outpost at Mpala in 1882. Almost immediately, Storms clashed with Lusinga. Storms wrote that 'All of Marungu recognises, if not his [Lusinga's] authority, at least his strength and perfidious nature.' Furthermore, Lusinga's 'reputation in all these lands [is] of being invulnerable. *He can change himself at will into a lion* . . . and if need be, can make himself invisible. This strength of Lusinga is not contested, and gives him unbelievable power. The fear he inspires sows panic.' Storms' conclusion was that 'to reduce this chief or to make him disappear would certainly be of the greatest benefit for the [people of the] Marungu.' Accordingly, Storms sent his men, bolstered by others from Paul Reichard's ill-fated Deutschen Expedition, to vanquish Lusinga. By subterfuge, they entered Lusinga's fortress, shot him dead and took his head.[34] Storms placed his ally, Ukala the Nyamwezi, as chief of Lusinga's lands. This Ukala was seemingly successor to the one Lusinga's forces had beheaded some years before; his selection by Storms was a calculated insult to Lusinga's people.

On the strength of his victory, Storms demanded recognition as 'chief of chiefs'.[35] Those quickly submitting to him included Manda, successor to the chief Lusinga had killed. By Storms' reckoning reflected in the 'treaty' signed by Manda, 'sovereign power over all the estates of Manda' was accorded to Storms for the AIA.[36] Thereafter, Manda proved his loyalty to Storms and his European successors by contributing men, often on his own initiative, to their various campaigns against those who must be recognised as Manda's *own* adversaries in the local-level, 'proto-colonial' context. Several of these were chiefs of Lusinga's Bushpig Clan who never willingly submitted to Storms or to those succeeding him and who considered Manda an intrusive element to be dispatched. Another adversary was Manda's own obstreperous sister's son, Katele.

Katele and Lusinga shared both goals and the means to attain them. Katele's ambition was directed to at least two ends: the slave trade and the local-level political arena in which he chafed as sister's son to Manda, a chief of greater stature by tradition and European reckoning. Katele's control of one of the few safe harbours along the central shore of Lake Tanganyika proved useful to his growing participation in the slave trade.

The coastal slaver Mohamed bin Xalfan, better known by his *nom de guerre* 'Rumaliza' ('The Finisher'), had long been operating from Katele's. Rumaliza's lieutenant, Mohamedi, established a fortress there in 1888. While he told the Europeans he meant to traverse the Marungu Massif from Katele's to obtain copper in LiTabwa (northeastern Zambia), it was apparent that Mohamedi's (hence Rumaliza's) broader aspiration was to claim the Massif itself. Coastal slavers operating in the far interior of the Congo Basin were alarmed by German colonial forces marching on their entrepot of Ujiji. According to a contemporary observer, Rumaliza hoped to claim the lands surrounding Lake Tanganyika and be named master or governor of the region, as his colleague Tippo Tip had been at Stanley Falls. Such a strategy would preserve his power in the nascent colonial context he perceived to be forming.[37]

Meanwhile, in 1885 Storms had ceded his fort at Mpala to the Missionaries of Africa, the White Fathers. Two years later, an ex-papal zouave named Leopold-Louis Joubert joined the Fathers to oversee matters of state in the 'Christian Kingdom' they had begun at Mpala. As Cardinal Lavigerie, founder of the White Fathers, is reported to have said one day during a reunion at Algiers, 'If Joubert wants the title of king, we shall give it to him'.[38] The 'king' of this *de facto* polity administered justice and capital punishment, and protected the bounds of his 'territory' with a standing army. One of Joubert's first acts was to seize canoes and slaves kept for Mohamedi by Katele. In 1890, after repeated hostilities, Joubert attacked and drove Katele from his fortress, settled there himself and renamed the surrounding village Saint Louis (Moba).[39] In less than fifteen months, several thousand refugees from slave-raiding formed seven new villages in the plain surrounding Saint Louis.[40] These, and many others over the thirty-seven years Joubert would live there, came to enjoy the peace the captain offered and imposed.

Rumaliza intensified his slaving along the shores of Lake Tanganyika, north of the 'Christian Kingdom' at Mpala. By one estimate, ten thousand people were killed or enslaved there in the first four months of 1892 alone.[41] Another White Fathers' mission at Kibanga or 'Lavigerieville' (founded 1883), just south of the Ubwari Peninsula, was isolated and particularly vulnerable.[42] Joubert was preoccupied with defending Mpala and too distant to assure the security of Kibanga, nor could the scant forces of the Congo Free State lend support. With regret, the Fathers closed Lavigerieville Mission in February 1893. 'Baudouinville' (Kirungu) would be established on the wide plateau above Joubert's Saint Louis, to accommodate the Lavigerieville refugees. Tabwa chiefs Manda, Kipili and Kibwire each had fifteen to twenty villages in which the priests might proselytise; the former, in particular, was of long-standing fidelity to Joubert like Storms before him.[43] Yet, although it appeared to the Europeans that there was room for all on the Kirungu Plateau, the founding of Baudouinville Mission was disruptive to local-level politics in ways the Fathers would feel very soon, but perhaps never *understand.*

The incidents of deaths by 'lions'

Case 1: lions and Nyamwezi intrusives[44]

Early in 1886, lions killed ten people in palisaded villages of 'the territory' of Mpala. A large, clawless male lion with only two teeth was caught soon thereafter: 'this proves African travellers' reports that only old lions that can no longer seize their prey, attack man'. As the mission scribe added, however, the lion might be dead, but not the affair of its ten victims. A month later Ukala, a Nyamwezi chief and close ally of Storms (who had ceded the fort at Mpala to the priests only months before) came to the missionaries saying he wanted their permission to wage war in the Marungu Massif, since he knew who had sent the lions to kill at the village of his fellow Nyamwezi, Chura: As Ukala told the priests, 'if you allow the deaths of these ten to go unpunished . . . it no longer will be possible [for us] to live in this land, as all of the Marungu [inhabitants] will send us their lions'. The scribe found 'the proposition grave, the argument well made, and yet we laughed.' The missionaries' mirth was, however, tempered:

> We see in all these requests to make war the result of the mockery we hear daily. When someone here [at Mpala, where cosmopolitan Lakeside Tabwa were joined by Nyamwezi and many other non-Tabwa immigrants] wants to insult a fellow,, he calls him a Mwenye Marungu. This means that the person has no positive qualities, only defects, and this because he is from the Marungu. Unfortunately, very often one doesn't stop with words, but proceeds to action. For all the savages around here who consider themselves superior to them, the Marungu people are true meat for the butchering; all they are good for is killing, and to do so seems a good act.

DISCUSSION OF CASE 1

Ukala the Nyamwezi was introduced above: a loyal follower of Storms and the White Fathers who succeeded him at Mpala. Chura (sometimes written 'Kyula') was another Nyamwezi who sent his men to join those of Ukala and Storms' Tabwa allies in the lieutenant's battles with recalcitrant inhabitants of the

Marungu. Villages were razed and, in a counter-gesture, Storms' own fort at Mpala was burned just as he was preparing to cede it to the Fathers. Both Ukala's and Chura's villages in lands conquered and given to them by Storms were attacked then as well, but attempts to set fire to them were thwarted.[45] Ukala was acutely aware of the menace to the fragile authority he and Chura enjoyed, now as men of the as-yet-unproven priests; and he understood the nature of the lions attacking the Nyamwezi villages. The underlying political message was clear: covert attacks were being made by the despised of the Massif, *now that Storms was gone*.

The mountain-dwellers, removed from active participation in the lakeside commerce of ideas as well as goods, were merely 'meat for the butchering'. Belittled and preyed upon by ruthless slavers who held every advantage in arms and strategy, they nonetheless had clandestine means to attack their foe. The priests, intrusives themselves, correctly recognised theirs to be a precarious position at best, and yet they could only laugh at Ukala's grave (and accurate) assessment of the deaths. The scribe's remarks are directed at Ukala's motive for requesting permission to attack the Marungu people: this was yet another example of the mockery, the seeking of any provocation, no matter how preposterous, to attack the loathed hinterland inhabitants. Yet the brutality so well captured by the scribe was the motive for the clandestine attacks of 'lions' 'sent' by Marungu enemies. The Fathers missed the obvious and logical connection between the 'lion' attacks on outsiders owing their security to Storms (and even then, to the soldiers Storms had left behind), and those who had grievances they accurately described.

Case 2: Katele and the Lions[46]

Throughout 1887, the men of Mohamedi, an 'Arab' slaver based at Katele's (at the present site of the town of Moba), attacked Katambwa, a chief living near what would become the mission village of Lusaka. The following year, the men of another coastal slaver, Makutubu, did the same.[47] In May 1888, Katambwa visited the missionaries at Mpala to say his people were being killed by 'lions'. The priests told him to trap them, but Katambwa held that this would be futile as these were men, not lions. As the scribe noted, 'here where people always look for some pretext to wage war and pillage, one so quickly accuses another . . . Poor savages, they still believe certain *sorciers* have the power to change into lions'. Katambwa returned four months later to say lions were still killing people, despite his having trapped two lions as the priests had suggested. The scribe recorded this with sarcasm, 'So Katambwa, these were men in lionskins, right?'

DISCUSSION OF CASE 2
These fragments, in the context of our other cases and corroborative detail, lead one to speculate that the same strategy, used in Case 1 by mountain people seeking both revenge and to panic those living with intrusives, could be directed toward different ends. Katele (as in Case 4 as well) acted within the Tabwa cultural idiom to attain his goals. A hypothesis here would be that he and his men, in assisting Mohamedi and Makutubu, sent 'lions' to sow panic. In the resulting melee, the slavers could snatch and pillage, without wasting precious powder.

Case 3: Lusinga and the lions[48]

Kipili, a Leopard Clan chief, married Kaimba of the Bushpig Clan. One day he discovered some Bushpig men flirting with her, pursued them in anger, fought, was wounded and died. To avenge his death, Kipili's clanfellows murdered Kaimba to bury with him. When Lusinga of the Bushpig Clan learned this, he said 'here is a story that makes men's ears prick up'. Lusinga avowed he wanted no war, at least no overt one. Since that time, there have been lions in the land, but these are not the lions of days gone by. These lions are sent by the Bushpig Clan (against the Leopard Clan of Kipili and his kin).

Case 4: Lusinga and Katele[49]

Lusinga of the Bushpig Clan and Kipili of Manda's Leopard Clan, became friends. In the jesting associated with their interclan joking relationship (*ubishi wa utani wao*), Lusinga said that it was their elders who had kept them from being closer friends. Lusinga suggested that Kipili murder his 'mother's brother', Manda, while he would do the same to his own, Kansabala. Lusinga then invited Kinsunkulu, a Luba chief, to come and kill Manda. Kinsunkulu's men came at night, hid behind Mount Murumbi for three days while Manda and his people drank beer, then during the third night when Manda went outside to urinate, they beheaded him. Manda's people awakened and panicked: the men hid, the women were taken as Kinsunkulu's captives. Manda's head was impaled at Kipili's. The latter was aghast and led his men after the Luba enemies (*maadui*),[50] but to no avail. Another kinsman, Kyengo, wanted to attack Lusinga but was prevented from doing so by Kipalapata, a Leopard-clan chief who said Lusinga was his in-law and besides, the killers of Manda had not passed that way. Kyengo returned and succeeded (*kupyana*) Manda.

Manda Kyengo died soon thereafter, and was succeeded by Mukeya Loba. This latter decided to abandon his two villages and move to where his ancestors were buried, at Mukula. He told both Katele, the chief of his smaller village, and Kaloko, chief of the larger, that each would succeed him at his death. Manda Mukeya Lobe died before too long, and Kaloko quickly succeeded him. Katele, taken by surprise, was greatly angered; he told his people they would go and practise *kisama* (interregnum pillaging) at Manda Kaloko's. Katele already had WaNgwana mercenaries with him, among them Kisukulo. They first raided Kitandere, killing people, razing the village and taking livestock which they returned to Katele's to eat; then to show Manda Kaloko his arrogance (*kiburi*), Katele went to the village of Manda's dependent, Kilaba, killed him and razed his village. When Manda Kaloko heard of this, he went to Mpala Mission with his niece, Zongwe Maria, and offered her as a *lilobo* in exchange for Captain Joubert's going to war against Katele. Joubert agreed and after three days set out with moonlight to pursue Katele.

DISCUSSION OF CASES 3 AND 4

These two cases may be considered together, since the contradiction of Case 3 and the first part of Case 4 indicate a long-standing and problematic relationship between Lusinga and Kipili that may be characterised differently by different narrators with different aims in the telling. Kipili is of the Leopard Clan by the primary recognition of matrilineal descent, but the present bearer of the name, like the ones he has succeeded, stresses he is a 'child of the TuSanga' or Bushpig Clan. It is they who granted him his lands (perhaps in appreciation for the common attack upon Ukala, as mentioned above). The joking relation (*utani* or *baendo*) between Lusinga and Kipili would be informed by this, as unpleasant services (especially burial) are exchanged along with oft-biting sarcasm (*ubishi*).

Those burying a chief are referred to as *vimbwi*, 'hyenas', or *vitambwa*. Both terms are multi-referential. Another name for the spotted hyena is *lupula-nkalamu*, literally, 'the one that begs food from an eating lion'. In a figurative sense, the chief is the 'lion' in question, but the joking partners (often real or classificatory grandchildren as well) call *themselves* lions, too. *Vitambwa*, the other name, is reported by Van Acker to refer to servants and slaves of the chief who are killed at his death to be buried with him, 'after they have engaged in diverse brigandage'. The root verb is **tamba*, 'to look far away',[51] as hyenas are deemed capable, their *malosi* or extended ken allowing them to find carrion no matter how distant.[52] The word **tamba* also means 'rampant', as plants may be, and a derivative verb refers to pursuing an animal in the chase, stooping over as one runs.[53] One story told to me was of *visanguka* observed as they left a village, walking and then trotting, upright and then stooping as they ran till they dropped to all fours, transformed to 'lions'. This in turn, through metaphor and pun, links the joking partners, the 'hyenas', to the 'lions' they serve, and to the 'lions' they may become in that service.

According to Storms, Lusinga reputedly could 'change himself at will into a lion' (as above), and in Case 3 'pricks up his ears' in an ominous fashion, when affronted by the assassination of his kinswoman. In Case 4, it is no coincidence that the narrator chooses a story form known to all Tabwa children: the trickster tale of Kalulu, the wily hare, and the oft-duped lion. Kalulu suggests he and Lion kill their mothers, to demonstrate their mutual friendship. Of course, only the lion's mother dies, while Hare feigns the homicide and hides his. So Lusinga tricks Kipili, and Manda is assassinated. In a prologue to this account, the narrator explains that Manda married Kaimba, the promiscuous woman who is murdered in Case 3. In these two myths, then, Kipili in Case 3 and Manda in Case 4 are interchangeable characters, of a paradigm, their *individual* identities less important than their *structural* (clan) identity, as foils for the wiles of Lusinga.

In the second part of Case 4, the problematic relationship between Manda as mother's brother, and Katele as sister's son, is developed. On the one hand, Manda Mukeya sets his successors against each other, surely knowing full well the consequence of his telling each that *he* will succeed, and not the other. Then the tables turn, as they always do in the Tabwa metaphysic, each force, social or natural, having its opposite and equivalent, each assuring its other's demise, then its re-ascendance. Katele, to show his displeasure at being bested in the succession, engages in 'diverse brigandage', the *kisama* associated with interregnum.

If the chief 'is' the moon, as mentioned briefly above, interregnum 'is' the tenebrous two or three nights between final and first crescents. Just as lions and other dangerous beasts 'wander about excessively' during *kamwonang'anga*, the dark of the moon; so during *kisama* the defunct chief's 'hyenas' and 'lions' are rampant. They 'run through the village yelling and shouting. They take all that they can lay their hands upon, outside the houses: picks, hoes, chickens, goats, children and even adults. Everyone hides at home'.[54] According to my informant Nzwiba, the *baendo bitambwa*, the stooped-over (hence bestial), rampant (hence man-like) joking partners, would wear crimson camwood (*nkula*) on their faces to indicate and increase their ferocity.[55] These *were* lions, Nzwiba insisted, but their rage would end in grieving as appeasing offerings were made by kinsmen of the deceased, and the fierce ones joined the mourning.

Nzwiba asserted that such 'lions' might destroy goods and kill livestock, but persons apprehended would be held for ransom. Katele acted within the idiom of *kisama*, and in its guise. In his outrage at being denied the succession, he pursued activities already begun in concert with WaNgwana. He killed and took slaves.

The tale is completed in Case 4 by Manda going to Captain Joubert, and offering him a sister's daughter (whom Joubert accepted and who was later married to Charles Faraghit, the Malta-trained Hausa medic at the mission), in exchange for his services in war against Katele. Joubert agreed but waited three days for moonlight before setting out; this was not only a practical measure, but in the story, one of symbolism, making Joubert the new moon rising after three nights' darkness, to signal and create a new order.

The four preceding cases better allow speculation concerning the killings from 1894 to 1896, as described in the White Fathers' journal of the Baudouinville Mission.

Case 5: Lions, missionaries, and Manda

In January 1894, a lion killed someone near the mission, and those living nearby asked the priests to pronounce an edict banning people of the Marungu Massif from transforming themselves into lions. The scribe's only comment was 'Poor people! When will they leave their superstitious ideas to walk in the light of truth?'[56]

In May, a village near the mission decamped when three were killed by lions. In June, Leopard Clan chief Kibwire, whom Joubert had chased from the area for slaving with Katele, returned from the Marungu saying there were too many there who could change into lions 'to get rid of their enemies'. He was allowed to settle at the foot of Mount Murumbi, where despite the paucity of his following, 'he still thinks himself as grand as the Czar of all the Russias', and wore the medal Monsignor Charbonnier gave him before he fell into Joubert's disfavour.[57]

In December 1894, lions tried to break down the doors to the mission village. Hearing a scratching at his door, a man thrust a brand outside to see what might be about and was clawed on the same arm by an unseen beast. Lions killed more than ten near the mission in 1894, and priests decided to build a stockade around their village. All inhabitants of two villages in close proximity to the mission fled because of these incidents, as did those of ten other villages within a day or two's walk.[58]

Flight from the missionaries was on everyone's mind, as lions killed within the priests' very village. The Fathers were told again that these were *not* lions but men. The missionaries hoped to kill several lions, as such a state of panic would ruin the mission. They obtained permission from the Congo Free State to prohibit people from deserting mission environs; village chiefs were armed with muskets to fend off lions, and Joubert sent men to bring back those taking flight.[59] Commandant Deschamps, the Free State representative, ordered that small villages be consolidated, ostensibly to render defence against lions the more effective. Monsignor Roelens remarked that 'these measures will be very useful to our work, for the more people are grouped, the easier we can instruct them'.[60]

As 1895 began, more lion-related deaths occurred around the Mission. Chief Manda was reported as living with a few men at the foot of Murumbi, where his people had been attacked repeatedly by lions over the previous two years. 'His people, crazed with fear, have fled, one after the other.' When lions killed yet another woman, Manda came to seek refuge with the priests at Baudouinville. The same night a lion was heard prowling around the mission compound. People living nearby complained that Manda had brought the lions, and threatened to flee should he be allowed to remain amongst them. 'To avoid stirring up these all-too-deeply-rooted superstitions of our savages', the priests told Manda to go and live with a kinsman of his, a day's walk from the mission. The next report of Manda, two months later in March 1895, was that he had died.[61]

In April of 1895 a man who had fled from the vicinity of the priests returned voluntarily and pleaded loyalty to the mission. Monsignor Roelens had the man bound in chains: 'Let this example serve as a lesson to all our neighbours, and diminish the attraction for travels!' Through April and May the killing continued. On 26 April, a woman was taken near the Mulobozi, the river near the mission; only her intestines were found. On 15 May, a young man who had regularly attended catechism was killed near the missionaries' residence. On the 16th, a woman carrying building stone for the cathedral was killed in broad daylight; no traces of the lions were found, and others carrying stone went on strike, On the 18th a woman and three men were killed near the mission. The priests' attempts to poison the lions with strychnine were in vain. At the beginning of June, a young and 'assiduous mission loyalist' was taken a hundred paces from the priests' village; a child taken ten minutes from the mission was 'clawed' and died of loss of blood; and there was news of others dying elsewhere close by. 'Everywhere terror reigns and villages threaten to flee *en masse*', especially the members of Manda's Leopard Clan.[62]

The killing persisted through the summer. In September 1895, people left the missionaries, seeking safety in Rhodesia. Some were flogged for trying to 'desert', and women apprehended were made to 'pay dearly for their escapade! We try to show them their mistake with "striking arguments", then reclothe them with a collar of iron'; that is, they were beaten and thrown in chains by the priests' men. Lions were about again in December 1895, and the 'only real witnesses are the three people that have been devoured on the other side of the Kyzie', a stream that flows next to the mission. More were killed near the priests' residence as 1896 began, including one of Joubert's men. Those who sought to flee were brought back and flogged. Lions heard prowling about the mission compound and village could not be seen nor was any caught. The priests implored Joseph, patron saint of the Baudouinville Mission, to rid their land of lions.[63]

And yet the killing continued. In March 1896, a lion was heard prowling about the perimeter of the priests' garden, but in the dark of a moonless night, they could hear but not see it. The next day a man was attacked and killed after stepping outside the village gate to defecate. A predator killed a sheep and a goat of the mission; a poisoned carcass left as bait was taken by the lion, but when found again, it had been hidden beneath a bush with a leg removed from the unpoisoned portion. In April 'our lions are no longer lucky, when they kill a person, someone always arrives to take their prey from them, and so they are now thin and weak'. People nearby killed three emaciated lions 'dead of hunger'. With a few more references to deaths through the end of 1896, the missionaries reported few from lions or other predators till 1903.[64]

DISCUSSION OF CASE 5

With the preceding cases in mind, one can speculate that 'lions' were being sent by inhabitants of the Massif to attack mission loyal. Given Manda's past performance in support of Storms and Joubert against recalcitrant Bushpig Clan chiefs, and Katele's fury at losing the succession to Manda Mukeya's chair, there are several possible bad guys in the who-done-it, and every likelihood that more than one faction was employing the same strategy of clandestine and violent politics, at the same time for different reasons. The new Lusinga and his people, as well as other Bushpig chiefs (whose cases have not been reviewed here) had plenty of reason to seek revenge against Manda and his Leopard Clan. So did the slaver Katele, who in so many ways on so many occasions was reckless and bloodthirsty.

Monsignor Roelens at the Baudouinville Mission wielded heavy-handed authority, yet I think it would be incorrect to assume that these attacks by 'lions' (and I assume them to be *visanguka* lion-*men*, as did those subject to their depredations) were simply or primarily in resistance to and rebellion against the new European presence. They *were* this, but only secondarily so.

At first the priests supported the same Tabwa and intrusive chiefs as had Storms, but then they shifted their favour. Concurrent with the change in the missionaries' regard for Manda was the rising influence and ordination in 1893 of Victor Roelens as Vicar Apostolic for the Upper Congo. From his arrival at the Lake Tanganyika Missions early in 1892, Roelens was critical of all those who had preceded him, secular and missionary personnel alike. He was irked by Joubert, who was 'an excellent man, a good Christian and a brave soldier, but a bad administrator . . . he is too good, too soft, too patient and hence too weak in the view of his people'; 'The niggers do whatever they want and the good captain shuts his eyes.' Roelens, to build his own authority, demanded obedience from the Africans residing with Joubert, even as he 'managed' the Captain himself. Joubert and several missionaries complained of Roelens' severity with the Africans, but Roelens prevailed.[65] Specifically, Tabwa chiefs who had enjoyed the support of and had lent support to Joubert and Storms before him, were denied Roelens' favour.

Manda's weakened position was recognized by Bushpig Clan members, long-standing opponents to his having usurped lands and prerogatives they felt theirs. Their 'lions' harried him, as did those of several chiefs of Katele's line. If mission support for Manda was on the wane, his humiliation could be increased by inducing his followers to flee, in mortal fear of the 'lions' pursuing him. A chief without people has no power. By the years of the First World War, Manda had so few subjects that he left the Belgian Congo altogether, abject, and was replaced by the priests with a loyalist from north of Lake Tanganyika who had accompanied them when the Lavigerieville Mission was closed.[66]

Scores in the local-level arena were being settled, then, and killing individuals identified with the priests who supported their adversaries was a means of indirect attack. The same strategy was used in opposing the priests themselves. As Roelens' and the Baudouinville Mission's influence grew, so must have the resentment of local chiefs whose powers were eroded commensurately. Tabwa and others were prevented from 'deserting' the mission, and were encouraged to provide labour for the immensely intensive task of building the new mission and its cathedral. Should those closest to the priests be seized with panic and flee, the missionaries, too, would no longer be 'chiefs', despite their being the ones with the guns. Manda Kabunda, humiliated by Roelens and the White Fathers, may be suspected of deploying a strategy used against him by others, seeking revenge through 'lions'. A lack of data makes this conjectural, but logically consistent with cases better known.

Conclusions and beginnings

The discussion above suggests as many beginnings for further inquiry as it does conclusions. The deaths by 'lions' *did* abate in 1896 when, causally or coincidentally, depending upon one's perspective, several raggedy lions were caught. It may be that some of the deaths were due to lions, and that lion-men 'followed in their tracks' and so disguised their own deeds.[67] I have suggested the following motives for the cases presented: (1) resistance to Nyamwezi intrusives; (2) the frightening of villagers while taking slaves; (3) revenge for the affrontery of homicide; and (4) participation in *kisama* interregnum or as a strategy in succession disputes. These are consistent with Joset's summary of reasons for leopard-man incidents in the region of Beni, several hundred miles due north of Tabwa country.[68] Most importantly, the theriomorphic strategy was available for use by a variety of actors in many different circumstances, among which might be resistance to European colonisers. As often, lion-men killings continued *in spite of* European colonisers.

Such terrorism is in logical opposition to the essence of political power: the recruitment and maintenance of a stable residence group. *Visanguka*, 'lion-man', is derived from a verb meaning 'to become' or 'change oneself into', but also 'to be unfaithful', as someone is who leaves one chief to live with another. The 'ones who change themselves into' and thus 'become' lions, in so doing cause people to 'be unfaithful' by changing affiliation from one chief to another. 'A chief is his people.' If his people are dispersed, he is no longer chief except in name. Within the group, particular individuals may pursue conflicts with particular adversaries by means of sorcery on the one hand, and sorcery accusations on the other. A history is known, and misfortune explained in terms of it: affliction is deserved due to laxity in following culturally prescribed behaviour, or at least comprehensible when divination defines its context and points a finger at the envious sorcerer.

Chiefs are 'sorcerers' in collusion with sorcerers of their communities; yet unlike 'true' sorcerers, they rarely use their highly potent medicines aggressively (except against other chiefs or, these days, government officials); nor do they eat victims' flesh brought to them by sorcerers. Rather, they sell the meat back to lesser sorcerers than they. Chiefs are their people, *all* their people. They are identified with the land (*mwine kyalo* or *mwenye inchi*). They transcend the particular. They have no shame, as to have it would be a particular regard or recognition. Their justice is *for* all, their wrath *against* all. The impunity with which *visanguka* lion-men deployed by or associated with chiefs transgress what seem the most essential of social rules (the preservation of life) and philosophical principles (that man lives in an ordered universe, in which acts and events make some ultimate sense) can be understand only be reference to the concept of chiefship for Tabwa.

Anguish is the greater when it is known that there will be victims as a conflict evolves, but that they will be random. A lion stalks a herd of its prey, its calculation awesome, as wrote Tabwa author Stefans Kaoze;[69] but *which* of the herd it will pounce upon is often a matter of chance. Aged or young, disabled or defenceless prey may fall first to real or theriomorphic lions, but only when, from the victim's viewpoint, they happen to be in the wrong spot at the wrong time.[70] This must be a monumental insult to the particular actors in the conflict who cannot combat their adversary face to face or even through magical medicine to magical medicine. They are powerless, as victims are plucked from their destinies without particular reason. Instead, as an *anioto* leopard-man implicated in the mid-1930s hecatomb near Beni remarked, 'a cadaver demands a cadaver';[71] and so a vendetta is carried forth, the victims third parties. Direct assault would make the affair finite. In this protracted torment, the adversary and his dependents, the would-be victims, are not permitted the security of social order. To follow Tabwa metaphors, were the moon to shine or the lion to be seen, one could escape harm, or hope to do so. But this is utter darkness, utter fury, utter chaos.

Among issues yet to be resolved is the missionaries' reluctance to recognise human actors in the 1890s killings, despite constant, straight-forward explanation to the effect by African residents. This subject merits more detailed coverage than is possible here, yet a last comment may be offered.

One can deduce from the writings of the scribe at the Mpala Mission cited above that the people of the Marungu Massif, 'true meat for the butchering' according to Nyamwezi and other lakeside dwellers, reacted to the double ignominy of insult and injury by sending 'lions'. Victor Roelens would prove little kinder in his assessment of the same Tabwa. In a number of contexts the Monsignor wrote of Tabwa 'psychology', and found them 'atavistically depraved', marked by an 'insouciant lack of reflection' and 'bestial frenzy' in their sexual life. 'One must have seen our pagan Blacks in their primitive state as I have; one must have sounded the depths of their soul and have analysed the sentiments that animate and direct them, to know the sort of pigsty of depravity in which the human beast can wallow.'[72]

Over and above an ignorance of natural history as known now, according to which man-eating lions seem rare, it served an ideological purpose for the missionaries to disbelieve their 'innocents'. On the one hand, Roelens found 'this spectacle' (the 'pigsty') 'produces a sentiment of immense pity' among his pious and righteous priests; on the other, relations with Tabwa were taxing, as the Fathers were 'given no respite, day or night'.[73] However, as another wrote:

> to suffer, wasn't that what ought to be our part, we, missionaries of Africa? We knew it well, and far from complaining, we thanked Our Lord for allowing us to endure something for Him . . . These ordeals . . . have done great good for our souls. If we had suffered less, we would have prayed less, we would have been less detached from the things of this earth, less united with God.[74]

Suffering, whether from illness, separation from loved ones, or from constant exposure to the 'atavistic depravity' of the Tabwa, brought the priests closer to God, farther from Satan. When, in four months of 1904, more than thirty people were killed by 'lion' in the plain between Baudouinville and Lusaka Missions, a Father wrote stressing that he meant real lions, for 'our Blacks even believe – how naive! – that certain men have the power to change into lions' and are called *visanguka*. 'The poor Blacks, for goodness sakes (tout de même)!' The priest is reminded of 1 Peter 5:8 (as in this paper's epigraph), '"Awake, be on the alert! Your enemy the devil, like a roaring lion, prowls round looking for someone to devour." For, after all, the true and great lion that we must all fear is the Demon.'[75]

NOTES

1 Four years' anthropological fieldwork (late 1973 to late 1977) at Lubanda, village of Tabwa chief Mpala, was funded by the National Institute for Mental Health, (1-FO1-MH-55251-01-CUAN) the Committee on African Studies and the Edson-Keith Fund of the University of Chicago, and the Society of the Sigma Xi. Travel funds to attend the 9th Annual African Studies Symposium at the University of Illinois, Urbana, at which this paper was first presented, and secretarial support were received from Albion College. A Mellon Foundation Faculty Development grant at Albion College permitted research at the Central Archives of the Missionaries of Our Lady of Africa (White Fathers), Rome; special thanks to Father Rene Lamey, archivist, for helping me find relevant documents. Thoughtful comments have been offered by Elizabeth Brumfiel, Christopher Davis-Roberts, Omari Kokole, Daniel Moerman, the late Robert Notestein, Jonathan Post, Tobin Siebers, Jan Vansina, Father Joseph Vleugels ('Mabawa') and the participants of the 9th Annual African Studies Symposium at the University of Illinois. Despite the generosity of these agencies and individuals, I alone am responsible for the contents of this paper. In fond memory of Father Joseph Kimembe.

2 Such losses would be reported by priests more readily than would the deaths of persons with whom they had no contact at all.

3 Harold Schneider, 'Male–female conflict and lion men in Singida', in Simon Ottenberg (ed.), *African Religious Groups and Beliefs* (Meerut India: Archana for Folklore Institute, Berkeley, c. 1982), p. 96. The degree to which the ethnography of missionaries in Africa has been overlooked is outlined by Thomas Beidleman, *Colonial Evangelism* (Bloomington: Indiana University Press, 1982).

4 Birger Lindskog, *African Leopard Men* (Uppsala: Studia Ethnographica Upsaliensia, VII, 1954). P. Joset, *Les sociétés des hommes-léopards en Afrique Noire* (Paris: Payot, 1955).

5 As cited in Lindskog, *Leopard Men*, p. 52. In his recent study of the same era and area, Randall Packard makes no mention of leopard-men, even though this must have been an important strategy in the political competition he would describe. See his *Chiefship and Cosmology: An Historical Study of Political Competition* (Bloomington: Indiana University Press, 1981).

6 Emil Torday, 'The things that matter to the West African', *Man*, Vol. 13, no. 116 (1931), pp. 110–13.

7 Lindskog, *Leopard Men*, p. 84.

8 Eric Hobsbawm, *Bandits* (New York: Pantheon, revd edn, 1981), pp. 17, 20.

9 If one considers antiquated and obsolete senses of these two words, the phrase is even more appropriate. In antiquity, according to the OED, 'hero' was 'a name given (as in Homer) to men of superhuman strength, courage, or ability, favoured by the gods; at a later time regarded as intermediate between gods and men, and immortal', *The Compact Edition of the Oxford English Dictionary*, Vol. 1 (New York: Oxford University Press, 1982), p. 1296. 'Criminal' formerly might be used in reference to beasts especially 'savage, fierce, malignant', as in the fifteenth century when describing 'the most terrible and most crymynel dragon' or 'Bestes . . . so righte stonge & crymynell that no men dare approche them' (ibid., p. 604). 'Heroic criminal', then, *could* refer to an intermediate or transitional being, half-man, half-beast of superhuman strength and unbounded savagery; Tabwa lion-men were felt to be just that.

10 Hobsbawm, *Bandits*, p. 22.

11 These portentous phenomena and the social change they were felt to import are described in Allen Roberts, '"Comets importing Change and Times and States": ephemerae and process among the Tabwa of Zaire', *American Ethnologist*, vol. 9, no. 4 (1982), pp. 712–29. And *idem*, '"Fishers of men": religion and political economy among colonized Tabwa', *Africa*, vol. 54, no. 2 (1984), pp. 49–70.

12 See Allen Roberts, '"Insidious conquests": war-time politics along the southwestern shore of Lake Tanganyika', in Melvin Page (ed.), *Africa and the First World War*

(London: Macmillan, forthcoming); and Andre Marissaux, *Albertville, Note Historique* (Brussels: L. Cuypers, n.d.).

13 Mary Douglas, 'Introduction: thirty years after witchcraft, oracles and magic', in Mary Douglas (ed.), *Witchcraft Confessions and Accusations* (London: Tavistock, 1970), p. xx.

14 Although the historical context and political aims of actors are very different, the hypothesis suggested here is consistent with Eugene Walter's conclusions concerning the despotic use of terror by Shaka the Zulu. See his *Terror and Resistance: A Study of Political Violence, With Case Studies of Some Primitive African Communities* (New York: Oxford University Press, 1969). My thanks to Professor Jan Vansina for bringing this parallel to my attention.

15 Kanengo was a celebrated diviner-healer, and an adept of the apparently defunct Kazanzi anti-sorcery society; one of his exploits is described in Allen Roberts, 'Anarchy, abjection and absurdity: a case of metaphoric medicine among the Tabwa of Zaire,' in Romanucci-Ross, Moerman and Tancredi (eds), *The Anthropology of Medicine* (Amherst: Bergin *for* Praeger Special Studies, 1983), pp. 119–23.

16 George Schaller, *The Serengeti Lion* (Chicago: University of Chicago Press, 1972), p. 220. René Malbrant, 'Faune du Centre Africain Français (mammifères et oiseaux)', *Encyclopédie Biologique*, Vol. 15 (Paris: Paul Lechevalier, 1936), p. 130.

17 J. H. Patterson, *The Man-eaters of Tsavo and other East African Adventures* (New York: Macmillan, 1948). The popular accounts cited here are Roger Caras, *Dangerous to Man. Wild Animals: A Definitive Study of Their Reputed Dangers to Man* (New York: Chilton Books, 1964), pp. 16–17; and C. Guggisberg, *Simba, the Life of the Lion* (Cape Town: Howard Timmens, 1961), pp. 256–8.

18 William Arens, *The Man-eating Myth* (New York: Oxford University Press, 1979), p. 9.

19 Sketches of such implements are presented in Armand Van Malderen, 'Note sur la secte des "Visanguka", territoire des Baanza-Bazimba', *Bulletin des Juridictions Indigènes et du Droit Coutumier*, vol. 1 (1935), pp. 22–4. Similar instruments from elsewhere in Zaire are in the collections of The Royal Museum for Central Africa (MRAC), Tervuren, Belgium.

20 Auguste Van Acker, *Dictionnaire kitabwa-français, français-kitabwa*, Annales, serie V, Ethnographie-Linguistique, Musée Royal du Congo Belge, (1907), p. 58. No accents, stress or tone marks are indicated by relevant authors for this or other terms discussed in the text, except as indicated. White Fathers, *Bemba-English Dictionary* (Cape Town: 1954), p. 656.

21 Van Acker, *Dictionnaire*, p. 58. E. Van Avermaet, and Benoit Mbuya, *Dictionnaire kiluba-français*, Annales, Sciences de l'Homme, Linguistique 7, Musée Royal du Congo Belge, (1954), p. 575. White Fathers, *Dictionary*, p. 657.

22 Gustave De Beerst, 'Essai de grammaire tabwa', *Zeitschrift fur Afrika und Ocean Sprachen*, vols. 1–2 (1894), pp. 19, 23.

23 James Brain, 'Ancestors as elders in Africa – further thoughts', in William A. Lessa and Evan Z. Vogt (eds), *Reader in Comparative Religion* (New York: Harper & Row, 4th edn, 1979), p. 395.

24 This imagery is discussed at length in Allen Roberts, 'Heroic beasts, beastly heroes: cosmology and chiefship among the lakeside BaTabwa of Zaire' (unpublished PhD dissertation, University of Chicago, 1980); *idem*, 'Anarchy, abjection and absurdity'; *idem*, '"Comets importing change"'.

25 Roberts, 'Heroic beasts', p. 491.

26 The metaphor of the lion, especially as employed by Tabwa when discussing precolonial chiefship, is discussed in Allen Roberts: '"Perfect" lions, "perfect" leaders: a metaphor for Tabwa Chiefship', *Journal de la Société des Africanistes*, vol. 53, nos 1–2 (1983), pp. 93–105; for further information on *visanguka*, see Van Malderen, 'Visanguka'; 'F.R.', 'Crimes et superstitions indigenes: l'affaire des hommes-lions en Territoire de Moba, Chefferie Kayabala', *Bulletin des Juridictions Indigènes et du Droit Coutumier Congolais*, vol. 8, no. 9 (1940), pp. 257–63; and T. Alexander Barns, *Across the Great Craterland to the Congo* (London: Ernest Benn, 1923), pp. 190–205. Comparative

material from the southern Tabwa of Zambia is in Robert Cancel, '*Inshimi* structure and theme: the Tabwa oral narrative tradition' (unpublished PhD dissertation, University of Wisconsin, 1981); and Anon., 'Man-eaters at Mporokoso: extracts from a district notebook', *Northern Rhodesia Journal*, vol. 1, no. 2 (1980), pp. 71–74. On lion-men (*bacisanguko*) among neighbouring Bemba in Zambia, see E. Labrecque, White Fathers' Central Archives, Rome, 'Quelques notes sur la religion des Babemba et tribus voisines', Typescript 180/2, *c.* 1923, p. 18; reproduced with slight editing in Language Center (eds), *Bemba Cultural Data: Monographs*, part one, 'La religion du noir infidèle', by E. Labrecque (Chinsali, Zambia: The Language Center, 1982).

27 Andrew Roberts, *A History of the Bemba: Political Growth and Change North-eastern Zambia Before 1900* (London: Longman, 1973), pp. 151–60.

28 These 'other' Yeke have yet to be studied closely, but see Thomas Reefe, *The Rainbow and the Kings: A History of the Luba Empire to 1891* (Berkeley: University of California Press, 1981), pp. 172–80.

29 ibid., Chs. X–XII.

30 Joseph Thomson, *To the Central African Lakes and Back: The Narrative of the Royal Geographical Society's East Central African Expedition, 1878–1880*, Vol. 2 (London: Frank Cass, 2 vols, 1968), p. 48.

31 ibid., p. 51. See also Robert Rotberg, *Joseph Thomson and the Exploration of Africa* (London: Chatto and Windus, 1971), p. 84.

32 James Wolf (ed.), *Missionary to Tanganyika 1877–1888: The Writings of Edward Coode Hore, Master Mariner* (London: Frank Cass, 1971), p. 125; and Thomson, *Lakes*, Vol. 2, pp. 51–2.

33 Archives du Bureau des Affaires Culturelles, Division Regionale des Affaires Politiques, Lubumbashi. Anon., 'Unentitled administrative report', c. 1915.

34 Archives du Musée Royal de l'Afrique Centrale, Tervuren, Fonds Storms, 1883–1885, Emile Storms, 'Journal de la Station de Mpala fondée le 4 mai 1883 par le lieutenant Storms', entry for 26 December 1884, my emphasis.

35 White Fathers, 'Journal du poste de la Mission de Kapakwe', typescript copy consulted at the Archives of the Kalemie-Moba Diocese (original at the White Fathers' Central Archives, Rome), 26 December 1884.

36 Storms, 'Journal', Vol. 2, 8 January 1885.

37 Leopold-Louis Joubert, 'Letter to General de Charette', *Mouvement antiesclavagiste*, vol. 2 (1890), pp. 18–24; idem., 'Letter to his brother of 3 May 1891', *Mouvement antiesclavagiste*, vol. 4 (1892), pp. 68–72. This was a European perception at a time when early missions in the area were severely menaced by Rumaliza's men. No diaries or personal documents of Rumaliza are known, but from conversations with him reported in missionary sources, it is clear that he was, like Mirambo, Nyungu-ya-Mawe and Tippo Tip one of the dominant figures of late nineteenth-century east-central Africa. On Nyungu-ya-Mawe, see Aylward Shorter, 'Nyungu-ya-Mawe and the "Empire of the Ruga-Rugas"', *Journal of African History*, vol. 9, no. 2 (1968), pp. 235–59.

38 Cited in T. Houdebine and M. Boumier, *Le Capitaine Joubert*, Collection Lavigerie 29 (Namur: Grands Lacs, n.d.), p. 88.

39 White Fathers, 'Diaire de la Mission de Baudouinville,' typescript copy consulted at the Archives of the Kalemie-Moba Diocese (original at White Fathers' Central Archives, Rome), March 1892.

40 J. Verhoevan, *Jacques de Dixmude, l'Africain: Contribution à l'histoire de la Société antiesclavagiste belge, 1888–1894* (Brussels: Librairie Coloniale, 1929), p. 53.

41 ibid., p. 77. See also Roger Heremans, *Les établissements de l'Association Internationale Africaine au Lac Tanganika et Les Peres Blancs, Mpala et Karema, 1877–1885*. Musée Royal de l'Afrique Centrale Annales-Sciences Historiques 3 (1966); and François Renault, *Lavigerie: l'esclavage africain et l'Europe*. 2 vols. (Paris: Editions De Boccard, 1971).

42 Verhoeven, *Jacques*, p. 101. See also Nteziryayo Gapangwa, *La première évangelisation du Massanze et du Bubwari (1880–1892)*. Mémoire de Licence, Faculté d'Histoire Ecclesiastique, Université Pontificale Gregorienne, Rome (1980).

43   White Fathers, 'B'ville,' February 1893, 10 May 1893.
44   White Fathers, 'Diaire de la Mission de Mpala'. Typescript copy of all but first notebook consulted at Archives of the Kalemie-Moba Diocese; copy of the first notebook obtained from White Fathers' Central Archives, Rome. 14 February 1885, 12 March 1886.
45   Unpublished letter, Moinet to Eminentissime Seigneur (Cardinal Lavigerie), 26 September 1885 from Mpala c-19-245, White Fathers', Rome.
46   White Fathers, 'Mpala', 15 May 1888, 22 September 1888.
47   Leopold-Louis Joubert, 'Personal diaries, 1887–1914', incomplete photocopy consulted at the National Museum of Zaire, Lubumbashi; 23 May 1887, 2 August 1887, 15 May 1888, 6 November 1888.
48   Joseph Vleugels (ed.), 'Table d'Enquete sur les Moeurs et Coutumes locales des Batabwa', Sixième fascicule, Archives of the Kalemie-Moba Diocese (c. 1960), no pagination. There is no specific date for this incident, although if Lusinga moved to the shore of Lake Tanganyika in the 1870s, then this and the next case must refer to the next decade or so, prior to Storms' arrival in 1883.
49   Henri Kissi, 'Mafasario juu ya maneno ya inchi ipatikanayo katika Territoire ya Baudouinville, District du Tanganika, Etat du Katanga', Dossier Territoire des Bazimba, Archives du Bureau des Affaires Culturelles, Division Regionale des Affaires Politiques, Lubumbashi, (1961), pp. 20–3.
50   *Adui* means both 'enemy' and 'predator' in Lakeside Swahili; in the second case with specific reference to lions (and occasionally leopards).
51   Van Acker, *Dictionnaire*, pp. 130, 55, 65. It appears that the Tabwa term **tamba* is more closely related to the Bemba verb **tamba*, 'to look at), with derivatives 'to stare at' and 'to bewitch' (White Fathers, *Dictionary*, pp. 734–4) than it is to the Luba, *ntambo*, 'lion' (Van Avermaet and Mbuya, *Dictionnaire*, p. 668).
52   On hyenas in Tabwa thought, see Roberts, 'Anarchy, abjection and absurdity'.
53   Van Acker, *Dictionnaire*, p. 65, cf., White Fathers, *Dictionary*, p. 743.
54   Charles Delhaise, 'Chez les Wabemba', *Bulletin de la Société Royale Belge de Géographie*, 32, 173–227, 261–83. 1908: 211, describing events at Tabwa Leopard-clan Chief Mulilo's.
55   Comparison was made by Nzwiba to his having heard of the use of cannabis by American soldiers in the Viet Nam war, to effect a similar result: using *nkula*, one no longer has fear, and is willing to die.
56   White Fathers, B'ville, 20 January 1894.·
57   ibid., 26 May, 20 June 1894.
58   ibid., 1, 3, 15, 17 December 1894.
59   ibid.
60   Victor Roelens, 'Les missions dans le Vicariat apostolique du Haut-Congo', *Mouvement antiesclavagiste*, vol. 8 (1896), p. 241.
61   White Fathers, B'ville, 14 January, 27 March 1895.
62   ibid., 22, 26 April; 15, 16, 18 May; 8 June 1895. See also A. Engels. 'Baptêmes de nos petits rachètés (Haut-Congo)', *Missions d'Afrique*, vol. 19, no. 8 (1898), pp. 227–31.
63   White Fathers, B'ville, 15, 17, 22 September; 23, 24, 25 December 1895; 23, 25, 26, 29 February 1896.
64   ibid., 8, 9, 14, 15 March; 20 April; 5, 8 December 1896; 10 July 1903. In 1898 a person was killed and half eaten near Mpala Mission; when the missionaries poisoned the remains of the cadaver (in gross insult to Tabwa desires for burial) the 'lion' did not return to take the bait. Soeur Willibrord, 'Nouvelles de Mpala', *Missions d'Afrique*, vol. 20, no. 9 (1899), pp. 277–86.
65   White Fathers' Central Archives, Rome, unpublished letters, Roelens to Livinhac, 17 February 1892 from Karema and 16 September 1892 from St. Louis. c. 19. pp. 430–31. For an excellent review of Roelens's role at Baudouinville and a critique of his particularly negative view of his African parishioners, see Roger Herremans, 'Missions et Ecoles: l'éducations dans les missions des Pères Blancs en Afrique Central avant 1914' (unpublished PhD dissertation, Modern History, Université Catholique de Louvain,

1978), pp. 183–5 and *passim*.

66 See Roberts, '"Insidious conquests"', where this is discussed further.

67 Lindskog, *Leopard Men*, p. 73.

68 Joset, *Sociétés*, pp. 53, 65, 18, 89.

69 Stefano Kaoze, 'Le mukowa ou le clan', MS in private archive, *c.* 1930.

70 See George Schaller, *The Serengeti Lion: A Study of Predator-Prey Relations* (Chicago: University of Chicago Press, 1972), pp. 225–32. In a survey of predator kills reported in 1960, by Bruce Wright, male prey outnumbered female three to two; animals in their prime, those past it two to one; and those in good condition, not those eight to one. Roger Caras concludes that 'lions simply do not engage in conscious selection of prey but take that which is closest,' offers the best chance for a kill, and requires the least amount of exertion'. See his *Dangerous to Man*, p. 15.

71 Joset, *Sociétés*, p. 48.

72 Victor Roelens, 'Les rayons et les ombres de l'Apostolat au Haut Congo', *Grands Lacs*, vol. 61, nos. 4–6, n.s. (1946), p. 31.

73 ibid., pp. 31–2.

74 Paul Lesourd, *Les Pères Blancs du Cardinal Lavigérie* (Paris: Eds Bernard Grasset, 1935, Collection Les grands ordres monastiques et instituts religieux), pp. 220–21.

75 Anonymous White Father, 'Au Pays du lion à Saint-Jacques de Lusaka', *Missions d'Afrique* (1905), pp. 263, 266.

SECTION II

Banditry

4 Social bandits and other heroic criminals: Western models of resistance and their relevance for Africa[1]

RALPH A. AUSTEN

The purpose of this essay is to explore the problems and possibilities suggested by the application of Eric Hobsbawm's concept of social banditry to African history and popular culture.[2] The problems arise from two sources: first, the lack of clarity with which Hobsbawm presents this very stimulating idea in its original Western context; second, the realisation that social banditry is a European-American idea (with an important set of Asian variants not discussed here) which ultimately does not fit very much African historical experience.

Importing Hobsbawm into Africa still offers a number of rewards. First of all (as other essays in this volume reveal) it forces us to think about the social meaning of crime, a very real issue in Africa which has not previously received adequate attention. Secondly it indicates some interesting contours of divergence and convergence in African and European history which in so many other respects are both remote from, and intimately linked with one another. Finally, this exercise in, and partial exorcism of, a model with limited relevance to Africa may suggest some other directions in which to pursue African experiences and understandings of social order, deviance, heroism, and resistance.

In order to deal at all systematically with the historical meaning of crime it is first necessary to reformulate Hobsbawm's categories of heroic criminals. Two general considerations will inform this effort. The first is to treat separately the empirical instances of crime which may have some positive social meaning and the folklore which expresses this meaning. Without such an effort it is sometimes difficult to avoid the Hobsbawmian heresy of adding to bandit folklore rather than analysing it. We should also not assume that the social milieu of criminal events is always the same as that of the heroic legends which grow up around them.

Secondly, more explicit attention must be given than in Hobsbawm's writings to the role of the state. In African cases, even when they transcend Hobsbawm's residual category of 'tribal or kinship societies',[3] it is not always necessary for deviant groups to enter the category of criminals opposed to a general political authority. They can instead resort to, or be confronted with, a process of segmentary self-help. Thus, instead of 'cops vs robbers' we may have community pressure against individuals, feuds between community segments, or the separation of dissident groups to form an entirely new state.[4] In other words for criminals to become social heroes – or, more significantly, for heroes to become criminals – there must not only be poor classes who resent the rich but also a political system claiming hegemony over both the rich and the poor.

Despite his claim that 'social banditry ... is one of the most universal phenomena known to history, and one of the most amazingly uniform',[5] Hobsbawm has noted that it takes a variety of forms. The categories under which heroic criminals will be grouped here do not follow the schematisation of *Bandits*, although they are partially inspired by Hobsbawm. The main basis for the present distinctions is the relationship between social crime and the evolution of the state, beginning with a rural situation in which segmentary self-help still provides an alternative to official authority and concluding with the self-conscious use of criminal action to overthrow modern governments. The categories (to be explained as each is presented) are: the self-helping frontiersman; the populist redistributor; the professional underworld; the *picaro*; and (in abbreviated form) the urban guerrilla.

The self-helping frontiersman

The most common theme in all accounts of bandits, whether real or legendary, tells of a man who suffers a personal wrong and, in seeking to retaliate directly, is driven outside the legal bounds of society. The settings for such dramas range widely over Western and African history, but whenever they become the institutionalised definition of a criminal career they describe a frontier between centralised authority and a local society which has not entirely abandoned self-help as a means of administering justice.

Bandits of this kind are the central figures in the earliest recorded heroic criminal literature, in the European tradition: the outlaw sagas of medieval Iceland, apparently based on true cases.[6] Here the central figures are heads of landowning families who have escaped the authority of the newly centralised Norwegian state and come into conflict with even their own representative council, the Althing. In later Western versions of this theme the geographical setting is also important: the state is always based on densely settled, developed areas while the outlaws flourish in peripheral mountain and island regions or the cattle plains of North and South America.[7]

This kind of banditry can also be found in Africa, particularly in rural areas imperfectly integrated into early states. The role of *sheftenat* in Ethiopia is delineated in several other chapters in this volume. We also have very full accounts of bandits in the mountains of Algeria during the nineteenth and twentieth centuries.[8] The fact that no such cases are presented in any detail for precolonial tropical Africa is undoubtedly a function of difficulties in documentation. However, it may be argued generally that the frontiers of most indigenous African states were too open for banditry to develop. Segmentary politics could continue without any serious attention to the formal central authority, or dissidents could find cultivable land to move into and form their own political system.

Algeria also stands out as a unique case of self-help banditry functioning as a major African response to colonialism. Further south in the continent there is plenty of evidence of violent resistance to European rule and also of rural self-help in settling affairs among Africans as well as appropriating goods from Europeans. Information concerning such activity comes mainly from police sources and has not, as far as is known, generated any indigenous heroic legends.[9] The apparently subdued nature of this form of criminality will be discussed below in relation to the colonial state.

As already indicated in the introduction to this volume, little is added to the understanding of conflicts between European invaders and African states by labelling the latter as bandits. Indeed, this term was often used by Europeans to express contempt for the guerrilla tactics resorted to by African political leaders or entire societies.[10] We need to make a distinction between frontiers within a given social system and the situation of 'interpenetration between two previously distinct societies,'[11] which is the African colonial frontier. The former gave rise to banditry as one means of contesting the values that were to be deployed in controlling economic and political life. The latter produces 'primary resistance' in which no institutions or repertoire of values are shared between the two sides.

The role of heroic frontier criminals can further be understood by exploring a distinction which both Hobsbawm and more recent students of rural society in southern Italy (especially Sicily) have made between the bandit and the *mafioso*.[12] The bandit represents the dramatic and heroic version of self-help in this society because he defies the official authorities from an isolated (often peasant) position, is forced to flee individually or with a few followers to 'the hills', and concentrates his depredations upon the rich since the poor have little to offer except shelter and admiration. The purity of such bandit careers in popular perceptions is guaranteed by the inevitable (and yet inexplicable) tragic defeats at the hands of the authorities.

Mafiosi represent a more subdued form of illegitimate self-help, despite the legends of their wide-ranging conspirational networks. They are organised, albeit rather parochially, around wealthy men who defy the formal rules of the national political system but come to terms with its weak local representatives. *Mafiosi* live openly in rural settlements and are 'respected' by the poorer population for their ability to mete out rewards and punishments and their genuine invulnerability to threats other than those of competitors within their own system.

The common denominator linking banditry and the mafia as forms of self-help is the assertion of personal honour and violence against the official law codes and bureaucratic agencies of the central state. The bandit acts this opposition out in extreme form as a mythic positing of nature (the flamboyant individual in the hills) against culture. However an informally understood code of honour gives the *mafioso* his much cherished 'respect' and his interests are necessarily enforced by acts of violence. What both figures achieve is advancement to wealth and power through the use of personal local resources rather than the abstract, universalised values embodied in the modernising state.

Criticism of the social-bandit notion as applied to such frontier situations rests first on the empirical demonstration that bandits and *mafiosi* (or equivalent semi-feudal elites such as north-east Brazilian landlords) were frequently allied, particularly when the position of the latter was threatened by any serious peasant movements. Secondly, the values of honour upheld by both groups are not simply, as Hobsbawm suggests, populist expressions of archaic dissatisfaction with the contemporary situation, but also (perhaps predominantly) a support for the principles of local patron-client hierarchies.[13]

Bandits and oligarchs may lead or represent rebellions when they enter into the organisation or ideology of regional movements which assert themselves against central or foreign rule (these are Hobsbawm's bandits as *haiduks*, or *mafiosi* as relatively 'good' Sicilians). But such situations are external to the social relations within a region. They reveal more about the nature of nationalist rebellions or regional self-identity than about criminals as they operate or are perceived in their local roles.[14] Indeed, in the Algerian case the ideological adoption of mountaineer bandits by the post-Second World War nationalist movement directly contradicted the efforts of these figures in their own lifetimes to avoid conflict with either colonial authorities or local French settlers.[15]

In tropical Africa it may be useful to think of traditional chiefs incorporated into various forms of indirect rule as a kind of mafia, although here the initiative came from a colonial state which was not really interested in assimilating rural areas into a national entity. At the same time such regimes were sufficiently effective to discourage violent criminal action as a real or symbolic form of peasant resistance.[16]

The one documented case of frontier criminality by colonial Africans occurred on the north-western borders of the eighteenth-century Cape of Good Hope Colony. Here 'coloured' retainers of local white ranchers (themselves involved in considerable cattle raiding) broke away to form independent communities which functioned for some years as bandit gangs. But it is characteristic of even this region of relatively assimilative European expansion that the most successful of the indigenous outlaw leaders, Jager Afrikaner, and his descendents could eventually redefine their following as an ethnic-political force on a par with other local African states, especially after they migrated from the vicinity of the Cape into the still uncolonised Namibia.[17]

The populist redistributor

'He robbed the rich, he helped the poor'; this 'wild colonial boy' or 'Robin Hood of ——' seems to be at the core of the entire concept of social banditry. As a motif, redistributive action forms part of the legend of virtually every category of heroic criminal, but it also has a fuller development under historical conditions which provide the basis for a separate category.

Redistribution is certainly not the central theme of self-help criminality, which may keep out the threatening power of the central state but offers its constituency at best only the preservation of their existing rights and property. Some rural bandits, however, have identified themselves with populations not of oppressed peasants but rather of small-scale agricultural entrepreneurs. Their enemies are not 'the rich' in general but specific classes who threaten popular ambitions for increased prosperity.

Heroic criminality of this kind is far less common than either the self-helping frontiersman or several of the categories which follow. Particularly for the prototype of the entire genre, historical analysis has to be based upon the contextualisation of an orginal legend, because Robin Hood has not been successfully identified with any independently documented person.[18] The early Robin Hood poems do, however, link their hero with a historical class in the contemporary English countryside, the yeomen and lower gentry. In the late fourteenth and early fifteenth century when these works were composed, such groups had gained a significant degree of prosperity as opposed to the still-enserfed (and soon to be displaced) lower peasantry and the manorial-based feudal aristocracy. Their enemies, clearly identified in the Robin Hood ballads, are the Church, as a landholding and moneylending institution, and certain nobles who abused their ties to the monarchy, as represented by the notorious High Sheriff of Nottingham. If the central state is defied by outlaws like Robin Hood, it is not because of its existence but because of its corruption. It would be carrying things too far to read into the Robin Hood legends the later rise of the gentry through the distribution of Church lands in the sixteenth century and revolution against absolutist monarchy in the seventeenth[19] but certainly the posture of the bandit here suggests optimism about integration into a national system rather than the tragedy or conservative compromise of self-help as an end in itself.

Of all the bandits in later periods credited with some of Robin Hood's virtues, the ones who most clearly play a parallel role are some of the bank and railway robbers of the late nineteenth-century American West and the Australian bushrangers (especially Ned Kelly) during the same period. Again, the class from which the outlaws emerged consists of aspiring small landowners, now newly settled on the rural frontiers of a more obviously expanding capitalist society. Their enemies (bankers, railroads, large landowners) were better-capitalised competitors in the same arena. Only in Ned Kelly's case was banditry consciously tied to demands for legislative changes favouring the smallholders.[20] However the broad support enjoyed by the James, Dalton, and Younger gangs in Missouri and Oklahoma indicates some of the roots of the later populist movement in these same areas, and is linked to a myriad of other local political issues.[21] Here again criminality is ultimately a means for influencing the modern state rather than opposing it with archaic peasant values, as Hobsbawm insists.

The absence of the populist distributor type of bandit in African reality and folklore is not surprising, but still instructive. Under colonialism many agriculture zones were integrated into the world market by colonial railways. However, outside of Southern Africa, local small-scale entrepreneurs tended to be more often peasants than yeomen or gentry; that is, they limited their commitment to new forms of production in the interest of minimising risks to traditional sources of economic and social security.[22] Competition from more-capitalistic farming was seldom a sustained threat while the real commanding heights of the economy – railways and banks – were rather difficult to attack, given the power of the colonial state.[23]

On the other hand, South Africa presents some of the conditions which might foster a banditry of the Robin Hood-Ned Kelly variety. Rural Africans here were displaced by Europeans and we have lately learned a great deal about the extent of small-scale entrepreneurship among these black farmers.[24] However, available evidence suggests that heroic criminality did not present itself as a meaningful response to competition from better-capitalised whites. The commanding economic heights here were even more remote than in tropical colonial Africa, given the scale of the mining industry, and there was no hope that direct action would elicit any shift in state support from Europeans. Indeed the South African government went beyond representing the interests of white farmers, who were not always such eager or efficient competitors against Africans, and deliberately suppressed independent black cultivators as a potential threat to the total structure of European domination in the subcontinent.[25] The response of rural black South Africans to this dilemma was neither the acting out nor contemplation of bandit models but rather the formation of independent churches as a very different expression of resistance.[26]

The legends of Robin Hood-type bandits exaggerate their real deeds of social benevolence and link these to more-universal mythic themes, but the profile always remains distinct. Although only the later versions of the original Robin Hood story emphasise the giving of money and goods to the poor, this gesture is consistent with the social programme of his more contemporaneous presentation. Jesse James and other outlaws in the American West were regularly credited with the stylised, and entirely undocumented, action of providing widows and other unfortunate folk with the money to pay off their mortgages and then stealing the funds back again from the banks. One of the most universal traits of this type of bandit, his adoption of disguises (even woman's dress) and other forms of trickery to accomplish his ends, stands in sharp contrast to the flamboyant machismo of the self-helping 'bandit d'honneur'. Even the common denominator between these two types, the outlaw's escape from culture into nature, has a unique form here. Robin Hood, in his idealised greenwood, exercises a self-conscious parody of the aristocratic style. The lone, upright gunslinger hero of the western movie acts out his role as a mediator between the humble homesteader and the greedy local magnate, two poles of the settled civilisation which his entire persona symbolically opposes.[27]

African culture, both traditional and modern, is rich in myths but that of Robin Hood seems absent. The mass-oriented Onitsha chapbooks of eastern Nigeria may tell us something about this hiatus in their lumping together of Robin Hood, Jesse James, and similar characters with the thoroughly asocial gangsters and 'spaghetti western' machos of the local cinema.[28] The figure closest to Hobsbawm's true social bandit may turn out to be the least universal of all heroic criminals.

The professional underworld

For Hobsbawm and most of his critics the complete antithesis of the social bandit is the representative of the professional underworld, the individual for whom crime is a source of livelihood rather than an act of defiance, and who is as alien to the peasants as to the wealthier classes upon whom he may prey somewhat more frequently. Yet many of the heroic criminals cited in Hobsbawm's book fall into this category. Moreover in both their actions and legends they too represent a form of opposition to established order with its own social meaning.

Professional criminals arise in situations where a central authority has control over legitimate violence and the rules for market transactions. The very term 'underworld' suggests a marginal relationship to dominant institutions, but in this case its location is not on the 'frontier', i.e. the point of expansion into geographically isolated or remote zones. Instead 'criminals' are found in the interstices of an already developed and fully extended order. These often include the literal frontiers between states and forested areas protected by law, because smuggling and poaching are widespread underworld activities, although one in which the state is the only obvious victim and local 'above-ground' communities may be active collaborators.[29] The more distinctive base of the professional criminal is the realm of occupations and castes which are not in themselves illegal but still considered marginal because they are not rooted in stable majority communities, in the legitimate social hierarchy, and in the concepts of 'honourable' activity. For early modern Europe this milieu has been located among tinkers, tanners, entertainers, beggars, and prostitutes. It is simultaneously associated with at least two widely dispersed ethnic groups, gypsies and Jews.[30]

The existence of such an underworld indicates the limited ability of the state and the legitimate economy to control or absorb the universe, including the major metropolitan centres, over which it claims sovereignty. The underworld does not directly compete with the dominant order for control of major political or economic functions but in its own sphere of operation it is intimately and often very ambiguously linked to lower agencies of the state and petty merchants. Policemen, soldiers, foresters, and customs officials thus alternatively pursue, collaborate with, and become (or recruit themselves from) criminals; local merchants (particularly Jews in Europe) are at once the victims, the fences, and sometimes the violent perpetrators of highway robbery.

The elements of a professional underworld existed in much of precolonial Africa and even more so in the twentieth century, but what we now know about both their empirical reality and their legends, suggests significant variations from the Western pattern. Virtually the same occupations (metal workers, leather workers, minstrels) which were stigmatised in various ways in pre-modern Europe became the basis for socially and ritually segregated caste groups in the Sudanic zone of West Africa and in Ethiopia. These groups tended to be more geographically mobile than ordinary cultivators, as were merchants (whose 'diaspora' communities were also marked by a separate religion, Islam).[31] In addition slaves, who became especially numerous throughout Africa during the nineteenth century, formed another marginal community, in this case usually at the bottom of the social order.

Despite the various forms of alienation in relations between these groups, criminality seems relevant as a category of behaviour only to dissident slaves and even here it is somewhat misleading. Caste groups do not appear anywhere in Africa to have established a separate corporate base from which they could operate 'underground' activities. Instead they seem to have attached themselves piecemeal to various dominant groups in patron–client relationships which fixed the place of their 'secret' activities and even incorporated them into the ritual and cosmology of the dominant groups. Muslim merchants were treated similarly, although in their case the ties were on a more egalitarian basis, since the aliens controlled access to what was recognised as a materially, and perhaps spiritually, more powerful external domain.[32]

There is historical evidence (and probably more could be found) that members of all these African groups engaged in some criminal activities. However it does not appear that any of them behaved as members of, or were perceived as connected to a professional underworld. We thus know that in both West and East Africa caravan merchants (both Muslim and indigenous) sometimes robbed one another. The urban centres of the West African Sudan also supported prostitution, theft, smuggling and other unsanctioned activities. But these activities do not seem to have loomed very large among such communities. As Nadel says of prostitutes in the northern Nigerian city of Bida, crime of this kind generally remained 'semi-professional.'[33]

Runaway slaves are described by both contemporary European observers and modern scholars in various regions of Africa as undertaking banditry.[34] However studies closer to the original sources offer a different pattern. The alleged slaves-turned-bandits of Ningi, in nineteenth-century northern Nigeria, thus become the base of either a pocket state within the general political system of the Sokoto Caliphate or a military organisation dedicated to enslaving the free population of the surrounding areas.[35] Similar material from other portions of the West African Sudan and East Africa also suggests that escaped slaves tended to form or join political entities which competed with the states in which they were previously held.[36] In short, even for marginalised groups within quite highly stratified societies, precolonial Africa offered segmentation as an alternative to criminal-isation.

In colonial tropical Africa organised professional crime seems to have been a rather minor factor in the countryside, where the state and still-intact local communities inhibited any effective underground activity. Levantine and South Asian petty merchants formed a new marginal community, but one again too well-connected to a dominant external sector to undertake the risk of criminal activities other than smuggling.[37] In any case, there was relatively little to steal on the highways and far more in the government coffers, which were vulnerable to a different sort of criminality, discussed in the next section.

Modern African cities, of course, contain major peri-urban settlements whose 'informal economies' provide a broad base for criminal activities of various kinds. However, research on these sectors up to the present suggests that the scale of illegal activities is limited to fairly petty enterprises of theft, prostitution, and illegal brewing. The perpetrators are tolerated as long as they provide useful services and prey only on those outside their own communities, but theft within the community is summarily, and often very violently punished.[38] This is not a situation from which, as will be seen below, a folklore of heroic banditry emerges.

It is in modern Southern Africa that a criminal underground has emerged more fully in both rural and urban sectors. Clarence-Smith has provided an excellent account of bandit gangs in late nineteenth-, early twentieth-century southern Angola.[39] These groups were made up of ex-soldiers and escaped slaves who harassed both settler and peasant populations of the region, and the leading figures regularly switched roles with the police forces. But no such fully developed underground has been reported for other rural areas of even this region since the early nineteenth century, perhaps because the boundaries between dominant white minorities and marginalised black majorities are inherently too unstable to tolerate such institutionalised interstices.[40] In the modern Republic of South Africa little uncontrolled activity has been allowed to rural blacks except within the confines of 'homelands'. In Zimbabwe, Mozambique, and eventually even Angola, illicit accommodation to deprivation found its outlet in guerrilla nationalist movements whose ultimate and quite realistic goals were not crime (whatever norms may have been transgressed at given moments) but rather the takeover of the centre of the state.[41]

The urban landscape of South Africa presents the most fully developed underworld of the entire continent because here the very instruments of repression become the basis for forming illicit, if not immediately political African organisation. The best documented case of this kind is the 'Ninevite/Regiment of the Hills' movement which flourished in various cities of the Transvaal between 1890 and 1910. Its original base was the criminal underground which Europeans had developed as part of the expanding mining society of Johannesburg. But, as the movement became more African in its personnel and cultural identity, it operated (like such Western criminal organisations as the Neopolitan *camorra*) out of the local prison system.[42] The South African police were eventually able to destroy the Ninevites. But there are suggestions in yet-unpublished research on South Africa that criminal gangs survive with some degree of autonomy in black South African society and are one of the models for wider forms of youth gangs which, in both urban and rural areas, offer a degree of non-political resistance to the repressive state.[43]

Legendary accounts of the professional underworld both reinforce its empirical character and manipulate it for more general purposes. Marginal groups are expected to behave perversely and as a consequence move easily from the dishonourable to the illegal. At another extreme, folklore (and the self-advertisement of some bandits) may present them as populist champions. The classic case is the late-eighteenth-century German robber, Johannes Bückler, whose claim to concentrate his thefts only on unpopular Jewish merchants was widely accepted although his nickname 'Schinderhannes' ('Jack the Tanner') indicates his own marginal origins, and, as contemporaries probably did not know, he sometimes worked as a minor dependent of a much more powerful Jewish bandit gang, the Nederländer.[44]

Much of the ballad, chapbook and ultimate cinematic presentation of the underworld makes little pretence of transforming it into something different. The marginal criminal has his own mythic base in the universal figure of the trickster who uses deception, cunning, and violence to achieve his egotistical ends. Even here we must be careful, because there are trickster mythic cycles which deal exclusively with such disorderly behaviour and others which link their central figures to the major creative forces of the universe. The European folklore of the professional criminal underworld belongs to the former category; it is located not at the centre of power but rather in the humbler market places, roadways and boundary crossings of rural provinces and poorer urban neighbourhoods.[45] Nonetheless its protagonists are heroic, first of all because they assert themselves, even if they must then be suitably punished. Moreover even when their actions fail to follow any code of honour or social programme, they at least provide a certain levelling function, because they inevitably pose a greater threat to the rich (if not the very rich) than they do to the poor.

The African underworld is a subject of both popular and high-cultural expression, but given its prevalence in at least contemporary urban settings, far less prominent than might be expected. The apparent absence of a rural tradition of banditry partially explains this absence in recent urban songs, chapbooks, and photo novels.[46] On the other hand the trickster theme, which is very prevalent in precolonial African narrative, enters more modern genres only in the form of moralising frames: the actual story which appears between them is deliberately non-traditional and non-heroic.[47] Possibly the limited extant research on popular culture in contemporary African cities has concentrated too much on written texts. More attention to oral forms might produce greater continuity with the past, including themes of trickster-as-heroic-criminal.

In the printed chapbook literature as well as the work of more polished popular novelists such as Cyprian Ekwensi, the form of criminality most frequently represented is uncontrolled female sexuality, as symbolised by the prostitute, courtesan, 'good-time girl', or *femme libre*, rather than the masculine bandit or swindler.[48] This image is rather inconsistent with the realities portrayed by empirical research but the effect is instructively contrastive with the Western criminal figure. In social terms, the authors deplore rather than extol the levelling function of the urban underworld. It threatens a positively valued hierarchy of sex rather than a more questionable class structure. In cultural terms, the focus on females suggests a totally different realm of contested values than in Western criminal literature. It is reproductive power, not property, which is most essentially at stake and which will have to be considered as the basis for an African paradigm of heroic deviance in the conclusion of this paper.

The African literature focusing on males gives more emphasis to detectives than it does to criminals. The Western models imitated from films and novels allow plenty of self-indulgence on the part of the hero, but ultimately the emphasis is rather on unambiguous supporters of the official order than any confusion of categories or orientation towards disorder. Even the male criminals portrayed in a small part of this literature are presented as shallow, often surrealistic pastiches of images from imported films with little sense of how they link to the real African underground.[49]

The more sensitive portrayal of urban criminals in a few major works of African fiction only underlines the absence of the heroic. They are about victims, not perpetrators. In Alex La Guma's classic, *A Walk in the Night,* the criminal character ends a very unrewarding career by being killed for a murder committed by his alter ego, an honest worker driven to unemployment and senseless brutality by the racism of South African society.[50] Meja Mwangi's *Kill Me Quick* is the first novel I know of to explore fully the existence of petty male criminals in a modern African city, this time Nairobi. But Mwangi's two protagonists are anything but heroic. The underlying theme of the novel is that they fail even to make the transition from adolescence to manhood because they can never establish their own households. They ultimately find their greatest stability in recurrent terms in prison, which is closely linked to secondary school, their last point of incorporation into an ordered social existence.[51]

The professional underworld is both meanly parasitical and heroically radical. It operates most immediately against the material interests and shared values of all members of society but it also challenges the capacity of the state to maintain its official version of the social order. African accounts of the underworld stress its parasitical side because secular action of this kind does not create any middle ground between offering the state a direct political challenge and resigning to it and its elite allies as the only force controlling significant resources. We will thus find more heroic forms of African criminality only in direct relationship to the state and wider forms of resistance (to be discussed in the conclusion) in non-secular realms.

The picaro

The confidence man, living by his wits rather than by violence, is usually not considered a heroic form of criminal. Indeed, one of the classic modern 'Robin-Hood' ballads draws an invidious comparison between those who 'rob you with a six-gun' and those 'with a fountain pen'.[52] Nonetheless, when seen as an isolated individual facing the urban authorities rather than as a representative of these authorities among innocent countrymen, the con man may display heroic qualities. Under the term *picaro* he is, in fact, the centre of a major European-American genre of criminal fiction. Moreover, unlike the other categories of heroic criminal, the *picaro* is present throughout colonial Africa, although not often seen there in his own terms and thus made a hero.

For a confidence man to make some kind of heroic career he must be based in an urban setting which is an essentially static, hierarchical order, but allows mobility for those who can practise the proper duplicity. In Western society, careers such as this have been possible in many times and places but the high point of the *picaro* literary genre, the 'ancien regime' of Spain, France and Britain between the sixteenth and eighteenth centuries, is also a model for the situation in which historical picaresque careers could be pursued. The commercial revolution and the growth of central political power had greatly enlarged the urban sector of European society. These same factors along with the connected and open-ended opportunities of the newly acquired overseas colonies allowed major new avenues of mobility, but the formal social and political order at the centre remained frozen into the rankings of aristocracy and commoners and (at least on the European continent) the absolutist monarchy. The real-life models for *picaro* fictional characters were stupendously unsuccessful manipulators like the Scotsman John Law, various charlatans who made brief careers in fashionable salons, and adventurers who amassed and lost fortunes in the East and West Indies by various illicit means.[53]

Colonial Africa in the twentieth century presents a similar picture if on a more modest scale. New urban structures grew up around the trade and administrative centres throughout the continent. The literate African could find new positions here and move far more easily than in the past across ethnic, geographic, and even new political boundaries to practise his skills. Yet the colonial system was both a racial-social hierarchy and a political system closely resembling the ancien regime of Europe. Real advancement often seemed to require the illicit manipulation of the interstices of this system, something relatively easy for educated Africans who functioned most commonly as intermediaries, often literally as interpreters, between the ruling white stratum and the rest of society.[54]

More explicitly and more fully than the underworld criminal, the legendary conception of the *picaro* conforms to the archetypal myth of the trickster.[55] He lives not in a marginal world of his own but on the boundaries between worlds, which he must constantly cross. His persona is subject to constant protean changes, so that its intrinsic essence becomes lost. He is classically portrayed as an orphan, i.e. without lineage, and engages in a wide range of sexual exploits without socially meaningful procreation. His career, although episodic, takes him through major sectors of society and thus holds a mirror up to its internal corruption while endowing the protagonist with a kind of mock-epic heroism. He travels far, although his achievement is mainly to survive rather than to accomplish any ordered purpose. The mediating appeal of such narratives is not to honour, shock, or moral sympathy but to laughter: the *picaro* himself is often ridiculous, but so is much of the society against which he struggles.

The picaresque narrative in Europe has its predecessors in the late medieval trickster classics, *Le Roman de Reynard* and *Till Eulenspiegel*, which used often cruel comic tales to satirise courtly and urban society, but did not link escapades in these areas to realistic accounts of criminal careers.[56] The *picaro* himself does not have any redeeming social qualities, but his presentation in literature does serve a radical function of exposing both the amorality of egotism and the matching corruption of a society which collaborates in such behaviour. The conservative element in these presentations is that we laugh, thus attributing a kind of negative freedom to the *picaro*, and that he is inevitably forced to accept the limitations of his aspirations, thus implying that there is at least an idea of order which keeps the trickster from being the ruler.

The *picaro* criminal in the social if not the ideological sense is a central figure in modern African literature. The difference from the European tradition is that, with a few notable exceptions, he is treated not as a trickster who inspires laughter and a certain degree of empathy, but rather as a betrayer, the 'fountain pen' agent of truly dangerous forces, and the tone of the presentation is not humour but outrage and despair.

The exceptional works are both novels with greater or lesser autobiographical elements: Dugmore Boetie, *Familiarity is the Kingdom of the Lost* and Amadou Hampaté Bâ, *L'étrange destin de Wangrin ou les roueries d'un interprète africaine*.[57] Boetie claims to present an accurate account (but there is considerable evidence to challenge its veracity) of a life as a black confidence man and sometimes clerk and musician in the cities and prisons of South Africa from the 1930s to the 1950s. It could probably be read as another story of underworld victimisation like the work of Alex La Guma and it certainly sometimes takes this tone, particularly when the hero attempts to lead a 'straight' life. Most of the book, however, narrates the various ways in which Boetie manipulates the system, moving in classic picaresque fashion from one substitute parent to another, and from one identity to another (twice he transforms himself from 'Bantu' to 'Coloured'). Boetie does not tell us anything we do not already know about the corruption of South Africa nor is the ultimate failure of all his schemes of great moral significance, given the overdetermination of his abject situation by the racist system under which he lives. It is nonetheless impressive to find the picaresque used here not only as a practical weapon to survive in this society, but also as a device for retaining spiritual freedom while holding absolutely no illusions about the grim surrounding reality.

Bâ's Wangrin is an interpreter in French West Africa from shortly before the First World War to about the 1930s. In this case the African picaresque, whether or not imitating any European mode, is more self-conscious. Wangrin is dedicated in his youth to a trickster deity,[58] and the story of this rascally intermediary is framed like a Malian epic, originally recited, we are told, in the courtyards of a great Sudanic household (which is also where much of the action takes place), accompanied by the griot's guitar and interrupted periodically by set-piece songs. What Wangrin actually does, however, is ruthlessly and greedily to use his position in the colonial government to carry on elaborate frauds, moving boldly and craftily between the worlds of Europeans and Africans, Islam and traditional Bambara religion, and the various African ethnic groups within the region. Inevitably he meets his downfall through the duplicity of an even more unscrupulous and clearly more dishonourable European woman and dies in a ditch, a bankrupt alcoholic with no male heirs to carry on his name. Bâ carefully disclaims his story having 'the least thesis . . . political, religious or otherwise. It simply concerns the account of a man's life' (p. 9). Nonetheless he has caught the flavour of an aspect of the African past with great accuracy and done so in a form which conveys very profoundly the practical possibilities and limitations as well as the cultural meaning of making an African career in such a setting.

The much greater number of African novelists who write about corrupt officials in a harshly moralising tone have recently been criticised by Oyekan Owomoyela for adopting alien values and fostering a kind of cultural dependency.[59] However, one can understand why they would not be able to make picaresque heroes of their protagonists in the manner of Bâ.[60] The figures they attack are not, at least within Africa, the intermediaries between peasant and foreign rulers, but rather the rulers themselves who claim to be pursuing an anti-colonial revolution. The criminals who once represented a possibility of movement within an oppressive state order have now become the state. The tone with which these figures are presented in modern fiction is, as Owomoyela notes, extremely condescending and unsympathetic to the African cultural dimensions of much of their behaviour. But it is difficult to address this kind of criminality in any but cosmopolitan terms since it operates directly in contact with an international economic and political order. In terms of the criminal categories used previously it is as if the bandit with some kind of local roots had now become recognisable as a *mafioso* intimately allied with alien authorities. Thus the most robust African version of heroic criminality ultimately fades out against a vision of the state which is now more powerful than in the precolonial segmentary situation but, nevertheless, fails to enter the internal dialogue of popular action and culture.

The urban guerrilla

In the same era which immediately followed African attainment of political independence, various groups of young urban radicals in Western countries self-consciously took up the posture of social banditry as a tactic for creating revolutionary consciousness within their own societies. Hobsbawm, in the postscript to the 1981 edition of *Bandits*,[61] makes some insightful but very unsympathetic comments about these movements, whose social and ideological base was essentially that of a bourgeois intelligentsia, rather than the milieu of the popular masses whom they claimed to represent.[62] Despite Hobsbawm's repudiation, the urban guerrillas ('terrorists' to their enemies) of Latin America, the United States, and Western Europe deserve a place in the tradition of heroic criminals.[63]

Africa (but more so Afro-American settings in the United States and Jamaica) have provided some of the inspiration for the concept of 'righteous gangsters'[64] who could arouse the ultimate oppressed class, the urban lumpenproletariat, to rise against established authority and class structure.[65] However, there has been little echo of urban guerrilla action or ideology within Africa itself. *Tsotsis* (urban juvenile gangsters) were employed by black nationalists in South Africa to support essentially above-ground political work during the early 1960s but more recent violent actions have been undertaken by trained cadres entering the country from outside.[66] The role of Westernized Ethiopian students during the 1977 revolution against Haile Sellassie more clearly followed European-Latin American models of urban guerrillas, but ended in bloody repression by a Marxist military regime with its own student and lumpenproletariat support.[67]

If Hobsbawm is right in his denunciations, the urban guerrillas phenomenon may indicate a terminal point in the relevance of the social-bandit idea of Western history. While it also suggests a degree of relevance for Africa in Western radical thought, this variant also marks even more clearly than its predecessors the limited meaning of heroic criminality in the internal African context. But rather than ending on such a negative note, it may be useful here to pick up a few of the loose ends of the attempts to fit the Hobsbawmian model onto Africa and look for alternative African concepts which may be more historically appropriate analogues of the bandit figure.

Postscript: towards an African model of resistance

It is no coincidence that the most fruitful Western model for African heroic criminality is the colonial-era *picaro*, an individual relatively isolated from African society yet not entirely integrated into the European-dominated state. But to define marginal figures of resistance who operate in a more fully African context (whether precolonial or within sectors of colonial and post-colonial society less involved with the modern state) we need to consider an entirely different set of concepts. To do this seriously would require at least another entire essay. Instead I simply want to note here a vocabulary of African public deviance suggested by the present inquiry.

The root difficulty in utilising Hobsbawm-derived models for Africa is the centrality to them of the concepts of territorially defined state and market-defined private property. Banditry, in the Western world, represents a challenge to existing control of both territory and property but does not question the validity of the concepts themselves. In the African situations which have been treated here, states and markets are always present in some form. But they do not pervade social relationships to the same degree as in the West and are less the domain than the issue of both the actions and ideology of deviance.

The vocabulary of deviance in Africa is thus seldom state vs criminal or legitimate vs illegitimate acquisition of property but rather the relatively static communal social order vs the appropriation of powers which allows individuals to transcend the community. In the most general sense, these powers may be described as magico-religious. They are seen as a necessary element of any state system in Africa, but in the cases discussed here, they were evoked by African sources to explain such bandit-like phenomena as the leadership of the Ningi runaway-slave state in northern Nigeria, caravan robbers in Tanzania, a successful bankrobber in colonial Cameroon, and the prophetic leadership of independent churches in the settler-dominated South African countryside.[68]

The idiom which more specifically addresses the relationship of marginal figures to changes in authority and property rights is that of witchcraft. Witchcraft in its most potent African forms refers to the transformation of collective social power (usually expressed in terms of sexual reproduction and control of food) into individual political and material power. It is intimately associated with the state in a number of African cosmologies, i.e. the ruler, no matter how 'legitimate' also represents an essentially life-threatening force.[69] Witches not associated with a social role which also brings some collective benefits are more unambiguously feared and punished. The witchcraft idiom may be used as a general attack on property accumulation, which is opposed to communal values of sharing and reproduction.[70] But at a private level it also provides a key to the conception of figures who resemble the Western notion of the heroic criminal. Thus the obsession of modern African writers with independent urban women who quite literally transform their control of sexuality and food into market commodities. Likewise, the appropriation of Zulu kingship terms by the leader of the Johannesburg Ninevites takes on a fuller meaning if we also relate the homosexual practices of this twentieth-century gang to the mythic biography and male regimental organisation of the Zulu state founder, Shaka. A common theme in both cases is the diversion of masculine sexuality from reproduction to male bonding.[71]

The notions of magic and the much richer connotations of witchcraft can here be no more than suggestions for some later development of an African model of resistance. It does, however, imply a vindication of Hobsbawm's fundamental project of understanding criminal deviance as an expression of positive alternatives to dominant social values, although often in archaic terms. In a Western setting where capitalist modernisation has generally prevailed, Hobsbawm's concerns with such resistance may be dismissed as marginal and romantic. For Africa, where precapitalist social forms are still very much alive, and marginality is a central condition, such investigations may provide us with critical clues to the relationship between political economy, culture, and historical change.

NOTES

1 I would like to thank James Spiegler for considerable advice on this paper and the National Science Foundation for a CAUSE grant which supported some of the research and discussion upon which it is based.
2 E. J. Hobsbawm's, *Bandits* (New York: Pantheon, 1981, 2nd edn), p. 17.
3 ibid., p. 18.
4 The use of segmentation as a diagnostic for classifying African political processes is derived from David Easton, 'Political anthropology', in Bernard Siegel (ed.), *Biennial Review of Anthropology, 1959* (Stanford: Stanford University Press, 1959), pp. 210–62.
5 Hobsbawm, *Bandits*, p. 17.
6 J. DeLange, *The Relation and Development of English and Icelandic Outlaw Traditions* (Haarlem: H. D. Tjeenk, Willink, 1935); I strongly disagree with the way in which DeLange attempts to identify very different types of heroic literature as a single 'outlaw tradition', particularly based upon my own reading of his presentation of the Icelandic sagas.
7 Hobsbawm, *Bandits*, is very good on European mountaineers and islanders; see also Silvia R. Duncan-Baretta and John Markoff, 'Civilization and barbarism: cattle frontiers in Latin America', *Comparative Studies in Society and History*, vol. 20 (1978), pp. 587–620.
8 Jean Dejeux, 'Un bandit d'honneur dans l'Aures de 1917 à 1921: Messaoud ben Zelmad', *Revue de l'Occident musulman et de la mediterranée*, vol. 26, no. 2 (1978), pp. 35–54. (This article is more general than the title suggests).
9 There is very little literature on such criminality in Africa. For a valiant but ultimately confused effort see Stephan A. Lucas, 'Social deviance and crime in selected rural communities of Tanzania', *Cahiers d'Études africaines*, vol. 16, no. 3 (1976), pp. 499–518. (I cannot resist quoting here the Tanganyika African policeman who, in 1963, guiding me to an entire shack filled with weapons confiscated from feuding tribesmen, could find no more appropriate metaphor than: 'These Zanakis are just like your hillbillies'.)
10 See Chapter 8, below.
11 Editors' introduction in Howard Lamar and Leonard Thompson (eds), *The Frontier in History: North America and South Africa Compared* (New Haven: Yale University Press, 1981), p. 7; it is notable that there is no reference to criminality in the introduction or index of this work and the only substantive discussion even relevant to the topic deals with inter-ethnic cattle theft during the early stages of European occupation of the north-eastern Cape Colony (see Herman Giliomee, 'Processes in development of the Southern African frontier', in ibid., pp. 114–16).
12 Chapter III, 'Mafia', in E. J. Hobsbawm, *Primitive Rebels* (Manchester: Manchester University Press, 1959), pp. 30–56; Henner Hess, *Mafia and Mafiosi: The Structure of Power* (Lexington, Mass: Lexington Books, 1973); Anton Blok, *The Mafia of a Sicilian Village, 1860–1960: A Study of Violent Peasant Entrepreneurs* (Oxford: Blackwell, 1974).
13 Anton Blok, 'The peasant and the brigand: social banditry reconsidered', *Comparative Studies in Society and History*, vol. 14, no. 4 (1972), pp. 494–505. On the relations between landlords and Brazilian *cangaceiros* see Billy Jaynes Chandler, *The Bandit King: Lampiao of Brazil* (College Station: Texas A & M University Press, 1978); Linda Lewin, 'The oligarchical limitations of social banditry in Brazil: the case of the "good" thief Antonio Silvino', *Past and Present*, no. 82 (1979), pp. 116–46; Peter Singleman, 'Political

structure and social banditry in northeast Brazil', *Journal of Latin American Studies*, vol. 7, no. 1 (1975), pp. 59–83.

14 Hobsbawm, *Primitive Rebels*, pp. 40–44; *idem*, *Bandits*, Chs 5 and 7; on the legend of the *cangaceiro* as a form of regional self-assertion by impoverished north-eastern Brazilians forced to seek employment in the south, see Maria Isaura Pereira de Queiroz, *Os Cangaceiros, les bandits d'honneur brésiliens* (Paris: Julliard, 1968), pp. 195–6; Ronald Daus, *Der epische Zyklus der Cangaceiros in der Volkspoesie Nordostbrasiliens* (Berlin: Colloquium Verlag, 1969).

15 Dejeux, 'Bandit d'honneur', pp. 39–42; see also the analysis of the shifting image of Bahta Hagos in Eritrea in the contribution by R. A. Caulk (Chapter 13, below) to this volume.

16 Henry S. Wilson, *The Imperial Experience in sub-Saharan Africa since 1870* (Minneapolis: University of Minnesota Press, 1977), pp. 186–99.

17 Johannes de Bruyn, 'The Oorlams Afrikaners: from dependence to dominance', unpublished paper, University of South Africa, n.d.; Alf Wannenburgh, *Forgotten Frontiersmen* (Cape Town: H. Timmins, 1980), pp. 37–65.

18 R. B. Dobson and J. Taylor, *Rymes of Robin Hood: An Introduction to the English Outlaw* (London: Heinemann, 1976). There is a great deal more literature on the social meaning of the Robin Hood legend, much of it cited in Hobsbawm, *Bandits*, but is not particularly useful. For a specific attack on Hobsbawm's interpretation of Robin Hood along with a historical account of late medieval English criminality as a largely 'professional underworld' phenomenon, see Barbara Hanawalt, *Crime and Conflict in English Communities, 1300–1348* (Cambridge, Mass: Harvard University Press, 1979), especially pp. 201–16.

19 For a valuable summary of the arguments on these issues, see Immanuel Wallerstein, *The Modern World System*, Vol. 1 (New York: Academic Press, 1974), pp. 235–60. In later versions of the legend, Robin Hood and the Sheriff of Nottingham are commonly identified with, respectively, the 'good' King Richard the Lion-Hearted and the 'bad' King John, but this motif is too anachronistic and stereotyped to provide any insight into the social or political meaning of Robin Hood.

20 Pat O'Malley, 'Social Bandits, modern capitalism and the traditional peasantry; a critique of Hobsbawm', *Journal of Peasant Studies*, vol. 6, no. 4 (1979), pp. 489–501; John, McQuilton, *The Kelly Outbreak, 1878–1880: The Geographical Dimensions of Social Banditry* (Carlton: Melbourne University Press, 1979).

21 Richard White, 'Outlaw gangs of the middle border: American social bandits', *Western Historical Quarterly*, vol. 12 (1981), pp. 387–408.

22 For this interpretation of African peasantries see Goran Hyden, *Beyond Ujamaa in Tanzania: Underdevelopment and an Uncaptured Peasantry* (Berkeley: University of California Press, 1980); Frederick Cooper, 'Peasants, capitalists, and historians', *Journal of Southern African Studies*, vol. 7, no. 2 (1981), pp. 284–314; Sara Berry, 'Rural class formation in West Africa', in Robert H. Bates and Michael Lofchie (eds), *Agricultural Development in Africa* (New York: 1980).

23 The recent appearance of African bank robbery is stressed in Nwokocha Nkpa, 'Armed robbery in post-civil war Nigeria: the role of the victim', *Victimology*, vol. 1, no. 1 (1976), pp. 71–83.

24 Colin Bundy, *The Rise and Fall of the South African Peasantry* (Berkeley: University of California Press, 1979).

25 Stanley Trapido, 'Landlord and tenant in a colonial economy: the Transvaal, 1880–1910', *Journal of Southern African Studies*, vol. 5, no. 1 (1978), pp. 26–58; M. L. Morris, 'The development of capitalism in South African agriculture: class struggle in the countryside', *Economy and Society*, vol. 5 (1978), pp. 292–343.

26 Bengt G. M. Sundkler, *Bantu Prophets in South Africa* (London: Oxford University Press, 1964, 2nd edn).

27 Will Henry Wright, 'Six guns and society: a structural study of a modern myth' (Berkeley: University of California Press, 1975); see his 'Type 1'.

28   See especially the excerpt from J. A. Okeke Anyiche, 'Adventures of Four Stars', in Emmanuel N. Obiechina (ed.), *Onitsha Market Literature* (London: Heinemann, 1972), pp. 139–43; also Obiechina, *African Popular Literature: A Study of Onitsha Market Pamphlets* (Cambridge: Cambridge University Press, 1973), pp. 96–7; Wright, 'Six guns', pp. 145f, sees the spaghetti western as representative of a later development of the genre in which violence has become a professional rather than a moral issue.

29   E. P. Thompson, *Whigs and Hunters: The Origins of the Black Act* (London: Allen Lane, 1975); curiously the very 'social' criminal actions in the eighteenth-century English forests described by Thompson have left, according to his account, almost no records in contemporary folklore.

30   The best work I have found on this type of criminality is Carsten Küther, *Raüber und Gauner in Deutschland: das organisierte Bandenwesen im 18. und frühen 19. Jahrhundert* (Gottingen: Vandenhoek & Rurpecht, 1976). For a colourful but largely unanalytical presentation of more material on this theme see Andrew McCall, *The Medieval Underworld* (London: Hamish Hamilton, 1979).

31   Nicholas Hopkins, 'Mandinka Social Organization', in Carleton T. Hodge, *Papers on the Manding* (Bloomington: Indiana University/New York: Humanities Press, 1971), pp. 99–114 (on indigenous Muslim traders, casted artisans and patron-client relationships). On Ethiopian castes see Donald N. Levine, *Greater Ethiopia* (Chicago: University of Chicago Press, 1974), pp. 56–7, 62, 169–70, 185–97.

32   Emmanuel Térray, 'Long-distance exchange and the formation of the state: the case of the Abron kingdom of Gyaman', *Science and Society*, vol. 3 (1974), pp. 315–45; Thomas C. Hunter, 'The development of an Islamic tradition of learning among the Jahanke of West Africa' (unpublished PhD, University of Chicago, 1978).

33   Kea on Akan (Chapter 5, below) in this volume; Robert Launay, Northwestern University, private communication, 1 December 1982 (on Ivory Coast Muslim merchant-bandits); Michael Singleton, 'Why was Giesecke killed', *Cultures et Développements*, vol. 8 (1976), pp. 657–60 (on Nyamwezi caravan robbers in Tanzania); Horace Miner, *The Primitive City of Timbuctoo* (Garden City, N.Y.: Doubleday, 1965), pp. 267–71, *et passim*; S. N. Nadel, *A Black Byzantium: The Kingdom of Nupe in Nigeria* (London: Oxford University Press, 1942), pp. 152–5.

34   P. Staudinger, *Im Herzen der Hausaländer* (Berlin: Landsberger, 1889), pp. 282–3; for a modern inference that banditry was a widespread strategy of runaway slaves see Paul E. Lovejoy, 'Slavery in the Sokoto Caliphate', in Lovejoy (ed.), *The Ideology of Slavery in Africa* (Beverly Hills: Sage, 1981), p. 230.

35   R. A. Adeleye, *Power and Diplomacy in Northern Nigeria 1804–1906. The Sokoto Caliphate and its Enemies* (London: Longman, 1971), pp. 73–4, 111; Adell Patton Jr, 'Ningi raids and slaving in the nineteenth century Sokoto Caliphate', *Slavery and Abolition*, vol. 2, no. 2 (1981), pp. 114–45.

36   Frederick Cooper, *Plantation Slavery on the Coast of East Africa* (New Haven: Yale University Press, 1977), pp. 200–10; Paul Irwin, *Liptako Speaks: History from Oral Tradition* (Princeton: Princeton University Press, 1981), p. 145; Irwin, 'Chronique du Liptako au 19e siècle', *Notes et Documents Voltaiques*, no. 9 (1975/76), pp. 15–17.

37   William A. Shack and Elliot P. Skinner (eds), *Strangers in African Societies* (Berkeley: University of California Press, 1979).

38   Nici Nelson, 'How men and women get by: the sexual division of labour in the informal sector of a Nairobi squatter settlement', in Ray Bromley and Chris Gerry (eds), *Casual Work and Poverty in Third World Cities* (Chichester: John Wiley, 1979), pp. 283–302; Keith Hart, 'Informal income opportunities and urban employment in Ghana', *Journal of Modern African Studies*, vol. 11, no. 1 (1973), pp. 61–89 (especially p. 76).

39   Gervase Clarence-Smith, *Slaves, Peasants, and Capitalists in Southern Angola, 1840–1926* (Cambridge: Cambridge University Press, 1979), pp. 37, 57, 79, 82–8. It might be fruitful to connect this kind of African banditry in Angola with the local *demi-monde* of European and *mestizo* rural merchants, especially those dealing in semi-contraband spirits.

40 George Fredrickson, *White Supremacy: A Comparative Study in American and South African History* (New York: Oxford University Press, 1981), p. 69; see also note 13 above.

41 Basil Davidson, *The People's Cause: A History of Guerrillas in Africa* (London: Longman, 1981), pp. 119 ff; see also the contribution by Terence Ranger (Chapter 17, below) in this volume.

42 Charles van Onselen, 'The "Regiment of the Hills": South Africa's lumpenproletarian army, 1890–1920', *Past and Present*, no. 80 (1978), pp. 91–121. See also van Onselen's further research on the leader of the Ninevites, 'Crime and total institutions in the making of modern South Africa: the life of "Nongoloza" Mathebula, 1867–1948', unpublished paper presented at the Conference on the History of Law, Labour, and Crime, University of Warwick, September 1983. On the *camorra*, see Hess, *Mafia and Mafiosi*, p. 94; Hobsbawm, *Primitive Rebels*, pp. 53–6.

43 William Beinart, 'The Family, youth organization, gangs and politics in the Transkeian area' unpublished paper presented to the Conference on the Family in Africa, School of Oriental and African Studies, London, 1981. Private communication, Beinart to author, 28 September 1982.

44 Küther, *Räuber u. Gauner*, pp. 25–6, 111.

45 For an introduction to recent literature attempting to treat the trickster as a serious category of myth see Barbara Babcock-Abrahams, 'A tolerated margin of mess: the trickster and his tales reconsidered', *Journal of the Folklore Institute*, vol. 11 (1975), pp. 147–86; Robert D. Pelton, *The Trickster in West Africa* (Berkeley: University of California Press, 1980); see also the trickster versions of European robber tales in Antti Aarne and Stith Thompson, *The Types of the Folktale* (Helsinki: Accademia Scientarium Fennica, 1964), nos 950–69. Unfortunately, since this standard reference work does not deal with 'legends', i.e. stories about real or supposedly real figures, it is useless for pursuing most issues of the present paper.

46 Elizabeth Knight, 'Popular literature in East Africa', *African Literature Today*, vol. 10 (1979), pp. 177–90; Obiechina, *African Popular Literature*; Richard Priebe, 'Popular writing in Ghana: a sociology and rhetoric', *Research in African Literature*, vol. 9 (1978), pp. 395–432. The research in this area is so limited that the African Subcommittee of the Social Sciences Research Council was unable to arrange a suitable panel to review scholarly literature on popular culture for the 1982 African Studies Association Annual Meeting.

47 Obiechina, *Onitsha Market Literature*, pp. 19–21.

48 For data from popular literature see citations in the previous two footnotes. For more formal novels and short stories, a useful description (but little analysis) can be found in Kenneth Little, *The Sociology of Urban Women's Image in African Literature* (London: Macmillan, 1980).

49 Obiechina, *African Popular Literature*, pp. 96–7.

50 Alex La Guma, *A Walk in the Night and Other Stories* (London: Heinemann, 1968).

51 Meja Mwangi, *Kill Me Quick* (London: Heinemann, 1973).

52 Woody Guthrie, 'The Ballad of Pretty-Boy Floyd'.

53 There has recently been a spate of interest by literary scholars in the *picaro*; e.g. Alexander Blackburn, *The Myth of the Picaro* (Chapel Hill: University of North Carolina Press, 1979); Alexander A. Parker, *Literature and the Delinquent* (Edinburgh: Edinburgh University Press, 1967).

54 Apart from the literary examples cited below, see Henri Brunschwig, 'Interprètes indigènes pendant la période d'expansion française en Afrique noire (1871–1914)', *Proceedings of the Second Meeting of the French Colonial Historical Society, 1976*, pp. 1–15. This pattern of African colonial careers is also confirmed in the biographical data collected through interviews by myself in Tanzania and Cameroon between 1963 and 1975 and by James Spiegler in various portions of French speaking West Africa in 1962–3; (see J. Spiegler, 'Aspects of nationalist thought among French-speaking West Africans, 1921–39' (unpublished PhD thesis, University of Oxford, 1967).

55 See note 45 above.

56 J. F. Flinn, 'Litterature bourgeoise et le Roman de Rénart', in E. Rombauts and A. Welkenhuysen, *Aspects of the Medieval Animal Epic* (Leuven: University Press/The Hague: Nijhoff, 1975); Werner Wunderlich (ed.), *Eulenspiegel-Interpretationen: der Schalk im Spiegel der Forschung, 1807–1977* (Munich: Fink, 1979).

57 Dugmore Boetie, *Familiarity is the Kingdom of the Lost* (New York: Dutton, 1969); Amadou Hampaté Bâ, *L'étrange destin du Wangrin ou les roueries d'un interprète africaine* (Paris: Union générale, 1973).

58 I have been unable to identify this deity, Gongoloma-Sooke, with any figures in the cosmology of the Bambara, from which the author claims he comes. Possibly, given the description in *L'étrange destin*, pp. 21–3, Hampaté Bâ borrowed a Yoruba-Fon concept and grafted it onto his Sudanese setting.

59 Oyekan Owomoyela, 'Dissidence and the African writer: commitment or dependency', *African Studies Review*, vol. 34, no. 1 (1981), pp. 83–98.

60 The contrast with current and popular literature is illustrated by a highly successful recent Cameroonian novel, *Sur le Chemin de Suicide*, which has been labeled 'picaresque' but presents its hero as the unceasing *victim* of 'robbery and betrayal'. See Karen Keim, 'Naha Desiré and Popular Fiction in Francophone Cameroon', African Studies Association Meeting Papers, 1981.

61 Hobsbawm, *Bandits*, pp. 157 ff.

62 For a newer Marxist approach to popular Western resistance as expressed in the lifestyles of working class youth, see Dick Hebdige, *Subculture: The Meaning of Style* (London: Methuen, 1979).

63 James Kohl and John Litt, *Urban Guerilla Warfare in Latin America* (Cambridge, Mass: M.I.T. Press, 1974); John Bryan, *This Soldier is Still at War* (New York: Harcourt Brace, 1975), (closest work to a scholarly account of the SLA); G. Louis Heath, *Off the Pigs! The History and Literature of the Black Panther Party* (Metuchen: Scarecrow Press, 1976); Aleszandro Silj, *Never Again Without a Rifle* (New York: Karz, 1979); Jillian Becker, *Hitler's Children: the Story of the Baader-Meinhof Terrorist Gang* (Philadelphia: Lippincott, 1977).

64 Heath, *Off the Pigs*, p. 16.

65 Percy Henzell, *The Harder They Come* (film on Jamaica, 1973); Lewis Nkosi, 'On South Africa (The Fire Some Time)', *Transition*, no. 38 (1971), pp. 30–34; Tom Wolfe, *Radical Chic and Mau-Mauing the Flak Catchers* (New York: Farrar, Strauss & Giroux, 1970); Irene L. Gendzier, *Frantz Fanon: A Critical Study* (New York: Pantheon, 1973); Peter Worsley, 'Frantz Fanon and the "lumpenproletariat"', in Ralph Miliband and John Saville (eds), *The Socialist Register 1972* (London: Merlin Press, 1972), pp. 193–230.

66 Gail Gerhart, *Black Power in South Africa: the Evolution of an Ideology* (Berkeley: University of California Press, 1978), pp. 223–5 and ff.; for an ideological evocation of *tsotsis* as a radical force see Nkosi, 'On South Africa' (Nkosi's ideas were articulated while in exile and reflect more of the contemporary North American 'radical chic' [see Wolf in previous footnote] than internal South African consciousness).

67 Marina and David Ottaway, *Ethiopia: Empire in Revolution* (New York: Africana Publishing, 1978), pp. 99–148; René Lefort, *Éthiopie: la révolution hérétique* (Paris: Maspero, 1981), pp. 241–79.

68 Staudinger, *Hausaländer*; Singleton, 'Giesecke'; Johannes Ittmann, 'Der kultische Geheimbund djengu an der Kameruner Küste', *Anthropos*, vol. 52 (1957), p. 18; Sundkler, *Bantu Prophets*, especially pp. 253 ff.

69 Wyatt McGaffey, 'The Religious Commissions of the Bakongo', *Man*, vol. 5, no. 1 (1970), pp. 27–38; Axel-Ivar Berglund, *Zulu Thought-Patterns and Symbolism* (Uppsala: 1976), pp. 246 ff; and Chapter 3, above, by Allen F. Roberts.

70 The classic studies of this ideology deal with Latin America: George Foster, 'Peasant society and the image of the limited good', *American Anthropologist*, vol. 67 (1965), pp. 293–315; Michael T. Taussig, *The Devil and Commodity Fetishism in South America* (Chapel Hill: University of North Carolina Press, 1980). For its expression in a widely

popular contemporary African attack upon capitalism see Ngugi wa Thiong'o, *Devil on the Cross* (London: Heinemann, 1982).

71   Van Onselen, 'Crime and total institutions'. I do not mean to explain Shaka psycho-historically as a homosexual personality à la Max Gluckman, 'The individual in a social framework: the rise of King Shaka in Zululand', *Journal of African Studies*, vol. 1, no. 2 (1974), pp. 140–41. However, it is also impossible to accept Shaka's structural opposition to reproductive sexuality as simply a functional necessity of population control under conditions of demographic stress, following Jeff Guy, 'Ecological factors in the rise of Shaka and the Zulu kingdom', in Anthony Atmore and Shula Marks (eds), *Economy and Society in Pre-Industrial South Africa* (London: Longman, 1980)

5 'I am here to plunder on the general road': Bandits and banditry in the pre-nineteenth-century Gold Coast[1]

RAY A. KEA

'Not to mention the multitudes of desperate villains and robbers which commonly pester the ways.'

John Barbot, ca. 1682[2]

Who were these 'desperate villains and robbers' of the late-seventeenth-century Gold Coast? And whom did they 'pester' on the roads? What social conditions made the emergence and activities of such persons possible? These questions lead inevitably to a consideration of banditry. According to Eric Hobsbawm, banditry is a universal feature of agricultural societies, such societies consisting largely of peasants and landless labourers ruled, oppressed, and exploited by someone else. It is 'a form of individual or minority rebellion within peasant societies', and is symptomatic of crisis and tension in these societies.[3] Were Barbot's 'desperate villains and robbers' drawn from the ranks of peasants and landless labourers by contradictory and antagonistic social relations, or did they emerge from other classes and occupational groups? In short, what was the nature of Gold Coast banditry?

The literature on banditry documents its occurrence in many societies throughout the world over very long periods of time.[4] A few pioneering studies have examined various forms of outlawry in late-nineteenth- and early-twentieth-century Southern Africa.[5] However comparable investigations of the pre-nineteenth-century social formations of tropical Africa have not been carried out. For example, no one has conducted research on the rise and development of brigandage in the Gold Coast, or, for that matter, in any other part of West Africa. Although the data for such research exist, scholars have so far failed to use them for an historical analysis of banditry, whether as social and political protest or as a means to personal aggrandisement, nor have they described and analysed the individuals or groups associated with it.[6] The reason for this state of affairs is essentially ideological rather than methodological.

This study consists of three main parts. The first considers relevant source materials and the terms and descriptions they use to designate bandits and their activities. The second outlines the changing political economy of the Gold Coast and the material and social conditions in which banditry emerged and flourished. Part three provides specific examples of banditry, from the late fifteenth century to the early nineteenth century. I have also included in this part references to urban criminality.

European documentary sources and oral traditions combine to provide a rich body of information on Gold Coast banditry. They indicate quite explicitly that banditry was a persistent phenomenon in the region before 1800. The European sources use various terms to designate bandits: *ladrões, struikrovers, røvere, banditti,* and 'pirates'. And to cite John Barbot once again: the woods and trade routes were 'everywhere pestered with robbers.'[7]

What activities were the brigands engaged in, according to the Europeans? The sources specify, for example, that in 1702 the ruler of the coastal Fante polity had two men imprisoned in Cape Coast Castle, because they habitually attacked merchants on the roads.[8] In 1705 officials in the port town of Elmina publicly hanged two men because both had violated the public roads.[9] Ludwig, Rømer, a factor in the service of the Danish West Indies and Guinea Company from 1739 until 1750, refers to Antufi, an Asante bandit or highwayman, who, with an estimated 2000 followers, operated in the southern districts of metropolitan Asante. Antufi attacked trading caravans and successfully opposed the authority of the Asante government for a period of about twenty years.[10] Rømer claims that the Asantehene Opoku Ware offered Antufi whatever he wanted on condition that he cease committing acts of brigandage against Opoku Ware's subjects.

Oral histories also provide useful data. One example of these is *beyee-den*, a praise poem of the Asantehene. Recited by royal bards on public occasions, the poem details the rulers and other prominent persons who were executed or slain in warfare by eighteenth- and nineteenth-century Asante sovereigns. Three lines are appropriate here:

You slew Ntiriwaa Kwodwo.
You slew Kwame Antwiwa.
You slew *werekyerewerekyere* and made the paths safe.[11]

Here we are dealing with brigands. Ntiriwaa Kwodwo was a 'legendary' bandit; Kwame Antwiwa was a bandit; and *werekyerewerekyere* is a generic name for a bandit.[12] The poem clearly establishes both the presence of brigands within the confines of the Asante state, and the state policy of summary punishment to violators of the public roads.

The sources are not ambiguous about the nature of banditry, and the actions of brigands. Banditry involved acts of predation and violence on the arterial trade routes. Merchants and their caravans were the principal victims of these acts. By impeding caravan traffic, bandits violated the security of the public roads. The public roads and their security were symbols of an established political order, which was solely responsible for their maintenance. Consequently by its very nature banditry was a form of rebellion against an established political system: brigands violated both the roads and the merchants. Through their acts of predation bandits repudiated symbols of political order and authority.

In the modern Akan language various terms are used to designate a bandit: *werekyerewerekyere, okwanmukafo, odwowtwafo,* and *owudifo*.[13] Seventeenth-century Akan vocabulary lists contain two words with the same meaning: *dwowfo* and *akwanmu*.[14] The second word was glossed 'murder' or 'cut-throat' (*mordre*, i.e., *meurtrier*) by Barbot. It carried the meaning of one who committed murder on a public road (*okwan*, 'way', 'road'; *mu*, 'in', 'within'), and is a cognate of the modern terms *okwanmuka* ('occasional highway robbery') and *okwanmukafo* ('highway robber'; 'robber'). The first term, *dwowfo*, doubtless derived from *adow* ('to kidnap'; 'to panyar'), and maybe an archaic form of *adwowtwafo* ('a habitual robber'; 'brigand'; 'freebooter'; 'highwayman') and the related *adwowtwa* ('open and habitual robbery'). The modern *owudifo* ('highwayman'; 'murderer') appears in Müller's vocabulary list as *udifo* and is glossed 'murderer', but not 'bandit'.[15] Consequently Akan-speaking peoples could have used at least three different words to refer to brigands in the seventeenth century: *dwowfo*, *akwanmu* and *owudifo*. The modern *werekyerewerekyere* does not appear in any of the pre-nineteenth century vocabulary lists.

The seventeenth-century evidence clearly shows that banditry was an established phenomenon in the region. When did it first appear? Regrettably, the available data are not fully informative on this point. Nevertheless, I would venture the view that banditry emerged on a significant scale in the second half of the fifteenth century, and may very well have existed earlier. After around 1480 it increasingly became a generalised and regular feature of the rural order, and by the seventeenth century there were 'multitudes of desperate villains and robbers' operating in the countryside. This major social development can be attributed to the regional political economy in the fifteenth century. These changes first produced the conditions necessary for the appearance of widespread brigandage; and then established the material and social basis for its persistence through to the nineteenth century.

For analytical purposes the history of the region from about 1250 to 1800 can be divided into three distinct (but overlapping) periods:
Period I: The age of agrarianism, thirteenth century–fifteenth century;
Period II: The age of mercantilism, fifteenth century–early eighteenth century;
Period III: The age of militarism, early eighteenth century–early nineteenth century.
The emergence of banditry coincided with the rise of mercantilism.

The identification of distinct periods is based in part on eighteenth-century oral historical traditions recorded by the Danish factor Ludwig Rømer. These traditions were collected by Rømer in the 1740s and were related by several Ga-speaking notables who lived in the coastal towns of Labadi, Osu, and Teshie. Though focusing on events in and around the Accra plains during the preceding three to four centuries, they nonetheless provide a unique and informative eighteenth-century ideological perspective on the historical transformations which had occurred in other parts of the region as well. The three periods can also be defined in terms of the chronology of social wealth appropriation and accumulation. During the age of agrarianism, for example, land and agriculture were the basis of wealth accumulation for the dominant classes. With the age of mercantilism a significant shift occurred. Trade assumed paramount importance in the process of surplus appropriation, and social wealth was realised in the form of merchant- and money-lending capital, while the products of agricultural production ceased to be an indicator of economic and social status. The age of militarism was characterised by the importance of warfare and territorial conquest as a means of wealth accumulation. Further, the periods can be differentiated with respect to political organisation. During the first period, the main form of political organisation was the captaincy; in the second it was the city-state, and in the third the territorial state. These political changes are indicative of changes in production relations and forms of productive property.[16]

The oral history of the eighteenth-century Ga notables covers the first periods outlined above and the beginnings of the third.[17] It begins with a distant, idyllic era, which predated the arrival of the Portuguese. Local upper-class families ruled the region and each family had symbols of honour denoting its high social status. The wealth of each family head depended on his children and slaves. The notables describe this era as one of peace, fertility, and prosperity. Their traditions recall the period as marked by a spartan frugality in aristocratic household affairs. There were no laws against theft, because theft and thieves did not exist. Indigent families did not have to steal in order to acquire their daily provisions, for the wealthy elite families provided them with food whenever they asked for it.

Beneath this ideological and highly idealised picture we may perceive an agrarian social formation dominated by high-status landholding and slave-owning families. The circulation of currency and merchandise was not essential to their way of life.

There were relatively few towns in the forest- and coastal-based captaincies at this time. Such towns as there were probably functioned either as administrative seats of local rulers or as agricultural depots of elite families rather than as artisanal and commercial centres. However, archaeological data indicate that specialisation in production had taken place in some towns, thereby indicating that the captaincy regimes promoted the separation of agricultural and craft production.[18] If the oral history is correct in minimising the importance of long-distance trade to the captaincy economies, we may surmise that the separation of production and trade had not occurred. Archaeology, also, indicates the presence during the thirteenth and fourteenth centuries of nucleated rural settlements, some of which were up to a kilometre in diameter.[19] This evidence we can harmonise with the oral tradition if we view the majority of upper-class landed families as rural-based, residing in village communities characterised by the combination of agricultural and craft production. We can interpret the eighteenth-century oral history in such a way as to attribute to this world of captains and agrarianism its own moral economy, which linked social notions of obligation and prestige with social relations of control and subordination.

The oral history attributes the first signs of social and political disunity in the region to the establishment of Portuguese trading posts along the coast; and, indeed, the fifteenth century did usher in a new era. We can view this change in terms of the emergence of new forms of social production.

The rural, agrarian order was succeeded by an urban, mercantile one.[20] A new dominant class appeared on the historical scene, the *abirempon* (sing. *obirempon*) or the rich, and with it a new political organisation, whose typical form was the city-state (*oman*). Within the contexts of rapidly expanding commerce the *abirempon* emerged as the agents of change and the shapers of society, and the city-state and its associated systems of material production became the dominant mode of historical development throughout much of the region for a period of about two centuries. For the *abirempon*, trade was paramount. The *obirempon* economic system was largely built on the profits of commerce, and these profits fed a political and social order. Accumulation of property, lavish consumption and the public display of wealth marked the *obirempon* way of life. The *abirempon* promoted gold mining and 'mercantile slavery'. They purchased the many thousands of slaves who were imported into the region from other parts of western Africa and employed them in such tasks as the production of gold and of food crops. As earlier, peasant cultivators continued to provide labour services and revenue payments in kind, but, in contrast to the captaincy regimes, the city-states of the *abirempon* also imposed a land tax payable only in gold. The earliest reference to the *abirempon* comes from a 1479–80 description of trade along the Gold Coast seaboard.[21] The world of the *abirempon* centred around ports and markets, gold mines, trade routes, and the workshops of urban artisans. Gold became the principal currency throughout much of the region, replacing in most districts the earlier iron currency. The commercial achievements of the *obirempon* system culminated in the Akani trading organisation, which encompassed virtually all of the region from the early sixteenth century to the 1690s.[22]

From the 1490s onwards the sources combine to reveal a new kind of rural settlement and new forms of rural ownership: the *obirempon* economic complex. Typically, a complex included gold mines, the fortified residence of an *obirempon* and his household, settlements of unfree gold miners, hamlets and villages of free and/or unfree families engaged in land clearance, food-crop production, and the rearing of livestock.[23] Archaeology has established the presence of comparable sites throughout the region, most of them dating from the fifteenth and sixteenth centuries, but concentrating in the auriferous districts on hilltops.[24] Such settlements were not the appurtenances of the peasant agrarian economy; they were part of the burgeoning urban and commercial sectors of the regional economy.

The site of one apparent *obirempon* complex, a 'promontory fort', is located on the banks of the lower Tano river near modern Nkara. Originally it was an entrenched, fortified settlement over a kilometre in length, whose inhabitants mined gold, worked metal, made pottery and engaged in other crafts as well.[25] Doubtless, the settlers of this site also farmed and reared livestock. Nkara very likely had at least several hundred permanent residents. The entire place, together with its hinterland, would have been under the jurisdiction of an *obirempon* family.

The Nkara settlement broke the scale of earlier peasant villages. Consider such a complex and view it simply in terms of labour. How systematic must the appropriation of surplus wealth have been in order to rear a settlement on that scale? By contrast, see what any farm had managed to become by the efforts of a single peasant family after many decades of labour. Clearly, it did not assume the dimensions and organisation of the Nkara complex. With Nkara we have a visible manifestation of power, wealth, and authoritative command and a new level of productive capacity. *Obirempon* rural practice was not that of the peasants.

Turn to the sixteenth- and seventeenth-century coastal towns where a number of *obirempon* families constructed for themselves large, elaborate residences, some with gardens and walkways.[26] These residences, too, were a visible display of power, wealth, and authority, and concrete expressions of the new social system. They stood in stark contrast to the single-room cottages of the peasantry and humble dwellings of the urban poor. Where the rural *abirempon* organised their economic complexes for production, the urban *abirempon* organised their residences for consumption.

One example of rural *abirempon* organisation can be found within a 50-km radius of modern Oda in the upper and middle Birim River valley. Here are the remains of over twenty settlements of the fifteenth and sixteenth centuries, each of which had trench systems, embankments, and deep ditches. Some settlements were over 1.5 km in circumference, suggesting the seats of dominant *obirempon* families. Crafts, farming, and gold mining were the principal economic activities associated with the Birim earthworks.[27] These settlements indicate the *abirempon*'s ability to mobilise and control a large labour force.

Consider another example: the transformation of the approximately 800 sq km Nyanaoase district in the Akwamu state. Around the middle of the sixteenth century this heavily forested district, situated on the eastern edge of the Birim earthworks complex, was inhabited by families of hunters and peasants. They had lived there for decades in scattered hamlets and villages, apparently organised in captaincies. Between 1570 and 1660 the district was transformed. Over twenty towns appear including Nyanaoase, the 15-km-long Akwamu capital. The population of the district grew to exceed 75 000, and transformed much of the forest land into arable. The district became a major exporter of produce to the coastal ports.[28] Generations of hunters and peasants had not transformed the district on this scale. The *abirempon* system wrought the transformation. But consider what surpluses the Akwamu *abirempon* had to appropriate in order to effect its changes in ninety years.

The *obirempon* age was one of widespread urbanisation. For example, in and around the coastal Accra plains the pattern of settlement changed during the course of the fifteenth and sixteenth centuries from one of small and medium-sized villages to one of towns occupying several square kilometres of land and with populations in excess of 10 000 inhabitants.[29] By the late sixteenth century more than thirty towns dominated the region economically and politically, many being the foci of autonomous city-states.

The emergent towns were multifunctional settlements, serving as craft, cultural, marketing, political, and religious centres. The city-state system promoted the division of labour, thus accelerating a process already in evidence under the captaincy regimes. Crafts were urbanised, through the settling of artisans either in the towns or in specialised craft villages around them. This separated agricultural and craft production in peasant communities and established market relations between town and country. Rural economic dependence on the town was enhanced when peasants began to pay their land tax in gold. In order to acquire gold the peasants had to sell a portion of their harvests in the urban markets. In this way, the *obirempon* system fostered petty-commodity production by the peasants.[30]

Changes in the division of labour were evident in another sphere. The expansion of trade and the rise of a wealthy merchant class separated commerce from craft production in the towns, and led to the emergence of distinct guilds for merchants and craftsmen. A 'confraternity of nobles' appeared proclaiming an ideology of accumulation and conspicuous consumption. This urban-based confraternity was an *obirempon* association open only to the rich and the 'well born'. Its members enjoyed a wide range of rights and privileges that were denied to non-members. The confraternity was found in every city-state, and every *obirempon* belonged to it.[31]

Who built the towns and who supplied them with produce, raw materials, and labour? Who paid the social cost of urbanisation and who maintained the elite's life of pomp and privilege? Did the *abirempon* initiate a dramatic increase in the rate of exploitation of peasants and slaves? Did much of their wealth come from rising rates of exploitation? A partial answer may be found in a Portuguese report of 1572. The report is explicit: when the most powerful wanted gold they took it from those who could least afford to lose it.[32] The most powerful were *abirempon*. Behind expanding trade and urbanisation stood a social relation of surplus extraction.

What of the others, those who could least afford to lose their gold? When they appear in the European sources, they are 'the common people', 'the poorer sort of people', 'the poor', 'the mob', 'the meaner sort': in short, the lower orders, the labouring classes. An *obirempon* only addressed them in a voice of command and authority.[33] Akan vocabulary lists of the seventeenth century identify the labouring classes by two collective nouns. The first, *anihumanifo*, was used at the beginning of the century and, presumably, in the sixteenth as well. The Dutch equated it with 'poor' (*arm*). The second, *adofo*, was in use in the second half of the seventeenth century, and also was equated with 'poor' (German: *arm*). Both words conveyed the same general meaning: 'the vulgar people', 'the poor', 'persons of low social condition', 'persons of no rank'.[34] Within the *obirempon* world the *adofo/anihumanifo* constituted the disadvantaged and the underprivileged, and they were numerous. *Animafo bebrifi* literally means 'the numerous poor'; similarly, *adofo* literally means 'the people who multiply' or 'the numerous people'.[35]

We can identify yet another group of poor. In the towns, particularly the coastal ports, there was a small, but visible group of persons whom the sources called idlers, rogues, scoundrels, and vagabonds. Unlike the *adofo/anihumanifo*, who look like the 'respectable poor', this group were the 'unrespectable poor' and, constituted the beginnings of a pre-capitalist lumpenproletariat. Vocabulary lists of the seventeenth century have three Akan terms that refer to members of this group: *konkonsafo*, *mantemantanni*, and *oduaduafo*.[36] Although these persons belonged to the larger *adofo/anihumanifo* grouping, they were distinct from it. Economically and socially displaced, they lacked corporate ties to the means of production and the means of subsistence. They were destitute. They would not have been very different from the 'young men' (*jongelingen*) of the coastal town of Kormantse who, it was reported in 1640, travelled from port to port in order to earn their livelihood.[37]

According to tradition, needy families of the pre-*obirempon* received provisions in times of dearth from wealthy families. Under the *abirempon* this informal, personal treatment gave way to a formal, less personal treatment of a distinct and separate class of poor. Already in the sixteenth century certain city-states developed systems of public poor relief to furnish the more necessitous of the *anihumanifo bebree* with their means of subsistence. Gold, consumer goods, produce, meat, and drink were distributed among them, and the rulers of the towns provided the elderly, the physically handicapped, and unemployed young men with menial tasks in the towns whereby they could earn their livelihood.[38] Such persons were the casualties of a changing rural economy on the one hand and the unequal distribution of money, goods, and produce in the towns on the other. The *obirempon* socio-economic system of the sixteenth and seventeenth centuries produced rural and urban pauperism. The moral economy and patriarchal ideology of an earlier time were no longer applicable.

Mercantilism and the *obirempon* system created conditions for widespread banditry. While late fifteenth- and early sixteenth-century Portuguese sources suggest bandit activity in some coastal districts, seventeenth-century data provide ample, direct evidence of brigandage. A Portuguese report of about 1618 is particularly striking and informative. In addition to its observation on banditry it offers a perceptive view of rural social life under the *obirempon* regimes:

> Because of the many commodities the Dutch have brought and are bringing, all have abandoned farming and have become and are still becoming merchants. Those who cannot pay become robbers of other merchants. There are no farms and no agriculture, and all of the neighboring coastal Kings are in despair and lament that they are lost and ruined, and that they will lose everything, and that they will die of hunger.[39]

Dutch trade along the Gold Coast began in 1593, so the social changes described here had twenty-five years of increasing mercantile activity behind them. But they also had behind them over a century of *obirempon* domination.

That some segments of the Gold Coast peasantry faced pauperisation and economic crisis early in the seventeenth century is beyond doubt. Consider the marked disparities in wealth. Around the middle of the seventeenth century the yearly gross incomes of commoners resident in the ports – canoemen, carpenters, smiths, day labourers, woodcutters, water carriers, etc. – ranged from 144 dambas to about 4000 dambas in gold. Peasants' incomes from the sale of produce in urban markets were on average between 300 and 2000 dambas a year. All of these earnings were subject to a variety of taxes, which could appropriate as much as one-half (or more) of a commoner family's income.[40] In the coastal towns the 'numerous poor' families each earned appreciably less than 500 dambas a year. Meanwhile prosperous merchants and landlord brokers of the ports each realised an annual trade turnover of between 200 000 and one million dambas in gold.[41] Admission into the confraternity of nobles cost a total of nearly 39 000 dambas in the early seventeenth century, compared with which commoner incomes become negligible. The admission fee alone exceeded the total annual earnings of 270 wood carriers, which averaged yearly around 144 dambas in gold each in the 1640s.[42] Moreover, peasant families were often in debt to prosperous townspeople. The levying of land taxes in gold frequently obliged them to borrow gold against the security of future harvests in order to pay the revenue collectors of this particular city-state.[43]

In the ports some of the destitute became in the eyes of the authorities a source of crime and disorder. Here there was an abundance of consumer goods and gold and large numbers of people travelling through the streets with objects of market value. Burglary and theft thrived throughout the seventeenth century. In the port of Elmina an enterprising thief from the pauper class co-operated with some slaves to steal gold and other goods from the homes of the well-to-do. In 1646 this 'delinquent rogue' was imprisoned, and shortly afterwards hanged himself.[44] The involvement of slaves in burglaries which were carried out by indigent commoners suggests the existence of a criminal social network consisting of 'persons of low condition' – paupers, vagabonds, and slaves. They were *awifo*, thieves.

Another example of urban crime concerns the militia of Mankessim (or Great Fante), the capital of the coastal Fante polity. In 1653 several militiamen waylaid a small caravan after it had left the capital. They seized all of its merchandise, but did not injure the traders. Through contacts they had the stolen goods placed on sale in a number of the town markets in Fante. A vigorous investigation by the Mankessim administration failed to recover the goods or to apprehend the militiamen.[45] The militiamen behaved like *odwowtwafo*, 'highwaymen'. Their criminality differed from that of the Elmina *awifo*.

Consider also the fishermen of the port of Mouri, a leading gold-marketing centre in the Asebu polity. In June 1645 they rioted for two days in protest against an increase in the fish tax. The port's authorities were forced to seek refuge in the Dutch fort in order to escape the wrath of the fishermen.[46] The tax was not increased. This kind of activity was sufficiently common to be reflected in the Akan language of the seventeenth century. Müller notes that a riot was known as *agyesem* and rioters as *agyesemfo*.[47] The fishermen carried out their protest as *agyesemfo*, a culpable action in the eyes of the state. Urban lower-class criminality had its counterpart in the countryside, banditry. Both were produced by the same social system.

Bandits were particularly active in several districts. A 1629 Dutch report recounts the expedition of two prominent merchants from the port of Little Komenda, who travelled to the gold-producing Egwira polity in the middle Ankobra River valley. They returned rich in gold.[48] The report considered the undertaking to be dangerous, because the route was plagued by *struickroovers*, highwaymen. The route would have carried the merchants past a number of towns and mining settlements in the Adom state, which *obirempon* families had founded and developed in the fifteenth and sixteenth centuries. The report does not locate the bandits precisely, but one suitable base of operations close to the route would have been the southern part of the heavily forested, sparsely settled Hunu hills on the border of two provinces in Adom, and between their main population centres.

Traders from the rich gold-producing Mampa province of Adom were subject to regular attacks when they travelled to the coastal ports of Axim, Akwafo-Busua, and Butri. Banditry on the roads to these ports often compelled the Mampa merchants to direct their trade to Little Komenda via a comparatively safe sub-coastal route which passed through Yabiew.[49] *Akwanmuka* on the routes linking the Mampa capital and the coastal ports may have been present already in the late sixteenth century. A 1602 Portuguese map identifies Ampago, the capital of Little Nkassa, and notes that the Nkassa destroyed 'petirces' in the Axim hinterland and along the west bank of the lower Ank⁒ ʰra River. Ampago was located between fifteen and thirty kilometres to the north-east of Axim, and was a bitter commercial and political rival of that port. The word 'petirces' has so far not been identified; it might refer to caravans, settlements, farmlands or to something else altogether. Still, it is clear that the destruction of 'petirces' by the Nkassa disrupted local caravan traffic. The Nkassa 'destroyers' may have been armed retainers from the town of Ampago, acting on behalf of the local *abirempon*; or they may have been a peasant force in rebellion against the *obirempon* systems throughout the entire area,[50] in conformity with the Portuguese report of 1618 which we discussed above.

Brigands were also active in the lower Pra River valley, an important arterial route. The origins of banditry in this area are uncertain, but it is possible that bandits first appeared on a significant scale around 1600 following a war between the coastal Eguafo and Yabiew polities. The victorious Eguafo army destroyed the Yabiew farming and fishing communities on the east bank of the river, thus forcing large numbers of peasants, fishermen, and others to settle on the west bank. Some of these displaced persons probably turned to banditry as a way of life, as the 1618 Portuguese report strongly suggests. Bandits were certainly present in the area a century later. In 1727, the Dutch factor at the main Yabiew port near the mouth of the Pra made an agreement with several Adom notables, which asked them to build a settlement on the west bank of the river 'in order to give safety to the *mercadors* [merchants] against all robbers'.[51]

In the first half of the seventeenth century traders from Axim were cautious about venturing to the inland gold-producing places. Likewise, inland traders from Adom to Axim had to adopt a circuitous route across the Ankobra River and through the city-states of Egwira and Jamu. The problem was rural folk in the lands neighbouring Axim who regularly murdered and plundered merchants. These unidentified people had neither kings nor a supreme central governing authority. Instead, they were organised in 'republics'.[52] Were the 'republics' of the Axim hinterland surviving remnants of the pre-*obirempon* captaincies, resisting mercantilism and incorporation into adjacent city-states? Whatever the case, by the 1660s the 'republics' and their brigandage were no longer a major problem to merchant caravans.

Banditry flourished in the lands east of the Ankobra River from the 1640s onward. Freebooters became active in the districts bordering Egwira and Jamu, thus restricting the movement of caravan traffic to Assini and Albine. They were also active along the coast road that linked Assini and Axim, which were over 130 km apart. The growth of Cape Apollonia, a port about midway between Assini and Axim, and the development of trading towns like Abuma and Sumane in the neighbouring coastal hinterland provided lucrative opportunities for bandit predation. Around 1656 the authorities of Jumore, the westernmost province of Axim where Cape Apollonia was situated, made a particularly strong effort to suppress brigandage.[53] Nevertheless, *okwammuka* remained an intermittent problem until at least the end of the century.

It would seem that from the sixteenth century onward many peasant villages in the lands between the Pra and Tano rivers experienced the same changes and pressures which the 1618 Portuguese report ascribed to the coastal village communities east of the Pra, where the *obirempon* system was older and more firmly established. If acts of banditry are evidence of rural discontent, one might conclude that the social order created by the *abirempon* between the Pra and Tano was bitterly contested by the peasantry. The new order tied peasants to the urban markets via the revenue system, thus securing their economic and political subordination to the hegemonic classes; it placed slaves on the land; and it established a new organisation of rural production in the form of the *obirempon* economic complex.

In March 1636 bandits were active in Abrem, a southern forest polity. Two major routes traversed this state, linking the coastal ports with the inland towns in the Pra-Ofin basin. The principal travellers along these routes were Akani traders. They supplied more than one-half of the gold that was received at the coastal European trading stations at this time. They were the bearers of an *obirempon* tradition that dated from the fifteenth century. Unidentified Abrem bandits seized gold and trade goods from Akani caravans. The Akani traders reacted swiftly. With a large body of armed retainers they carried out a vigorous military-police action and by May had eliminated bandit pillaging along the Abrem section of the arterial routes.[54] For the remainder of the century Abrem had little or no banditry.

Akani caravans also faced brigands along the routes north of the forest. These routes linked Bighu and other commercial centres in the valleys of the Black and White Volta rivers and the towns of the Pra-Ofin basin. Throughout the first half of the seventeenth century they were endangered by 'great multitudes of highwaymen', and Akani caravans were frequently attacked by outlaws of unknown social origins. By the 1650s banditry on the northern roads had been largely suppressed and caravan traffic to Bighu and the other towns increased substantially.[55]

Between the 1670s and the early 1700s bandits were active in the coastal territories east of the Volta. A late-seventeenth-century report mentions their presence in Keta, a coastal polity on the edge of the *obirempon* system. 'There are few wealthy men among the Cotos [Ketas], and the generality being very poor, many of them turn strolling robbers about the country, and do much mischief.'[56] The wealthy men were traders and landlord-brokers resident in the port of Keta, who realised large profits from the import-export trade. The rise of banditry amongst the numerous poor of Keta coincided with Keta's emergence as an important trading centre.

One group of Keta bandits operated along the coastal overland route that linked Keta and Ofra, the main port of the Ardra kingdom, about 130 km away. The bandits were active between 1675 and 1683. The leader was named Aban, and he had about fifty followers. Aban and his men pillaged caravans and murdered travellers, particularly official messengers who travelled from one port to another.[57] They disappear after 1683, but others replace them. One unusual band of piratical brigands specialised in pillaging sloops and other small boats that traded at the ports between the Volta River and the coastal town of Whydah. In 1686 they were described as 'a parcel of runaway rogues', which suggests that they were slaves and/or bonded servants who had fled from their masters. At Keta they plundered an English boat and killed many of its crew; and at the port of Popo, about 80 km to the east, they joined the slaves aboard the English sloop *Charlton* and 'killed ye commander and all they could and took all the gold and slaves.'[58] In contrast to Aban and his followers the 'runaway rogues' did not, as far as we know, attack the caravans of merchants or official travellers along the overland routes. This gang, like Aban's gang, drops from the record and we have no references from the 1690s to 'rogues' plundering coastal boats in the Keta-Popo area. However, other groups of 'strolling robbers' continued their acts of violence and predation on the trade routes east of the Volta through the eighteenth century.[59]

Akwamu oral traditions tell of bandit activities in the hinterland of Great Accra. From the 1540s until its destruction in 1677, Great Accra was one of the wealthiest and most important commercial centres in the Gold Coast region. For more than a century traders and caravans from all directions, from far and near, converged upon it. In 1572, Great Accra was described as a large, populous, and prosperous city full of rich merchants, and Ga historical traditions of the nineteenth century credited the city with a population of 40 000 to 50 000 people on the eve of its destruction.[60] To enterprising bandits the roads leading to Great Accra offered the prospect of quite considerable quantities of loot.

Akwamu traditions identify two bandit groups, both of uncertain social origin and uncertain date. The first group was founded and led by a man named Okwanmu (i.e., 'the bandit'). This name is a contraction of *okwanmu-ka ntiri na me woho*: 'I am here to plunder on the general road' or, alternatively, 'I am here for the sole purpose of plundering the passers-by.' By all accounts Okwanmu was a successful highwayman: 'Gradually he succeeded in getting many followers, which he formed into *asafo*, and they later on became the main army of the Akwamu Kingdom.'[61] This *asafo* is a military company. Okwanmu's bandits had become professional soldiers. The second group was equally successful. The traditions relate: 'A certain man who was related to the Akwamu established himself on one of the roads. He, too, said he was there solely for plunder: *Oko nti na me wo ha.*' This bandit leader later organised his followers into a military unit that 'was also recognised in the Akwamu realm.'[62]

Nyanaoase was the Akwamu capital, about 25 km north-west of Great Accra. Its growth and development were closely tied to the economy of that city. It was also well sited with respect to other important *obirempon* towns, all of which pre-dated Nyanaoase's founding in the 1570s. The early transformation of the 800 sq km Nyanaoase hinterland may be assigned to the same period. The rise of Nyanaoase disrupted the settlements and economy of the pre-existing rural population.[63] It is not unlikely that the bandits of Akwamu tradition emerged during this time of rural crisis and urban extension. Were Okwanmu and the others originally peasants facing great economic change and social displacement? Were they among those who could not buy their way into the ranks of the Akwamu or Great Accra *abirempon*, but who had broken with their old world and wanted their share of the new? In short is this situation comparable to the one described in the 1618 Portuguese report, with merchants and bandits emerging together out of the ruin of peasant villages? Certainly, Akwamu traditions and European reports, both of the early eighteenth century stressed the importance of merchants and skilled military men in the build-up of the state, if not the role of bandits in this process.[64]

The Akwamu evidence draws attention to a close link between banditry and professional soldiering in the Gold Coast region. Agona Asafo, a polity on the borders of Akwamu and Great Accra, provides additional evidence. A 1675 report relates that in this estate there were men who not only robbed merchants, but who also hired themselves out as mercenaries. Here we have men who were both brigands and professional warriors, a phenomenon that was not unique to the Gold Coast.[65]

The preceding examples of banditry would suggest that this activity was the preserve of commoners and slaves. This was not the case. The wealthy and privileged *abirempon* freely engaged in banditry when circumstances demanded. Thus, not all freebooters were peasants, 'runaway rogues', and the like. Upper class involvement was predicated on different personal and social motivations from those of the *anihumanifo bebree*, the 'desperate villains', and 'strolling robbers'. If peasant banditry was rooted in economic and political relations which intervened between the land and the peasants – the systems of land tenure, revenue collection, money-lending, and marketing – *obirempon* banditry was rooted in a different set of relations. Marx described their structural terms:

> The operation of merchant capital in pre-capitalist modes of production with limited surplus products generally concentrated in the hands of the ruling classes lends itself to manifestations of violence. Violence arises from conflicts between merchant capital and ruling classes over distribution of the surplus product between them.[66]

Obirempon banditry intervened in the process of the accumulation and distribution of surplus wealth in order to direct, if not actually to control this process. The *abirempon* directed their brigandage towards the forcible seizure and accumulation of wealth, often in the form of merchant capital, and brigandage engendered conflict among *obirempon* families and warfare between their regimes.

The earliest reference to upper-class brigandage dates to 1510, and data from the seventeenth century indicate that it was not unusual for important notables to employ their armed retainers in 'acts of robbery' against merchants and their caravans.[67] One such notable was Takyi, the ruler of the coastal state of Eguafo. In November 1646 he despatched a 'party of highwaymen' to the road that linked the port of Elmina and the inland town of Afutu in order to pillage caravans.[68] The 'highwaymen' were armed retainers in Takyi's service. In justifying his act Takyi claimed he was not opposed to trade or merchants, but was merely protecting himself from his political enemies, who were conspiring to drive him out of office. Indeed, Takyi had had a chequered political career. In the late 1630s he was a wealthy *obirempon* in open conflict with his equally wealthy brother, Edwan, the then ruler of Eguafo. Takyi overthrew his brother in 1641 or 1642 but was in turn overthrown in 1644 and driven out of the capital with his dependents. In the early months of 1645 Takyi's retainers pillaged villages which were the revenue-paying dependencies of his political antagonists. Takyi regained the kingship in late 1645 or early 1646 by using his accumulated wealth to bribe certain important *abirempon* in Eguarfo and some of the neighbouring coastal states. But in 1647 he was again forced from office and driven out of the capital.[69] Violence and pillaging were freely employed by Takyi in his efforts to further his political career. At the same time Takyi was heavily engaged in trade, particularly with the Dutch factors of Elmina, from which he realised considerable profit. However, commercial operations did not give him access to the surplus production of peasants and other revenue-paying commoners. This he could only achieve when he occupied a high political post in Eguafo, hence his protracted, often violent, struggle to control political power.

Another notable who employed violence to advance his political career was Ahen, the *brafo*, or chief military-police and third-ranking official, of the coastal Afutu state. In 1662 he and his armed retainers seized the Swedish trading station in Cape Coast, the principal port of Afutu, and confiscated all of the trade goods and 150 pounds-weight of gold. Shortly afterwards Ahen was appointed *fetere* by the King and the Afutu state council.[70] The post of *fetere*, or chief administrator of the state, was the second highest political position in Afutu. Ahen's sudden and violent acquisition of wealth played no small role in his climb up the political ladder.

A final example comes from the early eighteenth century.[71] Jan Kango, a ruler in Mampa province of Adom, rebelled against his overlord, and with his armed followers established himself as an independent force in the Ankobra River valley. He pillaged caravans and blocked the arterial routes. On one occasion he and his men crossed the Ankobra, invaded Abokro, in Axim and the state of Egwira and pillaged several riverain communities. Kango's marauding activities eventually led to his defeat and capture by a confederated military force. In 1707 he was executed. But prior to that Kango and his retainers had engaged in large-scale freebooting for several years. Here was no 'parcel of runaway rogues' or 'strolling robbers' but an *obirempon* and his armed force engaged in the systematic accumulation of wealth through plundering. The careers of Takyi, Ahen, and Kango exemplify the violence that was inherent in the conflict between merchant capital and officeholders over the distribution of the social surplus between them.

The framework of the *obirempon* city-state system could not resolve this conflict. The system began to break down in the late seventeenth century, and by the 1730s its viability had gone. It was replaced by militarism and large political formations. In the age of militarism the accumulation of wealth was realised through territorial expansion organised by the state. It was also a time of widespread de-urbanisation and, for certain districts, massive depopulation. In the eighteenth century the region was more rural than it had been in the sixteenth and seventeenth centuries.

The *awurafram*, 'the masters of firepower', replaced the *abirempon*. The *awurafram* were military men for whom the firearm was an instrument of production. In times of war an *owurafram* commanded a force of several thousand musketeers, and through successful campaigns he gained fame and wealth. He 'produced' slaves who were sold to European factors in exchange for military goods and other commodities. Thus, although the *awurafram* generally held professional merchants in disdain, they did not eschew trade but actively participated in it. Nevertheless, among the leading *awurafram*, commercial success took second place to success in combat. The soldier's musket replaced the merchant's gold weights as the symbol of the age.[72]

The dissolution of the *obirempon* system and the rise of the *owurafram* social order were directly linked to the emergence of the expansionist, militarised states of Akwamu and Denkyira and their mass armies of conscripted peasants, mercenaries, and slaves. From the late 1670s onward the Akwamu and Denkyira ruling classes launched a series of military campaigns which resulted in the subjugation or destruction of *obirempon* regimes over a wide area and the establishment of two imperial systems. Both systems were extensions of powerful and wealthy urban centres: Denkyira was a by-product of the rise of Abankesieso; and Akwamu was an extended revenue-producing hinterland of Nyanaoase. In short, they were city-state empires, and represented the first phase in the age of militarism. Far-reaching reforms in their armies, such as the introduction of mass conscription, strikingly enhanced their military capabilities, and made possible their territorial exansion. Through warfare the Akwamu and Denkyira ruling classes externalised their mode of surplus extraction and expropriated the surpluses of other polities. Through successful campaigns the Akwamu and Denkyira elite transformed themselves from *abirempon* into *awurafram* ('masters of firepower'). They initiated a military revolution in the region.

Denkyira was conquered in the years between 1699 and 1701, and the Akwamu imperial system collapsed in 1730 following a succession of internal revolts that coincided with an invasion by several of its neighbours. However, territorial conquests did not end here. They continued under Asante and Fante, both of which were territorial states organised around confederations of towns. Between them they came to dominate the entire region during the course of the eighteenth century. Their expansionism marked the second phase of the age of militarism. It also coincided with a dramatic increase in the number of slaves exported from the region.

Wealth and labour flowed into the metropolitan districts of the expansionist states, and were concentrated in the hands of the possessing classes. Wealth from booty and tribute supported the luxurious living of the *awurafram*. It was spent on the construction of urban and rural residences and was invested in the acquisition of land, subjects, and slaves, in moneylending, and in the financing of mercantile activities. The triumph of the 'masters of firepower' left the old *obirempon* system in ruins. Consider the Denkyira army commander Agya Ananse who was known as Obooman, 'the destroyer of towns'; or the Akwamu general Amputi Brafo, whose boast was that he had conquered all 'the lands of the blacks'.[73] The social experiences behind Ananse's and Brafo's careers are clear: forcible seizures, new kinds of ownership, altered political and social structures, reorganised trade routes, new titles of importance, and new symbols of authority.

Viewed as a social process, military expansionism in the Gold Coast was a form of surplus extraction developed from conditions of production rooted in the age of mercantilism. For example, in the various *obirempon* city-states, artisans and peasants engaged in petty-commodity production. Urbanisation had helped to separate craft and agricultural production, while the peasants had had to sell some of their produce in town to get gold to meet their revenue obligations. Under the *awurafram*, de-urbanisation led to a decline in market production and to the ruralisation of crafts. Peasants no longer paid their taxes in gold, but in kind and in labour services, mainly military. Throughout much of the region craft and agricultural production were re-established in village communities. These communities enjoyed a degree of economic self-sufficiency which their counter-parts of the *obirempon* age had not had. Petty-commodity production was not realised in the village community, but at the level of the militarised ruling classes. Through their ownership of the means of production, in this case military equipment, and their control over labour in the form of armies, these classes 'produced' slaves for overseas markets and received in exchange military wares and articles of consumption. The *awurafram* combined the functions of 'production' and trade through their control of the state apparatus. Hence, in contrast to those of the *abirempon*, the urban seats of the *awurafram* were essentially administrative in character.

And what of banditry during this age of the triumphant *awurafram*? The evidence is explicit. *Akwanmuka*, robbery on the main roads, did not end; it persisted throughout the eighteenth century. Several European reports of the early eighteenth century relate that inland traders en route to the seaboard ports were regularly 'robbed, killed, and plundered' by the rulers of the coastal and southern forest polities, who employed their retainers in these acts. The rulers were guilty of 'always violating the public roads', according to a 1704 report, and in 1712 it was noted that 'the small powers occupy themselves with robberies' on the main roads.[74] How do we account for increased brigandage by political authorities who were responsible for the security of their own roads? The authorities in the coastal and sub-coastal polities faced mounting economic and political problems. Military defeats meant a loss of political autonomy and reduction to tributary status. This, in turn, meant a loss of control over surplus appropriation and a decline in disposable incomes. Under the circumstances the rulers turned to brigandage to stablise their losses and, at the same time, to challenge the suzerainty of their political overlords. Jan Kango, whose marauding activities were outlined above, was one outstanding example of an officeholder turned bandit during this time of terminal crisis for the *obirempon* regimes.

Others were involved in highway robbery in the early eighteenth century. In 1701 Asante conquered Denkyira and incorporated it as a tribute-paying province. Following this, large numbers of Denkyira soldiers – including armed retainers now without masters, and displaced peasant conscripts – turned to Asante caravans passing through the province. These bandits were particularly active between 1706 and 1708. By the latter year the Asante government had sent 'militia' units into the province to stamp out this brigandage; the units were generally successful.[75] Denkyira banditry in the early eighteenth century was clearly a form of political opposition to Asante rule. In this respect it was similar to the freebooting of contemporary rulers of the coastal and sub-coastal states, which we have just considered.

By mid-century Asante dominated most of the region and Asante trading caravans regularly faced the problem of *werekyerewerekyere*, highwaymen, in the tributary provinces through which they passed. For example, in the 1740s Fante freebooters pillaged Asante caravans travelling through desolate and depopulated districts in the coastal hinterland on their way to Elmina and the western ports, and sold the traders to European factors.[76] In the 1740s and 1750s local bandits regularly pillaged in recently conquered lands bordering the lower Volta River. Indeed, this area had already witnessed bandit activity in the 1710s and 1720s when it was under the hegemony of Akwamu.[77] The Volta brigands attacked Akwamu and riverain settlements and, with the passing of the area under Asante dominance from the 1740s onwards, refocused their operations against Asante caravans. The lower Volta valley remained bandit-infested until well into the nineteenth century, when one observer offered the following informative comment on the situation there:

> The banks of the Volta on both sides are occasionally infested by a banditti, sometimes daring enough to plunder the smaller crooms [villages]. The insurgents are mostly Aquambu [Akwamu] negroes, incited to rebellion against their conquerors the Ashantees by other branches of the tribe who fled before the invading army and have established themselves on the Dahoman shore; yet in peaceable times they are considered subjects of Ashantee.[78]

As in the case of Denkyira, so the pillaging of Akwamu bandits represented a form of political protest against Asante rule.

To the immediate west of the lower Volta valley, banditry was a serious problem for traders throughout the second half of the eighteenth century. An important trade route linked the market town of Asuchary and the Accra ports. The route ran near the densely populated Krobo hills, a habitat of 'lawless banditti', who seized caravan goods and murdered traders 'with impunity' whenever they pleased. The 'banditti' were Krobo peasants, young men for the most part, who spent some of their time in food-crop production and the remainder in robbery on the main roads. Caravans passing through what European factors termed the Krobo 'republic' were obliged to have contingents of armed guards to protect them from bandit attacks. Local authorities, like the Akwamu ruler, a tributary of Asante after 1742, were responsible for law and order in the lower Volta basin and sought to suppress Krobo brigandage by destroying the farms and villages situated at the foot of the Krobo hills. However, the main Krobo settlements were located on the summits. The hills themselves were virtually immune to direct assault because of their natural and man-made fortifications.[79] As a result, even the systematic destruction of the Krobo peasants' means of agricultural production, had no lasting effect. Banditry in the area always managed to reassert itself, and only ceased to be a problem in the early nineteenth century when the economy of the area was transformed. At that time Krobo peasants and traders embarked on the large-scale production of palm oil for sale and purchased extensive lands from neighbouring towns and villages. Commercial agriculture promoted the growth of the great Krobo market towns which flourished during the nineteenth century.[80] A district of peasant communities had become a major producer of palm oil for overseas markets. But behind the transformation of this economy lay more than fifty years of banditry. If Krobo banditry emerged in the 1740s as a form of protest against Asante rule, by the 1760s it had become a major means for the accumulation of social wealth. This wealth provided the capital that contributed to the emergence of a Krobo trading class by the 1770s. We may also assume that it was the basis for the later development of palm oil production.[81]

A related development occurred with the founding of Kpong around the turn of the eighteenth century. Originally an island settlement in the Volta River, Kpong was a refuge for escaped slaves, debtors, and criminals who engaged in river piracy, pillaging barges and canoes loaded with salt, produce, and trade goods. Occasionally, these brigands operated against merchants and their caravans.[82] Kpong did not long remain a mere settlement of outlaws. By the 1850s it was a major market for the river trade, as the missionary J. Zimmerman noted:

> Though inhabited by a mixed and rough set of people, the town is thriving, having the best river canoes, the greatest share of the salt and palm oil trade and enjoying the neighborhood of the rich and thriving Krobo country . . .

Nevertheless, Zimmerman thought little of its social character: 'Kpong is a rotten town, a free city where runaway slaves and other rabble settle.'[83] Yet, some fifty or sixty years earlier, the brigandage of escaped slaves and 'other rabble' had laid Kpong's foundations as an important trading centre, much as the Krobo peasantry had earlier done.

Not all brigands experienced social changes of this sort. The Akwamu bandits of the early nineteenth century remained predators and insurgents. In the 1770s Asante and other authorities were periodically obliged to suppress brigandage on the arterial routes, 'to clear the paths of pirates' in the words of the factors, in order to safeguard the movement of trading caravans.[84] In these cases bandits remained *akwanmukafo*, violators of the public roads, opposers of the Asante government and *awurafram*. They were the *werekyerewerekyerefo* whom the Asante ruler regularly suppressed for the sake of law and order. Like many of their sixteenth- and seventeenth-century predecessors they came from the ranks of the rural *anihumanifo bebree*, and were no different from the Krobo bandits of the 1750s and 1760s. Some may have resembled Kwame Abe, a famous bandit of the nineteenth century, of whom we have the following description:

> The most notorious highwayman in the nineteenth century was probably Kwame Abe, who haunted the paths through Adansi, Denkyira, Wassa, and Assin. To this day old people mention his name with awe and trepidation. He was known to kill lone travellers and seize their goods, and occasionally abducted any beautiful girl who seized his fancy. So successful was he that many believed he could command magic powers. He could appear in an instant, and disappear at will if danger threatened. By all accounts this legendary figure met a sticky end; some say he was surprised when sleeping and cut up into tiny pieces, others that while taking his bath he was despatched by an ungrateful concubine who poured a pot of scalding soup over his head.[85]

The story of the Kwame Abes has yet to be written.

Not all of the *adofo* ('numerous poor') harboured thoughts of rebelling and pursuing the career of a Kwame Abe. Many would have shared the views of Kwasi, a bricklayer's apprentice, employed at the Danish fort of Christianborg in the 1730s and 1740s. When asked if he would like to be a king, Kwasi replied: 'No, that is not possible, for I know that as long as I have been on this earth I have been a slave; and as long as I shall remain here, I must be a slave.'[86] Kwasi was neither an urban thief nor a rural brigand. He did not protest against the material and social conditions of his life; he accepted them. Kwasi represented that large social mass upon which the political stability of the *awurafram* regimes depended. Banditry appealed to only a small minority of the population.

As we have seen, this minority was active in the eighteenth century and later. It was involved in banditry. It was also involved in particular forms of outlawry which had no visible precedent in the social order of the *obirempon* mercantilists. For example, the letters and reports of early-eighteenth-century factors refer to persons who kidnapped villagers from the forest in the manner of highwaymen and sold them at the European trading stations on the coast.[87] Carl Christian Reindorf describes the general insecurity of the later eighteenth century:

> In those days rivalry among the Dutch and Danish, as well as the Dutch and English merchants, manstealing and scarcity of provisions were in the highest stage in the country. The kidnappings by the Agonas had been checked by Dade Adu, but Akra women were not safe at Mlafi and other Volta towns, when travelling there to buy corn. As pillaging and plundering during that period

were general, the farmers of Akra could not make their farms more inland; so scarcity prevailed nearly every year, that people were forced to travel to Krobo, Ningo and such places for food, where they were never safe.[88]

Kidnapping and plundering generally arose out of disputes over the ownership of property between office-holders and the rich. Militias had to be organised to curtail this activity. Reindorf relates that 'a company for defensive warfare was organised by all the iron-hearted men of Akra' and 'through their operations a stop was put to the inroads of the kidnappers.'[89] Nevertheless, the problem persisted:

But in consequence of the unsettled state of the country by the incessant kidnapping and plundering of the Obutus and Akuapems, the [Akra] farmers were unable to cultivate the land as they should have done, until the robbers and plunderers of both places had been checked. Some even were killed, such as one hunter Nseni of Obutu ... and several others who shared the same fate from the Akra palm-wine carriers and the iron-hearted company known as 'Odshofoi'. The palm-wine carriers formed a most powerful body in those days, as they defended the country from such robbers.[90]

The Odshofoi militia and the Accra palm-wine carriers were pitted against kidnappers and robbers from the 1790s until the first decade of the nineteenth century. It was only in 1815–16 that an Asante army finally suppressed these outlaws.

The ongoing conflicts among wealthy families in such towns as Accra and Obutu over land, goods, and money brought about the two decades of open outlawry in their rural hinterland. The victims of the kidnappers and robbers were generally not merchants and their caravans nor Asante functionaries, but peasants and hunters, who were political dependents of the feuding, faction-ridden towns. Urban conflict reached into the countryside, where it manifested itself as kidnapping, robbery, and the destruction of farms.[91] The social basis of these activities was quite distinct from that of the *werekyerewerekyerefo*, who pillaged traders and caravans. Here we have a different phenomenon, one that can be linked to the rise of the *awurafram*.

We can trace back a hundred years the rural outlawry which plagued the hinterland of the coastal Accra towns at the turn of the eighteenth century. The sources associated this activity with the operations of gangs in the states of Akwamu, Agona, and Akron. The gangs comprised 'young men' who were described as clever, enterprising, unmarried, and without regular occupations. The young men of a gang were known collectively as *sika den*, 'black gold'. Each gang functioned in the same way. It went into a neighbouring land at night and placed itself near farms close to the paths, where it awaited peasants going to their farms. The peasants were attacked, bound, and gagged, and then transported to the coast, where they were sold to European factors as slaves.[92] The young men did not act on their own initiative but were sent by traders and office-holders resident in the towns to collect unsettled debts, owed by traders and office-holders in other towns. The creditors recouped their losses by hiring the disadvantaged urban poor to carry out acts of predation and violence against rural cultivators who were under the political authority of the debtors.

The practices of the *sika den* became institutionalised in the Akwamu state in the late seventeenth century. According to one report, the Akwamu ruler, Akonno (1705–25), had 1000 'clever young men' in his personal service.[93] However, the *sika den* not only ventured into lands beyond Akwamu, but they also, on occasion, 'seized their own countrymen' and sold them to the Accra brokers, who sold them to the European.[94] The Akwamu *sika den*, most of whom were drawn from the capital Nyanaoase and other Akwamu towns, directed their operations almost solely against the rural peasantry. It was no longer a question of settling old debts but a way of acquiring slaves who could be exchanged for a wide range of commodities at the European trading stations. For the rulers and high-ranking officials of Nyanaoase, the *sika den* served as an instrument in the process of wealth accumulation.

The collapse of the Akwamu imperial system in 1730 did not spell the end of the *sika den*. In the second half of the century the term was synonymous with kidnapping and robbery. The officials of a town frequently sent 'their young men on *sikkeding* in order to kidnap for old debts.'[95] The rural Accra districts of the turn of the eighteenth century well illustrate the socially disruptive nature of the *sika den*. By the third decade of the nineteenth century the *sika den* had generally disappeared. The breakdown of the *awurafram* socio-economic order and the decline of slave exports brought about this development.

For more than a century the rich and powerful of the towns organised lower-class youths into gangs of marauders who preyed on the rural population, forcibly expropriating their labour and converting it into a commodity for sale in overseas markets. The age of the *abirempon* had seen the rise of *akwanmuka* and the *akwanmukafo*, forms of rural violence directed against merchant caravans. The age of the *awurafram* saw this phenomenon continued but added to it the *sika den*, institutionalised urban violence directed against rural cultivators. With *sika den* we have a new form of upper-class domination of the countryside, one which involved the active participation of disadvantaged urban males. In the Gold Coast context this form of predation was a means to 'open' to the towns the relatively 'closed' countryside. It was a salient feature of the political economy of militarism during the trans-Atlantic slave trade.

In the foregoing pages I have surveyed the history of banditry and other forms of outlawry in the Gold Coast region down to the nineteenth century. I have tried to characterise these phenomena and to define the varying social experiences that gave rise to them and allowed them to flourish. Many questions remain unanswered and many relevant themes await examination. These themes and questions call urgently for systematic and thoroughgoing research. Banditry and outlawry on the Gold Coast remains a viable and tantalising topic.

Hitherto scholars of the region have relegated to the dustbin the thoughts and actions of the many. We have discovered them as 'multitudes of desperate villains and robbers', as 'a parcel of runaway rogues', as 'escaped slaves and other rabble', and as 'black gold', the *sika den*. They all appear fleetingly in the historical record. As Antonio Gramsci aptly stated: 'Every trace of independent initiative on the part of subaltern groups should therefore be of incalculable value for the integral historian.'[96]

NOTES

1   In this study, the Gold Coast region refers to the forest and coastal zones of southern Ghana.
2   John Barbot, *Description of the Coasts of North and South Guinea* (London, 1746), p. 191.

3 Eric Hobsbawm, *Bandits* (New York: Pantheon, 1981 revd edn), Ch. 1.

4 ibid., passim.

5 See Allen Isaacman, 'Social banditry in Zimbabwe (Rhodesia) and Mozambique, 1894–1907: an expression of early peasant protest', *Journal of Southern African Studies*, vol. 4, no. 1 (1977), pp. 1–30; W. G. Clarence-Smith, *Slaves, Peasants and Capitalists in Southern Angola 1840–1926* (Cambridge: Cambridge University Press, 1979), Chapter 6; E. P. Makambe, 'The Nyasaland African labour "Ulendos" to Southern Rhodesia and the problem of the African "highwaymen", 1903–23: a study in the limitations of early independent labour migration', *African Affairs*, vol. 79, no. 317 (1980), pp. 548–66; Charles van Onselen, *Studies in the Social and Economic History of the Witwatersrand 1886–1914*, Vol. 2 (New York, Longman, 1982, 2 vols), Ch. 4.

6 References to banditry in West Africa are fairly numerous. See for example, N. Levtzion and J. F. P. Hopkins (eds), *Corpus of Early Arabic Sources for West African History* (Cambridge: Cambridge University Press, 1981), p. 67; Harold Courlander, *The Heart of the Ngoni* (New York: Crown, 1982), p. 15. A final example comes from the kingdom of Gonja in what is now northern Ghana. When Jakpa Lanta (1622/3–66/7), the King of Gonja, appointed his grandson as ruler over the town of Bole and its surrounding environs, his instructions to him included the admonition 'never to resort to highway robbery'. Mahmoud El-Wakkad, trans., 'Qissatu Salga Tarikhu Gonja: The Story of Salaga and the History of Gonja', *Ghana Notes and Queries*, no. 3 (1961), p. 12.

7 Barbot, *Description*, p. 187.

8 Public Record Office, London (Hereafter PRO), T 70/1463. Memorandum Book Kept at Cape Coast Castle from 13 January 1703 to 2 January 1704: entry for 1 September 1703.

9 'W. de la Palma to the Assembly of Ten. dd. Elmina, 5 September 1705' in Albert van Dantzig (tr. and ed.), *The Dutch and the Guinea Coast, 1674–1742: A Collection of Documents from the General State Archive at the Hague* (Accra: Ghana Academy of Arts and Sciences, 1978), p. 112.

10 L. F. Rømer, *Tilforladelig efterretning om kystem Guinea* (Copenhagen, 1760), p. 182.

11 Kwame Arhin, 'Asante military institutions', *Journal of African Studies*, vol. 7, no. 1 (1980), pp. 22–3. Also pp. 26–7.

12 ibid., p. 29, notes 31, 38 and 54. In a personal communication, Professor Ivor Wilks of Northwestern University has suggested that Kwame Antwiwa and Rømer's Antifu are the same person.

13 See J. G. Christaller, *A Dictionary of the Asante and Fante Language* (Basle: Evangelical Missionary Society, 1881).

14 W. J. Müller, *Die africanische auf der guineische Gold-Cust gelegene Landschafft Fetu* (Graz: Akademische Druch- u verlagsanstalt, 1968 repr. of 1676 Hamburg edn), Ch. 17; Gabriel Debien et al., 'Journal d'un voyage de traite en Guinee à Cayenne et aux Antilles fait par Jean Barbot en 1678–1679', *Bulletin de l'Institut Fondamental de l'Afrique Noire*, vol. 40, no. 2 (1978), p. 343.

15 Müller, *Die africanische*, Ch. 17.

16 For a fuller discussion of Gold Coast historical periodisation see R. A. Kea, *Settlements, Trade, and Polities in the Seventeenth-Century Gold Coast* (Baltimore: Johns Hopkins University Press, 1982), pp. 6–7.

17 See Rømer, *Tilforladelig efterretning*, pp. 16–17, 26–7, 110, 112–26, 132–3.

18 James Anquandah, *Rediscovering Ghana's Past* (London: Longman, 1982), pp. 68, 115, 116.

19 ibid., pp. 69–70; P. Ozanne, 'Notes on the later prehistory of Accra', *Journal of the Historical Society of Nigeria*, vol. 3, no. 1 (1964), pp. 15–20. cf. Pieter de Marees, *Beschryvinghe ende historische verbael van het Gout Koninckrijcke van Gunea anders de Gout-Custe de Mina genaemt liggende in het Deel van Africa*, ed. by S. P. L'Honoré Naber (The Hague: Nijhoff, 1912), p. 45.

20 Kea, *Settlements*, passim.

21 'Voyage à la côte occidentale d'Afrique en Portugal et en Espagne (1479–1480)', ed. R. Foulche-Delbose, *Revue Hispanique*, no. 4 (1897), p. 182. *Abirempon* is rendered *berrenbucs* in this source.
22 For a description of the Akani trading system see Kea, *Settlements*, Ch. 7.
23 Valentin Fernandes, *Description de la côte d'Afrique de Ceuta au Sénégal*, ed. and trans. P. de Cenival and Th. Monod (Paris: Larose, 1938), p. 87; Antonio Duarte Brasio (ed.), *Monumenta missionaria Africana*, Vol. 3, (Lisbon: Agência Geral do Ultramar, 1952–), p. 94.
24 Anquandah, *Ghana's Past*, pp. 68–9, 97 (map 8.2), 114 (map 9.1); O. Davies, *West Africa before the Europeans* (London: Methuen, 1967), pp. 283–90; David Kiyaga-Mulindwa, 'The earthworks of the Birim valley, southern Ghana' (PhD dissertation, The Johns Hopkins University, 1978).
25 Davies, *West Africa*, p. 290. Also C. T. Shaw, 'Archaeology in the Gold Coast', *Première conference internationale des africanists de l'ouest*, Vol. 1 (Dakar: 2 vols, 1951), pp. 494–5.
26 Müller, *Die africanische*, pp. 147–9; India Office Library and Records, London, E/3/26. Treaty between the East India Company and Afutu, dd 18 January 1660. Also Anquandah, *Ghana's Past*, p. 133.
27 Anquandah, *Ghana's Past*, pp. 68–9; Davies, *West Africa*, pp. 287–90; Kiyaga-Mulindwa, 'Earthworks'.
28 See Ray A. Kea, 'Administration and trade in the Akwamu empire, 1681–1730', in B. K. Swartz, Jr and Raymond E. Dumett (eds), *West African Cultural Dynamics: Archaeological and Historical Perspectives* (The Hague: Mouton, 1980), pp. 371–5.
29 For a general discussion of urbanisation in the region see Kea, *Settlements*, Chs. 1 and 2.
30 ibid., pp. 16–21, 43–50, 108–9, 173–6, 182–4.
31 ibid., pp. 101–4.
32 Brasio (ed.), *Monumenta missionaria*, Vol. 2, p. 109.
33 cf. V. de Bellefond, *Relation des costes d'Afrique appellées Guinée* (Paris, 1669), p. 218.
34 De Marees, *Beschryvinghe*, p. 254, where *anihumanifo* is transcribed as *animafo bebribi*: Müller, *Die africanische*, Ch. 18, where *adofo* appears as *odofu*.
35 *Bebri* = modern *bebree*, 'numerous', 'many'; *adofo* is derived from the following elements *do*, 'to increase', *dodow*, 'multitude', and *fo*, 'people'.
36 Müller, *Die africanische*, Ch. 17; Debien et al., 'Journal d'un voyage', p. 342.
37 Algemeen Rijksarchief, The Hague (Hereafter ARA), The First West Indies Company (Hereafter OWIC) 13. Register der contracten . . . met de naturellen in Guinea: entry dd 26 August 1640.
38 For a discussion of the poor relief system see Kea, *Settlements*, pp. 303–7.
39 '1516–1619. Escravos e Minas de Africa' in Luciano Cordeiro (ed.), *Viagens explorações e conquistas dos Portuguezes*, Vol. 6 (Lisbon: 1881, 6 vols), p. 24.
40 Kea, *Settlements*, pp. 46–7, 307–13.
41 Dierick Ruiters, *Toortse der Zee-Vaert, 1623*, ed. by S. P. L'Honoré Naber (The Hague: Nijhoff, 1913), p. 74. See also Kea, *Settlements*, Chs 6, 7 and 8, passim.
42 Kea, *Settlements*, p. 308.
43 De Marees, *Beschryvinghe*, pp. 112–13.
44 For an account of the arrest and imprisonment of this person see K. Ratelband (ed.), *Vijf dagregisters van het kasteel São Jorge da Mina (Elmina) aan de Gout Kust, 1645–1647* (The Hague: Nijhoff, 1953), pp. 183–4 entries for June 7 and 8 June 1646, p. 184 entry for 9 June 1646, pp. 184–5 entry for 11 June 1646, p. 192 entry for 27 June 1646. For slaves involved in theft see ibid., entry for 8 September 1646. For an account of crime in Elmina over a century-and-a-half later, see Ch. 1 above.
45 ARA Aanwinsten 1898 XXII. Journal gehouden by my Louys Dammaert: entry for 28 April 1653.
46 Ratelband (ed.), *Vijf dagregisters*, p. 55 entry for 12 June 1645.
47 Müller, *Die africanische*, Ch. 17.
48 See ARA, Collectie Leupe no. 743. Caerte des Lantschaps van de Gout Kust in Guinea, dd Mouri, 25 December 1629.

49 Barbot, *Description*, p. 188. See also Ratelband (ed.), *Vijf Dagregisters*, p. 206 entry for 28 July 1646; Koninlijk instituut voor taal-, land-en Volkenkunde, Leiden (Hereafter KITLV), Leiden msc. H65a. Report by Jan Valkenburgh, dd Elmina, September 1659, pp. 424–9. The sub-coastal route ran parallel to and ten to fifteen miles inland from the littoral.

50 The 1602 Portuguese source is a map by the Portuguese cartographer Luis Teixeira. For the map by Luis Teixeira see Armando Coresão and Avelino Teixeira da Mota (eds), *Portugaliae Monumenta Cartographica*, Vol. 3 (Lisbon: 1960, 6 vols), pp. 67–70. An earlier Portuguese description of the area in 1573 contains no references to bandit activities. See '1516–1619: Escravos e minas de Africa', in Cordeiro (ed.), *Viagens explorações*, Vol. 6, pp. 18–19. Little Nkassa was conquered by Axim probably between 1610 and 1620, and it remained a tributary, though rebellious, dependency well into the eighteenth century.

51 'Letter from van Bosch, dd. Shama, June 14, 1727' in van Dantzig (tr. and ed.), *The Dutch and the Guinea Coast*, p. 223. For the Eguafo-Yabiew war see de Marees, *Beschryvinghe*, p. 94.

52 Olfert Dapper, *Naukeurige beschrijvinge der africaensche gewesten van Egypten, Barbaryen, Libyen, Biledulgerid, Negroslant, Guinea, Ethiopien* (Amsterdam, 1676, 2nd edn), p. 109. Also KITLV, Leiden msc. H65a. Report by John Valkenburgh, dd September 1659, p. 421.

53 'Vertoog of deductie, 1656', in J. K. J. de Jonge (ed.), *De oorsprong van Neerlands bezittingen op de Kust van Guinea* (The Hague: 1871), pp. 55–6. See also the various treaties and agreements between agents of the Dutch West Indies Company and local political authorities in the area. ARA, Archief der nederlandsche bezittingen ter kust van Guinea, The Hague (Hereafter ANBKG), 222 passim.

54 ARA, OWIC 51. To the directors of the chartered West Indies Company from Pieter Reyersen, dd Mouri, 5 May 1636. For Akani trade in the seventeenth century see Kea, *Settlements*, Ch. 7.

55 Dapper, *Naukeurige beschrijvinge*, pp. 109–10. For Bighu and other urban centres north of the forest see Anquandah, *Ghana's Past*, Ch. 7 and pp. 92–100; and Kea, *Settlements*, Chs 1 and 2, passim.

56 Barbot, *Description*, p. 324.

57 Aban and his men also destroyed the Dutch trading station at Popo. ARA, Second West Indies Company, The Hague (Hereafter WIC), 1024. To the director-general from the factor, dd Ofra, Ardra, 29 December 1680. cf also ARA, WIC 1024. To the director-general from J. Bruyningh, dd Ofra, Ardra, 11 March 1682.

58 Letter from J. Carter, dd Whydah, 22 November 1686 in D. P. Henige (ed.), 'Correspondence from the Outforts to Cape Coast Castle 1681–1699. A Guide to Rawlinson c. 745–747 (unpublished)' (Madison, 1972).

59 See, e.g., ARA, ANBKG, 84. Journal of St. George d'Elmina: entry for 1 January 1717: letter from J. Snoek, dd Accra, 24 December 1716; Rigsarkivet, Copenhagen (Hereafter RA), Generaltoldkammerets Archiv. Sager til Guineiske Journaler 1780. 1794–97. Letter from Hager, dd Christiansborg, 14 March 1793. PRO, T70/1565 (2). To A. Dalzel from A. P. Biørn, dd Christiansborg, 6 May 1792 and 12 May 1792.

60 C. C. Reindorf, *History of the Gold Coast and Asante* (Basel: the author, 1895), p. 12; '1516–1619. Escravos e Minas de Africa', in Cordeiro (ed.), *Viagens explorações*, Vol. 6, p. 19; Brasio (ed.), *Monumenta missionaria*, Vol. 3, p. 111.

61 Oheneba Sakyi Djan, *The Sunlight Reference Almanac of the Gold Coast Colony and Its Dependencies* (Aburi, Gold Coast, 1936), pp. 131–2.

62 ibid., p. 132.

63 Ozanne, 'Notes on the early historic Archaeology of Accra', *Transactions of the Historical Society of Ghana*, vol. 6 (1962), p. 70.

64 E. Tilleman, *En liden enfoldig beretning om det landskab Guinea og dets beskaffenhed langs ved sø-kanten* (Copenhagen, 1697), pp. 113–14; J. Rask, *En kort og sanferdig rejsebeskrivelse til og fra Guinea* (Trondheim, 1754), pp. 151–2.

65 'Description of rivers, capes, places, and towns in Africa' in W. N. Sainsbury (ed.), *Calendar of State Papers. Colonial Series: America and West Indies, 1675–1676* (London, 1893), p. 329. A similar phenomenon has been noted in Ethiopia, see Chapter 6, below by Donald Crummey.

66 K. Marx, *Capital*, Vol. 3, ed. by F. Engels, trans. by S. Moore and E. Aveling. (Moscow, 1961), p. 331.

67 For the 1510 reference to *obirempon* brigandage see D. McCall, 'Who was the Xarife on the Costa da Mina in the sixteenth century?', paper presented at the Conference on Manding Studies (London, 1972), pp. 10–11. For the seventeenth century see Ratelband, *Vijf dagregisters*, passim; ARA, OWIC 81. Dagregister, passim; 'Correspondence', passim.

68 Ratelband, *Vijf dagregisters*, p. 263 entry for 17 November 1646.

69 See ibid., passim; ARA, OWIC 11. Letter from J. van der Well, dd on the frigate Eendracht in the São Tome road, 18 March 1647; M. Hemmersam, *Reise nach Guinea und Brasilien, 1639–1645*, ed. by S. P. L'Honoré Naber (The Hague: Nijhoff, 1930 repr.), pp. 69–70.

70 Müller, *Die africanische*, pp. 14–15. Müller does not refer to Ahen's promotion. This interpretation is based on documents of the Danish African Company which are to be found in the Danish national archives.

71 'WIC 124: Minutes of Council, Elmina: entry for 16 September 1707' in van Dantzig (tr. and ed.), *The Dutch and the Guinea Coast*, p. 132.

72 For a discussion of the military revolution see Kea, *Settlements*, Ch. 4.

73 For Agya Ananse see Kwame Y. Daaku, 'History in the oral traditions of the Akan', *Journal of the Folklore Institute*, vol. 8 (1971), p. 118. For Amputi Brafo see RA, Vestindisk-Guineisk Kompagni (Hereafter VGK). Døkumenter vedk. diverse K'cbenhavnske Retsager II, 1720–1728. Letter from F. Boye et al. to his Royal Majesty, King of Denmark and Norway, dd Christiansborg, 9 October 1711.

74 See van Dantzig (tr. and ed.), *The Dutch and the Guinea Coast*, Chs 2 and 3, passim.

75 See e.g., ARA, Aanwinsten 1902 XXVI no. 115. Dagh register gehouden by den oppercoopman President Pieter Nuyts: entries for 31 July 1706: letter from J. Landman, dd Axim, 28 July 1706; 16 August 1706: letter from J. Landman dd Axim 14 August 1706; 16 April 1707: letter from J. Landman dd Axim 11 April 1707.

76 Rømer, *Tilforladelig efterretning*, p. 111. See also RA, VGK. Breve og dokumenter fra Guinea 1746–1750, passim.

77 ARA, ANBKG 84. Journal of St George d'Elmina: entry for 1 January 1717: letter from J. Snoek, dd Accra, 24 December 1716; RA, Guineisk Kompagnie (Hereafter GK) Palaberbog 1776–77. 'Account of the 1730 Akwamu war by Athee, broker at Osu', J. Giønge, dd Christianborg, 11 December 1776.

78 J. Dupuis, *Journal of a Residence in Ashantee* (London: Frank Cass, 1966, 2nd edn), pp. xxxi–xxxii.

79 Danish trading company records of the second half of the eighteenth century contain numerous references to Krobo banditry. See also 'Biørns beretning 1788 om de danske forter og negerier' in *Archiv for Statistik, Politik og Huus-holdnings Videnskaber*, vol. 3, ed. F. Thaarup (Copenhagen, 1797–8), p. 210.

80 M. Johnson, 'Migrants' progress', *Bulletin of the Ghana Geographical Association*, vol. 9, no. 2 (1964), pp. 21–3.

81 Information on Krobo traders comes from Danish trading company records.

82 Johnson, 'Migrants' progress', pp. 8, 22.

83 ibid., p. 22.

84 See e.g., PRO, T70/978. James Fort Day Book: entries for 18 January 1770 and 12 August 1779.

85 Timothy F. Garrard, *Akan Weights and the Gold Trade* (London and New York: Longman, 1980), p. 40.

86 Rømer, *Tilforladelig efterretning*, pp. 104–5.

87 See 'WIC 124, Minutes of the Council Meetings, Elmina: entry for April 4, 1710' in van

Dantzig (tr. and ed.), *The Dutch and the Guinea Coast*, pp. 155–6.

88 Reindorf, *History*, p. 100; also pp. 274, 314.

89 ibid., pp. 149–50.

90 ibid., p. 276.

91 For one account of factionalism in the coastal and sub-coastal towns see Rømer, *Tilforladelig efterretning*, pp. 83, 271.

92 ibid., pp. 122, 126.

93 ibid., pp. 143–4. The Akwamu *sika den* did not own their means of production (firearms, ammunition, etc.). These were held by the king who employed the *sika den* to 'produce' slaves for sale at the European trading stations.

94 ibid., p. 125.

95 RA, GK. Palaberbog 1776–70, 1773–77. Report by Giønge, dd Christiansborg, 30 September 1775. See also RA, GK. Sekretprotokol, 1767–68, passim; RA, GK. Kystdokumenter 1768 I. To the directors from Quist, dd Christianborg, 20 April 1768. For a case concerning the seizure of one family at Abrekuma, a village about ten miles from the coast, see RA, GK. Palaberbog 1767–70, 1773–77. To Stadlander at Bereku from Aaerstrup, dd Christiansborg, 28 August 1776.

96 Antonio Gramsci, *Selections from Prison Notebooks*, ed. and trans. by Q. Hoare and G. N. Smith (New York, 1978), p. 55.

6 Banditry and resistance: noble and peasant in nineteenth-century Ethiopia[1]

DONALD CRUMMEY

Eric Hobsbawm has placed bandits firmly on the scholarly agenda.[2] In explaining how bandits excite and fascinate contemporary audiences and scholars, Hobsbawm develops the notion of the 'social bandit', someone who transcends, if only partially, his criminal calling through avenging the injustices suffered by the rural poor. Acting out the role created by the prototype Robin Hood, the social bandit 'takes from the rich and gives to the poor'. Such bandits are to be distinguished sharply from 'mere' criminal outlaws.[3] Hobsbawm's concept of the social bandit has come in for vigorous criticism, some of it damaging.[4] However, there have been at least two notable gaps in the discussions of banditry: examinations of the role of bandits within a feudal mode of production; and on the continent of Africa.[5] The latter gap exists in spite of Hobsbawm's drawing attention to it in the introduction to both editions of *Bandits*. Other contributors to this volume explore banditry in a number of parts of Africa. The introduction to the 1981 edition of *Bandits* highlights the Mesazghi brothers of Eritrea, thereby tapping the rich lode of Ethiopian banditry, a lode which provides material to help fill both feudal and African gaps.

Banditry was widespread in nineteenth- and twentieth-century Ethiopia. Several Ethiopian sources even catch the ethos of social banditry. However, while close examination of the evidence reveals many instances of outlawry, of armed defiance, and of lives based on plunder, played out within a self-conscious class context, it reveals few links of a progressive or socially redeeming nature between the peasants and the institution of banditry. The Ethiopian ruling classes dominated the institution of banditry, and moulded it to their own ends. They used banditry as a tool for career mobility. Banditry may also have articulated vicariously with peasant needs, thereby helping to reproduce a social order whose tensions it so eloquently bespoke. However, the Ethiopian ruling classes turned the social challenge of banditry into a form of political competition for office. In so doing they strikingly revealed the criminal undercurrents of all forms of state power.

The Amharic for banditry is *sheftenat*, a word derived from the root verb *shaffata* (to rebel), whence also derives the term *sheftā* (bandit or rebel). Tasammā Habta Mikǎ'ēl provides a nice definition. A *sheftā*, he writes, is 'one who stirs up trouble, while taking to the forest or the bush, departing from the king, the government, rule [*gezāt*], instituted order [*ser'at*], and the law.[6] The term *sheftā* occurs commonly in both European and Ethiopian sources for the nineteenth century, as does the verb in Amharic. Modern Ethiopian usage is ambiguous. While *sheftā* is used to extol certain historical figures, it is equally used to dismiss

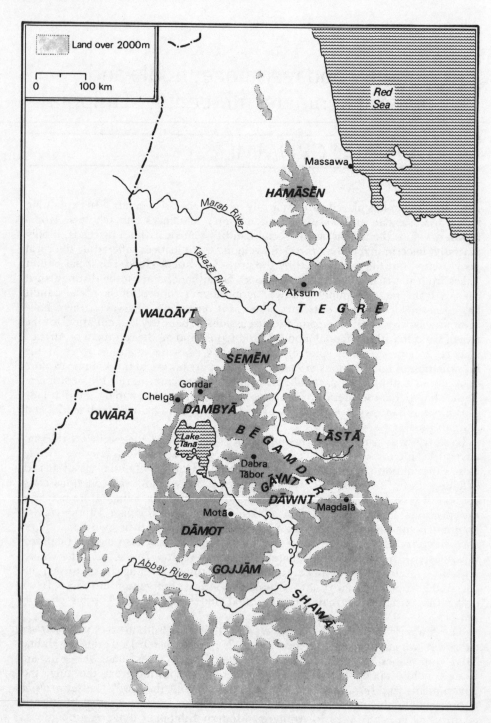

Map 3: *Northern highland Ethiopia*

such bitter enemies of the current regime as the Eritrean liberation movement.[7] The prevalence of banditry in the Ethiopian highlands during the twentieth century has contributed the term *shifta* to the English spoken in north-east Africa, where, from the Sudan to Kenya, it refers to any armed band at odds with the state.

The meaning of *sheftenat* deserves closer attention. The term covers both rebellion and banditry, in a sense which the English term 'outlawry' catches rather well. Both rebel and robber place themselves against established law and order, the rebel doing so in a more political way, the robber in a criminal one. Both rebellion and banditry entail the violent defiance of authority. However, although as early as the 1840s we have evidence that *sheftenat* meant both rebellion and banditry, and can therefore dispense with Levine's belief that the latter meaning is a recent 'vulgarisation',[8] nonetheless the relative incidence of the political and criminal poles of this institution did vary through the nineteenth and twentieth centuries.

As rebellion, *sheftenat* flourished down to the last quarter of the nineteenth century, the period being one of what Hobsbawm calls 'feudal anarchy'.[9] During most of that century the term *sheftā* covered many prominent noblemen as well as venal highwaymen, and Ethiopia's first two modern monarchs, Tēwodros and Yohannes both came to the throne through *sheftenat*. Such a combination of rebellion and establishment, of criminality and law, raises rather fundamental questions. It renders overt and pervasive the violent basis of all class societies, and challenges the meaning of law.[10] In the earlier twentieth century, as a loose central government slowly put itself in place, *sheftenat* narrowed towards banditry and increasingly became a provincial matter.[11] The Italian invasion of 1935 and the following six years of occupation brought a flowering of *sheftenat* and its merging with the patriotic resistance to Fascist rule.[12] The association of rebellion, banditry, and resistance to oppressive rule lasted beyond the ousting of the Italians in 1941. Bandits played important roles in the 1943 Wayānē revolt in Tegrē.[13] However, with the elaboration of Haile Sellassie's autocracy, *sheftenat* became more and more criminal and less and less political, as the emperor drew the rural nobility into the outer reaches of his administration.

In nineteenth-century Christian Ethiopia peasants were generally drawn into banditry and rebellion as followers of individual nobles. Banditry and rebellion were thus without class meaning for them. Instead, these forms of political behaviour were as oppressive to the peasants as was the normal ordering of society. Indeed, next to merchants, peasants were the main victims of banditry. As a result the peasants as a *class* resisted directly, and class warfare sometimes took open and literal forms. This means that if we are interested in peasant resistance and responses to oppression we must not confine our investigation to *sheftenat*, but must also look at this question directly.

The following survey of rebellion and resistance embodies three dyads: chronological, thematic, and perceptual. Chronologically, it embraces two periods. The first runs from roughly the 1810s to the 1850s and is marked by decentralisation. The second, from the 1850s to the early 1870s, features two attempts to revive monarchical authority. Thematically, each section first deals with rebellion and banditry and then looks at peasant behaviour. Finally the discussion embodies two viewpoints. Contemporary Ethiopian literature forms the main point of departure, with European travellers being drawn on for supplementary insights.

The chronicler Takla Iyasus characterises our first period in terms of class conflicts. He tells how God punished Ethiopia, 'by causing the nobles and peasants to destroy each other.'[14] He further explains how 'any son of a deputy [*meslanē*], or even of a village chief [*cheqā*], would kill an ox of his uncle and enter the forest.'[15] This pithy account captures the pervasiveness of rebellion, reinforces our image of an institution dominated by rulers, and introduces us to a common Ethiopian metaphor for *sheftenat*, 'entering the forest'. However, the verb *shaffata* and its derivatives appear infrequently in indigenous sources for the period down to the 1850s, since those sources are in the classical Ge'ez, not Amharic. However, *sheftā* does appear as a loanword in both Ge'ez and Tegreññā texts referring to this period, testimony to its aptness. Tegreññā traditions of the early twentieth century use *sheftenat*;[16] and several figures in the Ge'ez chronicles are called *sheftā*.[17] To deepen our understanding of the meaning of the term during this period we must turn to European observers.

A few European sources actually use the term *sheftā*, while others use bandit or brigand. During the 1810s Pearce wrote of his contemporary Goju, that he had turned *sheftā*: 'such being the name of a prince or powerful man, who maintains himself by plundering from place to place.'[18] Thirty years later Plowden used the term frequently, confirming that *sheftā* could be noblemen engaged in political struggle.[19] His *sheftā* inhabited wastelands: either the forested river valleys, with their seasonal malaria; or the high, cold barrens.[20] One striking passage referring to late 1844 portrays the *sheftā* as common bandit. Recounting a trip into central Gojjām via the mountains lying south of Motā, Plowden wrote: 'this range of dreary fens, where no cultivation is to be found above a certain level, with extensive caverns in almost impregnable situations, offers a shelter where the *schifta* . . ., setting his chief at defiance, subsists by plunder . . .'[21] His testimony is seconded by Beke, who scarcely two years before was warned off these same mountains: 'on account of some sort of Rob Roy people who infest the neighbourhood levying contributions [? original reads "consultation"?] on passengers, and on the farmers below [? original reads "former belo ad" ?].'[22]

Other European sources reinforce the image of *sheftenat* as both noble rebellion and banditry through use of such terms as brigand or bandit. Several authors note how every significant change of rule brought forth plundering bands of robbers or rebels.[23] The sources also confirm the endemic character of *sheftenat*, its prevalence on the inhospitable margins of the settled highlands, its political slant, and its dominance by the nobility. References cover most part of the Christian highlands, from Hāmāsēn in Eritrea in the north to Gojjām in the south.[24] Particularly notorious haunts were the Marab valley on the Eritrean–Tegrē border and the south-eastern corner of Lake Tānā, where the Abay River marked the boundary between Gojjām and Bēgamder;[25] while one Bēgamder village is said to have been entirely devoted to brigandage.[26] Shawā, lying south-east of the main territory inhabited by the Amharic- and Tegreññā-speaking peoples, enjoyed a reputation amongst European travellers as the securest part of Christian Ethiopia.[27] Here a firm monarchical rule equally held in sway both nobles and brigands. Nonetheless, even this haven had its pocket of endemic banditry.[28]

For these earlier years our most detailed accounts of banditry and rebellion refer to Tegrē in the 1830s and 1840s. They underline the political context of *sheftenat* and its dominion by the nobility. *Dajjāch* Webē, ruler of Tegrē in these years, was generally firm, but several times his hold slipped. In 1841 he ventured south of the Takazzē on a major expedition and disorder rose in his wake. It intensified in 1842 on the arrival of news of his defeat.[29] Webē re-established himself, but four years later the Tegrē brigands flourished again as Webē was besieged by his nominal overlord. Two accounts of the political disorders in Tegrē raise considerations of social class through their references to the peasants. The French travellers Ferret and Galinier witnessed how in 1841 'a crowd of malcontents and evildoers came forth from all sides, to beat the country and ransom the peasants.' Their servants fell foul of two young noblemen who robbed them of their cash and mules. Seeking redress they turned to *Dajāzmāch* Demtu, father of one of the culprits, and himself 'a bandit'.[30] Six years later the missionary de Jacobis complained that his stations near Addigrāt were continually exposed to 'incursions and to the pillagings of brigands'. Twice, however, the nearby villages had spontaneously defended the Catholics, so deep was their hostility to brigandage.[31] Shake-downs and plundering were not normally characteristic of Webē's rule, as they were of some of his contemporaries.[32] At the end of the decade, Plowden could write of Webē's lands that he believed that in them 'the roads were as safe as in any part of England'.[33] *Sheftenat* in Christian Ethiopia in the 1830s and 1840s was thus, to some degree, a function of political stability.

Our only encounter with the peasants has seen them take up arms against bandits. In other incidents they also directly resisted oppression. Ethiopian sources offer several models of oppressive rule. Takla Iyasus singles out *Rās* Hāylu who 'uprooted the inherited lands of the poor and established soldiers called *gwoynā* over them as rulers.'[34] Another chronicler pointed the finger at *Rās* Māryē, recording that 'in his days there was no one who did not have his house sacked and his wife raped.'[35] Plundering was one of the commonest forms of oppression. Although as Pearce pointed out this was a characteristic *sheftā* activity,[36] it was also much used by established, legitimate rulers. In 1842 Beke noted how *Dajjāch* Gwoshu was about to establish his protégé in office 'by burning down the villages and ravaging the country!'[37] But plundering was only one of several points of friction between the soldiers and peasants, another common one being the peacetime practice of billeting.[38] And relations between soldiers and peasants were but an expression of a broad set of feudal social relations, which were perceived as oppressive in a number of ways. For example, the custom of *dergo* required the peasants to house and feed anyone travelling in the name of their ruler. Petty officials sometimes abused this custom, creating peasant resentment.[39] Corvée obligations were ill-defined and a recurrent source of friction.[40] Peasants also contested both the amount of particular taxes and the rights of certain rulers to levy them.[41]

The peasants coped with these problems in different ways. Some must have taken to the bush. On other occasions they vigorously protested their rights, as Beke witnessed: 'Never was there such a row, all bawling out at one time, at the utmost pitch of their voices.'[42] Peasants also used sullenness, evasion, and outright refusal.[43] And they took up arms. Plowden noted that some districts always resisted billeting *en masse* and that 'the troops dare not venture'.[44] Pearce describes how some villagers responded to excessive demands from billeted troops by giving 'a general alarm, and rais[ing] the neighbouring villages to their assistance, and many lives are lost on both sides.'[45] The people might depose, and even kill unpopular governors.[46] And we have several instances of popular revolt in the 1840s, but few details allowing us to explore them.[47] Finally, warfare affected the peasants directly.

Warfare was the main means of conducting politics in our period. Yet major battles were infrequent. Warfare tended to be characterised by pursuit and evasion. Pursuing armies attacked their enemy's civilian base hoping to force submission. This made plundering into a political weapon. But it was also a daily necessity for armies on the march in alien territory.[48] Ethiopian supply systems were elementary and it was the most a ruler could do to restrain his army from plundering his own people. Enemy subjects were fair game.

Ethiopian peasants adopted various responses to plundering. They located their villages in obscure and defensible positions.[49] They developed secret underground storage pits for their grain.[50] They used the protective asylum of churches to harbour grain and cattle.[51] Evacuating both people and cattle, they fled from advancing armies.[52] And sometimes they fought those armies with ferocity. Foraging and plundering were dangerous activities for the soldiers, since armed peasants regularly harassed them.[53] When battles loomed, peasants would gather at strategic points in the vicinity to prey on the losers.[54] They might wreak their vengeance through slaughter, or through holding their victims for ransom. Not even the most exalted reaches of society were free from this threat. After the battle of Dabra Tābor in 1842, peasants harassed the fleeing *Rās* Ali, and a peasant band captured the nominal ruler, *Asē* Yohannes.[55]

Broad and vigorous as were these peasant responses, our sources describe none as *sheftenat*. *Sheftenat* they apply at this stage solely to the activities of the gentry and nobility. Labels aside, the peasants of the first half of the nineteenth century were active moulders of their own destinies, and were far from the passive ciphers which many modern Ethiopians have held them to have been. As peasant problems mounted in the third quarter of the century, so too did their response.

Our second period runs from the later 1840s to about 1870. The most famous *sheftā* in Ethiopian history dominated these years. He began as *Lej* Kāssā and ended as Emperor Tēwodros. His active years were turbulent; and their record contains many references to *sheftā*. Some observers have tried to make a social bandit of Tēwodros. Such a search for a progressive leader has drawn attention away from the people, while a close examination of the 1850s and 1860s reveals a vigorous continuation of popular resistance. By the later 1860s this resistance was directed against the emperor himself. Any understanding of the events of this period, and of the role of banditry and rebellion within them, must begin with Kāssā-Tēwodros.

Lej Kāssā entered the chronicles dramatically in September 1846 as a *sheftā*.[56] For the following six years he was active on the western borderlands, now enjoying official appointment, now rebelling.[57] In 1852 he made his last great act of rebellion, one which led him to the throne in 1855. His coronation marked the beginnings of the restoration of the monarchy. However, his career up to 1852 illustrates the *sheftā* as both highwayman brigand, and as a rebellious aspirant nobleman. Although it was his activities in these years which gave rise to the aura of social banditry, the aura dissipates as we examine it.

Kāssā was nobly born: the half-brother through his father of *Dajāzmāch* Kenfu, a marcher lord of the 1820s and 1830s.[58] By about the age of ten he was of sufficient standing to be used by others in staking political claims.[59] He attended a monastic school. He bore the honorific '*lej*'. As a child he had suffered poverty, and his mother had for a time been a street pedlar. This fact, well known in his lifetime, has misled many into believing him of humble birth. He did not come from the people nor did he exercise his power in the people's cause.

Kāssā's father was from the border province of Qwārā. This connection gave him free rein to plunder in the western lowlands beyond the border, at the same time as he cultivated his claim to legitimate rule in Qwārā. His main victims were either traders or the lowland cultivators. Culturally and religiously these victims were either Muslim or the adherents of indigenous African faiths known pejoratively to the highlanders as *Shānqellā*.[60] Several times he raided the Shānqellā for slaves;[61] other Shānqellā he reduced to the serflike status of *gabbār*.[62] He seems generally to have avoided jeopardising his appeal to the peasants of Qwārā, although he did pillage soldiers fleeing from his sometime overlord.[63] Kāssā entered a bandit subculture, marrying the daughter of another *sheftā*, and acting in solidarity with still others, notably a Sudanese called Edris.

Yet the image of social bandit is not easily shaken from Kāssā. In part it rests on an incident from the early years before 1845–6. His chronicler tells of the ruin suffered by Qwārā and of Kāssā's response. He gave the peasants 'lots of money' to buy digging tools, and he and his men set to work clearing the forest and turning over land, with the result that once again Qwārā prospered.[64] The incident is unusual, but not unique. Pearce reports how *Rās* Walda Sellāsē directed his men to clearing land in 1813; and Rosen writes of the conspicuous involvement of *Rās* Mengesha Seyum in development projects and manual labour on the eve of the Ethiopian Revolution of 1974.[65] Neither Walda Sellāsē nor Mengesha is a candidate for social-bandit status. Incidents aside, Kāssā's reputation as social bandit seems more the product of the deep and widespread need within class societies to make heroes of the defiant, a need which lies at the root of the phenomenon discussed by Hobsbawm. Sahle Sellassie has given us a recent statement of the social-bandit myth in his novel about Kāssā entitled *Warrior King*. But an earlier statement had already been sketched out by an anonymous chronicler, who claimed: 'The peasants of Qwārā loved him . . . all the peasants became *sheftā* with *Lej* Kāssā, and young men joined him from as far away as Dambyā and Chelgā';[66] and went on to describe how the peasants co-operated with him. But the source is a late one, and provides us, at this point, with evidence of what some people in the following generation were making out of Kāssā, and not of what he had actually been. Our best information makes Kāssā out as either a highwayman or the model of what he hoped to become – a ruler – for one of the main functions of *sheftenat* in our period was to serve the career aspirations of the nobility.

Our earliest accounts of Kāssā's activities in the late 1840s and early 1850s emphasise his unruliness. It seems that perhaps a year or so before his dramatic entry into the chronicles in 1846 he had already received appointment as governor of Qwārā.[67] So his career henceforth was no longer that of a simple bandit (if it ever had been), but of a recognised noble. At the beginning of 1847 he 'sacked many villages in Dambeyā'. A few weeks later just to the east he carried off the cattle of the peasants. During the rains of that year his troops proved very unruly. Towards the end of that year he sacked grain from an area he was supposed to be guarding for his overlord.[68] A year later his troops brought 'famine' to the town of Gondar by spending the rains 'sacking the villages and the city and devouring what they found.'[69] And in 1850, we learn, he returned to the borderland to raid the Shānqellā for slaves.[70] The mutation from brigand to noble was slow, but was perhaps helped by the fact that both roles called for repression.

By 1848 Kāssā the rebel had become the object of rebellion. He dealt with it severely. Around Easter he punished some *sheftā* by hacking their hands and feet off. It foreshadowed the repression of his rule. Yet neither at that time nor later did severity prove a lasting deterrent. In September 1851 some of his own soldiers became *sheftā*.[71] Then, in the following year, he embarked on the final stage of his own *sheftenat* with the act of defiance which was to carry him to the throne in three years.[72] There was little in his record to that point to suggest either revolutionary or reformer. Instead we find a career bounded by the role of *sheftā*: brigand and noble. We hear overtones of social banditry, but only by standing far back. As we approach they vanish. In short we find a noble using the institution of *sheftenat* in pursuit of his own ambitions. The scantier evidence relating to Yohannes IV supports similar claims.

Although Yohannes has never been viewed as revolutionary or reformer, yet he too rose as a *sheftā*, and the stages of his career strikingly parallel those of Tēwodros. He first lived in the bush with a few followers and later vied openly for power. He also shared his predecessor's given name and honorific. Around 1866, a chronicler tells us: '*Lej* Kāssā rebelled [*sheftaw*] and entered the forest.'[73] His brother and mother were imprisoned for keeping a *sheftā* in the forest, but he stayed there until Tegrē fell into the hands of a rival to the emperor. During the next few years Kāssā made and broke alliances with abandon, until he reached the throne in 1871. Another source fleshes out this rough picture. It portrays a quarrelsome man unable to get along with his siblings.[74] He eventually rebelled and entered the forest. But this was no ordinary *sheftā*:

> Though he had left the order of the King and became an outlaw [*sheftā*], he never took unjustly the belongings of the poor or their cattle or sheep or goats, nor did his soldiers take unjustly any bread from the houses of the farmers or ripe grain from the plant, or provisions from the needy on the way.[75]

According to the chronicler such measures were unnecessary since tribute flowed in voluntarily. This whitewashed image contrasts not only with what we know of the grubby world of the *sheftā*, but also with Kāssā-Yohannes' later úse of force. It also contrasts with the same author's explanation of Kāssā's subsequently adopted praise-name, *Abbā* Bazbez: 'Taker of booty day after day.'[76] *Abbā* Bazbez, the plunderer, is a fitting name of an Ethiopian ruler of the nineteenth century, for such rulers vaunted themselves on their ability to separate the producers from the fruits of their labour. It is also a fitting name for a *sheftā* emperor.

Both Kāssā, *sheftā* that they had been, proved strict rulers. The case of Tēwodros raises intriguing paradoxes. By about 1854 the crown was within his reach: the law he aspired to uphold implicit in his earlier outlawry. Unruly as his own troops had been, it is said that one of his major reforms was to have affected the army, which he sought to discipline and restrain from plundering.[77] This reform failed. It seems that the restraint on plundering was directed to clothes, cattle and 'serious things', not to grain and the like;[78] and from the beginning Tēwodros used plundering as a political weapon. Moreover, he never succeeded adequately in disciplining his troops. The missionary Waldmeier, sympathetic and close to the emperor, offers a picture of relations between Tēwodros's soldiers and people indistinguishable from the picture we have already seen: 'the Abyssinian soldiers, when passing through the land, put their horses, mules, and donkeys, into a prosperous harvest-field until they have ruined the crop; and often I have seen the owners imploring the intruders, with tears, to remove their animals, but they would neither move nor pity them.'[79]

Yet law, its maintenance and renewal, was central to Tēwodros' vision of his imperial role. We have vivid descriptions of how he dispensed justice, and popular paintings attest his lasting reputation in this regard.[80] As early as 1853 he severely suppressed robbers and thieves, 'hacking off their hands and feet', in order to make secure a road which he had previously himself threatened.[81] An alleged coronation edict states the interest in law and order:

> Ploughman, plough! Noblemen and ladies, holders of inherited lands and holders of fiefs! I confirm to you what your father held! Trader, Trade! If you have been plundered, pile up stones in the place, and come to me. Thieves and brigands! Give over! Submit! Woe to his limbs, he who robs![82]

And two sources summarise the establishment of Tēwodros' rule in terms of law and order:

> serpents deserted the roads, and brigands left for monasteries . . .
> . . . he began to settle down while the *shefta* fled. At this time, because of the strictness of his punishment there were no brigands [*wambadē*] to be found in the forest, nor thieves [*lēbā*] in the villages.[83]

Our accounts are not without irony. Consider the story about the village of brigands. One of the king's edicts guaranteed to everyone the legacy of their fathers. The brigands responded with an appeal to be allowed to continue thieving. Tēwodros asked them to come in full force to hear his response. On the appointed day he had them slaughtered.[84] His fierce justice is one point of the story. But another is the ambiguous nature of law and order: if law rests on custom, and if custom includes brigandage and plunder, then law must embrace its opposite.

Tēwodros failed to suppress robbery and brigandage. He failed because of the central paradox in *sheftenat* as a means of achieving political power: it is a means dialectically opposed to its end, and a tool which can destroy as quickly as it creates. Kāssā-Yohannes transcended this paradox. Kāssā-Tēwodros did not. *Sheftenat* plagued his reign and brought it down. The Ethiopian sources give full play to the emperor's rebellious noble opponents, sometimes organising their material into chapters headed *sheftoch* (plural of *shefta*).[85] Yet the actions of these men were so overtly political, and their social standing as nobles so unequivocal, that an examination of them would add little to our understanding of *sheftenat*. Here we need simply note that rebels emerged against Tēwodros in the year of his coronation, and waxed in strength pretty steadily until at last they held most of the country in their grasp.

Under the umbrella of noble rebellion, robbery again flourished. The rebellious nobles themselves acted as brigands, and the English consul Cameron airily dismissed one of them, claiming that his movement had 'no political significance', and that its leader was 'a mere robber', the son of 'a noted brigand'.[86] An Ethiopian writer caught well how oppressive the disorders of the later years of Tēwodros' rule became. In Tegrē, he claimed, 'robbers and bandits' (*gamānāñā* and *sheftā*) had precluded the movement not only of merchants but also of the peasants in certain areas. 'Whoever has strength,' he lamented, 'has no other occupation save robbery.'[87] Reinforcing our image of *sheftenat* as an activity dominated by the Ethiopian nobility, these accounts lead us away from, not towards, the people. If we are to learn about popular action, we must do as we did during the preceding period, look directly at the peasants themselves. They did not allow these turbulent years to pass by without making their mark.

Two themes run through the story of peasant resistance during the 1850s and 1860s. On the one hand it exhibits some new features. We find our first *sheftā* of peasant origins at this time. We have rather more evidence of direct, political action by the peasants. And we find the intensity of resistance at a new peak in the 1860s, the verb *shaffata* (to rebel) being used to describe collective actions by the peasants. But on the other hand the general patterns of resistance show continuity with the preceding decades. It is those general patterns at which we will first look.

The peasants' relations with the state entered a new phase with the coronation of Tēwodros in 1855. The emperor promised to end the oppression caused by disorder, by the rebellious nobility, and by bandits. So far it was in the interests of the peasants to support him. Yet Tēwodros ruled much as previous emperors had ruled, through extracting surplus from the peasants, and for this reason it was in their interests to resist him. Moreover, his prospects of success were uncertain, and the peasants had continually to take into account the demands of the rebellious nobles. Consequently they acted in many ways.

Sometimes they took direct action. Takla Iyasus tells us that when the rebel, Tadlā Gwālu, tried to make certain appointments, the people rejected them, one of them because he threatened the rights of the humble people. This case of resistance failed, since Tadlā imposed his men by force.[88] Peasants also used evasion. Several travellers recorded the refusal of villagers to pay *dergo*, the customary support levied by the state for its itinerant officials and guests. Dufton saw the acts as rebellious.[89] And peasants used flight. One case around 1860 involved Dāwnt, from which people fled with their cattle over a very wide area.[90] Finally, soldiers continued to be a major burden for the peasants, even during times of peace. The sources of Tēwodros' reign speak of billeting (*tasari*). Billeting must have occurred earlier, but it is particularly mentioned in the 1850s and 1860s. Dufton caught the problem when he noted that billeted soldiers 'have a great licence in robbery of cattle, mules, and articles of food.'[91] An Ethiopian source tells us how the introduction of billeting into Gāynt led to the death of a peasant. The king's response to an appeal for justice was terse: 'Soldiers eat; peasants provide.'[92] Thereafter billeting was frequent, and peasants had continually to be on their guard lest its hardships worsen. Sometimes they had their rewards. In August 1851 the people of Gondar scored a signal victory over the restless troops of Kāssā-Tēwodros.[93]

The reign of Tēwodros provides us with a number of examples of the peasants taking up arms for political reasons, examples generally lacking in the preceding period. Down to about 1860 the trend of such action was against the king's opponents. But from the mid-1860s it turned against the king himself. One colourful episode is attributed to the era of Tēwodros's rise. In 1853 following his victory at Ayshāl, he passed through Dagā Dāmot in Gojjām. The peasants left off ploughing. They rigged up weapons by fastening their wives' hairpins (*wasfē*) to sticks, and went out *en masse* to plunder the soldiers of their horses, mules, and money, ambushing and killing many. They were defeated by a ruse. Knowing their avidity for horses and mules Dajāzmāch Kāssā sent them a large number, and, whilst they quarrelled over the spoils, he had them surrounded and killed by musketeers.[94] One point of the story is peasant thick-headedness. A very similar point is made by a different author about our only *sheftā* of peasant origin, Gembāro Kāssā. Sometime in the 1860s Gembāro led an attack on Galmo, the governor of Qwārā, who was feasting when news of the attack arrived. Galmo abandoned his feast to the rebellious peasant, but returned to overwhelm his enemies while they glutted themselves.[95] Ethiopian authors shared widespread prejudices in attributing greediness and stupidity to peasants. However, the stories they tell also imply a degree of audacity and resolve on the part of peasants which we were unable to glimpse directly during the preceding decades. That the peasants of Dāga Dāmot were not alone in attacking a victorious imperial army appears from a brief claim that in 1874 the peasants of Walqāyt and Sagadē had defeated an army of Yohannes, capturing 500 muskets and seizing thirty imperial dignitaries.[96]

Down to about 1861 Tēwodros enjoyed some active support from the peasants. The period from 1858 to 1861 offers several incidents in which they acted against Tēwodros's rebellious nobles, whom the sources call *sheftā*. In 1858 Lāstā peasants caught some *sheftā*, who included *Wāgshum* Gabra Madhen and *Behtwaddad* Berru, and delivered them to the king.[97] The following year peasants of Dambeyā showed similar loyalty when an armed band of them successfully opposed the invasion of rebels from Tegrē. After an extended encounter, we are told, over three hundred rebels were left on the battlefield, and 318 were captured.[98] Early in 1860 local villagers on the edge of the Eritrean plateau helped frustrate attempts by a French emissary to establish contact with the rebel Negusē.[99] And the following year, Negusē having spent the interval in flight, yet other peasants in Tegrē seized him and turned him over to the king.[100]

The 1860s were a terrible decade in Christian Ethiopia. Natural disaster brought famine. Political oppression mounted. Rebellion strengthened, and the imperial army marched back and forth. Punishment rose until wholesale massacres took place. As the rebels eluded him the emperor took out his frustration on the peasants, whom he pillaged and murdered. They resisted.

1866 was a crucial year. Already by January an Amharic source used the verb *shaffata* to announce that several large districts had rebelled and passed out of control.[101] The same source, towards the end of the following year, informed his correspondent that: 'All the peasants of Bēgamder have rebelled', again using *shaffata*.[102] Other sources, of which Blanc is the most coherent, provide a few details of the rebellion of the Bēgamder peasants. It began in 1866 when the peasants resisted demands for higher imposts.[103] The king turned to plundering. But the peasants were now aroused: 'Joined by the deserters they fought in their own way, cut off stragglers, sent their families to distant provinces, and for miles around Debra Tābor ceased cultivating the soil.'[104]

In March 1867 Tēwodros set off to raid Qwārātā, a commercial town, but the peasants prevented him. For a year they had been on their guard against him, and signalled his moves by beacon fires. He turned to local sorties in the Dabrā Tābor region, but met armed replies. And when he ventured into nearby Foggarā, the peasants again foiled him, by burying their grain and removing their cattle and dependents as far away as Lāstā. At a village called Godu the peasants 'fell suddenly on the soldiers, killed a good many and took several prisoners.'[105]

In June the king turned on Balassā. By now the peasants had become subversive indeed. Walda Māryām tells of how the people of Bēgamder began to slip off at night, gather on the hilltops, and hurl insults at the king. When he came to Balassā the peasants took up slings to oppose his firearms. Tēwodros uprooted their crops, but they retaliated by destroying their own houses to deny him their use as shelter or firewood.[106] Peasant defiance to his very face outraged the king. It was a rare act indeed. Yet the circle of rebellion widened. Desertions from the royal army reached high levels, and as the contingents reached home 'they called upon the peasantry to arm.' In answer, royal appointees were deposed and despoiled.[107] The rebellion of Bēgamder and its peasants did not end until October 1867 when Tēwodros finally left the province, toiling eastward to Maqdalā, there to meet a British expeditionary force of whose landing he had just learned. In April 1868, faced with a new overwhelming foe, he died by his own hand.

After the trauma of Tēwodros's rule a degree of order returned as the long-standing rebels each consolidated their local position. But a new imperial order did not fully emerge until 1878 when the ex-*sheftā* Yohannes IV forced vassaldom on his rival, Menilek. The central provinces, Bēgamder included, enjoyed some respite. It was not prolonged. Toward the end of the 1880s Mahdist forces from the Sudan raided Bēgamder, and when the province reappears to European vision in the early twentieth century, it had become a backwater. No longer plagued by large-scale rebellion, no longer the prey of a ruler whose demands outstripped the bounds of conventional forms of surplus extraction, it was now the haunt of bandits.[108] Bēgamder's *shefta* in the twentieth century were more modest in their social origins and ambitions than their celebrated predecessor Tēwodros. Largely drawn from the gentry and lower nobility they flourished under the indolent rule of *Rās* Gugsā during the 1920s and joined with relish the patriotic struggle of the later 1930s. Not even the clergy lay beyond the reach of this social institution. Stories of the bandit priest, *Qēs* Yelmā circulated in Dambeyā in the late 1960s. They contained no socially redeeming episodes, the cleric exemplifying the predatory nature of banditry, a nature carried over easily into noble rebellion.

Sheftā in Amharic catches the wealth of the English bandit, and goes beyond it fundamentally to question the nature of law, order, and rebellion. The nineteenth-century record reveals an institution of *sheftenat* which embraced two poles, robbery and rebellion. The ruling class dominated this institution, as it dominated society at large. Such a judgment is possible in spite of the narrow class vision of the sources, none of which wholly excludes the people from its purview. We have no reason to believe that the meaning of *sheftenat* has changed in the last century and a half. But we have noted that the balance between criminal banditry and noble rebellion did shift through time, and that the political activities of Haile Sellassie I marginalised the *sheftās*, minimised the incidence of *sheftenat*, and removed *sheftenat* as a device for coming to power. Banditry and rebellion ceased to be options of interest to the Ethiopian ruling classes. However, the same forces which ended the era of noble rebellion paved the way towards the era of popular revolution. The deceptive stability of Haile Sellassie's regime during the 1950s and 1960s laid the foundations not of rebellion, or *sheftenat*, but of *abyot*, revolution. However, in parting let us return to the richness of Amharic. At times we catch glimpses of social banditry, in our one peasant *sheftā*, in the occasional attempts by *sheftā* to address peasant concerns, and in our acts of peasant rebellion. In some sense the very notion of social banditry is embedded in the primary sense of the verb *shaffata*, to rebel, as Camus may be read to have said.[109] But it is equally true that the meaning of *sheftenat* cuts in the opposite direction. If many Ethiopian leaders of the nineteenth century rose through *sheftenat*, their example reminds us of the criminal origins of all state power. Law demands the support of the state and cannot avoid emphasising its opposite, outlawry, nor embracing it.

NOTES

1 Support for the research embodied in this chapter came from the National Endowment for the Humanities and the Research Board of the Graduate College of

the University of Illinois. In addition to the Urbana symposium on Rebellion and Social Protest, this paper was also presented at the Seventh International Conference on Ethiopian Studies, Lund University, May 1982; and a lengthier version has appeared in the conference proceedings: S. Rubenson (ed.), *Proceedings of the Seventh International Conference of Ethiopian Studies* (Addis Ababa, Uppsala, East Lansing, 1984), pp. 263–77.

2 E. J. Hobsbawm, *Bandits* (Harmondsworth: Penguin, 1st edn, 1972; 2nd edn, 1981). See also his *Primitive Rebels* (Manchester: Manchester University Press, 1959).

3 Hobsbawm, *Bandits*, pp. 38–9 (pagination same in both editions).

4 ibid., postscript to 2nd edn.

5 ibid., 1st edn, p. 14; 2nd edn, pp. 11–15. But see here, F. Braudel, *The Mediterranean and the Mediterranean World in the Age of Philip II* (London: Collins, 1972–3, 2 vols) as indexed. Also A. Isaacman, 'Social banditry in Zimbabwe (Rhodesia) and Mozambique, 1894–1907, an expression of early peasant protest', *Journal of Southern African Studies*, vol. 4, no. 1 (1977), pp. 1–30; and E. J. Keller, 'A twentieth century model: the Mau Mau transformation from social banditry to social rebellion,' *Kenya Historical Review*, vol. 1, no. 2 (1973), pp. 189–205.

6 *Kasātē Berhān Tasammā*, p. 334. See also Ignazio Guidi, *Vocabolario Amarico-Italiano* (Rome: Istituto per l'Oriente, 1901); and Joseph Baeteman,. *Dictionnaire Amarigna-Français* (Dire-Daoua, 1929).

7 *Africa News*, 8 February 1982; *New York Times*, 21 February 1982.

8 D. N. Levine, *Wax and Gold. Tradition and Innovation in Ethiopian Culture* (Chicago: University of Chicago Press, 1965), pp. 243–4. This same passage well captures the complexities of peasant attitudes towards authority and banditry.

9 Hobsbawm, *Bandits*, p. 95.

10 See D. Shulman, 'On South Indian bandits and kings,' *Indian Economic and Social History Review*, vol. 17, no. 3 (1979), pp. 283–306, for stimulating thoughts on the essential connections between banditry and monarchy in South India.

11 See below. Chapter 7.

12 See Hobsbawm, *Bandits*, for the parallel role of the *haiduks*.

13 Gebru Tareke, 'Rural protest in Ethiopia, 1941–1970: a study of three rebellions' (unpublished PhD dissertation Syracuse University, 1977); *idem*. 'Peasant resistance in Ethiopia: the case of *Weyane*', *Journal of African History*, vol. 25, no. 1 (1984), pp. 77–92.

14 Institute of Ethiopian Studies, Addis Ababa, Ms. 254, *Alaqā* Takla Iyasus, 'YaGojjām Tārik,' f. 47.

15 ibid.

16 J. Kolmodin, *Traditions de Tsazzega et Hazzega* (Upsala: Appelberg, n.d.), Chs 92, 230, and 270 on pp. 66, 162, and 190 of the translation.

17 Stadtbibliothek, Frankfurt-am-Main, Ms. Rüppell Ib, f. 191v; translation published by C. Conti Rossini, 'Nuovi documenti per la storia d'Abissinia nel secolo XIX,' *Atti della Accademia Nazionale dei Lincei*, S8, vol. 2 (1947), p. 374; Bibliothèque Nationale, Paris, *Mss. Ethiopien-d'Abbadie 167*, f. 169r; translated by Conti Rossini, 'Nuovi documenti', p. 388.

18 N. Pearce, *The Life and Adventures of Nathaniel Pearce, Written by Himself, During a Residence in Abyssinia, From the Years 1810 to 1819*, Vol. 2 (London: Colburn & Bentley, 1831, 2 vols), p. 50.

19 W. C. Plowden, *Travels in Abyssinia and the Galla Country with an Account of a Mission to Ras Ali in 1848* (London: Longmans, Green, 1868), pp. 239, 277.

20 ibid., pp. 432, 199. See also R. Austen's chapter (Chapter 4, above) for these as classic bandit haunts.

21 ibid., p. 201.

22 British Museum, *Additional Manuscripts* 30251, C. T. Beke, 'A Diary written during a journey in Abyssinia in the years 1840, 1841, 1842, and 1843', p. 410.

23 C. W. Isenberg and J. L. Krapf, *Journals of C. W. Isenberg and J. L. Krapf Detailing their*

Proceedings in the Kingdom of Shoa and Journeys in Other Parts of Abyssinia in the Years 1839, 1840, 1841, and 1842 (London: Seeley, 1843), p. 439; Samuel Gobat, *Journal of a Three Years' Residence in Abyssinia* (London: Seeley, 2nd edn, 1847), pp. 302–3; and A. von Katte, *Reise in Abyssinien im Jahre 1836* (Stuttgart: Cottasche, 1838), p. 85.

24 Katte, *Reise*, p. 32; Gobat, *Journal*, pp. 70, 239–40; T. Lefebvre, *Voyage en Abyssinie éxécuté pendant les années 1839, 1840, 1841, 1842, 1843*, Vol. 3 (Paris, 6 vols, 1845–48), p. 26; Isenberg and Krapf, *Journals*, pp. 343, 440–41, 456, 459–60, 463–4; J. Bell, 'Extract from a journal of travels in Abyssinia in the years 1840–41–42', *Miscellanea Aegyptiaca*, vol. 1 (1842), pp. 10–13; Plowden, *Travels*, p. 167 and as cited in notes 19 and 21, above; Edmond Combes and Maurice Tamisier, *Voyage en Abyssinie, dans le Pays des Galla, de Choa and d'Ifat*, Vol. 4 (Paris: Desessart, 1838, 4 vols), p. 57; BM, *Add. Mss.* 30251, Beke, 'Diary', p. 40; Arn. d'Abbadie, *Douze Ans de Séjour dans la Haute-Ethiopie* (Vatican City, 2 vols. 1980), Vol. 1, p. 29; Vol. 2, p. 226 *inter alia.*

25 For the Eritrean–Tegrē borderland and the Marab valley see: Katte, *Reise*, pp. 50–51, 53; W. P. E. S. Rüppell, *Reise in Abyssinien*, Vol. 1 (Frankfurt-am-Main, 2 vols, 1838–40), p. 324; J. L. Krapf, *Travels, Researches, and Missionary Labours during an Eighteen Years' Residence in Eastern Africa* (London: Trübner, 1860), p. 443; Isenberg and Krapf, *Journals*, p. 507; and Plowden, *Travels*, p. 370. For Lake Tānā: Plowden, *Travels*, pp. 195, 251, 252.

26 Combes and Tamisier, *Voyage*, Vol. 4, p. 57.

27 BM, *Add. Mss.* 30250A, Beke, 'Diary', pp. 155–6. See also Isenberg and Krapf, *Journals*, p. 349.

28 BM *Add. Mss.* 30250A, Beke, 'Diary', p. 228; W. Cornwallis Harris, *The Highlands of Aethiopia*, Vol. 2 (London: Longman, 3 vols, 1844), pp. 321–2.

29 Lefebvre, *Voyage*, Vol. 1, pp. 319, 352, Vol. 2, pp. 19, 91–6; Pierre Victor Ferret and Joseph Germain Galinier, *Voyage en Abyssinie dans les provinces du Tigré, du Samen et de l'Amhara*, Vol. 2 (Paris: Paulin, 3 vols, 1847), pp. 134–44, 475 *et seq*; Maison Lazariste, Paris, 'Giornale di Giustino de Jacobis', Vol. 2, p. 78; BM, *Add. Mss.* 30252, Beke, 'Diary', pp. 246–7.

30 Ferret and Galinier, *Voyage*, Vol. 2, pp. 135, 138–44, 149.

31 Letter of 10 July 1847 in *Annales de la Congrégation de la Mission*, vol. 13 (1848), pp. 75–6.

32 Gobat, *Journal*, pp. 71, 102–3; Plowden, *Travels*, p. 402.

33 Plowden, *Travels*, p. 378.

34 Takla Iyasus, 'YaGojjām Tārik', f. 48, also f. 50; *yadahewn rest naqqalannā gwoynā mālat gult gazhē watādar sarābat.*

35 Conti Rossini, 'Nuovi documenti', p. 367.

36 See above, quotation referenced to note 18.

37 BM, *Add. Mss.* 30251, Beke, 'Diary', p. 25.

38 See, for example: Pearce, *Life and Adventures*, Vol. 1, pp. 183–4; BM, *Add. Mss.* 30251, Beke, 'Diary', p. 27; Ferret and Galinier, *Voyage*, Vol. 2, p. 338; Kolmodin, *Traditions*, pp. 20–21.

39 Pearce, *Life and Adventures*, Vol. 1, pp. 164–5; Lefebvre, *Voyage*, Vol. 2, p. 10; BM, *Add. Mss.* 30251, Beke, 'Diary', p. 184.

40 BM, *Add. Mss.* 30252, Beke, 'Diary', p. 35; d'Abbadie, *Douze Ans*, Vol. 1, p. 33.

41 Combes and Tamisier, *Voyage*, Vol. 4, p. 168; Rüppell, *Reise*, Vol. 1, pp. 321–2; Pearce, *Life and Adventures*, Vol. 2, pp. 130–31; Kolmodin, *Traditions*, p. 120, Ch 179; d'Abbadie, *Douze ans*, Vol. 1, p. 391.

42 BM, *Add. Mss.* 30251, Beke, 'Diary', p. 89; Combes and Tamisier, *Voyage*, Vol. 1, p. 215.

43 BM, *Add. Mss.* 30251, Beke, 'Diary', pp. 439–40.

44 Plowden, *Travels*, p. 135.

45 Pearce, *Life and Adventures*, Vol. 1, pp. 183–4. See also Mansfield Parkyns, *Life in Abyssinia* (London: Murray, 2nd edn, 1868), p. 128; and d'Abbadie, *Douze Ans*, Vol. 1, p. 96, Vol. 2, p. 160.

46 Pearce, *Life and Adventures*, Vol. 1, p. 101; BM, *Add. Mss.* 30252, Beke, 'Diary', p. 128; Isenberg and Krapf, *Journals*, p. 366.

47 Beke records a revolt by the Shināshā people of the Blue Nile lowlands: BM, *Add. Mss.* 30251, 'Diary', pp. 555–7; and Beke, BM, *Add. Mss.* 30248, 'Journal of travels in Southern Abyssinia', pp. 47–50. Several sources refer to revolts by different districts in Eritrea during the rule of *Daj*. Webē: for Sarayē see BM, *Add. Mss.* 30252, Beke, 'Diary', p. 337; for Qohāyn see Lefebvre, *Voyage*, Vol. 2, p. 44 and C. Conti Rossini, 'Vicende dell'Etiopia e delle missioni cattoliche ai tempi di Ras Ali, Deggiac Ubie e Re Teodoro, secondo un documento abissino', *Rendiconti della Reale Accademia dei Lincei*, S5, vol. 25 (1916), p. 466.

48 For the general significance of plundering, see d'Abbadie, *Douze Ans* Vol. 1, pp. 302, 371–2.

49 Lefebvre, *Voyage*, Vol. 3, p. 221.

50 Pearce, *Life and Adventures*, Vol. 1, pp. 206–7; d'Abbadie, *Douze Ans*, Vol. 2, p. 176; L. Fusella (ed.), *Yātē Tewodros Tārik* (Rome, 1959), p. 4; trans. as 'La cronaca dell' Imperatore Teodoro II di Etiopia in un manoscritto amarico', *Annali dell' Instituto Universitario Orientale di Napoli*, ns, vol. 6 (1957).

51 Combes and Tamisier, *Voyage*, Vol. 3, pp. 284–5; BM, *Add. Mss.* 30251, Beke, 'Diary', p. 11; Lefebvre, *Voyage*, Vol. 2, pp. 290–91; Conti Rossini, 'Vicende', p. 486.

52 Rüppell, *Reise*, Vol. 2, pp. 266–7; Bell, 'Extract', p. 19; Isenberg and Krapf, *Journals*, pp. 352–5, 487.

53 Pearce, *Life and Adventures*, Vol. 1, pp. 205–9, 245; Plowden, *Travels*, p. 134; de Jacobis, 'Giornale', Vol. 2, p. 114; Gobat, *Journal*, pp. 264–5; Combes and Tamisier, *Voyage*, Vol. 1, p. 219; d'Abbadie, *Douze Ans*, Vol. 2, pp. 176–7.

54 By far our most graphic source here is Arn. d'Abbadie: *Douze Ans*, Vol. 1, pp. 299, 459, 475; Vol. 2, p. 175.

55 For *Rās* Ali see Lefebvre, *Voyage*, Vol. 1, p. 361; for Yohannes see Vatican Library, 'Carte d'Abbadie', Cartone VII, p. 675. See also d'Abbadie, *Douze Ans*, Vol. 2, p. 130.

56 Conti Rossini, 'Nuovi documenti', pp. 388–9. The earliest reference to Kāssā as *sheftā* which I have found in a European source is: de Jacobis, 'Giornale', Vol. 4, pp. 125–6, April 1849.

57 S. Rubenson, *King of Kings. Tewodros of Ethiopia* (Addis Ababa: Haile Sellassie University/Nairobi: Oxford University Press, 1966).

58 ibid., Ch 15, especially pp. 26–7.

59 Conti Rossini, 'Nuovi documenti', p. 374; Rubenson, *King of Kings*, p. 29.

60 Zanab, *YaTēwodros Tārik* (Princeton, 1902), published by E. Littman, pp. 7 ff; translation by M. M. Moreno, 'La cronaca di re Teodoro attribuita al dabtarā "Zanab"', *Rassegna di Studi Etiopici*, Vol. 2 (1942), pp. 143–80; Henry A. Stern, *Wanderings Amongst the Falashas in Abyssinia* (London: Wertheim, 1862), p. 65; Henry Dufton, *Narrative of a Journey through Abyssinia in 1862-3* (London: Chapman & Hall, 1867, 2nd edn), p. 122; and Henry Blanc, *A Narrative of Captivity in Abyssinia* (London: Smith, Elder, 1868), p. 2.

61 Zanab, *Tārik*, pp. 8, 11.

62 ibid., pp. 9–10.

63 ibid., p. 10.

64 ibid., p. 11; Rubenson, *King of Kings*, p. 36.

65 Pearce, *Life and Adventures*, Vol. 1, pp. 125–6; C. Rosen, 'The governor-general of Tigre province: structure and antistructure', in H. G. Marcus (ed.), *Proceedings of the First United States Conference on Ethiopian Studies* (E. Lansing: Michigan State University, African Studies Centre, 1974), pp. 171–83.

66 Sahle Sellassie, *Warrior King* (London: Heinemann, 1974); Fusella, *Yātē Tēwodros*, p. 3.

67 Zanab, *Tārik*, p. 11.

68 Conti Rossini, 'Nuovi documenti', pp. 390–91, 394, 395. See also Vatican Library, 'Carte d'Abbadie', Cartone XIII, pp. 284, 475–77.

69 Conti Rossini, 'Nuovi documenti', p. 396: original text at Bibliothèque Nationale,

Paris, *Mss. Eth.-d'Abb.* 240, f. 243.

70 Conti Rossini, 'Nuovi documenti', p. 400.

71 ibid., pp. 391, 396, 400, 403.

72 Walda Māryām, *Chronique de Théodore II Roi des Roi d'Ethiopie (1853–1868)* (Paris: Librairie orientale & américaine, 1904), published with a French translation by C. Mondon-Vidailhet, references to the Amharic text. Walda Māryām's chronicle opens with Kāssā's *sheftenat*, or act of rebellion, in 1852; but p. 2 refers to an earlier episode.

73 Lamlam, 'History of *Asē* Takla Giyorgis and *Asē* Yohannes', Bibliothèque Nationale, Paris, Mss. Eth. 259, f. 15. For a general study, see: Zewde Gabre-Sellassie, *Yohannes IV of Ethiopia. A Political Biography* (Oxford: Clarendon Press, 1975). See also Chapter 13, below, for an example of yet another Ethiopian ruler whose career began with *sheftenat*.

74 Bairu Tafla, *A Chronicle of Emperor Yohannes IV (1872–89)* (Wiesbaden, 1977), pp. 34–5, 40–41.

75 ibid., pp. 58–9.

76 ibid., pp. 64–5.

77 Rubenson, *King of Kings*.

78 Zanab, *Tārik*, p. 19.

79 Theophilus Waldmeier, *The Autobiography of Theophilus Waldmeier, Missionary* (London: Partridge, 1886), p. 15.

80 ibid., pp. 16–17; Walda Māryām, *Chronique*, pp. 26–8/32–5; Dufton, *Narrative*, p. 139.

81 E. Fenzl, 'Bericht über die von Herrn Dr. Constantin Reitz . . . auf seiner Reise von Chartum nach Gondàr in Abyssinien gesammelten geographisch-statischen Notizien', *Denkschriften der Kaiserlichen Akademie der Wissenschaften Mathematisch-Naturwissenschaftliche Classe*, vol. 8 (Wien, 1854), p. 14.

82 Fusella, *Yātē Tēwodros*, p. 12; see also an expanded version on p. 23; and Dufton, *Narrative*, p. 136.

83 Fusella, *Yātē Tēwondros*, p. 23; Walda Māryām, *Chronique*, p. 19. See also Dufton, *Narrative*, p. 137; Stern, *Wanderings*, pp. 108–9, 128, 129; and Takla Iyasus, 'YaGojjām Tārik', f. 66.

84 Walda Māryām, *Chronique*, p. 19; Stern, *Wanderings*, pp. 129–30.

85 Fusella, *Yātē Tēwondros*, p. 16. But see also Walda Māryām, *Chronique*, pp. 22–3 and following; Zanab, *Tārik*, p. 33; Takla Iyasus, 'YaGojjām Tārik', f. 68; and C. Conti Rossini, 'Epistolario del debterà Aseggachègn di Uadlà', *Rendiconti della Reale Accademia dei Lincei*, S. 6, vol. 1 (1925), pp. 449–90; translated by L. Fusella, 'Le lettere del debtarā Assaggākhañ', *Rassegna di Studi Etiopici*, vol. 12 (1953), pp. 80–95; vol. 13 (1954), pp. 20–30. This last source I cite hereafter as Assaggākhañ, 'Lettere', with primary reference to the Italian translation.

86 Public Record Office, Kew, FO 401 1, No. 811; Cameron to Russell, Aksum, 1 January 1863. The individual in question was none other than Kāssā Goljā, for whom, and many others of his ilk, see R. A. Caulk, 'Bad men of the borders: Shum and Shefta in Northern Ethiopia in the 19th century', *International Journal of African Historical Studies*, vol. 17, no. 2 (1984), pp. 201–27.

87 Assaggākhañ, 'Lettere', No. III, dated Halay, 1858 Eth. Cal. (September 1865–September 1866 AD). Assaggākhañ wrote from the affected areas.

88 Takla Iyasus, 'YaGojjām Tārik', ff. 62v–63r.

89 H. Steudner, 'Herrn Dr. Steudner's Bericht-Reise von Adoa nach Gondar. Dec. 26, 1861–January, 1862.' *Zeitschrift für Allgemeine Erdkunde*, NF, vol. 15 (Berlin, 1863), 58; Dufton, *Narrative*, pp. 191–2.

90 Staatsarchiv des Kantons Basel-Stadt, *C. F. Spittler Privat-Archiv 653*, D3/1, Letter of Bender from Chachaho, n.d.

91 Dufton, *Narrative*, p. 61.

92 Zanab, *Tārik*, p. 32; *watādar belā, bālāgar abelā*. Stern records the same incident and judgment: *Wanderings*, p. 132. Walda Māryām once mentions *tasari*: *Chronique*, p. 18.

93 Conti Rossini, 'Nuovi documenti', p. 403.

94 Takla Iyasus, 'YaGojjām Tārik', f. 61v.
95 Walda Māryām, *Chronique*, p. 30.
96 Assaggākhañ, 'Lettere', p. 30.
97 S. B.-S., *Spittler 653*, D3/5, Kienzlen to Flad, Suramba, 3 June 1858. Blanc, a rather later source, attributes the capture of Gabra Madhen to his successor Tafari, but his account may not wholly contradict Kienzlen: Blanc, *Captivity*, p. 263.
98 S. B.-S., *C. F. Spittler 653*, D3/9, Saalmüller to Schneller, Magdala, 12 October 1859.
99 Letters to Biancheri, Massawa, 25 January 1860, and Delmonte, Halay, 14 February 1860: *Annales de la Congrégation de la Mission*, vol. 26 (1861), pp. 46–50, 64–74.
100 Dufton, *Narrative*, p. 153.
101 Assaggākhañ, 'Lettere', p. 85. The Amharic has *shaffatu wayānē hona*.
102 Assaggākhañ, 'Lettere', Amharic text, p. 467. A later Amharic account combines the two notions here and informs us: 'All the Bēgamder peasants became rebels and bandits', *yaBēgamder bālāgar hulu wayānē sheftā hona*, Fusella, *Yātē Tēwondros*, p. 38.
103 Blanc, *Captivity*, p. 315.
104 ibid., p. 316.
105 ibid., pp. 316–22. See also India Office Records, London, *Abyssinia Original Correspondence*, Vol. I, p. 315, Blanc to Merewether, Magdallah, 30 April 1867.
106 Walda Māryām, *Chronique*, p. 34; Blanc, *Captivity*, pp. 229–30.
107 India Office Records, *Abyssinia Original Correspondence*, Vol. I, p. 389, Blanc to Merewether, Magdala, 18 June 1867.
108 See Chapter 7, below.
109 Albert Camus, *The Rebel* (New York: Knopf, 1954).

7 Social mobility and dissident elites in Northern Ethiopia: the role of bandits, 1900–69

TIMOTHY FERNYHOUGH

Throughout history banditry has manifested itself in diverse forms. Most commonly it has appeared as a form of peasant resistance to class oppression and injustice and above all as a challenge to the ruling classes and the state, the upholders of the social order. Fernand Braudel has suggested that banditry was a latent form of the *jacquerie*, the product of poverty, hunger, overpopulation and urbanisation.[1] The bandit's very existence was an act of resistance. Braudel posits that brigandage first became widespread with the rise of rural and urban poverty in Europe in the early sixteenth century. The poor and oppressed, with little recourse to protest against their humiliation, and few opportunities to alleviate their condition, turned increasingly to modes of resistance which the ruling elites defined as criminal and dangerous. Brigands and vagrants proliferated in early modern Europe and many were carried across to the New World.[2] Yet Braudel has stressed that sixteenth-century outlaws only rarely exhibited or expressed class consciousness. While he argues for a close relationship between brigandage and social distress Braudel cautions against pointing to a single place or period for the first appearance of bandits. The extant literature on medieval and early modern brigandage in Languedoc, Provence and Catalonia also indicates that bandits may exist at various times and in very different societies.[3]

Eric Hobsbawm has extended Braudel's analysis by suggesting that a few 'social' bandits may consciously exact class vengeance and redistribute wealth to the poor.[4] Though he notes the universal aspect of brigandage, Hobsbawm asserts that increases in the incidence of social banditry correspond to changes in classic modes of production, especially when rural society is forced to adapt to the needs of agrarian capitalism and the modern economy. In consequence, the search for brigands has focused on Latin America, Australia and Europe in the nineteenth and early twentieth centuries.[5] However, Hobsbawm believes that social banditry may also increase when a communally based society is replaced by one founded on class and state. Hence Braudel and Hobsbawm differ not about the existence of banditry in pre-capitalist societies, but on the conditions in which it flourishes. Braudel ties the rise of brigandage generally to social misery, Hobsbawm specifically to fundamental changes in the economy. If their views are reconcilable, at least for the sixteenth-century Mediterranean, the activities of brigands in parts of the ancient world, in medieval and early modern Europe and in colonial Latin America testify to the resilience and longevity of this kind of crime and suggest that widespread banditry may occur in all societies which have developed beyond kinship organisation to the stage at which classes appear. Certainly, by the twentieth century there was very little that was new about banditry. While it may have increased, in some cases dramatically, in societies in rapid transition from precapitalist to capitalist modes of agrarian production, it clearly could exist where such development had not occurred.[6]

If scant attention has been paid to pre-nineteenth century banditry elsewhere, only a few glances have been directed to Africa in any period. Those few have been concerned mainly with banditry in African societies where colonial rule and the introduction of capitalist relations of production have gone hand in hand.[7] Yet banditry existed in parts of Africa before the arrival of Europeans. Ethiopia presents an interesting case study because, with the brief interlude of the Italian occupation between 1936 and 1941, it was neither colonised nor incorporated to any significant degree into the world economy. Before the mid-1950s only very limited capital penetration and urbanisation had occurred. The domestic and international market was small and land was still held and worked in time-honoured ways.[8] Land, labour and agrarian wealth had not been transformed into commodities. Ethiopian society more closely resembled a feudal order than any other. Only under the impact of economic changes which began in the mid-1950s was a small modern agricultural sector created by foreign companies and a few Ethiopian entrepreneurs. By and large though, capital penetration occurred in the immediate vicinity of new arterial roads and existing market towns. Only in Italian Eritrea was the intrusion of capital more advanced and methods of landholding touched by colonial administration.

In this essentially pre-capitalist society outlaws played a very definite role. Bandits were not merely disruptive elements reflecting wider societal tensions, but they also served as agents by which such strains were reduced. The prevalence of banditry in nineteenth- and early twentieth-century Ethiopia was matched only by its integration within society. Banditry, or *sheftenat*, encompassed different types of behaviour, ranging from highway robbery, extortion and kidnapping, to political and personal conflict, often so protracted that it became indistinguishable from blood feuding. However, categorisation along these lines tends to obscure the fact that most Ethiopian bandits engaged in more than one, if not all, of these activities and that *sheftenat* in its diverse forms had an important social function. Banditry in Ethiopia provided a means by which low men might rise to positions of authority, while less-successful members of the nobility sought in brigandage a way of asserting their power. The absence of rigid social stratification within Ethiopian society has been noted elsewhere.[9] I would suggest that a factor contributing to social mobility in both directions between the peasantry and nobility was the existence of brigandage.

By offering alternative avenues for mobility within the social order brigandage also provided a way in which class conflict could be diffused. In seeking to advance their careers few Ethiopian bandits encouraged their followers to challenge the nobles and gentry who ruled their strikingly hierarchical society. Rather, as Anton Blok's studies of the Sicilian *mafia* suggest, brigands tended to seek official or noble protection.[10] In Ethiopia *sheftenat* reinforced vertical bonds within a society based on patron–client relations. For the Abyssinian nobility and for a few talented peasants, banditry was an important step on the road to office and political power. Amongst the ruling elite banditry was largely a matter of political power. For dissident nobles, especially for noble offspring for whom there was insufficient land, the achievement of office could be used to exert political influence, and ultimately to acquire land. The nature of landholding in Ethiopia meant that once an official position had been attained, the holder's effectiveness at litigation allowed him to activate a far wider range of land-use rights than would otherwise have been possible. For individual peasants *sheftenat* also offered a chance of real social mobility, a move away from landlessness or impoverishment.

Those who became bandits were usually nobles, officials and peasants intent on raising their status, gaining access to higher office, and increasing their power. Others have explored the social context of banditry in nineteenth-century Ethiopia. Their research emphasises the frequency with which prominent noblemen engaged in *sheftenat*, and discloses how two emperors of the period, Tēwodros (r. 1856–68) and Yohannes (r. 1871–89), came to the throne through banditry and rebellion.[11] Crummey has also argued that Ethiopian banditry changed over time. As the centralisation of government under emperors Menilek and Haile Sellassie eroded noble pre-eminence in society, it also weakened noble control of *sheftenat* as a road to political power.[12] The link between rebellion and banditry carried over into the twentieth century, as the participation of *sheftā* in successive rebellions demonstrates. Nevertheless, the incorporation of the gentry into the administrative hierarchy under Haile Sellassie further diminished the scope of political *sheftenat* and highlighted its criminal element.

After 1900 ambivalent and noncommittal attitudes to *sheftenat* remained common. In Tegrē *sheftenat* was widely accepted as a way of bringing grievances before the authorities.[13] In Bēgamder bandits who tried to rob the compound of a station of the London Society for the Propagation of Christianity amongst the Jews in Jandā in the late 1920s had previously been supplied with food by the villagers. After the raid one wounded *sheftā* was cared for by his local relatives and by the missionary Baur. The American traveller, Hermann Norden, explained the attitude of the local people by suggesting that these villagers had little to fear from brigands. They were often friends and relatives and were primarily a menace to merchants and caravans, not to the inhabitants of the village.[14]

Sheftā were not only drawn from all social ranks, but they appeared with regional variations in most parts of Ethiopia during this period. They flourished in frontier regions and in the more inaccessible parts of the highland plateaux. This was where there was often a confusion of authority and overlordship which *sheftā* exploited. After the 1890s, devastating raids by Christian highlanders on southern Ethiopia produced communities of brigands along the border with the British East Africa Protectorate.[15] The Sudanese border was also a region of lawlessness and illegal enterprise. An illicit market thrived here for guns, slaves, and all kinds of contraband traffic, including alcohol.[16] Banditry prevailed throughout the north-western borderlands, especially in Semēn, Bēgamder, and the subprovince of Walqāyt.[17] At the turn of the century the most famous bandit in this region was Kidāna Māryām of Nuqāra who operated between Qallābāt and the Setit River.[18] During the Mahdiya, Kidāna Māryām was leader of a large *sheftā* band in 1898 had over 2000 followers with 500 rifles. In 1900 he was particularly active in the border province of Qwārā where his band established a base.[19] Kidāna Māryām and his lieutenants, Barihun and Hāgos, carried out raids within Ethiopia as well as incursions into the Sudan and Eritrea. Barihun and Hāgos continued to operate after Kidāna Māryām surrendered to *Rās* Waldē, a personal friend, and was removed to Addis Ababa where he was eventually promoted. In 1902 Hāgos was still active in the rough, deserted, terrain east of Qallābāt.[20] In 1906 his activities were eclipsed by another powerful frontier bandit, Hāyla

Māryām, who led a slave raid exceptionally deep into the Sudan, on Abū Qulud, in April of that year which involved 350 riflemen. The raiders killed soldiers of the Sudanese battalion and carried off many prisoners and cattle.[21]

Their position on the periphery of the Ethiopian state had its advantages for *sheftā*, who readily exploited government crises in Addis Ababa. Political instability in the capital was reflected in the borderlands. In 1916, during fighting between *Rās* Walda Giyorgis and the recently deposed Emperor, *Lej* Iyasu, *sheftā* bands immediately appeared throughout the north-west and especially in the frontier regions of Semēn, Walqāyt and Wagarā.[22] The incidence of banditry on the Ethiopian–Sudanese border remained at a high level through the 1920s. In the middle years of the decade four sons of Barihun who had been deprived of their rights by the governor of Bēgamder, *Rās* Gugsā, ostensibly on the grounds of their illegitimacy, also turned to banditry and assembled a force of over 500 men who devastated large areas near the Sudan border, with the exception of their home village of Chelgā.[23] In February 1929, at least one band of 200 *sheftā* was operating between Lake Tānā and the Sudan border. The road between Qallābāt and Dāngilā in Gojjām was especially vulnerable to *sheftā* attacks. Bandits closed the road completely in 1930 while most provincial governors were in Addis Ababa attending the coronation of Haile Sellassie.[24] Subsequently travel on this route was subject to the depredations of a notorious bandit called Sheguti (pistol).[25] Further south, the western frontier of Wallaggā was similarly afflicted.[26]

Banditry thrived in these frontier regions because they were not vital either to the Italians, Ethiopians or the British. The Bēgamder–Sudan border was remote from power centres in Addis Ababa and Khartoum and held little of interest to them. Neither the absolutist regime in Ethiopia nor the young colonial administration in the Sudan possessed the resources to hold their peripheries under firm control. If the British hand rested lightly on their side of the border, Menilek was largely indifferent to the turn of events in an area where he exercised only nominal sovereignty.[27] Moreover, the western borderlands were not merely an area in which the authority of one set of rulers was exchanged for another, for this was a zone of profound transitions. From the high Ethiopian plateaux to the desert plains of the Sudan, changes in physical ecology underlined fundamental cultural, religious and demographic differences. For three centuries, even while Gondar was imperial capital, the western periphery was more than a political frontier. It represented an area in which rulers were loath to act and would do so only under the gravest provocation. The swift response on both sides of the border to the Abū Qulud raid is a case in point.[28]

In the first decade of this century confusion over the actual line of the border with the Sudan accentuated the problem. Until the May 1902 treaty between Britain and Ethiopia the boundary remained uncertain, while the Ethiopian–Italian treaties relating to the Eritrean frontier, signed in July 1900 and May 1902 were widely ignored, even in principle, by local officials. Demarcation remained a complex issue. On the Ethiopian side of the border constant changes of provincial governors aggravated the situation. Walqāyt, for example, had at least six rulers and four overlords between 1898 and August 1919.[29] Actual lines of authority were yet more difficult to define. Ethiopian involvement in Nuqāra operated on at least three higher levels, from the central government in Addis Ababa, through the provincial governor in Bēgamder, down to the sub-provincial administration of Walqāyt. In these conditions *sheftā* proliferated, exploiting the ineffectiveness of government and confusion over the frontier to assert their power. Periodically they attempted to take over whole provinces and districts.

In contrast to the conflicts of authority which facilitated the rise of brigandage along national borders, provincial boundaries within the Ethiopian state were so rigidly observed that *sheftā* again benefited. None of the authorities of one province, or sub-province, had the right to enter another without permission. These restrictions applied whether the authorities were in pursuit of bandits, or for any other reason. Thus outlaw bands were numerous around Laka Tānā, whose shore was divided between the provincial governorships of Gojjām and Bēgamder and the municipal administration of Gondar. Brigands used these borders to get beyond the reach of one set of authorities and into the jurisdiction of another. Banditry was endemic in the unhealthy marshland of the south-western shore of the lake, where also lived the pacific, pastoral Zallān. The villages of Qunzelā and Dangal Bar had an especially sinister reputation for brigandage.[30] Nearby swamps and reedbeds offered protection to *Grāzmāch* Babil of Maqal, a *sheftā* who controlled brigandage on the western side of the lake in the 1930s.[31] Whenever government forces were sent against him, he simply took to his *tankwā* (reed-boats) and disappeared, accompanied by a large force of riflemen.

The high incidence of banditry for which the Lake Tānā basin was notorious by the early twentieth century extended along its eastern shore. Hundreds of bandits were active here in the 1930s.[32] The road up the escarpment to the highlands of Libo and Dabra Tābor was especially unsafe. Brigands were so audacious in the rough terrain here that they often came into the town of Ifāg, which dominated the coastal plain and which was one of the most important market centres in Bēgamder. In Lake Tānā itself secluded islands like Bēt Manzo, Angāro and Bergedā Māryām made ideal retreats for lawbreakers.[33] Very early, Jelu island, named after a powerful bandit chieftain, had been strongly fortified. A rock wall surrounded the whole island at the water's edge. In the nineteenth century a *sheftā* called Berru Haylu took refuge on the island of Mesrahā, from which he was dislodged by the Emperor Tēwodros and a fleet of reed-boats.[34] In the 1930s a rebellious *fitāwrāri* emulated Berru by fleeing to Bēt Manzo to escape the wrath of *Rās* Gugsā of Bēgamder. Such *sheftā* often expected religious recluses who lived on the islands to supply them with food from their scarce stores. With no means of resistance, many unarmed monks were driven to find protection in larger monasteries whose flocks and granaries outlaws tended to respect. Occasionally monks served as intermediaries between *sheftā* and villagers, arranging sufficient local tribute to keep the robbers away.[35]

Degrees of inaccessibility were naturally relative to the forces which the authorities could send against *sheftā*. Nevertheless, in this period bandits were to be found in mountainous areas, where lines of communication were tenuous and slow, and where travel was a cumbrous affair. Brigands with an intimate knowledge of the terrain could easily prey on unwary travellers. Peasants and disaffected nobles were very active in the highlands of Tegrē. Gebru Tareke has suggested that, in the unusually favourable conditions for brigandage in the 1940s, there were as many as 5000 armed *sheftā* in north-eastern Tegrē alone.[36] In these conditions not only travellers but whole settlements were vulnerable. A favourite tactic of *sheftā* in Tegrē in the late 1940s was to set fire to villages and loot them in the ensuing panic. After 1946 this kind of brigandage spilled over into Eritrea as many Tegreññā-speaking bandits crossed the river Marab.[37]

While mountainous terrain offered protection to *sheftā*, it could not alone provide the wherewithal to live. In Semēn and Walqāyt in the 1920s brigands were often villagers pursuing peaceful occupations, who only raided strangers passing through their districts or seized them for ransom.[38] In Dagā Dāmot in central Gojjām, *sheftā* also tended their *rest* lands, secure in the knowledge that they could slip off into forested ravines if the undermanned police force should venture out of Faras Bēt, the local government seat.[39] All *sheftā* relied not only on secure lines of retreat, but also on access to major trade routes. Thus banditry flourished along natural divides, between highlands and low-lying districts, near river gorges, and in the terraces and ravines of the escarpments.[40]

The Eritrean and Tegrean escarpments had an especially long association with banditry, blood-feuding and raiding. When the Italians first occupied Asmarā in 1889 they took vigorous action to deter *sheftā*. In this they achieved considerable success, though at the turn of the century the high plateau of the Qohāyto in Eritrea still harboured bands which preyed on caravans between Asmarā and Aksum. Merchants, and even priests, were attacked between Addi Qwālā and the Geshiarqah pass. *Sheftenat* in Eritrea was never extirpated by the Italians and continued in the vicinity of the escarpment in the 1930s. It reappeared openly as a form of resistance after the Ethiopian defeat by the Italians in 1936 and in the disturbed post-war conditions in Eritrea under British administration. In Tegrē the road south from the river Marab to Adwā, a distance of about fifty kilometres, was subject to attacks by bandits who descended from cave hide-outs in the virtually impenetrable mountains lining both sides of the route.[41]

In Bēgamder banditry also prevailed in mountain areas adjacent to trade routes or important market towns, and along escarpments. An especially perilous route was that between Lālibalā and Gondar, but it was close to the latter town that most bandit attacks occurred. In the early 1900s brigands were to be found in the Cherē mountains and near the Gannat river, a day's ride to the north of Gondar.[42] In the 1920s an English traveller, Rosita Forbes, was robbed near the village of Talla, a day's march from Gondar.[43] In Bēgamder forested ravines, lying below the escarpments, also offered ample protection to *sheftā*. The notorious Sheguti and his band lived in the uninhabited forest between the western edge of the plateau and Qallābāt. This kind of bush at the foot of the escarpment, near Afete, two day's march from Matammā, also harboured the bandits whom Hermann Norden witnessed raiding a caravan *en route* from the Sudan to Gondar.[44] A Gojjāmē proverb suggested a similar situation in that province. 'Better the Tumcha Forest than a bad master,' it went, referring to the heavily wooded escarpment near Dambachā town.[45]

The rivers which lay in such forested ravines clearly had strong associations with *sheftenat*. Travelling caravans were especially vulnerable to attacks and arbitrary levies when they descended into gorges and forded rivers at the few crossing points.[46] In Tegrē the area around the Warri River was notorious for banditry, as was the region near the Gundar Wehā River in Bēgamder. In the early 1920s a band of *sheftā* with seventy-two rifles operated regularly near the Takkazē River, while bandits were perennially active along the Abāy (Blue Nile) between Gojjām and Bēgamder, and around its tributaries, which often delineated sub-provincial boundaries. In the 1930s the confluence of the Abāy and Mugar in Shawā had a reputation for banditry, as had the village of Wanzagē further upstream. Near the Abāy gorge there were immense caves, said to be the homes of *sheftā* who robbed caravans crossing the river. As late as the 1960s brigands assaulted successive expeditions to navigate the Abāy. In 1962 a night attack on a Franco-Swiss expedition near Shogalin left two expedition members dead; in 1968 a British military expedition repelled two well-organised *sheftā* attacks near the confluence of the Abāy and Tami.[47]

Topography and the proximity of political jurisdictions cannot alone explain the appearance of bandits in northern Ethiopia. Other factors played an important role. In particular, pestilence, agricultural crisis and war reinforced peasant hardship. From the late 1880s the northern provinces suffered a combination of all three: cattle plague, harvest failure and foreign invasion.[48] Thousands of cattle, sheep and goats perished in the rinderpest epidemic which spread southward from Hāmāsēn and Tegrē to Bēgamder, Lāstā, Gojjām and Shawā between 1887 and 1889. An estimated 90 per cent of Ethiopian cattle died in the outbreak which destroyed entire herds and left many pastoralists destitute. Rinderpest not only deprived peasants of an immediate source of wealth, but the shortage of livestock also left much land unworked. In the northern provinces, where successive failures of the annual rains between 1888 and 1891 had already inflicted drought conditions, the lack of oxèn for ploughing or for the import of grain from areas of surplus aggravated a worsening situation. Unusually large locust swarms and a rare appearance of caterpillar pest compounded the crop disaster. The resultant famine of 1890–91 affected the whole of Ethiopia, though the northern provinces were hardest hit. In Tegrē and Bēgamder social taboos were ignored as the famished ate mules, horses and dogs. Throughout the north there were acute food shortages; prices of scarce grain and livestock soared. Starvation and migration left whole regions depopulated and lands untilled. Outbreaks of influenza, cholera, typhus and smallpox decimated victims already weakened by famine.

This savage affliction of dearth and disease enhanced the conditions in which *sheftā* operated and hungry bandits were quick to pillage travellers from stricken regions.[49] The Great Famine was a disastrous blow to a social order already suffering the ravages of foreign invasion at the hands of Sudanese Mahdists. Bēgamder bore the brunt of this invasion. The region west of Lake Tānā, especially near the rivers rising in the highlands to the west of the lake, the Dambeyā and Ifāg plains and the subprovince of Walqāyt were all invaded by Sudanese Mahdists. Subsequently all had strong associations with banditry. The Mahdist expeditionary force which penetrated Bēgamder was probably the largest army to invade Ethiopia in the nineteenth century.[50] It defeated *Negus* Takla Hāymānot of Gojjām in January 1888, and indirectly caused his rebellion against Yohannes later in the year, provoking Yohannes' imperial army to ravage Gojjām mercilessly between September 1888 and February 1889. In Bēgamder members of the Duchesne-Fournet expedition (1901–3) repeatedly noted the depopulation and ruin of the countryside, especially around Lake Tānā where the land was still largely unrestored in 1902, well over a decade later.[51] The Italian traveller, Maurizio Rava, passing through the region in 1908, also commented on the deserted and desolate countryside on the north-eastern side of the lake, between Ifāg and Dabra Tābor. These conditions prevailed around Ifāg as late as the 1930s.[52]

The western shore of the lake also suffered. Qunzelā lost five-hundred young men who had been recruited to fight the Mahdists by *Negus* Taklā Hāymānot and was also ruthlessly pillaged.[53] The ruination of the Qunzelā plain shocked the Duchesne-Fournet expedition, who also remarked the coincident rise of brigandage. Near the battlefield, close to Delgi Māryām on the north-western shore of the lake, where the *Negus* was so heavily defeated by the Mahdists in January 1888, the land remained uncultivated for the next twenty years. Only a tiny hamlet existed where before there had been a large, thriving village. Indeed, along the greater part of the west coast of Lake Tānā virtually all the churches remained ruined even in the 1930s.[54] Amidst this devastation only *sheftā* flourished. The British naturalist and ethnographer, Powell-Cotton, who passed along the western side of Lake Tānā in 1900, needed an armed escort to protect him from bandits who robbed travellers between Gondar and Dangal Bar. So predatory were *sheftā* here that they drove many villagers away from the lake. In 1899 the inhabitants of Abanu fled when *sheftā* burnt and looted their houses and in 1900 Dangal Bar itself, once a flourishing town, lay in ruins. At the turn of the century abandoned fields and deserted houses lined the western shore of the lake.[55]

The Mahdist battles also directly affected the peasants of Tegrē, because the Emperor Yohannes recruited and provisioned his troops from that region. Tegrē also suffered most from the political struggles between the sub-provincial chiefs after Yohannes's death, besides being the major theatre of war in the late 1890s, first between Yohannes's son, Mangashā and the emperor Menilek and subsequently against the Italians. The high level of banditry from Adwā down to the Warri River, after the 1890s, can be directly linked to the depredations suffered by the town and by the peasantry of the surrounding districts. Adwā was looted and partly burnt by *Dajāzmāch* Ambassā in December 1889; plundered by troops under General Baratieri; and poorly treated again by the Emperor Menilek. In 1893 only one-third of the houses in the largely ruined town were occupied and two years later the weekly market, where formerly thousands had congregated, was only visited by a few hundred people with meagre supplies. In the late 1890s the wide Faras May valley, between Adwā and the Chalanqo district, was a scene of blackened villages, which had suffered the depredations of Italian troops when they tried to conquer Tegrē and of Menilek's forces while at Adwā. After the battle of Adwā the hills around the town, and around Aksum, became the refuge of many *sheftā*. The high Chalanqo plateau became known for bandits there who daubed their faces with white and red mud to disguise themselves.[56]

The correlation between conflict and the rise of brigandage is clear. The Italian invasion of 1935 legitimised the kind of guerilla warfare practised by *sheftā*, and there was a very fine distinction between *arbaññā* (patriots) and bandits during the Italian occupation. In Wallaggā a *sheftā* called Olenga Hengle became a leading *arbaññā*.[57] *Arbaññā* who advanced personal as well as patriotic interests were common and banditry continued unabated after the Italian occupation ended. On the southern edge of the plateau, near Ankobar in Shawā, the almost legendary patriot, *Fitāwrāri* Fardē had been a bandit long before the Italians invaded. After the occupation, in the 1950s, the escarpment continued to provide hide-outs for *sheftā*, as it had earlier for patriots.[58] In the north banditry increased in the early 1940s because the government in Addis Ababa was unable to fill the power vacuum left by the Italians. In 1942 Tegrē became the first test of Haile Sellassie's restored regime, and by the end of the year the province was in open revolt, with many Tegrean towns under *de facto* control by bandits. Many *sheftā*, like Hāyla Māryām Raddā and Yekuno Amlāk Tasfāyē, gave their support to the Wayānē Rebellion and were elected by village assemblies (*garrab*) and by the coordinating assembly, the *shango*, as leaders of the rising. Other *sheftā* served as armed followers.[59]

The Wayānē rising demonstrates that their familiarity with guerrilla tactics made *sheftā* natural resistance leaders. Yet if banditry tended to arise from conflict, it could also help foment it. During successive revolts against new taxes in Gojjām in 1942, 1944, 1951 and 1968, village assemblies elected brigands as rebel leaders. These rebellions were all sparked off by attempts to change tax laws, and in the latter two cases at least, by a widespread belief that the planned registration of land was a prelude to an alienation from the peasants of their rights to individual holdings. Thus it is significant that the *sheftā* leadership in these risings was largely of peasant origin.[60] In the 1940s the patriot leaders, like Balāy Zallaqa, who rebelled against Haile Sellassie, were often bandits or sons of bandits. In 1968 a hero of the rising was a one-eyed peasant *sheftā* from Bechanā called Bāmlāku.

Throughout Ethiopia disbanded soldiers were a prime source of bandits. During the battle of Adwā thousands of men saw military action and subsequently returned to their home districts. The appearance of so many brigands bearing the title of *fitāwrāri*, which retained a stronger military assocation than either *dajāzmāch* or *rās*, reinforces the idea that bandits were often ex-soldiers. The elder Barihun was a *fitāwrāri*, as was the powerful bandit Shebeshi who was active in Semēn and Walqāyt between 1916 and 1919.[61] Other bandits, like Sheguti, had probably served in the Sudanese police militia. Many of the armed retainers seen by the Italian, Rafael di Lauro, around Ifāg in the 1930s were ex-*askāri* (native troops) of the Libyan Battalion.[62] Walda Gabre'ēl Masāzgi, the renowned Eritrean bandit from Marata Sabana, served in Libya where many other Eritreans perished. In 1935, when Italy was about to invade Ethiopia, the Italian administration in Eritrea introduced conscription, which raised more than 65 000 *askāri*, many of whom were used by the Italians as shock troops.[63] All the Masāzgi brothers served in the *askāri* at this time, and Walda Gabre'ēl and Bayyana, the eldest two, were stationed in the Gondar region. Like thousands of *askāri*, they returned home to find there were no jobs and only small cash savings. More important for their later careers, they had learnt how to use weapons and had acquired military skills. Similarly there were disbanded Ethiopian patriots who after five years of *sheftā*-style guerrilla resistance found it difficult to readjust to rural life.

Political conflict also resulted in an increasingly wider dispersion of firearms from the mid-nineteenth century on. Firearms became an essential prerequisite for attaining political power, and also for aspiring *sheftā*. By the reign of Emperor Tēwodros (1855–68) there were thousands of firearms in Ethiopia, and especially in Tegrē.[64] Emperor Yohannes's victories over the Egyptians in the 1870s resulted in the capture of thousands of breech-loading rifles and more than a dozen heavy guns. By the end of the 1880s, the future emperor, Menilek, was importing large quantities of rifles via Assab.

After the death of Yohannes at Matammā in 1889, the disintegration of his army aided the diffusion of arms in Tegrē. Thousands of well-armed soldiers frustrated Menilek's initial challenge to Yohannes's successor, *Rās* Mangashā. Nevertheless, Menilek was swift to re-equip his forces, issuing improved rifles to his troops in the early 1890s. Most of *Rās* Mangashā's retainers at this time received one good rifle, but often they owned other firearms and cartridges used by sons not obliged to carry firearms for the *rās*. With the rifles earlier obtained by the Tegreans, several tens of thousands were in Ethiopian hands before the battle of Adwā, where Menilek commanded an army well-equipped by contemporary African standards. In November 1894 there were an estimated 82 000 rifles and five-and-a-half million cartridges in the Ethiopian countryside. This was probably a conservative estimate. At most, though, there were probably no more than 110 000 rifles and an undefined number of machine guns.[66] At the battle of of Adwā the Ethiopians captured a further 11 000 rifles.

Despite Emperor Menilek's formidable authority, large numbers of firearms and soldiers remained outside his control at the turn of the century. Before the battle of Adwā he started to issue surplus weapons to trusted subordinates. This process was accelerated after 1896.[67] The British Legation in Addis Ababa estimated there were at least 500 000 modern Gras and Remington rifles in Ethiopia by 1900 and at least eighty pieces of artillery.[68] As Memilek became increasingly incapacitated after 1908, those who assumed power in Addis Ababa on behalf of *Lej* Iyasu distributed even more arms. In mid-1910, after an abortive rising by a cousin of Yohannes, Abreha Arāyā, the government ended restrictions on the open sale of weapons.

The flood of imported arms after Adwā and the guerilla tactics of the *sheftā* temporarily redressed the local balance of power in favour of banditry. Diverse opportunities opened for a *sheftā* to enrich himself and clearly a number of villages, like Chankar in Dambeyā, prospered from their association with brigandage.[69] The theft or expropriation of cattle was probably the least sophisticated form of *sheftenat* and occasionally the sale of such livestock may have been important in local markets. Until the 1930s rustlers were especially active around the market town of Dāngilā in Gojjām and stolen livestock were a constant source of local litigation.[70] In Bēgamder *sheftā* operating near Tagosā, on the western shore of Lake Tānā, exacted such heavy tributes in cattle and so frequently demanded a share in butchered meat that herdsmen moved north to escape their predators.[71] In Tegrē rustlers were greatly feared and cattle theft was common. Caulk has shown how, in the late nineteenth century, Tegrean *sheftā* distributed stolen livestock, or portions of slaughtered meat, in an elaborate order of precedence.[72] Local notables, minstrels, and priests among Christians, came before members of the band; within it the *sheftā* leader, lookout, butcher and cook received additional heifers.

Slaves were more valuable commodities seized by bandits. Although Emperor Menilek issued proclamations against the sale and purchase of slaves and against the passage of slaves through customs houses, these were rarely enforced.[73] *Rās* Tafari's edict of March 1924, which regulated the emancipation of slaves, also had little initial effect. Raiders continued to ravage rich and populous south-western provinces, seizing thousands of slaves for sale in the great markets at Jemmā, Gorē and Addis Ababa. Long processions of chained slaves headed north to Gojjām and Bēgamder and east via Dassē in Wallo province to the Gulf of Tajurah for export to Arabia and Persia. About ten thousand slaves reached north-western Ethiopia each year in the mid-1920s, of whom approximately one-tenth were sold in Gojjām. As late as 1927 traders sold five hundred slaves at Naqamtē in Wallaggā and despatched a further eight hundred to Gojjām and the Sudan. On the eastern route in 1925 and 1926 dhows carried another 2500 slaves annually across the Red Sea for public sale in Aden.[74] Nevertheless, as *Rās* Tafari strengthened imperial edicts against slaving, especially by the appointment of officials to keep close watch on the main slave routes, the traffic in human cargoes declined. At first slave caravans took to moving by night. In Gojjām slave transactions, once held openly in town markets, now occurred in nearby houses. As penalties for slaving were enforced, sales shifted to deserted or lonely villages or to the forest.[75] By the late 1930s these too had virtually ceased.

Bandits displayed an interest in slaves. *Sheftā* raided for slaves across international frontiers and within Ethiopia. The major outlaws on the north-west frontier, like Kidānā Māryām, Hāgos and Barihun, repeatedly raided into the Sudan for slaves, like their illustrious forebear, Tēwodros. The Abū Qulud raid in 1906 took more than 142 Sudanese as slaves, and incursions into the Sudan continued on a yearly basis until the third decade of the new century. Brigands also raided inside Ethiopia, often razing whole villages and carrying off the inhabitants. In 1924 the villagers around Balaya mountain near Laka Tānā fled to Qallābāt to escape enslavement by the district governor, *Fitāwrāri* Zallaqa, a frontier *sheftā* who commanded four hundred men and whose father-in-law and a son had been killed in earlier expeditions to the Sudan.[76] However slave raids within the northern provinces declined rapidly after 1925, though the traffic in slaves from further south continued.

Constant changes of provincial governors in the south-west after Menilek's death ensured that each in turn collected and sent north thousands of slaves during their brief tenure of office.[77] Predatory southern officials, like *Dajāzmāch* Kabbadā, governor of Gorē before 1919, and *Dajāzmāch* Hāyla Māryām, governor of Shawā Gamirā in the 1920s, welcomed *sheftā* and slave traders from northern Ethiopia to assist them acquire and market slaves. ·Indeed, in these recently conquered southern provinces so many brigands and slave raiders came from further north that the term *tegrē* was used to describe all *sheftā*. In 1919 one European observer characterised *Fitāwrāri* Dastā's regime in Māji as a robbers' roost, and by the 1920s Māji, like other south-western provinces, lay ruined and deserted except for brigands and small communities of Abyssinian officials and soldiers. Often lacking tenants to provide for them, the unpaid soldiers also turned *sheftā*. Like the provincial governors, and usually at their behest, renegade soldiers raided for slaves across provincial borders and returned to share their loot with their patrons in relative safety. In addition to such raids *sheftā* also enslaved travellers. By 1922 kidnappings had become so common in the forests of eastern Kaffā that on market days merchants and peasants journeyed together in large parties, shouting and firing guns to deter brigands.[78] Where *sheftā* seized travellers and unwary villagers they escorted their newly branded captives north by night for a highly profitable sale in Gojjām or Bēgamder, or sold them to traders in Gorē and other slave markets. *En route* north from Gorē, slave traders were themselves occasionally victims of *sheftā* attacks. In Gojjām James Baum, a member of the 1926–7 Field Museum Abyssinian expedition, witnessed the robbery of slaves from more than one pack-train.[79]

As pressures against slavery increased, bandits renewed interests elsewhere and in more easily marketable commodities. Most commonly, *sheftā* preyed on unprotected merchant caravans, often posing as customs officials. In Walqāyt several deposed chiefs in the 1930s took to the bush to exact arbitrary levies from caravans, confident that the provincial government would come to terms with them.[80] Travellers in the vicinity of the Abāy in Gojjām frequently found it difficult to distinguish between customs officials and *sheftā* who set their own dues.[81] Europeans also reported the theft of mules and horses. Baum saw *sheftā* take a dozen mules from merchants near the Abāy, and the Duchesne-Fournet expedition also lost two horses in Gojjām. *Sheftā* stole Rosita Forbes's best mules near Gondar, while an attempted theft of mules from Major Cheesman in January 1932 led the authorities to act against the culprit, the notorious Sheguti. Once captured, he was flown, bound hand and foot at the pilot's insistence, to Addis Ababa where he was tried for the murder of at least nine men – those whose names he could remember – and hanged on Entotto ridge.[82]

Trivial items were often the objects of highway robbery. Letter carriers enjoyed a customary immunity from interference, which was not always respected. Correspondence usually attracted no interest and money was rarely sent by courier. Often the messengers' small food parcels were taken and occasionally even their clothes.[83] In contrast, the *sheftā* who penetrated Ifāg were attracted by its importance as a market centre. In 1931 alone, the official customs revenue amounted to 12 000 Ethiopian dollars, while *sheftā* levies applied on the trade routes leading to the town were probably equally high.[84] Travelling caravans were also subject to occasional attack, though so close to Dabra Tābor such a strong deterrent to trade would have attracted wider official attention than most brigands wished and might have reduced their extra tolls. Few bandits would wish to attract government troops to defeat them or be so predatory that trade shifted elsewhere. For this reason *sheftā* maintained sophisticated intelligence services to identify and anticipate the moves of potential victims, often keeping travellers under surveillance and communicating between themselves by a system of bird and animal calls.[85] Kidāna Māryām's band kept close watch on the Powell-Cotton expedition from hilltops. The *sheftā*'s followers made enquiries about Powell-Cotton in local villages and even sent spies into his camp by night.[86] In Dambeyā brigands kept detailed information about travelling caravans, assessing the number of men and animals and the nature of the cargo.[87] Near the Sudanese frontier, *sheftā* were bolder and merchants employed stronger guards. Even encampments of 50 men and 250 animals were attacked by *sheftā*. Raided caravans were usually loaded with currency (Maria Theresa dollars), coffee, cotton goods, raw cotton, spices, incense and mules, the most common items of trade on the border. When pack-trains were fortunate enough to reach the frontier unscathed they still attracted criminal interest. Smugglers often attached mules laden with contraband, usually highly valued alcohol forbidden in the Sudan, to legitimate caravans of high-ranking Ethiopians or Europeans so they would be carried through with only cursory inspection.[88]

Viewing brigandage, and especially highway robbery and the levying of arbitrary dues, from a European perspective and through European sources may be misleading. Europeans were largely exempt from local exactions and were only attacked by brigands occasionally, especially if they dressed like Ethiopians. Europeans were feared by bandits, not least because they could bring down the wrath of higher authorities, as in the case of Cheesman and Sheguti. This particular incident is interesting because when Sheguti's men realised they were taking the mules of a European, they immediately returned them, saying that Sheguti had ordered them to rob only Ethiopian merchants and not to interfere with European caravans.[89]

In the 1920s *Rās* Hāylu of Gojjām took vigorous action against *sheftā* who attacked foreigners. In his province, as elsewhere, Ethiopian travellers and merchants often attached themselves to foreign pack-trains for protection.[90] When di Lauro's courier was robbed in May 1932, he was allowed to keep the letter he was carrying and was otherwise unharmed. *Sheftā* respect for the correspondence of the Italian consul at Dabra Tābor mitigated against a more sinister fate. When Europeans were robbed, they were usually recompensed by the local governor, who had to wait to catch the *sheftā* before he could recoup his outlay. In Dagā Dāmot in Gojjām those foreigners attacked by bandits were often compensated by members of the community in which the robbery occurred.[91] However, Europeans were not always so fortunate. Greek merchants were robbed and killed on the road from Asmarā to Adwā, while the attack on the London Society Mission at Jandā was unusual and probably occurred because of the missionary's temporary absence.[92] In contrast to most of the incidents involving foreigners, after one of the Masāzgi brothers was killed by an Italian carabinieri in December 1948, the two eldest brothers avenged his death by hunting Italians throughout the Eritrean highlands, killing at least eleven and harrassing many more.[93] In the context of the uncertainty of the future of Eritrea, the Masāzgi emerged as anti-Italian resistance heroes far beyond their own locality.

Economic reasons for banditry were important, but by no means overriding. By turning *sheftā*, peasants, ex-soldiers, escaped prisoners and state outlaws, evaded not only poverty, but also the authority of overlords and of the state.[94] In Wagarā it was common for a peasant who had committed a murder to become a brigand to escape official retribution.[95] Although bandits rejected their rulers' laws, they often remained close to their natal communities. Peasant brigands maintained connections with their villages, relatives and friends, and were occasionally fed and assisted by them.[96] Yet their revolt against an oppressive society did not make social bandits of these outlaws, for few *sheftā* expressed their rebellion in class terms. Lacking political organisation and ideology, *sheftā* were unlikely to initiate concerted peasant action to change their society. Brigandage usually benefited peasants only by offering a slim prospect of social mobility to an ambitious few. The low corporate consciousness of the Ethiopian nobility made it quite feasible for a talented individual to rise above his station and gain office and land.[97] The powerful bandit Kidāna Māryām, for example, rose from peasant origins to the rank of *fitāwrāri*. Later, as a *dajāzmāch*, he married a relative of Empress Tāytu and was given overlordship of Semēn and Walqāyt.[98] Kidāna Māryām's rise owed much to his ability to raise an army which was larger than most provincial governors could assemble, while his official appointment was intended to keep other *sheftā* in check. Hāylā Māryām, who organised the Abū Qulud raid, also rose from a humble background to become a *qañāzmāch* under *Dajāzmāch* Makonnen of Walqāyt.[99] Other *sheftā*, like the peasant leaders of robber bands active in 1906 north of the River Takkazē, achieved no such distinction.[100]

For most Abyssinian cultivators banditry was an ambiguous form of protest. Dissident nobles and gentry dominated *sheftenat* and even brigands of lowly origin plundered indiscriminately. In Tegrē, as elsewhere, peasants fought back.[101] North of Adwā in the 1920s the inhabitants of one district joined together to wipe out a particularly troublesome band of *sheftā*.[102] By highlighting such instances, though, the sources suggest they were infrequent. Probably the most common peasant response was to pay the tribute *sheftā* required, even at the cost of official reprisal. In 1900 the inhabitants of Qunzelā clearly preferred to conciliate local *sheftā* and risk the more distant authority of *Negus* Takla Hāymānot, even though on one occasion this had resulted in the dismissal of their chief, the imposition of a heavy fine, and the pillaging of the village by the *negus*'s troops.[103] The only alternative to maintaining good relations with *sheftā* was for threatened villagers to move away. When Kidāna Māryàm demanded too much from Amalar in western Bēgamder the inhabitants abandoned their village and headed south.[104] The arbitrary levies imposed by *sheftā* and the dislocation they caused undoubtedly increased hardship, especially when brigandage was combined with excessive violence. For most peasants, *sheftenat* was a burden of tribute and fear. With rare exceptions, it weakened collective peasant resistance and inhibited class unity. By offering the prospect of advancement to a few talented individuals, *sheftenat* eroded horizontal ties in this highly deferential society and diverted the peasantry from challenging their rulers directly. Indeed in southern Ethiopia when peasant villages were looted and lands and goods seized, the dispossessed survivors often joined soldiers and *sheftā* in further raids on their own people.[105]

Thus while *sheftā* bands were occasionally led by peasants, banditry could depend for its leadership on a reservoir of disaffected nobles or *bālābbāt* excluded from office and access to land by political rivalries and feuds. Banditry provided a means whereby a noble could gain status, power and political office or, if he were incompetent, might sink in position or be killed by the authorities. An ambitious chief might become a brigand as an act of rebellion. In 1929 one of *Rās* Gugsā's retainers did precisely this.[106] Other chiefs became *sheftā* after disputes with their overlords. In the 1890s an early recruit to Kidāna Māryām's band was a *shum* (chief) called Kāssā who quarrelled with his superior over taxes.[107] Subsequently he became commander of a brigand troop operating in the district he formerly governed. Captured and chained by *Negus* Takla Hāymānot, Kāssā's career turned full circle. On his release Kāssā became an officer in *Rās* Dārgē's army, his abilities either as *shum* or *sheftā* too valuable to waste. For similar reasons many autonomous *meslanē* (district chiefs) in isolated parts of Tegrē and Bēgamder were also former bandits.[108] One such, a *Grāzmāch* Balāy Tafari, had over 250 armed men in May 1934. In Walqāyt a noble *sheftā*, *Lej* Webnah Tasammā, became a *meslanē* under *Dajāzmāch* Ayalew Berru. Like Kidāna Māryām, *Lej* Webnah was given a mandate to arrest other *sheftā*. Yet other enterprising noble brigands sought positions as administrators of customs revenues. In Walqāyt and Dabāreq, *Bālāmbārās* Shebeshi and *Grāzmāch* Abitu, both former bandits, were appointed as respective customs chiefs in the early 1930s. Elsewhere, officials were not merely former bandits who had made good. Many continued to behave like outlaws after attaining a title and office.

The distinction between office-holding and *sheftenat* was often unclear. In the 1930s the governor of Sallamt and the district north of the River Engo was an important brigand for whom the local chiefs had great respect.[109] Like *Dajāzmāch* Makonnen of Walqāyt, he supervised all brigandage in his area and offered escorts to travellers. It was widely believed in Bēgamder that *Rās* Gugsā tolerated banditry because he found it profitable. The elder Barihun was one of his retainers, while two *sheftā* bands operating around Qāllābāt in April 1906 were associated with him.[110] Two months later, the *rās* was noticeably reticent to detain *Lej* Yegzaw, a *sheftā* from Wagarā. His eventual capture and hanging was the work of *Dajāzmāch* Gassasa of Semēn who also executed Hāyla Māryām.[111] Even when he did capture brigands, *Rās* Gugsā often extracted graft from them in return for their release.[112] Lesser officials emulated the *rās*. In the Bēgamder town of Talla, local chiefs, villagers and bandits were in league with each other, while on the south-western shore of Lake Tānā, *Qañāzmāch* Negusē of Forē shared his authority at the turn of the century with a brigand called Dastā. Dastā enjoyed the protection of one of *Negus* Takla Hāymānot's principal officers and evaded capture more than once.[113] Thirty years later, *Grāzmāch* Babil of nearby Maqal was involved in thefts of sugar and other commodities from mail-runners who passed through his territory.[114]

In Gojjām the regimes of *Rās* Hāylu (Governor, 1908–32) and *Dajāzmāch* Sahay Enqo Sellāsē (Governor, 1960–68) made extensive use of bandits in official positions.[115] *Rās* Hāylu protected *Fitāwrāri* Zallaqa and sponsored at least one of his Sudan raids for a share of the spoils. *Dajāzmāch* Sahāy regularly appointed *sheftā* to higher posts in his administration, most notably Damessē Alamyerāw who rose to become sub-provincial governor of Bechenā and whose misrule, as Sahāy's henchman, was felt far outside the area of his own jurisdiction. Damessē began his career as a *sheftā* whose exploits were celebrated in local folk songs. He switched sides to become a thief-catcher, was appointed a *fitāwrāri*, and the subsequent hatred he provoked among the people exceeded their dislike of the governor. He was ably assisted by another former bandit, Gassasa Wadajē. *Dajāzmāch* Sahāy also made extensive use of *sheftā* in the *Nach Lebāsh* (the auxiliary police) and these especially fleeced peasants who had resisted them during their former bandit days. While this kind of institutionalised banditry, far heavier it would seem than legitimate taxation, was possibly acceptable under *Rās* Hāylu, a member of the royal house of Gojjām, it was far less so under the Shawān outsider Sahāy, and this probably contributed directly to the violence of the 1968 Gojjām rebellion.

Further north, *bālābbāt* also hovered on the uneasy line between banditry and office. Most of the leaders of the Wayānē rebellion were members of the *makwānnent* (nobility) of Tegrē.[116] Yekuno Amlāk was an alienated official, while Hāyla Māryām Raddā had been rebuffed in an attempt to gain high office. Among other things, the Wayānē Rebellion marked a distinct coalescence of Tegrē nobles and *sheftā* in a bid for power and position. The mixture of motives which caused revolt in Tegrē were matched in Eritrea where many bandits, like the Masāzgi brothers, became involved in the political struggles over the future of the province between 1948 and 1952.[117] Very clearly, prominent supporters of the union of Eritrea and Ethiopia approached bandit leaders and persuaded them to play their part in terrorising Eritrea into federation. Support from Ethiopia gave these political *sheftā* arms and money, and refuge across the frontier. Between 1948 and 1952 bandits played a crucial role in the politics of reuniting Eritrea and Ethiopia by persecuting the opponents of union. The bandits only lost their Ethiopian and Unionist support after the United Nations voted for federation in 1952.

This kind of office-seeking or office-holding bandit is distinctly Ethiopian. In the nineteenth century *sheftenat* operated at the highest levels in the political struggles between the most elevated contenders for power. Successful *sheftā* could become emperors if they possessed the ability, and their claims could be strengthened by an appeal to the blood of Solomon. By the twentieth century these political conflicts, which had once determined the destiny of the state, were now more firmly subordinated to it. Successful *sheftā* were less likely to be great princes. More often they were striving to achieve local power and appointment. As before, this kind of banditry prevailed where the authority of government was weakest and where local officials were themselves local men, lacking the means or inclination to diminish brigandage. *Sheftenat* still offered economic gain and political advancement, but not at so accelerated a rate as might have been possible before 1890. After 1900 banditry continued to exist, and was widely tolerated, because it acted as a mechanism by which the class structure of society was preserved. It provided one way in which social and political mobility could occur without distorting the whole social fabric.

Thus banditry in Ethiopia appeared as a form of rebellion at once more primitive and more sophisticated than that described by Eric Hobsbawm. More primitive, in that it rarely raised itself above feuding and political conflict; more sophisticated, in that in the hierarchical, though not rigidly stratified society of northern Ethiopia, it reinforced the Abyssinian perception of their society as a mobile one. When peasants defied their rulers and the law to become leaders of *sheftā* bands, they too could gain official appointment and noble status. The rare occasions when banditry raised peasant *sheftā* demonstrated the fluidity of class relations in Ethiopian society. However, these limited chances for social mobility were not without their drawbacks. By presenting an avenue of advancement, banditry may have obstructed peasant mobilisation, diminishing peasant solidarity and reducing class conflict. Moreover, as Lewin and Chandler have argued for nineteenth- and twentieth-century Brazil in Ethiopia the interdependence of ruling class elites and bandits prevented *sheftenat* from realising its full potential as an agent of social protest.[118] In Abyssinia noble faction and brigandage diffused mass dissent within vertical relationships. Limited in perspective by their mode of production – small, largely self-sufficient units – Ethiopian peasants tended to look to noble patrons for leadership. With a few notable exceptions, most peasant *sheftā* were followers of disaffected nobles. Only occasionally, by their statements of the social or personal injustice which caused their entry into *sheftenat*, by their peasant origin, or by their attacks on the rich and powerful, did Ethiopian bandits express a class identity and hostility to their rulers. As Singleman has suggested for bandits in north-eastern Brazil, the elements of protest in *sheftenat* were secondary, but by no means illusory.[119]

During the early decades of the twentieth century a period of *sheftenat* remained an effective way to office but by the late 1940s the number of individuals whose *sheftenat* led to official appointment surely declined. Hereditary provincial rulers were replaced, and the influence of central government grew. Routes upwards in society increasingly had to be found by acquiring education and technical skills. Banditry, which had always been a symptom of the stresses within Ethiopian society, no longer provided a mechanism by which those tensions might be treated.

Ethiopian banditry requires an extension of the analytical framework within which we examine rural criminality. In Eritrea, the transition from precapitalist modes of agrarian production to more-commercial forms of agriculture, combined with substantial unemployment and disorder after the Italian occupation of Ethiopia, produced a form of *sheftenat* similar to Hobsbawm's model. Eritrean bandits, like the Masāzgi, became resistance leaders. fine examples of Hobsbawm's *haiduks*, named after robber groups who waged guerrilla resistance against the Turks in the Balkans.[120] Similar figures also flourished south of the Eritrean border in resistance to the Italian occupation of Ethiopia. However, more commonly in Ethiopia a very different social and political context dictated very different kinds of banditry. While most *sheftā* enjoyed support, being regarded as heroic figures, they did not act in a class interest at all. They did not exert class vengeance, nor did they redistribute wealth. They usually robbed for their own gain. In a few instances villagers may have benefited from their association with banditry, but more often those who shared the illicit gains were chiefs and officials, who may even have been *sheftā* themselves. The kind of banditry found in Ethiopia was a feature primarily of a feudal society. Such banditry may increase and may be transformed as society is changed by new, capitalist relations of production. Yet even where social bandits do appear, they make up a very small proportion of the brigand population.

The *sheftā* was essentially a pre-capitalist phenomenon. *Sheftā* bands included ex-soldiers, escaped prisoners, and dispossessed or ambitious peasants, who, in a deferential society, tended to accept noble leadership. Though these noble figures ultimately sought reintegration within society, all bandits were rebels against higher authorities, and against the feudal state. *Sheftenat* thus helped create insecurity, especially in the peripheral borderlands. It destabilised provincial government, and occasionally affected the centre of the state. Ironically, in the last analysis, its very existence provided a mechanism by which the pressures imposed by the feudal order of society could be resolved.

NOTES

1 F. Braudel, *La Mediterranée et le monde Méditerranée à l'époque de Philippe II* (Paris: Colin, 1966, 2nd edn), pp. 75–90.
2 An excellent study of the spread of vagrancy and brigandage from Spain to colonial Mexico is offered by N. F. Martin, *Los Vagabundos en Nueva España. Siglo XVI* (Mexico City, 1957), especially pp. xii–xxi, 17, 41–64, 106–23. For further discussion of banditry in Spain and colonial Mexico see I. A. A. Thompson, 'A map of crime in sixteenth century Spain', *Economic History Review*, vol. 21 (1968), pp. 244–67; C. M. Maclachlan, *Crime and Justice in Eighteenth Century Mexico: A Study of the Tribunal of the Acordada* (Berkeley: University of California Press, 1974), especially pp. 28–41; and articles by Taylor and Archer in the special issue of *Bibliotheca Americana*, vol. 1, no. 2 (1983); J.-P. Berthe, 'Conjoncture et société. Le banditisme en Nouvelle Espagne', *Annales E. S. C.*, vol. 19 (1965), pp. 1256–8. For more general discussion of colonial vagrancy and banditry in Latin America see M. Gongora, 'Vagabondage et société pastorale en Amérique Latine (spécialement au Chili Central)', *Annales E. S. C.*, vol. 20 (1966), pp. 159–77.
3 P. Dominique, *Les brigands en Provence et en Languedoc* (Avignon: Aubanel, 1975), pp. 7–153; J. Llandonosa, *El Bandolerisme a la Catalunya occidentale* (Barcelona, 1972), especially pp. 3–13.
4 E. J. Hobsbawm, *Primitive Rebels* (Manchester: Manchester University Press. 1959), pp. 13–29; also *idem*, *Bandits* (London: Weidenfeld & Nicolson, 1st edn, 1969; rev. edn, 1981, to which reference is made here), pp. 17–29, and *idem*, 'Social banditry', in Henry A. Landsberger (ed.), *Rural Protest: Peasant Movements and Social Change* (London: Macmillan, 1974), pp. 142–57.
5 P. Vanderwood, *Disorder and Progress: Bandits, Police, and Mexican Development* (Lincoln and London, 1981). See also *idem*, 'Banditry in Latin America: an introduction to the theme', *Bibliotheca Americana*, vol. 1, no. 2 (1983), pp. 1–27; B. S. Orlove and G. Custred, *Land and Power in Latin America* (New York, 1980); R. Schwartz, 'Bandits and rebels in Cuban independence: patriots and predators', *Bibliotheca Americana*, vol. 1, no. 2 (1983), pp. 91–130; L. Perez, 'La Chambelona – political protest, sugar and social banditry in Cuba, 1914–1917', *Inter-American Economic Affairs*, vol. 21, no. 4 (1977), pp. 3–27; and articles by P. O'Malley, especially 'The class production of crime: banditry and class strategies in England and Australia', *Research in Law and Sociology*, vol. 3 (1980), pp. 181–9.
6 For banditry in Ancient Egypt see H. F. Lutz, 'The alleged robbers' guild in ancient Egypt', *Semitic Philology*, vol. 10, no. 7 (1937), pp. 231–42: For brigandage in precolonial Africa see the essays in this volume by Kea (Chapter 5, above) and Crummey (Chapter 6, above).
7 For instance, A. Isaacman, 'Social banditry in Zimbabwe (Rhodesia) and Mozambique, 1884–1907: an expression of early peasant protest', *Journal of Southern African Studies*, vol. 4, no. 1 (1977), pp. 1–30; and G. W. Clarence-Smith, *Slaves, Peasants and Capitalists in Southern Angola, 1840–1926* (Cambridge: Cambridge University Press, 1979), Ch. 6.

8 A general summary of landholding is offered by John M. Cohen and Dov Weintraub, *Land and Peasants in Imperial Ethiopia: The Social Background to a Revolution* (Assen: Van Gorcum, 1975). See also Allan Hoben, *Land Tenure Among the Amhara of Ethiopia: The Dynamics of Cognatic Descent* (Chicago: University of Chicago Press, 1973), pp. 1–28. For discussion of the effects of changes in Ethiopian society after the 1950s see, Patrick Gilkes, *The Dying Lion. Feudalism and Modernization in Ethiopia* (New York: St Martin's Press, 1975), Chs 4 and 5 (pp. 101–71); and John Markakis, *Ethiopia: Anatomy of a Traditional Polity* (Oxford: Clarendon Press, 1974), especially Part III, pp. 143–82.

9 D. Crummey, 'Abyssinian Feudalism', *Past and Present*, no. 89 (1980), pp. 115–38; see also Hoben, *Land Tenure*, pp. 182, 188–9, 198–203, and his 'Family and class in northwest Europe and northern highland Ethiopia', in H. Marcus (ed.), *Proceedings of the First United States Conference on Ethiopian Studies* (E. Lansing, Michigan State University, African Studies Center, 1973), p. 162. For specific, though brief, reference to social mobility and *sheftenat* see A. Hoben, 'Social stratification in traditional Amhara society', in A. Tuden and L. Plotnicov (eds), *Social Stratification in Africa* (London: Collier-Macmillan, 1970), p. 291.

10 A. Blok, *The Mafia of a Sicilian Village, (1860–1960): An Anthropological Study of Political Middlemen* (Amsterdam, 1972); see also his 'The peasant and the brigand: social banditry reconsidered', *Comparative Studies in Society and History*, vol. 14, no. 4 (1972), pp. 494–503, with Hobsbawm's 'Social bandits: reply' in the same issue, pp. 503–5.

11 See chapter by Crummey (Chapter 6 above); and R. A. Caulk, 'Bad men of the borders: *Shum* and *shefta* in northern Ethiopia in the 19th century', *International Journal of African Historical Studies*, vol. 17, no. 2 (1984), pp. 201–27.

12 Crummey, Chapter 6, above.

13 Augustus B. Wylde, *Modern Abyssinia* (London: Methuen, 1901), p. 176.

14 Hermann Norden, *Africa's Last Empire: Through Abyssinia to Lake Tana and the Country of the Falasha* (London: Witherby, 1930), pp. 204–6.

15 Public Record Office (PRO), Kew, FO1/40, pp. 197–200, Baird to Lansdowne, Adis Alam, 25 July 1902; PRO, FO 401/13, No. 15, Enclosure, Memorandum on the Southern Frontier of Abyssinia, Major Gwynn to FO, London, 27 July 1909.

16 Norden, *Africa's Last Empire*, pp. 217, 222.

17 Raffaele di Lauro, *Tre anni a Gondar* (Milan: Mondadori, 1936), p. 162 and *passim*.

18 P. Garretson, 'Frontier feudalism in northwest Ethiopia: Shaykh Al-Imam 'Abd Allah of Nuqara, 1901–1923', *International Journal of African Historical Studies*, vol. 15, no. 2 (1982), pp. 261–82.

19 P. H. G. Powell-Cotton, *A Sporting Trip through Abyssinia* (London: Ward, 1902), pp. 254–7.

20 Arthur J. Hayes, *The Source of the Blue Nile* (London: Smith, Elder, 1905), p. 55.

21 PRO, FO 371/2, p. 238, Cromer to Grey, Cairo, 20 April 1906.

22 Garretson, 'Frontier feudalism', p. 278.

23 James E. Baum, *Unknown Ethiopia. New Light on Darkest Abyssinia* (New York: Grosset & Dunlap, 1935), pp. 56–7, 332–3. This was previously published as *Savage Abyssinia* (New York: Sears, 1927). Reference here is to the 1935 edition.

24 PRO, FO 401/30, No. 4, Enclosure, Addis Ababa Intelligence Report for the Quarter ended 31 December 1930, Dangila District.

25 R. E. Cheesman, *Lake Tana and the Blue Nile* (London: Macmillan, 1936), pp. 43, 131–2.

26 Stuart Bergsma, *Rainbow Empire: Ethiopia Stretches Out Her Hands* (Grand Rapids: Eerdmans, 1932), p. 144.

27 PRO, FO 401/9, No. 80, Harrington to Cromer, Addis Ababa, 3 May 1906.

28 PRO, FO 401/10, No. 12 Enclosure, Heard to Wingate, Gedaref, 14 July 1906.

29 Garretson, 'Frontier feudalism', pp. 270–71, 279. For his discussion of levels of administration in Nuqara see p. 266.

30 J. Duchesne-Fournet, *Mission en Ethiopie, (1901–1903)* (Paris: Masson, 1908–9), p. 141; Powell-Cotton Museum, P. H. G. Powell-Cotton, Diary 20, book 3, entry for 7 April

1900; PRO, FO 1/39, pp. 160–64, 1901 General Report by Baird, Addis Ababa to Metemma.

31 Cheesman, *Lake Tana*, pp. 114–16.
32 Di Lauro, *Tre anni a Gondar*, pp. 34, 74–6.
33 Cheesman, *Lake Tana*, pp. 166, 198.
34 Takla Iyasus, 'Ya Gojjām Tarik', Institute of Ethiopian Studies, Ms. 254, f. 68, col. 2; Walda Māryām, *Chronique de Théodorus II Roi des Rois d'Ethiopie (1853–1868)* (Paris: Libraire orientale & americaine, 1904), with French translation by C. Mondon-Vidailhet, p. 47. For the *fitāwrāri* who emulated Berru in the 1930s see Cheesman, *Lake Tana*, p. 166.
35 Powell-Cotton, *Sporting Trip*, p. 268; Cheesman, *Lake Tana*, pp. 166, 198.
36 Gebru Tareke, 'Peasant resistance in Ethiopia: the case of *Weyane*', *Journal of African History*, vol. 25, no. 1 (1984), p. 87.
37 For brigandage in Tegrē in the 1940s see D. R. Buxton, *Travels in Ethiopia* (New York, 1967), p. 129; for banditry in Eritrea see Kidane Mengesteab, 'The Mesazgi brothers: banditry in 1941–1951 Eritrea' (unpublished BA thesis, Addis Ababa University).
38 Baum, *Unknown Ethiopia*, p. 71.
39 Hoben, *Land Tenure*, p. 37.
40 Crummey, Chapter 6, above.
41 For measures taken by the Italians to deter brigandage see J. Kolmodin (ed.), *Traditions de Tsazzega et Hazzega* (Rome, 1912), p. 283; For *sheftenat* in Eritrea see J. Theodore Bent, *The Sacred City of the Ethiopians* (London: Longmans Green, 1896), pp. 95–9, 128, 217; and Kidane Mengesteab, 'Mesazgi brothers', pp. 1–2; Wylde, *Modern Abyssinia*, p. 136.
42 Maurizio Rava, *Al Lago Tsana* (Rome: Reale società geografica, 1913), pp. 46–7,
43 Rosita Forbes, *From Red Sea to Blue Nile. Abyssinian Adventures* (London: Cassell, 1925), p. 299.
44 Norden, *Africa's Last Empire*, pp. 230–31; all references to Sheguti are derived from Cheesman, *Lake Tana*, pp. 130–32.
45 PRO, FO 1/39, pp. 144–5, Report by Baird.
46 For banditry around the Warri river see Wylde, *Modern Abyssinia*, p. 175; Hayes, *Source of the Blue Nile*, p. 176, for *sheftā* near the Gundar Wehā; and Forbes, *Red Sea to Blue Nile*, p. 282, for the band operating near the Takkazē in the 1920s.
47 For the caves near the Abāy see C. F. Rey, *The Real Abyssinia* (Philadelphia: Lippincott, 1935), p. 82; for *sheftā* attacks on Blue Nile expeditions, R. Snailham, *The Blue Nile Revealed* (London, 1971), pp. 195–200, 205–10.
48 This paragraph is based on R. Pankhurst, 'The Great Ethiopian Famine of 1888–1892: A new assessment', *Journal of the History of Medicine and Allied Sciences*, vol. 21 (January 1966), (Part I), pp. 95–124; vol. 21 (July 1966), (Part II), pp. 271–94.
49 ibid., pp. 167, 281; Taurin de Cahagne, 'Voyage dans le pays Galla', *Les Missions Catholiques*, 1896, p. 206.
50 R. A. Caulk, 'Firearms and princely power in Ethiopia in the nineteenth century', *Journal of African History*, vol. 13, no. 4 (1972), pp. 622–3; also see S. Rubenson, *The Survival of Ethiopian Independence* (London: Heinemann, 1976), p. 383.
51 Duchesne-Fournet, *Mission*, pp. 135–40.
52 Rava, *Lago Tsana*, pp. 77, 90–94; Di Lauro, *Tre Anni a Gondar*, p. 76.
53 Rava, *Lago Tsana*, p. 153.
54 Duchesne-Fournet, *Mission*, pp. 140, 161–2; Powell-Cotton, *Sporting Trip*, pp. 284–5; Rava, *Lago Tsana*, p. 142; Cheesman, *Lake Tana*, pp. 215–18.
55 Powell-Cotton, *Sporting Trip*, pp. 282–4 and (Powell-Cotton Museum) his Diary 20, book 3 entries for 26 April–5 May, 1900; PRO, FO 1/39, pp. 160–72, Report by Baird.
56 Wylde, *Modern Abyssinia*, pp. 174–5; Bent, *Sacred City*, pp. 116, 169.
57 Kidane Mengesteab, 'Mesazgi brothers', p. 16; Gebru Tareke, 'Peasant resistance', pp. 87–9.

58 Buxton, *Travels*, pp. 49–50; W. Weissleder, 'The political ecology of Amhara domination' (unpublished PhD dissertation, University of Chicago, 1965), p. 63.
59 Gebru Tareke, 'Peasant resistance', pp. 79, 87–9.
60 For the connection between the Gojjām rebellions and *sheftenat* see Gilkes, *Dying Lion*, pp. 181–6; Markakis, *Ethiopia*, pp. 386–7; Gebru Tareke, 'Rural protest in Ethiopia, 1941–1970: a study of three rebellions' (unpublished PhD dissertation, Syracuse University, 1977), p. 402.
61 For Barihun see PRO, FO 403/430, No. 25, Sudan Intelligence Report, 1912; for Shebeshi see Garretson, 'Frontier feudalism', p. 278.
62 Di Lauro, *Tre anni a Gondar*, p. 76.
63 E. de Bono, *Anno XIII, The Conquest of an Empire*, p. 31.
64 R. Pankhurst, *An Economic History of Ethiopia, 1800–1935* (Addis Ababa: Haile Sellassie I University Press, 1968), pp. 581, 594–602. This passage and the following paragraph also draw heavily on Caulk, 'Firearms and princely power', pp. 617–26.
65 Caulk, 'Firearms and princely power', p. 621.
66 These estimates are based on Rubenson, *Survival of Ethiopian Independence*, p. 398; for the number of rifles captured from the Italians at Adwā see Wylde, *Modern Abyssinia*, p. 212.
67 Caulk, 'Firearms and princely power', pp. 627–8.
68 PRO, FO 401/37, No. 23, Harrington to Salisbury, Addis Ababa, 22 May 1900.
69 Duchesne-Fournet, *Mission*, p. 145; Wylde, *Modern Abyssinia*, p. 304.
70 PRO, FO 401/30, No. 4, Enc., Addis Ababa Intelligence Report for the Quarter ended 31 December 1930, Dangila District; Cheesman, *Lake Tana*, p. 58.
71 Powell-Cotton, *Sporting Trip*, pp. 283–4.
72 Caulk, 'Bad men of the borders', pp. 217–18.
73 PRO, FO 403/451, No. 118, Dodds to Curzon, Addis Ababa, 14 September 1920; PRO, FO 403/454, No. 55, Enclosure 1, annex, Memorandum by Consul Walker on Slavery in Abyssinia, Gorē, 30 September 1923.
74 For slaves sent to Addis Ababa see PRO, FO 401/17, No. 15, Enclosure 1, Memorandum by Bullock, Addis Ababa, 9 October 1924. For the Gojjām and Wallaggā estimates see PRO, FO 401/17, No. 16, Memorandum by Home on the Slave Trade in North West Abyssinia, Dangila, 22 March 1924; and PRO, FO 401/22, No. 144, Enclosure 1, Memorandum by Home on Abyssinian Slave Raiding, Addis Ababa, 2 May 1927. For the numbers of slaves shipped across the Red Sea see PRO, FO 401/21, No. 60, Enclosure 2, Extract from reports of proceedings of HMS *Clematis* in the Red Sea during the period 15 April–30 June 1926.
75 PRO, FO 401/17, No. 16, Memo. by Home on Slave Trade, 22 March 1924; Cheesman, *Lake Tana*, p. 327.
76 PRO, FO 401/18, No. 102 and Enclosure, FO to Lugard, 14 July 1925; PRO, FO 742/17 Nos 139–54, Correspondence respecting Abyssinian raids and incursions into British Territory and the Anglo-Egyptian Sudan.
77 This paragraph draws on Henry Darley, *Slaves and Ivory: A Record of Adventures and Exploration in the Unknown Sudan, and Among the Abyssinian Slave Raiders* (London: Witherby, 1926), especially pp. 36–7, 64–84, 138–41; and Arnold Hodson, *Where Lion Reign: An Account of Lion Hunting and Exploration in S. W. Abyssinia* (London: Skeffington, 1929), especially pp. 25–7, 160–3. For references to *Dajāzmāch* Kabbadā see PRO, FO 403/450, No. 188 and Enclosures, Campbell to Curzon, Addis Ababa, 22 September 1919; and PRO, FO 403/454, No. 55, Enclosure 1, Memo. by Walker on Slavery, 30 September 1923. For *Fitāwrāri* Dastā see PRO, FO 403/450, No. 187, Enclosure, Summarised report of Maji Mission, 21 September 1919. For *Dajāzmāch* Hāyla Māryām see PRO, FO 401/22, No. 144, Bentinck to Chamberlain, Addis Ababa, May 1927. For use of *tegrē* to describe *sheftā* see the many references in PRO, FO 742/16 and particularly, PRO, FO 742/17, Nos 139–154, Correspondence respecting Abyssinian raids and incursions.
78 PRO, FO 403/453, No. 19, Enclosure, Extract from report of H. M. Consul at Maji, 1

February 1922.
79 Baum, *Unknown Ethiopia*, p. 107.
80 Di Lauro, *Tre anni a Gondar*, p. 132.
81 Cheesman, *Lake Tana*, p. 294; Baum, *Unknown Ethiopia*, pp. 117–27.
82 Baum, *Unknown Ethiopia*, p. 107; Duchesne-Fournet, *Mission*, p. 162; Forbes, *Red Sea to Blue Nile*, pp. 291, 299–300; Cheesman, *Lake Tana*, pp. 130–31.
83 Norden, *Africa's Last Empire*, p. 57; Hayes, *Source of Blue Nile*, pp. 82–3; Cheesman, *Lake Tana*, pp. 13, 43; Di Lauro, *Tre anni a Gondar*, p. 76. Hayes was informed that couriers were never stopped in Gojjām and Bēgamder, but Cheesman's mail-runners were repeatedly robbed in the early 1930s, especially near Qāllābāt. Near Ifāg Di Lauro's courier was robbed and stripped by *sheftā*, while on one occasion bandits stole the trousers of Cheesman's runners so they had to wait in the forest and enter Qāllābāt under cover of darkness.
84 Di Lauro, *Tre anni a Gondar*, p. 76.
85 Caulk, 'Bad men of the borders', p. 217–18.
86 Powell-Cotton, *Sporting Trip*, pp. 254, 270–71, and (Powell-Cotton Museum) his Diary 20, book 3, entry for 28 April 1900.
87 Norden, *Africa's Last Empire*, p. 206.
88 ibid., pp. 230–32.
89 Cheesman, *Lake Tana*, pp. 130–32.
90 Baum, *Unknown Ethiopia*, pp. 107–27, 232; Hayes, *Source of Blue Nile*, pp. 54–5; Powell-Cotton, *Sporting Trip*, p. 282.
91 Hoben, *Land Tenure*, p. 37.
92 For attacks on Greek merchants see Bent, *Sacred City*, p. 86; for the Jandā attack, Norden, *Africa's Last Empire*, p. 204.
93 Kidane Mengesteab, 'Mesazgi brothers', p. 10.
94 For state outlaws who reverted to *sheftenat* on their escape see PRO, FO 403/421, No. 18, Doughty to Grey, Addis Ababa, 27 July 1911.
95 PRO, FO 401/10, No. 12, Enclosure, Heard to Wingate, Addis Ababa, 14 July 1906.
96 Kidane Mengesteab, 'Mesazgi brothers', p. 6; Norden, *Africa's Last Empire*, p. 205; Buxton, *Travels*, p. 49.
97 Crummey, 'Abyssinian feudalism', pp. 137–8.
98 Garretson, 'Frontier feudalism', p. 271; Powell-Cotton, Diary 20, book 3, entry for 28 April 1900.
99 PRO, FO 401/10, No. 3, Extract from the Italian 'Tribuna'; PRO, FO 401/10 No. 12, Enclosure, Heard to Wingate, Addis Ababa, 14 July 1906.
100 ibid.
101 Buxton, *Travels*, p. 129.
102 H. C. Maydon, *Simen: Its Heights and Abysses* (London: Witherby, 1925), p. 32.
103 PRO, FO 1/39, pp. 160–64, Report by Baird.
104 Powell-Cotton, *Sporting Trip*, p. 269.
105 PRO, FO 403/454, No. 35, Enclosure 1, Hodson to Russell, Gemira, 6 June 1923.
106 Norden, *Africa's Last Empire*, p. 206.
107 Powell-Cotton, *Sporting Trip*, pp. 263–4.
108 The reference and the following passage is based on Di Lauro, *Tre Anni a Gondar*, pp. 132, 162. Bandits around Dambeyā in the 1930s were often disaffected officials (Norden, *Africa's Last Empire*, p. 206), suggesting a striking fluidity between officialdom and *sheftenat*.
109 ibid., p. 180.
110 Baum, *Unknown Ethiopia*, pp. 56–7; PRO, FO 401/9, No. 79, Enclosure 1, Wingate to Owen, Khartoum, 27 April 1906.
111 PRO, FO 401/9, No. 91, Cromer to Grey, Cairo, 18 May 1906.
112 Norden, *Africa's Last Empire*, p. 206.
113 PRO, FO 1/39, pp. 167–9, Report by Baird.
114 Cheesman, *Lake Tana*, pp. 114–16.

115 For *Rās* Hāylu and *Fitāwrāri* Zallaqa see PRO, FO 742/17, Nos 139–54, Abyssinian raids and incursions. For employment of bandits in official positions see Gilkes, *Dying Lion*, p. 184; Gebru Tareke, 'Rural protest', pp. 388–9; and Markakis, *Ethiopia*, p. 380.
116 Gebru Tareke, 'Peasant resistance', pp. 86–7.
117 Kidane Mengesteab, 'Mesazgi brothers', p. 12.
118 L. Lewin, 'The oligarchical limitations of social banditry in Brazil: the case of the "good" thief Antonio Silvino', *Past and Present*, no. 82 (1979), pp. 116–46; Billy Jaynes Chandler, *The Bandit King: Lampiao of Brazil* (College Station: Texas A & M University Press, 1978).
119 P. Singleman, 'Political structure and social banditry in northeast Brazil', *Journal of Latin American Studies*, vol. 7, no. 1 (1975), pp. 58–83.
120 Hobsbawm, *Bandits*, pp. 61–9.

8 Bushman banditry in twentieth-century Namibia[1]

ROBERT GORDON

Despite the claim by Allen Isaacman[2] that if academics persevere and turn over enough stones they will find examples of social banditry, it appears that the optimism and excitement which greeted the genre of resistance studies in the early 1970s has given way to a rather dour denial of their worth. There are many factors which have led to this decline, not least naive assumptions, sloppy scholarship and theoretical inadequacy, as well as changes in academic fashions. Populism, which provided a useful mother-lode for studies of the little man, is now apparently 'out' to be replaced by studies which emphasise macro-factors and industrialisation as the 'engine of change'.[3] P. O'Malley, in a comment on banditry studies in general, penetrates to the heart of the problem when he complains that they have 'generated a pseudo-liberationist rhetoric which barely conceals theoretical dogmatism and a cavalier disregard for the variability of capitalist relations of power'.[4]

We risk throwing the baby out with the bath water if we blindly follow the latest fashions. I hope to show how, if we combine historical research with an awareness of social-organisational arrangements derived from ethnographic studies, we can gain greater historical understanding. In his classic study of banditry, Eric Hobsbawm noted that hunter-gatherers supply a disproportionately large number of bandits when their society is incorporated into larger economies resting on class conflict.[5] This chapter focuses on one of the most heavily researched 'peoples' in Africa: the people known variously as San, Masarwa or Bushmen, perhaps the most academically famous hunter-gatherers in Africa.

There is some debate as to when the white colonisers at the Cape of Good Hope first labelled an entity of people as Bushmen and what precisely they meant by the term. 'Bushmen' as a label was only invented after the colonisers had been at the Cape for some time and had consolidated their rather precarious toehold in Africa. It was not surprising, according to Theal, that the earlier colonisers did not recognise any 'racial differences' between the Hottentots and the Bushmen because they were about as indistinguishable as Celts from Saxons. It was only in 1686 that colonial opinion began to emphasise racial differences.[6] Initially the settlers followed their pastoral allies, the Hottentots, in referring to the Bushmen as Saokwa or Sanqua Hottentots, the term Sanqua apparently meaning robbers and murderers. To these observers the major differentiating criterion was possession of livestock: Hottentots possessed sheep and cattle while the Bushmen did not. Who or what the Bushmen were was a topic which was to occupy some of the brightest minds at the Cape for many years. As late as 1854, Sir Francis Galton, who was later to found the Eugenics Movement, declared that the

Bushmen were simply impoverished Hottentots without cattle. Indeed the debate still rages in academic circles.[7]

A closely related debate concerns the origins of the term Bushman. Basically three theories are advanced to account for the word. These theories hold that the term was derived from: (1) the fact that the people so labelled lived in the 'Bush'; or (2) the Dutch 'Bosmanneken' and the perceived similarity between the Bushmen and the East Indian Orangutan; or (3) the Dutch 'Bossiesman' which means highwayman or bandit. Nienaber, perhaps the foremost etymologist of Afrikaans, gives the most credence to this last version and my own rather cursory reading of early Cape history also tends to favour this explanation.[8] The term did not drop ready made from some Calvinistic heaven but was rather first used in the context of 'Bosjesman-Hottentots'. It was a lumpen-category for all the dispossessed, and non-livestock owners; in short for all the outlaws and bandits. Accounts of the so-called Khoisan Wars of the seventeenth and eighteenth centuries suggest that contextually this is the only valid explanation.[9] Moreover, it is strange that all the early observers at the Cape, while capable of providing linguistic and tribal data on the Hottentots, were not able to do the same for the Bushmen.

It was only towards the end of the nineteenth century, after most of the Bushmen in the Cape had been exterminated, that scientists started to make internal differentiations among Bushmen, in sum to move from using the term simply as a lumpen-category to providing it with specific ethnographic content. This process began when Europeans reached into the Kalahari, especially into the Kaukauveld, the area in which this paper is grounded. The Kaukauveld, and its close neighbour, the Oshimpoloveld, stretch north of present-day Gobabis to past the Kavango River and into southern Angola, westwards well into Ovamboland and eastwards to Lake Ngami in Botswana.

Early travellers' accounts are almost unanimous that this was the heartland of the Bushmen, who were mainly of two large groupings, the Nama-speaking Heikom (who might conceivably have been impoverished pastoralists) and the Kung-speakers. In the Kaukauveld, white and black pastoralists had not yet made their appearance. The reason for this was so obvious that it was often ignored: stock diseases served to deter even the most aggressive cattle farmer.

But game was plentiful, and so from about the middle of the nineteenth century, numerous adventurer-hunters trekked into the area in search of ivory and ostrich feathers. Relations between these representatives of merchant-capital and the Bushmen were generally friendly. Gun-hunters appreciated the skills of the bos-hunters and hired many of them as 'shoot-boys'. When the great white hunter, van Zyl, who had declared the Kalahari north of Gobabis to be a veritable hunters' paradise, established his world record of 103 elephants killed in one day just south of Nyae Nyae, Victorian sportsmanship dictated that he omit mentioning the fact that he was helped by twenty armed Bushmen. Other great white hunters like Brooks were either so scared, drunk or lazy that they left all the elephant hunting to the Bushmen.

The fact that the Bushmen were armed and knew how to use firearms served as a strong deterrent against entering Bushmanland to many outsiders. In fact, the pocket Republic of Upingtonia was established by the Thirstland Trekkers at Grootfontein in 1884 at the instigation of William Jordan, an adventurer who hoped that by having a white community in the vicinity he would be able to take control of the immensely rich Bushman-operated copper mines. The plan failed.

Jordan was killed, and the Trekboers, instead of finding themselves in a Calvinistic utopia, found themselves in a Bushmen-created hell. By force of arms the Bushmen drove them out of the country into Angola. The Bushmen were responsible for the first successful Namibian war of liberation.

Success was short-lived, however. An early multinational company was formed to exploit Jordan's mineral claims and a short while later, in 1885, the area was proclaimed a German Schutzgebiet. Initially the colonisers did not have sufficient coercive powers to incorporate the indigenes forcibly into their society so the Governor signed a series of protective treaties with, *inter alia*, Buschmann-Kapitan Aribib who, in return for accepting German protection and promising not to fire the grass, was to receive a yearly retainer of 50 marks. By 1906 the situation had changed dramatically: the railway line had snaked its way up to the Grootfontein area and a crude but effective (if costly) vaccine had been developed for bovine 'lung-sickness'. The area around Grootfontein with its high rainfall became the centre of a veritable land rush. Indeed, despite having the highest land prices in the Territory, the district showed the most rapid increase in farmer-settlers. By the outbreak of the First World War, Grootfontein had the second highest number of settler-farmers. In contrast to their dealing with the pastoral Herero and Nama, the Germans refused to recognise any legal claims to land by the Bushmen, since nomads supposedly did not own land. Thus began one of the greatest unpublicized land thefts in Namibia.

Bushman retaliation was swift and devastating. By 1911 Governor Seitz, supposedly one of the more liberal members of the Colonial Service, acted with pragmatism. In a decree dated 24 October 1911 he proclaimed:

> Now that the Bushman's attitude towards other natives, white colonisers, officials and police has become so hostile that it has led to the death of one settler and one police officer, an end must be made of this danger. Therefore with regard to the use of weapons by the police in dealing with Bushmen I am amending the following:
> 1) When patrol officers . . . are searching Bushman areas breaking up their settlements or searching for cattle thieves and robber bands, they must have their weapons ready at all times to fire . . .
> 2) Firearms are to be used:
>   a. in the smallest case of insubordination . . .
>   b. when a (suspected) felon does not stop on command but tries to escape through flight . . .
> 3) If some of the male Bushmen who have been arrested are strong enough to work, they should be handed over to the district authorities at Luderitzbucht to work in the diamond fields.[10]

For a while influential German colonials even considered having the Bushmen declared to be non-human so that they could be shot like vermin. This was never put into law but it emerged in practice despite the appeal of the Grootfontein district commandant that extermination was not so much immoral as impractical as over half the farms in the Grootfontein district were dependent upon Bushmen for their labour requirements. By 1915 the situation had become so serious that the Germans felt compelled to detail an army company to the Grootfontein district to cope with the 'Plague'. This despite the fact that the South Africans were invading their colony in the south with an overwhelmingly superior force.

One of the troopers who participated in these Bushmen hunts has deposited his journal in the Windhoek Archives. It provides a sense of how policy was translated into practice: Bushmen armed with bows were shot on sight even when the Bushmen were unaware that whites were in the vicinity. Captured females were forced to carry supplies. On another occasion the troopers succeeded in capturing two women in a raid: 'We sent both women ahead and when they were five yards away, by arrangement Falckenburg and I shot them in the head from behind. Both did not feel their deaths.'[11]

Contrary to popular belief, the difference between the German and South African colonisers in their relations with the Bushmen was more ostensible than real. Variations in policy and action towards the Bushmen did not arise because of 'culture', 'race' or 'ethnicity', but were more the product of the phase of incorporation of the Bushmen into the capitalist world system. Thus we find, for example, that the court sentences imposed by the South Africans upon Bushmen convicts initially show no differences from those imposed by the Germans. Similarly, one finds that whites convicted of killing Bushmen were treated just as leniently by the South Africans as they were by the Germans.

German and South African policies towards the Bushmen, inarticulate and apparently contradictory as they are, shared a number of assumptions. Both agree that it was just a matter of time before the Bushmen would disappear off the face of the earth. This accorded well with the unilinear Darwinist doctrine of the time which saw the Bushmen as one of the lowest strata. Given the inevitability of their demise, policy towards them was based on 'practical' considerations.

Major Herbst, the Secretary of the Protectorate, drew on his experience in dealing with the Bushmen in the Northern Cape, and argued that 'the only policy . . . successful in overcoming the Bushman trouble is the settling of a European population in the area in which these raids occur. When this particular area . . . [is] more thickly populated the Bushmen will retire and seek fresh fields'.[12] Small wonder then that prospective white South African settlers believed that land could be had virtually for the asking in South West Africa in an area dubbed 'Bushmen land and Baboon country'.[13]

Yet colonial policy was not monolithic. While in neighboring Bechuanaland the British defined the 'Bushman problem' as one of ending the situation of slavery the Bushmen found themselves in *vis à vis* the Tswana people,[14] in South West Africa the 'Bushman problem' underwent various changes in emphasis over time. These changing emphases indicate in an approximate way the differing phases of the process whereby the Bushmen were incorporated into the dominant authoritarian capitalism. In this process the role of power and law was crucial. If we look at Grootfontein court cases over time, important transformations are immediately apparent. Initially most cases involve what one might term employment-related offences and most of the offenders are Bushmen, but the total number of offences is still relatively small. In Grootfontein this period lasted from 1896 to *c*. 1911 when stock theft started to increase as an offence, peaking in 1920 when 63 per cent of Bushmen convictions were in this category. It then shows a steady decline until in 1976 only 4 per cent of all Bushmen convictions were for stock theft. Stock theft shows a general decline and the role of the Bushmen in it also decreases. Thus in 1920 over 70 per cent of *all* stock theft cases involved Bushmen, but by 1976 this had dropped to 20 per cent. Clearly, judging by court records the proclivity of the Bushmen to engage in stock theft had been neutralised. This legal mechanism for forcing the Bushmen to settle down on white farms or move out beyond the police zone shows a constant if erratic application gradually serving its purpose and going into decline. The Master-Servant Ordinance which is the main legislation governing the relations between employer and employee shows a consistent increase in application before it was abolished in the early 1970s. What these cases indicate statistically is the Bushmen's gradual transition from relatively autonomous hunting and foraging groups to a situation of rural proletarianisation. In this regard the fact that 30 per cent of all Bushman cases in 1976 involved assaults is especially significant since they are almost exclusively concerned with assaults by Bushmen upon Bushmen; an indication of disintegration of old social forms.[15]

In sum, a study of the use of state power in the courts reveals a change of definition of 'the Bushman problem', according to overlapping, and at times, interweaving phases of incorporation. Initially there was a period of relatively minor labour problems. This was followed by a definition of the problem as one of banditry and stock theft and was typically associated with rapid white encroachment onto Bushman land. This phase effectively worked itself out by the Second World War. The Bushman had now been 'tamed', and had to be made into 'reliable' farmworkers. This constituted the next phase of incorporation. It is still current although on a vastly reduced scale. These three phases together constitute the first wave of incorporation into the colonial order, an order characterised by authoritian capitalism.

The second wave of incorporation took place after the Second World War and was characterised by the promulgation of a Bushman Reserve. It ended the autonomy of the last of the Bushmen forager-hunters by extending administrative control over them. In a sense this second wave was an insurance policy for the first wave, since one of its major purposes was to secure the supply of Bushman labour. The second wave also indicated a growing sophistication in attempts to regulate the life of the Bushmen. While in the first wave the 'Bushman problem' was tackled by changing the law, in the second wave the emphasis shifted from the legal realm to that of policy and planning. Laws, since they have to be promulgated in the Government Gazette, are subject to outside critical scrutiny; policy and planning is more private and hence less embarrassing. Nevertheless these statistics provide useful scaffolding from which to examine Bushman banditry more closely.

An illustrative case: the 'notorious' Hans's gang

Bandit gangs appear to have functioned from about 1906 to about 1930. Amongst them, the gang of Hans epitomises Bushman banditry at full bloom. This gang flourished during the heady days of uncertainty generated by the outbreak of the First World War when the Germans were mobilising to fight the South African invaders in the South. At its peak, Hans's gang had up to seventeen rifles and a large quantity of ammunition. The number of people associated with the gang fluctuated, peaking at between thirty to forty, mostly Bushmen and a few Damaras. But when the final capture came, the gang consisted of nine males plus women and children.[16] Hans, a Heikom, was reported to be a good hunter employed by a white German farmer named Wegener. When Wegener was mobilised, Hans collected a small gang. At Easter 1915, at Guntsas, Hans and his friend Max killed a German farmer named Ludwig for stealing Max's wife. Ludwig's offence was not a sudden irritation. Max had complained to the police at Nurugas about Ludwig's relations with his wife. The police apparently told her to return but she had refused.[17] After killing Ludwig, Hans and Max took to the bush and led a life of brigandage. They persuaded some other Bushmen employed on white farms to join them. As one of Hans's followers later explained in court; he had worked for Farmer Buccheim until 1913, but had not been well treated and so had left and resided in the bush. It was here that Hans had come to him and asked him to join him as he knew how to shoot. Hans claimed that the Germans would soon be chasing all the Bushmen in the vicinity and they would have to resist with arms.[18] Other members also joined out of fear for the whites, this was the dominant factor in determining Bushman behaviour in those parts: 'we ran away because we were frightened of the white man. White people shoot Bushmen so we were frightened of them.'[19] A short while later Hans shot and killed another farmer, Muller of Knakib, a farmer who used to boast about the number of his Bushman concubines.

Bushman fears of white retaliation were well grounded. Within a short period, special army units were delegated to deal specifically with the 'Bushman' problem, even though the country was on a war footing. However German army patrols were not especially successful because of the dense bush and scrub and in January 1917, the 'notorious' Hans was still reported to hold 'considerable sway' in the Gorobab and Choiganeb (now Otjituo) areas.[20] Farm labourers reported that Hans threatened to kill all natives who worked for whites.[21] The gang was reported to have killed Fritz, an Ovambo with marked pro-farmer sentiments.[22]

Walbaum, one of the soldiers sent to the Grootfontein District to deal with the 'Bushman problem' left behind a journal describing a state of siege amongst the farmers. They could not burn lights in their houses at night for fear that Hans and his band would shoot them out. Most of the farm labour in the area, especially the Bushmen working for unpopular farmers, had absconded with goats and cattle. The army and farmers responded with brutality.

In view of this brutality it was not surprising that Hans retaliated in May 1915 by attacking the farmhouse at Sus which was serving as a temporary police station in the operations against him. He attacked while the owner and his family were away visiting and the police were out hunting. With three armed men, Hans razed the farmhouse and killed three Herero labourers in their huts, but apparently did not touch the women. 'The whole house was demolished. Every window was broken. In the house nothing was left: suits, clothing, laundered children's clothes, food, tobacco, schnapps, and 200 marks in money: everything was stolen' was how Walbaum described the scene.[23] Hans's gang also took ten head of cattle and a Model 71 Mauser rifle.[24] In a later court case involving one of the participants in this raid, it emerged that the three black farm workers had caught Max and that to rescue him, Johannes had been compelled to shoot them.[25]

German retaliation was swift. Walbaum's patrol captured four males and two rifles in a surprise raid because the Bushmen were 'totally soaked from the schnapps'.[26] They also managed to capture the wife of Max who had been Ludwig's concubine. Because of her value as a hostage, Walbaum had a farmer, Reyelien (Regelen), personally deliver her to headquarters in Grootfontein. Despite severe and urgent warnings to be careful, Regelen was surprised by Max at dawn one morning 27 km from Grootfontein. Max killed him and rescued his wife. Under pressure from the Germans, Hans and his band retreated to Tsebeb water hole from where, in September 1915, they launched a raid which shocked the local white populace. Max ordered his lieutenant, Johannes, to go to the farm, Gorobab West and kill the owner, Eckstein, because he had recently taken away some of the Bushmen women from Tsebeb water hole to work on his farm, perhaps hoping that they would attract their men as well. Max did not go on this mission. Three Bushmen under Johannes, armed with two rifles, set off to exact justice. At the farm they spoke to various farm workers, including Andreas in the mealie fields, and told him not to accompany white men when they went into the bush and stole Bushwomen who were doing no harm. In a carefully laid ambush, two white farmers, Ohlroggen and Korting were killed. Eckstein was not with them. At his trial, Johannes took the blame for both killings, even though they had been committed with two different rifles.

During 1915 and 1916 Hans and his gang continued to harass the white farmers, scaring away their labourers and stealing stock. It was not indiscriminate harassment, however, being focused as it was on a few farms. Farmer Tributh of Sus is reported to have lost nearly all his livestock. Ackermann and Baumgarten reported that they had lost nearly a hundred head of cattle and Voswinkel, Wynack, Thomas, and Wilhelm also reported heavy losses. Farmers put a reward on Hans's head and the German government offered 200 marks for every rifle removed from a Bushman. Farmers organised unofficial posses and Bushmen continued to shoot at army patrols.

In October 1916 Hans was killed. Early one morning Hans and a small party of followers, including women and children, were walking along playing a long, bow-like musical instrument called a 'chas'. They only had one rifle which August, a loyal follower, was carrying. Hans had an infected foot and was hobbling along. Then galloping towards them came Feuerstein, an ex-post officer clerk who was interned at Sus for the duration of the war. The band scattered, except for the unarmed Hans, who because of his infirmity, stood still. Feuerstein charged up and emptied his pistol at him. He then proceeded to mutilate the body by cutting off Hans's head, so that as he later explained, he could claim the reward. The South African court which tried Feuerstein for this crime could find no extenuating circumstances and sentenced him to death. This sentence shocked the local expatriates, but Feuerstein managed to make a daring jail break and was not seen again in the Territory.

Max now appears to have taken over the leadership of the gang. But its existence was short-lived. South African troops wounded Max, who died in the bush, and captured the rest of the gang. For his services in making this possible, a Bushman informer was rewarded with the princely sum of one pound!

Motives and causes: attributed and otherwise

The image and reality of the 'other', of the Bushman as befits most social constructions, was, on closer inspection of an arbitrary and contradictory nature. This is well illustrated in an examination of the motives and causes which settlers attributed to Bushmen in their attempt to account for the behaviour of these shadowy 'others'. Farmers, abetted by some academics[27] put about the story that 'wild' Bushmen committed most of the stock thefts. Conveniently this absolved the farmers from being held accountable for such losses. Farmers were known on occasion to exaggerate their stock losses for self gain. It enabled them to claim various tax breaks, and relieved the pressure from the Land Bank to repay loans, etc. Farmers inevitably tended to blame the victim and in this they were sometimes assisted by government officials.

A popular version of this approach was to attribute the Bushman's behaviour to biological factors: Bushmen simply had to have meat. Echoing this widely held view, Brownlee observed that the 'tendency' towards cattle thefts could be understood as a consequence of game becoming more scarce. There was definitely no provocation involved although he had noticed:

> that certain thefts are wanton in nature and that more cattle or sheep are stolen at a time than can possibly be consumed by the thieves . . . [this] illustrates . . . a trait of barbarian character which is rather difficult to explain in words, but may be on account of an inherent lust for slaughter[28]

Other officials were not so naive and were more sensitive to some of the ploys which farmers would use. In 1925, the Gobabis magistrate reported: 'Most farmers bordering the hinterlands graze their stock on the crown lands and during the rainy season the stock stray and the cry of stolen stock by Bushmen is made too easily in order to get the Police assistance for its recapture.'[29]

Thus, in his Annual Report, the Director of Lands gave as his considered opinion that:

> the Bushman is generally dubbed as dangerous, both to the farmer and his cattle: they are said to steal great numbers for slaughter, and also to attack isolated farmers, but with regard to these depredations, I think that treatment hitherto meted out to these people is, and has been, the cause of most of the trouble. Personally, although I have met hundreds of Bushmen, I have never had any trouble.[30]

Twelve years later, with his characteristic bluntness, Hahn, the Ovamboland Native Commissioner, wrote that thieving was caused by hunger pure and simple and it was impossible for the Bushmen to exist on the food issued to them by farmers. Moreover, they were constantly being pushed off their own land. He was surprised that there was not more stock theft.[31]

Sometimes banditry was attributed to ill-treatment on farms. It appears, wrote Sergeant Coetzee from Steinhausen, that farmers who consistently chase the Bushmen from their farms are the only people who have trouble with them.[32] And in this he was supported by the venerable Vedder in the South African Senate.[33] Revenge had been recognised as a strong motive earlier by Lieutenant Hull in Grootfontein and was to appear as an explanation later as well, especially where Bushmen left meat or farm produce to rot in the sun.[34] The sergeant investigating the stock-theft complaints of Farmer Pleitz of Hayas Farm reported that Pleitz did not pay his workers and that it was 'strange' that he was the only farmer in the area suffering losses from stock theft by Bushmen.

The strongest motive for revenge was held to be when the farmers interfered with Bushmen women or took them as concubines. Fugitive Bushmen who had previously been farm labourers committed thefts after they had been badly fed and flogged, the Bluebook said, but the chief cause of all the trouble was when the farmers used Bushmen wives as concubines.[35] Lieutenant Hull was a little more candid: 'It seems that the Bushmen have lost all faith in the white man's methods [of justice], more especially as their women were being constantly interfered with by both farmers and police'.[36] The threat of revenge was recognised by farmers and served as a minimal control on the more outrageous abuses. This was, for example, the reason why Farmer Bohme of Kakuse West appealed to the Governor for assistance in 1915.

Perhaps the most thoughtful analysis of the causes of banditry was that made by Hauptmann von Zastrow in 1913. Most of the stock theft was committed by Bushmen who had experience of white farmers and who were treated badly in some other way. They generally belonged to the Gaikokoin Heikom and extending Bushman patrols to other areas would simply lead to the unrest spreading. Von Zastrow argued that harassment of Bushmen only led to increased stock theft because patrols only succeeded in arresting the aged and infirm. Instead, von Zastrow proposed to work more closely with the missions.[37] Elsewhere, in an article, he suggested that drought was a major factor and vengeance an important secondary factor. When the question was raised about how to explain the lack of banditry in previous droughts, von Zastrow showed a sensitivity to ecological factors and the reality of white encroachment by pointing out that raiding tended to occur in areas heavily populated by whites, for example at Nurugas, but not in areas like Ojtitix and Namutoni which had heavy Bushman settlements, but comparatively speaking fewer white settlers.[38]

The Bushmen tamed

So strong was Bushman resistance to the post-First World War influx of settlers from South Africa, that the administration was compelled to invoke new laws. In South West Africa the vagrancy law (1922), which was more draconian and under which it was easier to be convicted than in South Africa, was aimed specifically at containing the 'Bushman danger' rather than black migrant workers. Not only do we find direct allusions to this in official corresponsence, but structurally too such an argument appears reasonable. Bley points out, for example, that by 1908 over 90 per cent of the total black male population was employed, and of the pastoral Herero and Nama, only an estimated 200 males were not employed by whites.[39] There was little need to control Herero rural workers with a battery of laws since they were trying to replenish their livestock and were thus not so mobile as the Bushmen as they were dependent upon white farmers for grazing rights. In the 1920s and 1930s the other major potential labour sources for farmers, the Ovambo and Kavango, were still to be diverted from the mines. When the Vagrancy Act proved unsatisfactory, additional legislation was enacted to make the carrying of a 'Bushman' bow an offence. One could still carry a Herero or an Ovambo bow but not a Bushman bow. However when whites had Bushman bows they were not prosecuted, as these bows were then redefined as 'curios'.

Bushman 'tamed' in such a manner made good farm labourers. One of the most striking impressions I have after combing the relevant literature at the Windhoek Archives is the strength of the attachment Bushmen had to their traditional water holes. Workers had an almost fatalistic preference for working on farms which encompassed their traditional land areas. Indeed, even today very few Bushmen, less than 3 per cent, are to be found in the urban areas. While Bushmen had a reputation for 'unreliability', they were generally the cheapest and most exploitable labour around. As one magistrate put it:

> [Farmers] will get more value by employing Bushmen than if [they] were to employ other natives, because Bushmen are usually satisfied with wages in kind such as tobacco, beads, material, etc., and there need be no cash disbursements. A Bushman has no sense of value of these articles and will render the service required by his master for them without considering their intrinsic value.[40]

One Native Commissioner was more astute, complaining, to no apparent effect, that on the farms in the territory, 'Bushman wages, if they received any at all, are always less than those paid to Hereros or Ovambos and their food rations too are less and of inferior quality'.[41]

Despite this rank exploitation, Bushmen were working on farms in increasing numbers. Already in German times von Zastrow had pointed out that over half the farms in the Grootfontein district were run on Bushman labour.[42] These numbers increased as colonisers expanded their number of farms. By 1939 the magistrate could report that 'now there is hardly a farm in the district where Bushmen farm labourers are not to be seen'. These farms were mostly in the eastern and northern portions of the district, 'proving it seems that they yielded to the domination of the white man only where there was practically no further escape'.[43] A similar situation was played out in the Gobabis district. The importance of these two districts can be seen from the fact that they supply between 60 per cent and 75 per cent of the total commercial cattle in the country.

It was only after the Second World War that a supposedly clear policy towards the Bushmen was formulated. In 1950 a Commission on the Preservation of the Bushmen under the chairmanship of Professor P. J. Schoeman was established, Schoeman found evidence that the Bushmen were in danger of extinction, a situation which he attributed to their love of smoking tobacco!

According to Schoeman their love of tobacco was such that the Kung would trek up to the Kavango River where they would prostitute their wives to migrant workers returning from the mines for a bit of tobacco. This in turn led to them all becoming infected with venereal diseases for which they had little resistance.[44] Schoeman proposed that a Bushman reserve be created in the vicinity of Nyae-Nyae, a traditional Kung area, which was fortunately well-isolated from white farms. Troublesome Bushmen from other parts of the country, including the Kalahari Gemsbok Reserve in South Africa were to be resettled there.[45]

It was to be a good many years before Schoeman's recommendations were implemented. Why did it take so long? Was it because of bureaucratic muddles? I suspect that they were not acted upon until 1959 for the reasons indicated in an undated memorandum of the Chief Bantu Affairs Commissioner quoted by Olivier:

> It is being found that the Bushman farm labourers are deserting . . . from the Gobabis district to go to Bechuanaland. To replace this labour, farmers have to place orders for contract migrant labour and the waiting period is now four-and-a-half years. The reason for the desertions appears to be that the Bechuanaland authorities have appointed an official to deal with Bushman affairs over there, and apparently offer them protection. It is feared that unless similar steps are taken immediately we will lose many of our Bushman labourers as well as others. The matter of creating a Bushman reserve is thus seen as exceptionally urgent.
>
> It is not only the farm labour question which makes the creation of this reserve so urgent, there is also the question of control of our eastern border [with] Botswana where the penetration of Bechuanaland natives to hunt game still continually occurs and which usually resulted in the deliberate setting of bush and veld fires.[46]

This quotation is important because it stresses the importance of placing the Bushmen within the regional political economy. They have not been as isolated and independent as many believed, and moreover, we cannot understand the likely course of their future unless the economic basis of the larger social formation is considered because one thing should by now be obvious: Bushman policy did not drop ready made from heaven.

Patterns and forms of banditry

Moving from the world of mythic construct to that of empiric construct, clearly, ecological factors played an important role in determining the pattern of raids on livestock and migrant workers. Data from court records show that there was a close correlation between frequency of stock thefts and the lack of rainfall. This is true for the annual cycle as well as for longer drought cycles. In the annual cycle, for example, October was generally believed to be the peak month because that was the period just before the rains when the veld was dry, and stock theft occurred during the dry months immediately preceding the rainy season. During this time not only is game scarce, but it is easier for farmers to track lost or stolen livestock, so they are also more likely to be aware of thefts during this period. A further inducement to stock theft during this period is the fact that it is the most uneconomic one in the ranching cycle, so that farmers tend to be spare in their provision of meat to their work force. In the longer cycles, it is clear that drought brings larger than average numbers of Bushmen to the farms looking for work and subsistence, and that stock thefts thus increased. Indeed, even the attacks on Kavango labourers coincide with periods of drought and famine.

Perhaps the most remarkable feature of the pattern of stock theft and social banditry was the general adaptability of the Bushmen to changes in the environment and white attempts to control them. For example, they would steal horses belonging to Herero in Epukiro and use them for hunting, and once done with hunting, kill the horses to prevent the Herero from using them to track them down. They were also apparently aware of changes in local police strength and this had a significant impact on raiding. Captain Swemmer, the Grootfontein police commander complained that the Bushmen squatting at Police stations at Otjituo, Nurugas, and Auuns were not working but watching and observing police movements which they would report to their bandit friends.[47] This awareness also accounts for the reported resurgence of Bushman banditry during the two world wars when most of the rural police stations only had a token staff. Bushmen also would steal as much livestock as was possible, typically in the range of three to seven head of cattle, but on occasion they would take up to sixty head of sheep and herd these to a place 130 km from the farm. They were quite capable of waging a war of attrition against unpopular farmers. Thomas, the manager of Foxhof, reported that in six years his flock of 1500 sheep had been reduced to 70, and this was clearly not only the result of disease.[48]

During the initial period of colonialism, an entire herd might be driven off, but this pattern gradually changed to the occasional killing of a beast which would then be slaughtered and the meat carried off to the encampment. Some anthropologists have suggested that this is evidence of the survival or resilience of the Bushmen's 'traditional' methods of hunting, but given the environmental and social constraints such behaviour is eminently appropriate. To drive a herd of livestock into the Kalahari is not only to leave a trail as visible as the full moon, but also to slow down the speed of one's flight from expected pursuers. Moreover, killing one beast at a time lessens the possibility of discovery. Numerous stock-theft cases involved a number of cattle or sheep which were killed at irregular intervals of about a month. Police observers were at times amazed at how quickly stolen meat or white artefacts were dispersed or consumed: in short, placed into the exchange network. This should not surprise the knowledgeable person. It was given a decided stimulus by the fact that the police inevitably saw possession of white articles or a quantity of meat as proof of theft.

Adaptability is most clearly illustrated in the Bushman reaction to white punitive measures adopted against them. This ranged, inevitably, from being forced to move away, sometimes fleeing to Botswana,[49] to a pattern of behaviour which observers described as 'increasingly shy of contact with whites',[50] to behaviour which can best be described as downright saucy. Farmer Freiderichs complained that after his court case for inciting his 'tame' Bushmen to kill a cattle thief, 'the Kalahari have become so bold as to stone one of my cattle herds who attempted to prevent them driving off my animals so that he had to leave them in their hands whilst going for assistance'.[51] Farmer Weinrebe complained that the Bushmen had sent him a challenge telling him exactly when they were going to raid his cattle kraal and had then engaged in a bit of psychological warfare by building large bonfires around his kraal.

There was a rationale underlying many of these patterns of banditry: the desire to escape colonial domination. Whites did not believe the Bushmen capable of this motive. Yet in 1920 Deputy Police Commissioner Kirkpatrick visited Farmer Thomas of the Nurgugas area where he met a Bushman woman who had given birth to three boys while resident on the farm and had killed all three of them. 'She says she will not bring up boys in order that they have to work for white men! She has no grudge against white people, but she adopts the idea that Bushmen should not work for white people and is not going to have any child of hers so doing.'[52] Bushmen actions clearly involved more than just a few isolated individuals; indeed there is clear evidence of a larger common consciousness of being oppressed by the white farmers. Describing one of his efforts to arrest some Bushmen on the outskirts of a farm, Lieutenant Hull found that 'during our absence Acerkmann's tobacco garden had been looted and destroyed by another party of Bushmen. No doubt that the so-called tame Bushmen are in league with the cattle stealers and should be removed from their present abode. They have given warning to wild Bushmen as to the movements of white men'.[53] Bushman solidarity was frequently expressed in the following manner: 'Immediately the Police took action and arrested a number of Bushmen ... all the Bushmen in the employ of the complainant deserted his service, leaving him and his wife to do the caring of their stock themselves'.[54]

Perhaps the major reason for the ability of Bushmen to maintain their banditry for so long was their capacity to split up into small groups and to survive off the 'bush' by fleeing into the vast, largely waterless Kalahari where few outsiders could penetrate and hope to capture them. But at the same time factors which contributed to their rise as bandits were also responsible for their decline: solidarity was never strong enough to overcome the pervasive libertarian egalitarian ideology which bred informers at a remarkable rate especially from the women;[55] and, secondly, the continued, sustained government policy of driving 'troublesome' Bushmen into the wasteland beyond the Police Zone eventually took its toll.

Peripherality enshrined: the praetorianisation of bandits

In 1975, the Western Caprivi witnessed a large influx of Bushmen refugees, the *Vasekele* from strife-torn Angola. According to a recent South African army version, they had been living peacefully in south-eastern Angola until the late 1960s when guerrillas entered the area, abducted many men and forced them to undergo military training. The remaining *Vasekele* then fled and joined Portuguese counter-insurgency units. *Soldier of Fortune* magazine, on the other hand, claims that they worked as bounty hunters for the colonisers, and were paid on the basis of the number of ears brought in.[56] During a foray into Angola some South African Bushman trackers were supposed to have met with the fleeing *Vasekele* soldiers and their families: 'It took little persuasion to encourage the Vasekele to join' the South African military machine.[57] These refugees then became the core of the white-officered 201 (Bushman) Battalion. The South Africans claim that they volunteered for military service, but the *New York Times* was more forthright: Vasekele joined up 'not only because of the absence of employment in this virtually underdeveloped area [sic] but also because they fear that if the guerrillas are victorious, they will be massacred for having supported the Portuguese in Angola'.[58] Given this general psychological climate, it is rather difficult to talk about the Bushmen volunteering, although, of course, the relatively high salaries and well-publicised services they receive might create the delusion that it was an 'act of free choice'. Certainly their decision to enlist was not a well-informed choice:

> Lieutenant Wolff (the only White who speaks Bushman) admits that the Bushmen have 'no political sense' and know little about the causes in the war they are helping to fight . . . 'they do ask what is going to happen to them' in the future . . . 'At this stage I can't tell them anything' he says, 'I'm here for the fighting part, not the talking'[59]

A year later, the army acknowledged that the Bushmen refugees had very little choice in the matter. They either signed up or had to leave the region entirely.[60] No matter how 'adult' the Bushmen of 201 Battalion might be, in terms of international law they have unwittingly become mercenaries, since not by the longest stretch of the imagination can they be regarded as either Namibians or South Africans.

Given their power, how does the army perceive its wards? Army officers adduce *hate* as the major reason for the Bushmen joining the army. 'A Bushman's hate for SWAPO will give you the shivers . . . they hate SWAPO because they enslaved them and took their daughters for prostitutes' said Commandant Botes.[61] However these statements contradict the traditional army denial that SWAPO was very effective in Angola, let alone in Namibia, and its claim that SWAPO activities amounted mostly to planting a few isolated land mines. A subsidiary myth used to support this belief is that 'since earliest times they have been despised and persecuted by the other tribes',[62] that 'traditionally they were the drawers of water and hewers of wood for the blacks'.[63]

Another reason given by the military (and some anthropologists) for Bushmen joining the army, is that they have nothing to do: 'Traditionally the man here is in heaven. The only thing he does is plough fields [!]. The women do all the work. It's taken years to teach the man to do his part. But we are gaining. We do it through school – agriculture is a compulsory subject.'[64] What we have here is clearly a massive case of invented tradition; customs that are the products of the imagination of the colonisers and imposed upon those people labelled Bushmen.

To judge by the numerous tours on which it takes the press to view the Bushman base at Omega, the army is clearly thrilled with the performance of its Bushmen soldiers and their attempts at 'upliftment'. Part of the attraction is clearly derived from white myths that the Bushmen are supposedly the best trackers in the world. Their reputation as trackers meant that they were recruited for psychological purposes. Their presence was meant not only to inspire fear among the guerrillas, but also to serve as ideological talisman for the colonial troops: 'The Bushman's senses in the field are unbelievable. If a patrol has a Bushman with it, then it is unnecessary to post guards at night. The Bushman also goes to sleep. But when the enemy is still far away he wakes up and raises the alarm' according to one senior officer.[65] Another white soldier is quoted as saying: 'With the Bushmen along, our chances of dying are very slight. They have incredible tenacity, patience and endurance. They've taught me to respect another race'.[66] The Bushmen have become prisoners of the reputation placed upon them by the whites. Some newspaper reports also hint at this and anthropologists agree. Some of the most knowledgeable of them believe that all Bushmen born after about 1960 have lost the ability to track, and most of their veldcraft skills.

While Bushmen and white soldiers are supposedly fighting the war as compatriots and as equals, receiving the same wages on a symbolic level the Bushmen are constantly reminded of the structured inequality of the situation. This becomes apparent most strikingly in their insignia. The commanding officer explained the 'pied crow' emblem of the 201 Battalion in these terms:

> The black portion of the bird represents the Bushman population while the white breast represents the white leadership element (thus they accept that whites take the lead in their development process). The crow is the first bird which was let out of Noah's ark and did not return – this symbolizes the fact that the Bushmen too will not return to their previous customs.[67]

'The Defence Force does not only make war. On the contrary, the task of *civilising* . . . is probably *greater* than the military function'.[68] Despite its having a large and active Ethnology Section, drawn mostly from ethnologists trained at Afrikaans-language universities, the army's 'civilising programme' seems at best to be *ad hoc*, naive, and contradictory. Army officers rationalise their attempts at 'upliftment' by arguing that it will enable the Bushmen to protect and be proud of their identity. They hold a narrow and romantic view of Bushman culture as grounded on hunting, and to protect Bushman identity have modified the Western-oriented school curriculum so that school children can spend a few days in the bush every year in the belief that the children would then retain their basic hunting skills. Very little attention appears to have been given to what will happen to the Bushmen after the army leaves, as one day it will. On the contrary, prevailing military opinion tends to suggest that when the South African army leaves Namibia, the Bushman battalions will leave with them.[69]

So successful have the Bushman mercenaries of 201 Battalion been judged to be, that the South African army has enthusiastically extended recruiting to local Bushmen and even dropped the educational requirements. The favourite game of Bushman farm school children is now 'Army Army'. Within the space of five years the Bushmen have emerged with the dubious distinction of being the most militarised ethnic group in the world.

Conclusion

Given the phantasmic mix of reality and myth through which the colonisers saw the Bushmen, it is not surprising to observe that like many other African peoples, these shadowy peripheral people were incorporated by their rulers as a social control device. From the earliest days of white colonialism Bushmen were much prized as police trackers. So highly were they valued in this role that when a star Bushman tracker was charged with homicide for killing another Bushman, the police, in an unprecedented move, hired the best white lawyer in the territory to defend him. In recent years this control role of the Bushmen has, as we have seen, expanded significantly as the South Africans have waged a protracted 'low-intensity' war with SWAPO. Like many despotic regimes before them, the South Africans have learnt that the Bushmen, like 'outsiders' in other parts of the world, make ideal 'servants of power'. And as an added bonus, the Bushman 'upliftment' programme has great propaganda value, a value derived in part from academic research on the Bushmen. Undoubtedly one of the most influential theses generated by the extensive Bushman ethnography has been the notion of the 'original affluent society' as developed by Marshall Sahlins. In essence it argues that in pre-contact times hunter-gatherers like the Bushmen lived in a state of 'primitive affluence' because they had to work for a comparatively short period of time in order to meet all their basic requirements. While there is considerable controversy as to whether the original affluent society ever existed *in natura*, the ultimate paradox is that the model appears to be valid as a characterisation of Bushman bandit gangs both in the historic and present time.

NOTES

1   Research on this project was funded by a Walshe-Price Fellowship and by the University of Vermont. It was conducted in Namibia during 1983. Fuller details on banditry in Namibia and more detailed archival references are to be found in my forthcoming monograph entitled *The Bushmen: A Myth Explored*.

2   Allen Isaacman, 'Social banditry in Zimbabwe (Rhodesia) and Mozambique, 1894–1907: an expression of early peasant protest', *Journal of Southern African Studies*, vol. 4, no. 1 (1977), pp. 1–30.

3   See, for example, Gavin Kitching, *Development and Underdevelopment in Historical Perspective* (New York: Methuen, 1982).

4   P. O'Malley, 'The class production of crime: banditry and class strategies in England and Australia', in R. Simon and S. Spitzer (eds), *Research in Law and Sociology* (Greenwich: JAI Press, 1981).

5   E. J. Hobsbawm, *Bandits* (New York: Dell, 1969), p. 14.

6   G. Theal, *History of South Africa 1652–1795*, Vol. 1 (London: Swan Sonnenschein, 1897), p. 227.

7   See, for example, R. B. Lee, *The Kung San* (New York: Cambridge University Press, 1979); C. Schrire, 'An inquiry into the evolutionary status and apparent identity of San hunter-gatherers', *Human Ecology*, vol. 6, no. 1 (1980), pp. 9–31.

8   G. S. Nienaber, 'Die Woord "Boesman"', *Theoria* (Pietermaritzburg) (1950), pp. 36–40.

9   Shula Marks, 'Khoisan resistance to the Dutch in the seventeenth and eighteenth centuries', *Journal of African History*, vol. 13, no. 1 (1972), pp. 55–80.

10 State Archives, Windhoek, Namibia (hereafter SAW), Verordnung 26883/5391 of 24 October 1911.

11 SAW, Walbaum Journal, p. 47.

12 SAW, ADM 38238, Herbst to Gage, 25 November 1919.

13 J. Wellington, *South West Africa and its Human Issues* (Cape Town: Oxford University Press, 1967), p. 272.

14 J. Hermans, 'Official policy towards the Bushmen of Botswana: a review, Part I', *Botswana Notes and Records*, vol. 9 (1979), pp. 55–67.

15 L. Marshall, *The !Kung of Nyae-Nyae* (Cambridge, Mass: Harvard University Press, 1976).

16 SAW, Walbaum Journal.

17 SAW, SCC/1918, Rex v. Feuerstein.

18 SAW, SCC/1918, Rex v. Johannes Fritz.

19 SAW, SCC/1918, Rex v. Feuerstein.

20 SAW, ADM 3360, Report of 26 January 1917.

21 SAW, SCC/1918, Rex v. Feuerstein.

22 SAW, SCC/1918, Rex v. Massinah.

23 SAW, Walbaum Journal, p. 48.

24 SAW, SCC/1918, Rex v. Feuerstein.

25 SAW, SCC/1918, Rex v. Johannes Fritz.

26 SAW, Walbaum Journal.

27 C. M. Doke, 'The Qhung Bushmen of the Kalahari', *South African Geographical Journal*, vol. 8, p. 43.

28 SAW, ADM 3360, Report of 10 April 1917.

29 SAW, A50/25, 22 July 1925.

30 SAW, Annual Report, Director of Lands, 1919.

31 SAW, A50/67, Ovamboland Native Commissioner to Chief Native Commissioner, 5 September 1940.

32 SAW, LGO/3/1/4i, A50/25, Steinhausen 12/1930.

33 Hansard, Senate of the Union of South Africa, June 1951, p. 5620.

34 SAW, ADM 273, Hull to Military Magistrate, 24 December 1915.

35 Great Britain, Parliamentary Papers, Cmd 9146 (1918), *Report on the Natives of South West Africa and Their Treatment by Germany*.

36 SAW, ADM 1/13, Report of 6 November 1915. Space precludes a discussion of the key role of female exploitation in the subjugation process of the Bushmen.

37 SAW, Grootfontein Annual Report, 1913.

38 B. von Zastrow, 'Uber die Buschleute', *Zeitscrift für Ethnologie*, vol. 46 (1914), pp. 1–7.

39 Helmut Bley, *South West Africa Under German Rule* (Evanston: Northwestern University Press, 1971), p. 250.

40 SAW, A50/67, Magistrate, Grootfontein to Chief Native Commissioner, 15 April 1939.

41 SAW, A50/67, Hahn to Chief Native Commissioner, 5 September 1940.

42 Von Zastrow, 'Uber die Buschleute'.

43 SAW, A50/67, Meintjes Memorandum, 1939.

44 P. J. Schoeman, 'Weeskinders van Afrika', *Landbouweeklad*, 12 October 1971, pp. 10–15. Schoeman was not alone in his naivete. Isaac Schapera, the doyen of Southern African ethnography, also noted the decline in numbers over this period and attributed it to infanticide and malaria!

45 South Africa, Department of Welfare, *Die Boesmans in die Noord-oostelike dele van Suidwes-Afrika* (1962, mimeo).

46 M. J. Olivier, 'Inboorlingbeleid en-administrasie in die mandaatgebied van Suidwes-Afrika' (unpublished PhD dissertation, Stellenbosch University, 1961), p. 140. The translation is mine.

47 SAW, ADM 2; ADM 112, 25 October 1919.

48 H. J. K., 'A trip to the Okavanga', *The Nonggai*, December 1920, pp. 626–8; January 1921, pp. 2–5; February 1921, pp. 58–62; March 1921, pp. 115–20; April 1921, pp. 174–8.

49 J. Hermans, 'Official policy towards the Bushmen of Botswana: a review, Part I',

Botswana Notes and Records, vol. 9 (1979), pp. 55–67.
50 H. J. K., 'Trip to the Okavango'; L. Engelbrecht, 'Wagposte aan die Oekavango', *Die Brandwag*, vol. 13, no. 9 (1922), pp. 263–8.
51 SAW, ADM 3360, Report of 27 February 1923.
52 H. J. K., 'Trip to the Okavanga' (April 1921, p. 177).
53 SAW, ADM 273, Report of 24 December 1915, Hull to Secretary, Windhoek.
54 SAW, Annual report, Outjo, 1923.
55 See, for example, Lee, *Kung San*.
56 Article in *Soldier of Fortune*, February 1984.
57 Article in *Paratus*, February 1983, p. 28.
58 *New York Times*, 24 February 1981.
59 *Christian Science Monitor*, 19 March 1981.
60 *The Star* (Johannesburg), 24 November 1982.
61 *Sunday Tribune*, 1 March 1981.
62 *Eastern Province Herald*, 21 August 1980.
63 *Die Burger*, 8 February 1980.
64 *The Star* (Johannesburg), 9 October 1981.
65 *Die Burger*, 1 June 1982.
66 *Time*, 2 March 1981.
67 *Paratus*, May 1978.
68 *Die Volksblad*, 15 July 1981.
69 *The Star* (Johannesburg), 24 November 1982; *Windhoek Advertiser*, 10 November 1982.

SECTION III

Protest and resistance

9 Forms of resistance: songs and perceptions of power in colonial Mozambique[1]

LEROY VAIL and LANDEG WHITE

'Do you know, my son,' Papa spoke ponderously, and gesticulated a long time before every word. 'The most difficult thing to bear is that feeling of complete emptiness . . . and one suffers very much . . . , very very very much. One grows with so much bottled up inside, but afterwards it is difficult to scream, you know . . .'

Mama was going to object, but Papa clutched her shoulder firmly. 'It's nothing, mother, but, you know, our son believes that people don't mount wild horses, and that they only make use of the hungry docile ones. Yet when a horse goes wild it gets shot down, and it's all finished. But tame horses die every day – as long as they can stand on their feet.'[2]

These words are from Bernardo Honwana's 'Papa, Snake, and I'. That story describes with warmth and great subtlety the effects on a small boy of growing up in colonial Mozambique with the feeling, and then suddenly the clear knowledge, that his father is a weak, exploited man who does nothing to resist his exploitation. The setting, as for most of Honwana's stories, is one of those small administrative towns – with a church, a school, a *cantina*, a police post, a club, and an administrative building – that can be found throughout Mozambique. The father works as a minor civil servant in the Portuguese colonial administration, and the crisis in the story comes when the boy hears his father cursing under his breath a white Portuguese neighbour and demands of him, 'Why didn't you say that to his face?' At the end of a long evening of tension, father and son come to understand one another. The father realises that the time has come to admit his son to that adult world of powerlessness and humiliation from which he has so long sought to protect him, while the son, suddenly matured, accepts the responsibility of sharing his father's burden. The excerpt quoted above defines a kind of struggle in which the 'tame horses' who 'die every day' try to bring up their children with the knowledge of values other than those prevailing in colonial Mozambique.

The historiography of Africa during the 1960s and into the 1970s was dominated by studies of African resistance to European imperialism and to colonial rule. The rise of mass nationalism in postwar Africa led historians to ransack the past for earlier leaders who might have served as role models for the anti-colonial struggle, and resistance became nationalism's historical dimension.[3] African leaders who had attacked colonial armies were granted heroic stature, and Africans who had sought education and had attempted to foster modernisation and Westernisation were seen as shining apostles of African advancement. In this approach, unexceptional people, ordinary villagers were usually ignored. A book of cardinal importance in shaping this attitude among both scholars and African intellectuals was George Shepperson's path-breaking *Independent African*, published in 1958. In it, the Malawian peasants who failed to support the abortive revolt of the author's hero, John Chilembwe, in 1915 are brusquely dismissed as 'a violent vacillating lumpenproletariat'.[4]

At about the same time that nationalist historians were demonstrating the importance of resistance movements, students of imperialism also recognised the role of resistance in giving shape to imperial expansion. Imperialism and colonialism could not be understood without linking them to the specificities of African societies, and Ronald Robinson and John Gallagher even suggested that the thrust of expansion was called forth by 'African proto-nationalism' rather than propelled by any conditions within the European economy.[5] Common to both approaches was the tendency to concentrate on centralised political structures and to examine the histories of these movements by focusing on the biographies of their leaders.[6] Where genuinely popular uprisings existed in the historical record, they were usually given standing by the discovery that their effective leaders were scions of 'royal' houses. Counterpoised to the resisters in these early studies were those Africans who sought accommodation with the agents of capitalism and colonialism, the 'collaborators'. These, too, had a modern relevance: if nationalist historiography required the creation of a long line of heroic resisters, the disasters of the first years of independence could usefully be explained in terms of an equally long line of guilty collaborators. Again, however, the collaborators tended to be exceptional people – chiefs, mercenary soldiers, educated clerks – rather than ordinary villagers.

By the end of the 1960s, historians accepted that 'resister' and 'collaborator' were clumsy categories with little analytical power. Resistance and collaboration were recognised as alternative strategies open to African leaders as they sought to maintain or augment their power in the face of the colonial challenge. With this recognition came a greater sophistication in dealing with the politics of imperialism and colonialism.[7] Yet, despite their growing sophistication, resistance studies remain essentially concerned with the military and political history of members of the elite or relatively centralised political structures.

The shift in the mid-1970s away from the search for the roots of nationalism to the search for the roots of underdevelopment, as historians sought to account for the emergence of so many poverty-stricken and politically unstable countries, necessitated a change in the study of African reactions to colonialism. In examining, as the title of Walter Rodney's immensely influential book succinctly phrased it, *How Europe Underdeveloped Africa* (1972), there has been little scope for discussing African initiatives. The modernising and Westernising leaders of the earlier pantheon have become in this new historiography tarnished and irrelevant. But the peasantry, once viewed as obscurantist and backward, has been elevated to an 'authentic' stature, largely because of concerns within African countries themselves. In many parts of Africa the peasants in the past showed quick willingness to produce for the market. An influential group of specialists on Eastern and Southern Africa, defining peasants in terms of their exchange relations, charted the destruction of this efficient and market-oriented peasantry by the forces of colonial capitalism, a process assumed to have been completed by the 1920s.[8] Even as their work was being published, however, nationalist movements claiming a peasant ideology and mass peasant support were sweeping to power in

Mozambique, Angola, and Zimbabwe. The new 'political' peasants, figures linked charismatically with the revolutionary peasants of Vietnam and Latin America, have been superimposed on the old 'economic' peasants to create, in African countries where agricultural production is faltering and where the failures of the early nationalists are all too apparent, mass heroes of irresistible appeal.

Yet, in the process of elevating the peasantry to its new legitimacy, the vocabulary of the earlier historiography has not been abandoned. Since the early 1970s, resistance studies have been extended to include everything from foot-dragging and dissimulation to social banditry, arson, poaching, theft, avoidance of conscription, desertion, migration, and riot. The new resistance is, in short, any activity that helps frustrate the operations of capitalism or of capitalism's creations. With the argument moved from the rather uncertain political terrain of the 1960s to firmer economic grounds, much more room is allowed than hitherto for the variety of actual human behaviour, which must in itself be a welcome advance. But there are problems. Quite apart from the violence done to language by describing such things as desertion and emigration as 'resistance', the continued use of 'resistance' and 'collaboration' blurs analysis. Is alienation resistance? Is it collaboration to be involved in producing goods for sale on a capitalist market? What happens when tax evasion, desertion, and low worker productivity are practised by the peasantry in new revolutionary states, and such behaviour has to be returned once more to the categories of ordinary human vice? Is it simply a matter of whether the Left or the Right is in power?

Even more worrying is the naiveté of so much resistance psychology, with its accompanying political moralising. Human reactions to exploitation are enormously diverse. The moment we go beyond the kinds of organised resistance that are politically or militarily visible and try to deduce from people's behaviour their attitudes, perceptions, and cultural values, we find ourselves in areas where terms like 'resistance', 'collaboration', 'subjective consciousness', 'false consciousness', and the like lack the necessary nuance. This is particularly true when dealing with preindustrial societies, for, as George Rudé has cogently remarked, 'Such terms as "true" or "false" consciousness (that Marx had originally applied to the industrial working class) can have no relevance at all.'[9] A frustrated farm labourer in colonial Zimbabwe who misses work because of drunkenness may be viewed as undermining settler capitalism and, hence, as 'resisting'; a man who works hard, saves his money to educate his children, and finds his satisfactions in ferocious church-going may well be seen as selling out to the system and, hence, as 'collaborating'. But with reference to which theories of the human mind do these labels become the 'right' things to be said about such behaviour? Or, equally important, the most interesting? Any competent novelist would have no difficulty in devising twenty different narratives describing twenty different ways of coping with exploitation, and each of the twenty heroes and heroines – their days made up of a mixture of semi-resistance and semi-collaboration – would have an equal claim on our attention. Honwana's 'tame horse', the boy narrator's father, is presented with great sympathy and respect; the fiction would not be 'truer' for being angled as an attack on his 'consciousness'.

At the heart of this problem is a difficulty faced with special acuteness by social historians – namely, the difficulty of transcending the particular. Out of all the different case histories, all the variety of human experience, how can one establish historical themes? One solution is to allow available printed sources to operate as a kind of sieve, so that we examine not actual human behaviour but behaviour that has reached the attention of newspapers, law courts, governmental and non-governmental reports, and the like. The information so gleaned is more easily quantifiable since processes of selection have already occurred. Attitudes, motives, and cultural values can then be supplied either by reference to anecdotes or by extrapolation from the historian's own interests and social concerns. A second possibility is to observe, as anthropologists and political scientists have done, institutions such as chiefdom, kinship, the law, religious practices, and similar 'objective' phenomena and to deduce from them the nature of ideas, beliefs, and motives. At best, the conclusions are interesting and suggestive, but, just as a skeleton is not a living body, so the structures postulated from a study of institutions do not represent the living reality of African thought. At worst, this approach can amount to taking refuge in theory, accepting a version of society and history as correct and then noting selectively whatever fits the theory. Rarely has it been possible to demonstrate in any depth the part played by African perceptions in the unfolding of events. With the thoughts of the actors largely unregarded, the way has been left open for the elaboration of grand theories that lack human and material specificity and that treat people as objects while documenting their exploitation.

A third possibility, and one that is more explicitly 'subjective' than the first two, looks at the creation of the people themselves, the forms in which their own major concerns are expressed. Of these, the most accessible are songs and oral narratives, which often, either directly or implicitly, make symbolic statements about common experiences and concerns. Whole groups of people are able to identify with such cultural expressions; hence, these narratives and folk-songs reflect popular consciousness. To illustrate such an approach to writing social history 'from below', we shall discuss the content of some two hundred African songs recorded in three different areas of Mozambique.

There are difficulties in handling this kind of material for historical purposes. Dating is often difficult, and, even when the poetry is translatable, the words are frequently obscure. One must beware of assuming that every song springs automatically from 'the community' or from 'the people'. Each song is ultimately the work of an individual singer, and, although oral performance (and eventually oral transmission) must depend on some degree of communal acceptance, the link between individual and group opinion has in each case to be demonstrated – either by proving that the song concerned is popular over a wide area or by showing how certain themes and preoccupations keep recurring in large numbers of songs. Ideally, in using such material, one would wish to pay the closest possible attention to the actual meaning of the songs, supported by oral testimony on their meaning to the people who performed and listened to them, and supplemented by investigation into the social position of the singers, into the conventions of the forms they are using, into the context and contingencies of the performances, and into the importance of such songs in the oral literature and general culture of the area as a whole. This is a demanding programme, and in the case of some of the material we shall be considering is no longer feasible, leaving us to some extent dependent on interpretive criticism supplemented by intelligent hypothesis.

Set against such difficulties, however, are two enormous advantages. First, many societies in Central and Southern Africa, as elsewhere on the continent, take for granted that poetry is an appropriate medium for discussing the impact of power. The praise-poem in its many different forms seems to have arisen specifically to provide such a resource, and most of our examples deal in one way or another with the exercise of Portuguese colonial power. Second, it is a convention of the form that power may be openly criticised. Chiefs and headmen may be criticised by their subjects, husbands by their wives, fathers by their sons, employers by their workers, and colonial officials by their underlings in ways that the prevailing etiquettes do not otherwise permit. As one informant declared in speaking of a song directed against the field manager of a sugar estate, 'You could swear at him and he just smiled'; to attack him outside the song 'would be just insulting him . . . just provoking him,' but, as long as it is done through singing, 'there will be no case.'[10] The form, in short, legitimises the content: this 'free expression' is in many African societies not only tolerated but openly welcomed as a major channel of communication between the powerless and the powerful, the client and the patron, the ruled and the ruler. As a consequence, the kinds of Mozambican songs discussed in this essay provide unparalleled insights into popular perceptions and dissatisfactions during the period of Portuguese colonial rule. As a by-product, these songs make clear how insensitive and inappropriate is the habit of categorising African reactions in terms of 'resistance' or 'collaboration'.

The songs to be considered are taken from two large groups. The first is a collection of forty-six songs that Hugh Tracey compiled in the 1940s, taken from the Chopi musicians of southern Mozambique. As a musicologist, Tracey was attracted to the Chopi by their impressive orchestras of xylophones and, especially, by their spectacular *migodo*, annually staged entertainments made up of dances, songs, and orchestral music in presentations that lasted some fifty minutes each. In his *Chopi Musicians* (1948), Tracey examined seven complete *migodo* (singular, *ngodo*), six from Mozambique and one from the Witwatersrand, which he collected from 1940 to 1943. The second group consists of 187 songs we collected from 1975 to 1977 in the Quelimane district of Mozambique. The majority of these songs date from the 1940s and 1950s, although some are much older, and a few originated in the late nineteenth century.[11]

Tracey's *Chopi Musicians* is a minor classic. By paying serious attention to African music, Tracey was a generation ahead of his time, and in the Chopi orchestras he found a splendid subject. Surprisingly, given the infectious enthusiasm of his study, no one has followed Tracey's lead and investigated the *migodo*, either for their content or their performance.[12] It was the music of the *migodo*, particularly the instrumentation of the orchestras, that commanded Tracey's most informed attention. Neither as a literary critic nor as a historian was he equipped to plumb the meanings of the words of the songs he transcribed. Although his explanatory notes are frequently very illuminating, his general comments range from the whimsical and patronising to the completely erroneous. Two examples will suffice to demonstrate the need for a fresh look at his material. The first of Tracey's songs, the fourth section of the *ngodo* by Katini of Wani Zavala's village in 1940, runs as follows:

> It is time to pay taxes to the Portuguese,
> The Portuguese who eats eggs
> And chicken!!
> Change that English pound![13]

Tracey commented that the tax question 'is an oft-recurring theme in native songs', and his explanation of the fourth line, that migrant workers returning from South Africa resented having to change their sterling currency to Portuguese escudos at the border with a 10 per cent discounting, is helpful. On the song's central eggs-and-chicken metaphor, however, he had only this to say:

> The rightness of reference to the Portuguese as those who eat eggs and chicken will be wistfully admitted by every good Portuguese housewife in the district, of whom there may be a dozen or so. (The last census shows a total of 27 Europeans in the Zavala district.) Had they added fish from the lakes as well, the picture of the available choice of proteins would have been complete! But with the memory of my hostess vividly in mind I would gladly return any day to sample again the hundred and one delicious disguises she has conjured up for this culinary trinity.[14]

There is more in the same vein, and, although the song itself is almost forgotten, Tracey inadvertently helped confirm its central complaint – which is, of course, that the Portuguese consume everything (eggs *and* chicken) without regard for the future. Equally innocent is Tracey's comment on a song that attacks the use of the brutal *palmatoria*, a perforated paddle used to beat people: 'Here is a mystery, the Portuguese beat us on the hands,/Both us and our wives!'[15] Tracey's 'mystery' is a weak word for what ought to be translated as 'arbitrary and inexplicable', and his comment that this is a 'very light-hearted lyric' arising from the opinion of the Chopi that they instead should be beaten on 'the part specially fatted by nature for sacrifice' is typical of his general refusal to take the words of these songs seriously. There are real grounds, then, for re-examining his material, partly to illumine the realities of Chopi society and partly to provide comparative material for the larger group of songs that we collected.

Mozambique's Quelimane district, which is bounded by the Zambesi, the Shire and the Ligonha rivers, has for the greatest part of the twentieth century operated as an enclave economy.[16] From the 1890s to 1930, most of the district was under the quasi-governmental control of four large plantation companies. Sena Sugar Estates, founded with British capital and under largely British management, controlled three *prazos* (estates leased from the Portuguese crown) straddling the Zambesi River. The Companhia do Boror, founded with French capital, grew coconuts and sisal in a group of five *prazos* to the north and north-west of the town of Quelimane. The Société du Madal, based in Monaco, ran coconut plantations in four *prazos* to the south of Quelimane. Finally, the Companhia da Zambesia, the only Portuguese firm and based in Lisbon, grew coconuts and sisal in three separate *prazos*. Two further companies founded with Portuguese capital operated briefly until each was absorbed by the expanding Sena Sugar. The Companhia do Luabo grew coconuts and sugar along the Zambesi, while the Empresa Agricola de Lugella controlled three huge *prazos* bordering Nyasaland. These companies ran most of Quelimane district until 1930, when Salazar abolished the concession system. Administered from their head offices in London, Paris, Monaco, and Lisbon, these firms levied labour, collected taxes and customs revenues, coined money, monopolised trade, ran their own police forces, and distributed justice throughout their *prazos* – limited only by Portuguese 'agents of authority', whose salaries they paid. The sole exceptions to this regime in the whole district were the circumscriptions of Maganja da Costa and of Alto and Baixo Moloque, which remained under the control of the state and which after 1903 became suppliers of labour to the Witwatersrand.

After 1930, when state administration was extended to the whole district, Quelimane remained a single and self-contained economic unit. Although Salazar's 'reforms' were meant to 'Portugalise' the district's economy, the new administration's most pressing economic problem was to supply workers not only to the established plantation companies but also to new tea producers and to the concession holders who had exclusive control over the compulsory commodity production of rice and cotton. At first, the state employed European labour organisers, but, as the demand for workers grew, it decided to appoint African chiefs, who were paid to act as front men for the government, the plantations, and the concession holders. Legally, all Africans were required to pay a head tax, which corresponded to the minimum wage for three-to-four months' work, and could be obtained from any source. In practice, especially after the Governor-General's Circular of October 1942, all male Africans between the ages of 18 and 55 were obliged to prove that they had worked for at least six months of each year as labourers in the service of the state, private concession holders, or companies. In practice, too, labour on the cotton and rice concessions was supplied by women. No one legally left Quelimane district to work for the higher wages available elsewhere. Apart from a certain degree of temporary clandestine emigration to Manica-Sofala, to Southern Rhodesia, and to South Africa, and apart, too, from a substantial degree of permanent emigration to Nyasaland, which placed constraints on labour recruitment in border areas, Quelimane district was successfully administered as a labour reserve for the plantation companies and for the concession holders. Although the state constantly modified its labour strategies in the face of African reactions, not until April 1974 was the essential pattern of the Quelimane labour reserve broken.

Behind the apparent uniformity of these pressures on the African population, however, were two quite distinct historical experiences. When the plantation companies gained control of their *prazos* between 1890 and 1904, they acquired estates that had in most instances been in existence since the seventeenth century. Despite the repeated efforts of Lisbon governments from 1832 onward to reform the system of land tenure in the Zambesi valley, the *prazo* system had proved extraordinarily tenacious. This was principally because – as the wars of the 1850s and 1860s made clear – no Lisbon government could control Zambesia except in alliance with the *prazo* holders, whose interests were diametrically opposed to metropolitan liberal policies and whose power allowed them to ignore decrees and laws, to withhold troops from the government, and even to talk of secession. Only when international pressures, culminating in the British Ultimatum of 1890, forced Portugal to appear to be developing its colonies instead of just occupying them were the earlier attempts to abolish the *prazo* system abandoned. Instead, the Portuguese modified the system to attract international capital, offering longer-term investments and more stringent labour laws. At that point, the major plantation companies moved in. But, in taking over the *prazos*, they became heirs to that relationship between *prazo* holder and African cultivator, between landholder and tenant, between patron and client that had characterised the system since the late seventeenth century.

The African inhabitants of the *prazos* were accustomed to a system of administration that had as its ultimate head the *prazo*-holder, and they were used to paying taxes to the holder, both in money and in produce. They were even used to supplying tribute labour. But they were also accustomed to a pattern of reciprocity that allied their chiefs with the *prazo*-holders (if they were peasants) or that gave them the semi-protected status of one of the many categories of slaves. The *prazo*-holder was expected to provide military protection and to offer help in time of famine. Against any *prazo*-holder who exceeded his rights or failed in his obligations, the Africans applied the ultimate sanction of emigration. This rough *modus vivendi* was drastically modified and curtailed in scope during the long upheaval of the nineteenth-century slave trade. But it remained in the culture as a set of perceptions about the rights and duties of government, waiting to be employed by the workers and peasants against their new company overlords.[17]

This *modus vivendi* applied, however, only in those parts of the district that had historically been incorporated into the *prazo* system – that is, in the area south and west of Quelimane and along the banks of the Zambesi River. When troops of the Companhia da Zambęsia invaded the newly designated *prazos* of Milange, Lomwe, and Lugella in 1899, they came as an occupying colonial army into territory that had been completely independent. None of the company's demands – for the head tax, corvée labour, or a monopoly of commerce – could be comprehended as an adaptation of earlier systems of government. The reaction of the people of the area was, first, to attempt armed resistance and, when that was suppressed, to flee to coastal areas that the state had yet to bring under control, to fortified hilltop villages that had been their refuge during the slave trade, and in great numbers to British Central Africa across the border. Similar problems, though on a scale modified by the different terrain and by the greater distance of the border, were encountered by Sena Sugar and by the Companhia do Boror, which tried to impose their demands on inland areas, and by the state, which attempted to levy the head tax and to conscript migrant labour in the new circumscriptions of Maganja da Costa and Alto and Baixo Moloque. In none of these areas was it possible to imitate the kind of rule that had evolved in the older *prazos*, where by 1914 there had already emerged a version of company paternalism whose ultimate source was the idea held by the people themselves of what the *prazo* system was all about: the reciprocity of patron and client.

Instead, after decades in which migrant labour was obtained for the sugar, copra, sisal, and tea plantations by the old-fashioned methods of licensed raiding and in which mass emigration to Nyasaland continued unabated, Portuguese control of the interior of Quelimane district was achieved after 1945 through the agency of state-appointed chiefs and headmen. 'The chief was chosen by the government. When they came in a village like his one, they would gather everybody together and look for a clever one and name him a chief, and everybody would have to respect him for his title. Even a bad one.'[18] These chiefs became the administration's agents in tax collection, labour recruitment, and the enforcement of schemes for the compulsory production of rice and cotton. Though liable to ill treatment and dismissal if they failed to co-operate, the chiefs were paid handsomely and were hated by their new 'subjects' for whom they epitomised colonial rule. Thus, although the whole of Quelimane district operated as a single labour pool for the plantation companies, African reactions to their labour exploitation were far from uniform. The two extremes of response are represented by the Lomwe-Chuabo-speaking peoples of the interior highlands and by the Sena-Podzo-speaking peoples of the north bank of the lower Zambesi around Sena Sugar's head offices, first at Mopeia and then at Luabo. Elsewhere in the district, the distinctions were less clear cut; local circumstances and Portuguese pressures produced differing results.[19]

The Lomwe-Chuabo songs date mainly from the 1940s to 1960s, and they are dominated by two themes linked by a central preoccupation. The first and most obvious of these themes, expressed in more songs than we are able to quote here, is a bitter hatred of the state-appointed chiefs and headmen. Such hatred arose partly from the chiefs' official function, that of labour recruiter, responsible for supplying labour to the company recruiters or to the police, or for presenting the requisite number of workers at the administration building or at the roadside recruiting posts:

> The headman - ay - ay - ay - ay
> The headman harassed and seized one of my sons for Luabo,
> The other went to São Thomé and never returned.
> I'm going to bury the headman and build my house on his head!
> The headman - ay - ay - ay - ay![20]

Similarly, the chiefs and headmen were hated for their role as tax collectors:

> I'm being tied
> *I'm being tied up far from home*
> Tax, tax,
> *My heart is angry.*[21]

They were hated for their role in the compulsory cotton- and rice-growing campaigns. One of the women's songs derives from the Macuse division of the Azehna, Barbosa e Mendes rice concession:

> Because it's me, I'm being beaten,
> Because it's me, I'm being beaten,
> Because it's me, I'm being beaten,
> > Look, friends!
>
> I've been tied up, I've been tied up,
> I've been tied up, I've been tied up,
> I've been tied up, I've been tied up,
> > The chief – ay![22]

They were hated for their sexual abuse of the wives of men they had dispatched as migrant labourers (chiefs were permitted to retain 40 per cent of the recruitment tax of twenty-five escudos on workers contracted from their area who travelled *without* their wives). One supremely scornful song satirises the chiefs under a set of derogatory names: *aNankhollo* ('he-goat'), *aNamthullu* ('impotent bull'), *aMullalleya* ('court messenger'), and *aMukhopella* ('upstart') – all of which indicate the contempt the Lomwe women had for the nonentities appointed by the Portuguese to such positions. The nickname 'hoe-handle' refers to their unflagging licentiousness:

> aNankhollo, hoe-handle,
> aNankhollo, hoe-handle,
> It's not that I want to summon you!
> You chose my husband yesterday
> You sent him to the work place.
> Then you came banging on my door.
> You took me for a prostitute,
> > *Ali!*
> You took me for a prostitute!
> > *Ali!*
>
> aNamthullu, hoe-handle,
> aNamthullu, hoe-handle,
> It's not that I want to summon you

> You chose my husband yesterday
> You sent him to the work place.
> Then you came banging on my door.
> You took me for a whore,
>   *Ali!*
> You took me for a whore,
>   *Ali!*

> aMullalleya, hoe-handle,
> aMullalleya, hoe-handle,
> My heart is worn out.
> My heart is worn out!
> My husband came back on the day before yesterday.
> Today he's picked up again.
> I'm worn out!
>   *Ali!*
> Ay - ay - ay - ay - ay
>   *Ali!*

> aMukhopella, hoe-handle,
> aMukhopella, hoe-handle,
> It's not that I want to summon you!
> I too would like someone to work in my garden,
> Last year, when you picked him up,
> You took me for a whore,
>   *Ali!*
> You took me for a whore,
>   *Ali!*[23]

Above all, they were hated as figures who possessed a type of power that had no local precedent:

> You can do it
>   *You can do it*
> You can do it
>   *You can do it*
> You can find the chief and beat him up
>   *You can do it*
> But the chief is a powerful man![24]

Created in accordance with Portuguese notions of what African political relations were, the chiefs and headmen exercised their new authority arbitrarily and despotically.

The second theme of these songs is the situation not in the village under the headmen but at work on the plantations. The Lomwe-Chuabo songs are more sharply critical of the inequalities of the company system than any others from the district:

> I'm working, in hunger,
> The owners are full;
>   It's a bad sign,
>   It's a bad sign.[25]

And these songs express, more clearly and concisely than any others, the destructive impact of the companies' demands for labour:

This finished off the young men,
   The Company - ay,
This finished off the young men,
   The Company - ay,
This finished off the young men,
   Hair,
This finished off the young men,
   Since they started,
This finished off our grandfathers,
   Cunha,
This finished off the young men,
   Cunha,
This finished off the young men,
   The Company - ay.[26]

The company referred to is Sena Sugar, and the officials mentioned (Hair and Cunha) were plantation managers. A women's song from Macuse makes the same point about labour recruited for the Boror Company's sisal plantations at Namacurra:

Ay - ay - ay e-e ay - ay - ay
   e-e ay - ay - ay
Listen to me, all of you,
   My children are dying
My children are dying – hard work for Kokora![27]

Kokora, which means 'sweeping', was the African nickname for Nelson Saraivo Bravo, administrator at Namacurra. The background to this name and to the complaint expressed in the song is that the sisal plantations, given their location in the hot sandy plain inland from Quelimane and the prickly nature of the plant, were invariably unable to obtain labour except by the direct intervention of the administration's police.

Perhaps the most moving of all these songs, however, is a women's pounding song from Ile, which compares what is happening at home within the local cotton concession with what is happening 275 km away on Sena Sugar's Luabo estate:

I suffer, I do,
   Oyi - ya - e - e
I suffer, I do,
   I suffer, my heart is weeping,
What's to be done?
   I suffer, my heart is weeping,
I cultivate my cotton,
   I suffer, my heart is weeping,
Picking, picking a whole basketful,
   I suffer, my heart is weeping,
I've taken it to the Boma there,
   I suffer, my heart is weeping,
They've given me five escudos
   I suffer, my heart is weeping,
When I reflect on all this
   Oyi - ya - e - e

I suffer, I do,
   I suffer, my house is weeping,
My husband, that man,

> *I suffer, my heart is weeping,*
> He went there to Luabo,
> > *I suffer, my heart is weeping,*
> He went to work, work hard,
> > *I suffer, my heart is weeping,*
> He broke off some sugar-cane for himself,
> > *I suffer, my heart is weeping,*
> Leaving work, he was arrested.
> > *I suffer, my heart is weeping,*
> He was taken to the police,
> > *I suffer, my heart is weeping,*
> He was beaten on the hand,
> > *I suffer, my heart is weeping,*
> When I reflect on all this,
> > *Oyi - ya - e - e*
> I suffer, I do,
> > *I suffer, my heart is weeping.*[28]

In its broad details, this song is accurate. The singer received only five escudos for six months' work in the cotton fields. (She did not exaggerate. In 1940 the average annual income in Mulevala, where she came from, was eleven escudos for cotton pickers, and the district was eventually declared unsuitable for cotton.)[29] Meanwhile, her husband was beaten for eating a piece of sugar cane he was harvesting. (Again the complaint is justified. Sena Sugar claimed losses of hundreds of tons of sugar cane per year through theft, and the penalties were severe.[30]) Implicit in this comparison of what happens at home with what happens on the distant plantation is a comment on the whole unified economic system in which the singer was trapped. But the song's shape, as it moves from the singer's experience to that of her husband, stresses an even more important theme, that of the separation of husband and wife – the fundamental impact of this system of migrant labour on the family.

This concern is, indeed, the most striking feature of the Lomwe-Chuabo songs. Although they characterise more sharply and eloquently than any others from the district the injustice and brutality of the plantation and concession systems, they make no mention at all of Portuguese rule. None of these songs, on internal evidence alone, can be identified as coming specifically from Mozambique. Although the themes of the songs are representative of a fairly wide area of Quelimane district, the actual targets of protest are very local indeed. The songs attacked particular companies and individual chiefs and headmen. There is no larger colonial, or nationalist, or revolutionary, or even ethnic dimension. The central preoccupation is the family – what is happening to husbands, wives, and children and to the village farms. All the anger, grief, suffering, and loneliness of these songs ultimately arise from the destruction of a way of life centred on the rural homestead. One last song, sung by a woman from Maganja da costa where Lopes e Irmão ran the rice concession, makes the point poignantly and economically. The song begins with the message that her husband has reached Sena Sugar's Marromeu plantation:

> Marromeu has spoken
> > *He has arrived;*
> Marromeu has spoken
> > *He has arrived;*
> Marromeu has spoken

He has arrived, Marromeu.

I grow my rice
  He has arrived;
I am watching the road
  He has arrived;
The road is empty of people
  He has arrived, Marromeu.[31]

This concern in the songs is supported by oral testimony:

> When people were arrested and taken to the *Posto* and told they were going to
> work, they used to write their names down, and if one of them ran away, the rest
> of his family would be arrested – such as his wife and children. So, if you're on a
> journey, you don't run away because you know your family is going to suffer.
> The family in the village will suffer.[32]

What the Lomwe-Chuabo-speaking peoples desired above all else was to be left
alone, to be allowed to continue living and working undisturbed in their small-
scale village systems. They carried that desire over into the period of Frelimo's
liberation struggle, and it has largely coloured and ordered their reaction to
Mozambique's independence.[33]

The Sena-Podzo songs of the north bank of the lower Zambesi are quite different
indeed. For people living in the shadow of the country's largest capitalist company,
Sena Sugar Estates, the local chiefs and headmen were insignificant and slightly
comic figures. The people of this region had historically been part of the *prazo*
system, and the demands of the *prazo*-holder for taxes and labour were familiar –
even acceptable – so long as they did not exceed certain limits. The taxes paid and
labour supplied formed part of the bargain between rulers and ruled. As late as
1945, three years after the circular of 1942 had made six months' labour
obligatory, Sena Sugar still required only some three-and-a-half to four months'
work from its locally recruited workers to fulfil their tax obligations. Any attempts
to go beyond this resulted in widespread absenteeism and emigration, which made
the effort to get more than four months' work pointless. For most of this century
the company made good its larger requirements with migrant labour, which
proved much easier to discipline and control. It thus became a key part of the
system that the company obtained at fixed prices from local producers the
foodstuffs needed for the labour compounds in which the migrant workers were
housed. The exception to this rough and unequal *modus vivendi* between former
prazo-holders and former *prazo* inhabitants was the cotton concession, which Sena
Sugar held for the area from 1936 to 1961. Because cotton was a six-month crop
and because, to protect its labour supplies, Sena Sugar demanded that women
grow the cotton, for twenty-five years the company and its agents presided over a
system of oppression at least as bad as the one prevailing in the Lombwe-Chuabo
areas further north.[34] By this stage, however, the Sena-Podzo peoples were
inclined to put the blame not on the company but on the Portuguese state.

The Sena-Podzo songs are not about chiefs or taxation or the decay of village
and family life. These are not perceived as the main issues. Instead, they are about
the abuse of power and the breaking of implied agreements between patron and
clients. Exploitation is seen in entirely personal terms, and its human faces are
identified. A typical song begins with the name of the person who is attacked, often
repeated once or twice, followed by a denunciatory epigram that sums up why he
or she is especially hated. In the 1950s Fernando Braz Valezim, for example, was

an overseer on Sena Sugar's cotton concession:

> Varajin - ay
> > Ay - ay
>
> Varajin - ay
> > Ay - ay
>
> Varajin - ay
> > Varajin must go home, Varajin's too much![35]

Muripata was one of the African police attached to the Luabo police post and, like Varajin, had the reputation of confiscating babies to make their mothers work faster:

> Muripata
> I'm exhausted,
> > Ay - ay - ay, Ay - ay - ay
> I want to put the child on my back, Muripata,
> > Ay - ay - ay, Ay - ay - ay
>
> Muripata,
> I'm falling asleep,
> > Ay - ay - ay, Ay - ay - ay
> I want to put the child on my back, Muripata,
> > Ay - ay - ay, Ay - ay - ay
>
> Muripata,
> Witchcraft doesn't harm him,
> > Ay - ay - ay, Ay - ay - ay
> I want to put the child on my back, Muripata,
> > Ay - ay - ay, Ay - ay - ay.[36]

And Dona Anna d'Oliveira, a small planter and trader, was annually provided with labourers by the local *chefe do posto*:

> Dona Anna, Ay - ay,
> > Ay - ay - ay
> Ay - ay, Dona Anna,
> > Ay - ay,
> 'Go there!' telling us,
> 'Come here!' telling us.
>
> Dona Anna, Ay - ay,
> > Ay - ay - ay
> Ay - ay, We used to be tied up,
> > Ay - ay,
> 'Go there!' telling us,
> 'Come here!' telling us.
>
> Used to be tied up, Ay - ay,
> > Ay - ay - ay
> Ay - ay, Tied up with rope,
> > Ay - ay,
> 'Go there!' telling us,
> 'Come here!' telling us.
>
> Policeman: Dig, dig, *ari!*

Beat, beat, beat this person, *ari!* Beat her, beat her!
    (Sobbing noises)
Son of a bitch, you dog, son of a bitch!
You haven't worked, you lazy so and so.
I'll fuck you today, I'll fuck you today, I'll fuck you today, *ari!*
This woman wants to be fucked! I've caught you this time!

Woman: No, I don't want to be beaten.
            It's better to be screwed.

Policeman: Son of a bitch!

Dona Anna, father!
    Ay - ay - ay
Ay - ay, We used to be tied up, father!
    Ay - ay,
'Go there!' telling us,
'Come here!' telling us.[37]

In this third song, the actors in the inset drama were both women. One played
the part of the policeman sent by Dona Anna to recruit workers, and the other
the woman being recruited. The 'policeman' chased the 'woman' round the
circle of singers, shouting and beating her with a truncheon before enacting the
rape to screams of laughter from the audience.

    Some of the differences between these songs and those from the Lomwe-
Chuabo area are already evident. The Sena-Podzo songs are cruder, louder, less
decorous, less controlled, more peremptory, and more vigorous, and they are
driven by a fierce conviction of protest. The targets are individuals – the cotton
and rice overseers, the labour recruiters, the concession holders, the small
planters, the field *capitaes*, the administration police, company officials, and the
like. The subject is violence, the state- and company-sanctioned violence that is
at the heart of the system. In particular, the subject is violence against women,
for all three are women's songs. There are men's songs from the cane fields –
which share, and probably originated, this same form. Of the following songs,
the first is addressed to a Portuguese overseer on Sena Sugar's Luabo estate:

    Oso Romeya - ay
       O - o
          O
       O - o
          O,
    Oso Romeya - ay,
     I peeped and saw his mother's cunt![38]

The second is addressed to an African *capitão* on the same estate:

    Weno,
      It wants to touch you,
    Weno,
     O - ay - ay,
    Weno,
     O - ay - ay,
    Weno,
     His mother's cunt![39]

The third, however, is an all-purpose insult:

O Capitão,
O Capitão,
O Capitão,
    You mother's prick, Capitão, I'm tired![40]

When we recorded these songs from retired field workers, they were sung with tremendous relish; the *O - o*'s and *Ay - ay*'s built up much suppressed laughter before the explosion of the insult. They were obviously very satisfying to sing in the cane fields to overseers who had to put up with them, both because the prevailing etiquette dictated that they should and because the men worked faster when singing rhythmically. All the same, it is hard not to conclude that the Sena-Podzo men's work song, with its conventional and repetitive obscenities, is a somewhat limited form, especially when set along side the Lomwe-Chuabo men's work songs. It leaves us with an impression of broad dissatisfaction with labour conditions on the estates, but it supplies no details. One of the singers of *Oso Romeya* filled in the background. 'This is about suffering. While pushing trucks, loaded with sugar cane. And while you push, someone is hitting you on the back. On each truck there used to be four people to push.'[41] The comment is more informative than the song!

Only in the mouths and minds of the Sena-Podzo women did the work song, as danced in the village, breach the cramped confines of the name-plus-epigram format. A typical performance begins with the women standing in a circle, bending forward from the waist and clapping or shaking tin *machacha* as rhythmic accompaniment to the lead singer and the chorus. One at a time, they perform brief solo dances, backing slowly round the circle. After several verses, however, the song breaks off while an extemporised drama is performed, illustrating the song's theme. The stage is the circle of singers, and anyone can perform; the actors are frequently replaced part of the way through by women who feel they can do better. The audience consists of the remaining women, who encourage and applaud the caricature of bribery and beatings and rape. Thus, while the men insulted their overseers with uninventive obscenities, the women characterised in detail the whole range of the regime's impact on their lives. In one inset drama, part of a song that consists of 'he single repeated phrase 'They used to tie us up', two policemen from Luabo raid a village in search of illegal *Kachasu* stills:

First Woman: (Whispering) The white man's coming!

Second Woman: Sh! Sh! *Chefe da Policia!*
                Take care of the barrel.

First Woman: Who are they? Who are they? Joaqui?
                Who sent them?

Second Woman: He's from the administration. It's Blanket!

First Woman: But it's Sunday today.
                Our husbands went to look for sugar cane.

Second Woman: Let's pretend we're digging.

First Woman: Now today, should there be inspection today?
                Today, we're going to be arrested here.
                Run away with the children, with the women,
                    everybody, run away to the reeds.
                The people are going.
                It's Rodrigues, here he comes.

We're being tied up.
Five litres, ten litres, we're carrying the
    bottles in our hands.
We must sleep in hiding.
Now they're arresting us.

Second Woman: When we get there, we're going to be beaten up.
We're going to be held by the head, and have our heads
    thumped against the cement wall.
Getting beaten, after everything else.
Thump! Thump![42]

The policemen in this song are Joaquim Rodrigues and Agostinho Pires, nicknamed 'Blanket' because 'he used to beat people all over'.

The most important, and the most popular, of these Sena-Podzo songs, however, was originally a men's work song before it was 'taken to the village' to be performed in expanded versions by both men and women. The *Paiva* addressed in the song was originally José de Paiva Raposo, in whose name the *prazo* on which Sena Sugar first operated was initially leased in the late nineteenth century and who, as *prazo*-holder, had the right to levy labour. Other members of the Paiva Raposo family worked for the company until the 1950s. The family's association with the company helped keep the song alive by giving each generation of workers a fresh 'Paiva' to attack. But *Paiva* is more than the name of a person. It became the local name for the company itself as *prazo*-holder, the name under which Sena Sugar had appropriated the land and commandeered the labour. The song, then, is really addressed to the company itself. This version was sung by a male professional singer and composer in a village close to Luabo:

Paiva,
    Paiva, I've felt it
Paiva,
    Paiva, I've complained,
Paiva,
    Paiva, I've complained
Paiva,
    The mbira is my witness
Paiva,
    Paiva, father, I've wept
Paiva,
    Pàiva, going weeping
Paiva,
    Paiva, going weeping
Paiva,
    Paiva complained
Paiva,
    Paiva, I've fallen on a fire
            LUABO!

Paiva,
    Paiva, I've complained
Paiva,
    Paiva, I'm speechless
Paiva,
    Paiva, I'm speechless
Paiva,

> Paive, I've wept today
> Paiva,
> *Paiva, I've complained*
> Paiva,
> Paiva, I've fallen on a fire
> Paiva,
> Paiva, going weeping
> Paiva,
> Paiva, I've fallen on a fire
> PAIVA LUABO!
>
> Paiva,
> *O - o - o*
> Paiva,
> *O - o - o*
> Paiva,
> *O - o - o*
> *Paiva, I've killed his money for him, his penis!*[43]

Many different versions of this song exist in the villages that, until 1978, were subject to Sena Sugar's dominance. This particular version is fascinating in two ways. The concluding seven lines, which were sung by the people present many times after the soloist had completed the body of the song, represent the earliest version of the *Paiva*, dating from the 1890s. Thus, this simplest and oldest version has exactly the same form as the other work songs – the name, repeated, plus the epigram. And it was sung in this form in the cane fields from the 1890s to the 1950s. As an address to Sena Sugar, the song expresses an idea central to the Sena-Podzo people's perceptions of company and colonial over-rule. The complaint – that the workers have made *Paiva*'s money for him – is actually a complaint about appropriation without reciprocity. The phrase 'I have killed for him' (*Ndampera*) refers metaphorically to the hunting of wild game, which, in Sena society, is governed by a set of rules specifying the just distribution of the kill. *Paiva*, the company, is demanding too much and providing too little in return: this is not what the *prazo* system is supposed to be about. Beyond this, the hierarchy in and of itself is not challenged.

Indeed, in the full version of the song just quoted, the address is to *Paiva* as protector. After the abolition of the *prazo* system in 1930 and the arrival of the Portuguese administrators and *chefes do posto*, the people of the area continued to look to the company as patron. The extremes to which such attitudes could be carried are illustrated by one version of the state takeover in which the whole problem of violence is attributed to an error of judgement made by the Africans themselves:

> Now, when the government came, we were all called to the Company. They told us that the Company was no longer going to deduct the taxes from us, because now there was white government. 'You can choose, if you want to stay with the Company, or if you want to move over to the government. But we will no longer be responsible for any complaints.' And then, we all voted to go to the government. And then the company said to us, 'You've now chosen to go to the government, but I can tell you in future you are going to cry for the Company.' But the government was really bad! Our hands are all swollen with this palmatoria. Then we went to the Company and said, 'We're being beaten up.' And then the Company said, 'Didn't we tell you? You chose the government, now you go to the government. Don't come here to complain about it.'[44]

Hence, this version of the *Paiva* song continues to invoke the relationship of client to patron that the people of the area had, with some success, imposed on the company when it was first founded. *Paiva* is addressed as 'father', as though company and workers were bound by kinship relations, and the whole thrust of the song is an appeal to *Paiva* to have pity. In other versions, *Paiva* is addressed repeatedly as 'father' and 'mother' and, in one instance, as *Mbuya*, the 'Big One' or 'Elder'. Given such perceptions of the relationship of workers and their employers, all the other Sena-Podzo songs understandably complain about the misconducts of specific individual members of the official hierarchy and about conditions of living and working but pose no larger challenge to the colonial system. Just as, in the last resort, the Lomwe-Chuabo peoples wanted to be left alone in their rural homesteads, so in the last resort the Sena-Podzo peoples wanted the *prazo* system to continue, in a reformed, more equitable fashion with more adequate and responsible patrons but otherwise unchanged in its patterns of authority. This attitude, too, influenced their reactions to the anti-colonial struggle and has conditioned their response to the challenge of independence.[45]

The majority of the songs examined thus far have been work songs, intended originally to accompany such activities as pushing trucks or pounding grain into flour, and all of them were sung by the people actually doing the work – whether on the sugar or sisal or tea plantations, on the cotton or rice concessions, or at home doing domestic chores. They record first-hand experience directly, and most of them are in the first person. The forty-six Chopi songs that Hugh Tracey recorded come from a rather different context. The composers of the *migodo* he discussed were professional musicians – Katini, Gomukomu, Sauli Ilova, and Sipingani Likwekwe. And their orchestras of massed xylophones, which contained as many as forty-eight musicians, could not have been brought together without some system of patronage:

> Their orchestras are to be found in every large village. In the Zavala District alone each of the eight more important chiefs has his own *Ng'godo* of orchestra and dances . . . Katini is the leader and composer at the kraal of the Paramount Chief Wani Zavala, and Gomukomu holds the same office at the kraal of Filippe we Madumane Banguza, an important chief whose country, called Mangene, lies along the north-western boundary of the Zavala District.[46]

The orchestras are maintained by the chiefs, and the role of the orchestra leader is, in part, that of a court praise-singer. The songs themselves stress the specialist skills of the musician and the mystery of the composer's craft. 'To play the/mtimbila you must dream about it,' sings Gomukomu. Or, as Sipingani Likwekwe puts it, 'You must dream to compose music.' Katini's version is the most vivid statement of this theme:

> Wani Zavala!
> Hush, you people of Zavala,
> Cease your chatter
> At this court of chiefs!
>
> *Timbila* music is so moving it brings tears,
> This music of Katini's *timbila*
> Singing and dancing.
>
> Wani Zavala!
> Hush, you people of Zavala,

> Cease your chatter
> At this court of chiefs![47]

The social aspects of professionalism and patronage should not, however, be overstressed. Musicians had to cultivate their gardens and build their own houses like everyone else. Like everyone else, too, they were subject to taxation, to corvée labour and recruitment for the Rand mines, and to ill treatment at the hands of police and court messengers acting under the authority of the Portuguese administration. When they complained in the songs about such treatment, they were speaking not just for an elite but for the community as a whole. But the context of protest is very different from that of the Lomwe-Chuabo or the Sena-Podzo songs. When Gomukomu complained about the chieftaincy ('You elders must discuss affairs:/ The one whom the white man appointed was the son of a commoner'),[48] his grievance sounds very similar to the Lomwe-Chuabo complaint about the way the Portuguese manufactured 'traditional' authorities under their policy of indirect rule. Unlike the Lomwe-Chuabo, however, whose political systems were fragmented by the nineteenth-century slave trade, the Chopi retained a relatively centralised political life with strong chiefs and genuine chiefs.[49] Gomukomu, therefore, actually distinguished between those chiefs who were chiefs by legitimate succession and those who were mere puppet creations of the Portuguese – his own patron and brother-in-law, Filippe Banguza, would have been flattered by the comparison. Similarly, when Gomukomu protested about the use of the *palmatoria* by court messengers, his real outrage was not that '*I've* been beaten,' as in the Sena-Podzo songs, but that 'Even chiefs are beaten on the hands . . ./The arrogance of Julai in beating even the hands of chiefs!'[50] Such was the respect of these musicians for the popularly accepted legitimate hierarchy that Sipingani Likwekwe even turned on his own patron, Chugela Chisiko:

> You, Chugela, you are proud of your position, yet
>   you are only a chief made by the white men.
> Oh, the chieftainships of Nyaligolana and Chugela!
> Oh, the chieftainships of Nyaligolana and Chugela!
> It is a shame that should be hidden from Wani.[51]

'Wani', here, is the paramount chief, Wani Zavala, patron of Katini, whose music Sipingana Likwekwe also admired.[52]

Because the *ngodo* is an established genre, its set form imposes to some extent demands on the content of the songs. It must, in other words, fulfil certain roles and respect audience expectations. One very obvious limitation of the genre is its male domination. Not only do the *migodo* have no women participants, but in only two of the forty-six songs is any attempt made to represent women's experience. One is the brief opening song of Sauli Ilova's *ngodo*:

> My husband will tell me when he's drunk,
> My husband will tell me when he's drunk,
>   'You bitch!'[53]

The other, again a brief dance song, is by Gomukomu:

> I am most distressed,
> I am most distressed as my man has gone off to work,
> And he does not give me clothes to wear,
> Not even black cloth.[54]

These two themes, especially the second with its complaint about the migrant labour system, are very common in women's pounding songs throughout Southern Africa. Apart from these two moments, however, nothing in the Chopi *migodo* matches the kinds of concerns expressed in the Sena-Podzo and Lomwe-Chuabo women's songs.

Each *ngodo* contains certain established features. The order of the different sections in Katini's *ngodo* of 1940, for example, is as follows:

1.	MUSITSO WOKATA	First orchestral introduction
2.	MUSITSO WEMBIDI	Second orchestral introduction
3.	MUSITSO WORARU	Third orchestral introduction
4.	NG'GENISO	Entry of the dancers
5.	MDANO	Call of the dancers
6.	JOOSINYA	The dance
7.	JOOSINYA CIBUDO COMBIDI	The second dance
8.	MZENO	The song
9.	MABANDLA	The councillors
10.	CITOTO CIRIRI	The dancers' finale
11.	MUSITSO KUGWITA	The orchestral finale[55]

Allowing for slight variations in, for instance, the number of orchestral introductions, this is the basic shape of six of the *migodo* Tracey recorded. The demands of this structure are frequently referred to in the songs themselves: 'Mguyusa, my young brother, help me compose my music./ I have no *mzeno* for my *timbila*.'[56] Certain stock phrases recur and certain issues, such as the question of facial tattoos, are repeatedly aired as the composers maintain running arguments.[57] Quite clearly, the composers found these structural patterns satisfying, providing some unity in diversity. But the need to fulfil the requirements of form does affect the content of the songs, and no complete *ngodo* could be constructed from the kind of songs sung by the Sena-Podzo or Lomwe-Chuabo peoples. Just as there are sections where the orchestra plays alone, so there are others when the dancing takes precedence over the words, and *vice-versa*. The high points of each *ngodo* in terms of the lyric are the *mdano*, the call of the dancers, and the *mzeno*, the song, which is performed by the composer himself.

Tracey placed considerable emphasis on the overall artistic unity of each *ngodo*, and he explained the origin of this unity by describing how the sections were composed. The composer, he said, began with the words (presumably, although he did not say so, the words of the *mzeno*). The music was suggested by the flow of the words, as the Chopi language is tonal and 'the sounds of the words themselves almost suggest a melodic flow of tones.' This melodic kernel was then transferred to the xylophone, and contrapuntal melodies were devised. Those melodies then became the themes of both of the orchestral accompaniments to the songs and of the orchestral introductions. Finally, the dance leader listened to the music and began to devise dance routines for the other sections.[58]

What Tracey did not note, and what is somewhat obscured by his procedure of examining the forty-six songs one by one, is that each *ngodo* has a particular subject. Katini's *ngodo* of 1940, for example, opens (after the orchestral introductions) with the entry of the dancers to words we have already discussed:

> It is time to pay taxes to the Portuguese,
> The Portuguese who eat eggs
> And chicken!
> Change that English pound!

The next section, the call of the dancers, describes an incident when Katini and his wife were beaten up by Kapitini, the court messenger, for drinking cashew cider. The section that follows, the dance itself, is accompanied by these words:

> O - oh, listen to the orders,
> Listen to the orders of the Portuguese.
>
> O - oh, listen to the orders,
> Listen to the orders of the Portuguese.
> Men! The Portuguese say, 'Pay your pound.'
>
> Men! The Portuguese say, 'Pay your pound.'
> This is wonderful, father!
> Where shall I find the pound?
>
> This is wonderful, father!
> Where shall I find the pound?
>
> O - oh, listen to the orders,
> Listen to the orders of the Portuguese.[59]

By this stage, the subject of Katini's *ngodo* of 1940 is clearly established. Its theme is Portuguese colonial rule. The complaint thus far is about taxation, about official violence, and about the prodigality of the Portuguese settlers. The next section, the second dance, contains lines already noted about the use of the *palmatoria* – 'Here is a mystery, the Portuguese beat us on the hands,/Both us and our wives.' This is followed by the *mzeno*, which attacks the ignoring of legitimate procedures in the selection of state-appointed chiefs and deals with some of the problems that arise when the wrong man is appointed. The next section carries this point further, complaining about Portuguese interference in matters of Chopi succession. Running through the whole *ngodo* are words like 'orders', 'threaten', 'beat', 'sjambok', 'avenge', and 'make trouble' and lines like 'I heard them trying to hush it up' and 'Fambayane was brought bound before the judge'. The songs are full of peremptory instructions – 'Change that English pound!' and 'Pay your pound!' and 'Don't waste your time with *timbila*!' and 'If you come across Chimuke, greet him with a "Good Day!" Greet him well because he likes to be in amongst the chiefs.'[60] Chimuke was the district administrator, the source of all of these orders and the local representative of the colonial system, which is being criticised from so many different angles. Only in the final song of the *ngodo* does the mood relax, with a brief amusing song about the rather dilatory courtship by her suitor of Katini's sister-in-law, an attractive widow.

In Katini's next *ngodo*, however, performed in early 1943, the subject is quite different. There is no mention of taxation or *sjambok* or the *palmatoria* or the state-appointed chiefs. Kapitini, the court messenger, re-appears very briefly, but only to be ridiculed in the context of an amusing little scandal. The dominant theme of the *ngodo* of 1943 is not the nature of Portuguese colonial rule but pride in Chopi culture and in the triumph of Chopi music. The first song accompanied the entry of the dancers:

> Hey, Dawoti!
> Dawoti go and ask Madikise.
> He will tell you about our grandfathers.
> Chitombe, behold Madikise![61]

Dawoti was a court messenger attached to the office of Administrator Luiz de Vasconcelos (*madikise*, 'law-giver'), and Chitombe is one of the great ancestors of the Chopi people. The song's double irony is, first, that instead of being ruled by Chitombe the Chopi are now ruled by the Portuguese and, second, that for flunkies (like Dawoti) Chopi history and culture exist only through the antiquarian researches of the Portuguese administrator! This is a sarcastic beginning to the *ngodo*, but in the next section, the call of the dancers, Katini described how in July 1939 the Chopi musicians were summoned to entertain Portuguese President Carmona on his state visit to Mozambique:

> Come, you people of Zavala,
> Come, you people of Zavala, and go to Magule.
>
> The Song of Madikise,
> It is wanted by Ngundwana at Magule.[62]

Magule is the Chopi name for the place of the Gaza Paramount Ngunguyane's defeat at Lake Coolela in 1895 by the Portuguese general, Mousinho de Albuquerque.

The Gaza, known throughout Southern Africa as the 'Shangaans', from the name of their first chief, Soshangane, were old enemies of the Chopi. After a series of bitter wars, the Gaza defeated the Chopi in 1891. Now, at Magule, the Chopi and the Gaza are again in competition, this time before the Portuguese president and in terms of culture.[63] The encounter is described in the dance, which juxtaposes two quite distinct events. The first four and last two lines refer to Chopi experiences working as migrant labourers in the mines of the Witwatersrand and encountering prejudice on the part of the African 'boss boys' because they speak Chopi; lines 5–8, however, refer to the performance of Chopi songs at the meeting with the Portuguese president:

> Malanje says, 'You swear at me if I speak chiChopi',
> So I will speak chiSotho.
>
> Malanje says, 'You swear at me if I speak chiChopi',
> So I will speak chiSotho.
>
> Katini will come to Magule to play *timbila*.
> The President is glad to see the waChopi.
>
> The Shangaans are left to sing their 'Ho'ho siyana'
> Until very late for the President.
>
> Malanje says, 'You swear at me if I speak chiChopi',
> So I will speak chiSotho.[64]

To the delight of Katini, as lines 6–8 make clear, the Chopi defeated the Gaza at the very place where the Gaza were defeated by the Portuguese.

But, meanwhile, what happened to the Portuguese? This emphasis on entertaining the Portuguese president may seem pathetic, but Katini was interested in the nature of the encounter. Although the Portuguese now governed the Chopi, it appears from the *ngodo* thus far that they investigated Chopi history and were fast succumbing to Chopi culture, especially Chopi music. These triumphs were secured not by open confrontation and resistance, such as the Gaza had once attempted so disastrously at Magule. Rather, they were achieved by stealthy compliance of the type illustrated in the song by the vignette of ethnic relations on the Rand: when the 'boss boy' refused to listen when addressed in Chopi, Katini spoke to him in Sotho, a small concession where there were other, more substantial victories to be gained.[65]

With these points established, Katini reached the climax of his argument in the song (*mzeno*). The song's principal theme is the visit that Katini and his orchestra paid to Lisbon in the summer of 1940, when they performed in the celebrations of Portugal's tricentennial anniversary of freedom from Spain: 'We made new tunes for the *timbila* in the midst of the sea/As we passed foreign lands'.[66] Chopi ethnic pride could have had no greater triumph than this victory for music over politics and warfare, celebrated in Portugal itself![67] Although the events of the *ngodo* were three or four years out of date by the time it was performed in 1943, Katini did not, as Tracey suggested, comment haphazardly on topical events; instead, the composer pursued a complex argument about the nature of one type of response to the realities of Portuguese rule, drawing on recent Chopi history to make his point about the power of Chopi music to conquer the conquerors. Appropriately, therefore, this *ngodo*, in the section following the *mzeno*, contains the song 'Wani Zavala/Hush you people of Zavala' about the power of Katini's music to 'bring tears'. Appropriately, too, the *ngodo* closes with a demonstration of Katini's moral authority within the community, as he used the dancers' finale to deliver a stern warning about the attempted seduction of an under-aged girl.[68]

All of these considerations make the songs of the Chopi *migodo* very different from the Sena-Podzo and Lomwe-Chuabo songs. Because the *migodo* were performed by professional musicians rather than by labourers and cultivators and because they were devised for lengthy public entertainment rather than simply for accompaniment for communal or work-gang activity, the Chopi songs are much more thematically and aesthetically ambitious. It cannot be firmly established, on Tracey's evidence, that the songs of the *migodo* genuinely represent popular opinion, as some of the Sena-Podzo and Lomwe-Chuabo songs clearly do. Yet the very fact that they are the work of individual professional composers, working chiefly under patronage, provides greater scope for comment on the broad issues that are engaged. Scholars must, however, beware of romanticising work songs as automatically the 'voice of the people'. When the Sena-Podzo find in the *Paiva* song 'a map' of their experience, the reason is not simply that the song was originally a work song, sung by everyone, but more significantly that life in the overrule of Sena Sugar was stiflingly uniform. No one in a place where *Paiva* was 'the store, the factory, the railway line, the compounds, the cane-fields' could be said to be free from the intrusion of the company in their lives. In this sense, the *Paiva* song spoke for everyone.[69] Similarly, the Lomwe-Chuabo people, given the circumstances of their history, naturally focused much of their anger on the state-appointed chiefs and headmen who were the most local and most visible enemies of the rural household's integrity.

Chopi experience was considerably more diverse. In the nineteenth century, the groups that came to be called Chopi had lived under separate chiefs and had distinct political systems, despite their relatively homogeneous culture. Certain of these groups had benefited from a close alliance with the Portuguese, enabling them to keep the Gaza and other hostile groups at bay.[70] With such support, they had been able to take advantage of trading opportunities in the Delagoa Bay area and at Inhambane and had developed an active exchange economy. Initially, they traded large amounts of ivory and some slaves and then, after the opening of the

Suez Canal in 1869, they exported a wide range of products, including rubber, a variety of oil-seeds, beeswax, and copra.[71] In the 1880s, however, as the Portuguese stepped up their efforts to gain control of the interior, the balance of power altered. The Gaza, driven back from the banks of the Zambesi, attacked many of the Chopi chiefdoms; some declared themselves tributary to the Gaza, and others fought to maintain their independence.[72] Finally, in 1895, the Portuguese defeated the Gaza at Lake Coolela and solidified their power over southern Mozambique. In the first two decades of the twentieth century, the Chopi moved into labour migrancy both to the Witwatersrand and to Lourenço Marques, where they constituted a substantial portion of the local labour force. Their strategy was interesting. Accepting the inevitability of taxation and forced labour, they determined to maintain the integrity of Chopi society. By timing their trips to Lourenço Marques and the Rand so as not to conflict with the local planting season, by co-ordinating their absence with kinsmen or friends who agreed to do necessary agricultural work back home on their behalf, and by investing some of the money they earned in tools that improved productivity or in bridewealth, thus increasing the exploitation of female and child labour, the Chopi were able to preserve much of their rural economy as well as their access to purchased goods.[73]

After the First World War, however, pressures increased on the Chopi. Mozambique and Angola were granted a measure of local autonomy and their own budgetary responsibilities, and attempts were made to bring about economic growth. In Mozambique, especially in the southern part of the country, this took the form of encouraging the establishment of capitalist agriculture on the plantation model. Portugal hoped that growth in the local economy would help free Mozambique from its dependence on labour migration to South Africa for income.[74] In Chopi country, this departure involved the establishment of the Sociedade de Zavala, which was granted monopoly rights over the sale of peasant-grown *mafurra* oil, and a railway was built from Inharrime to carry the oil. Prior to the granting of the concession, the Chopi had exported the *mafurra* oil themselves, at rates of seven to eight thousand tons annually, through Indian traders acting either independently or as agents of various foreign trading houses. The pattern, in short, repeated that in Quelimane district in the 1890s, when peasant trading in copra through Indian intermediaries was taken over by the big copra companies; the difference in Zavala was the specific governmental support of Portuguese settler interests. In the event, the Chopi refused to sell their *mafurra* oil to the new company at fixed rates. Rejecting Portuguese rationalisations that they had been 'scandalously robbed' by the Indian merchants, the Chopi used the new railway line to facilitate labour migration to South Africa. The new company collapsed amid accusations that the high commissioner, Brito Camacho, had been bribed by the Rand Chamber of Mines to prevent it from succeeding. South Africa's interest in obstructing any large-scale investment in southern Mozambique lest it should affect labour supplies for the Rand was confirmed in May 1928. Portugal reached a new agreement with South Africa, permitting the recruitment of eighty thousand migrant workers annually.[75]

The Chopi, then, escaped the over-rule of a single company. Although they were, of course, subject to the same labour laws as the Africans of Quelimane district and they may be said to have belonged, in a sense, to the vast labour enclave of the Rand mines, the Chopi in fact had a comparative variety of outlets for employment. The variety lay not just in the choice of employers – the Incomati sugar estates, the rice and banana plantations of the Incomati valley, the Lourenço Marques railway, and of course the Rand – but more profoundly in the kinds of work available. We are speaking comparatively; we by no means are trying to suggest that the Chopi people had an easy time under the Portuguese. But there is a qualitative difference between working for decades as plantation labourers in a single, closed district and working in the city or encountering the full might of industrial capitalism on the Rand. These broader experiences demanded a more complex and extended poetical form. The world of the *migodo* is a relatively big place. That world includes the village and the city, the plantation and the mine, Mozambique and South Africa (and even, briefly, Lisbon). It takes account of two different versions of colonialism and of a variety of different kinds of Europeans – Portuguese, English, Dutch, Italians, and Germans. It records the contact with other African peoples, the Shangaans in Mozambique and the Sotho and Xhosa on the Rand. The Chopi could not, then, have had, as the Sena-Podzo and the Lomwe-Chuabo peoples did, any single enemy present at all these locations or among all these choices. The *migodo* contain no uniform explanation of why things went wrong. They do exhibit, however, a definite pattern of concern. Although the forty-six songs of the seven *migodo* Tracey collected are the work of at least five different composers, so that the first impression is one of variety and contrast, they all share one central and passionate preoccupation.

This point can be most effectively established by an examination of how these *migodo* from the early 1940s deal with the potentially radicalising experience of labour migration to the Rand. Little in these songs suggests that working in the mines was an unpleasant experience. 'I'll go to the mines to work for money so that when I come back I can buy cashew-cider to drink,' states one song. 'If we go to the city we see wonders as we pass Pretoria,' states another, from a *ngodo* actually recorded on the Rand. Only one line from one song – 'If I go to the mines, where shall I find the courage to get into the cage?' – gives us any impression of what it might have been like to work underground.[76] Most of the songs that refer to the Rand deal with the setting of the mining compound, and the subject is usually ethnic rivalries:

> Cast off your skins!
> There is no relish left, you Shangaans,
>   it has been eaten by the Sotho.
> It has been eaten by the Sotho and the Xhosa,
>   and we will not get it.[77]

On the Rand, the enemies tended to be not the Portuguese or the Randlords but other African peoples. For a fuller analysis of what is damaging about the migrant labour system, we must turn to Gomukomu's *ngodo* of 1942–3.

Gomukomu introduced the subject in the call of the dancers, apparently as a joke:

> It is Filippe's opinion
> That the girls also should sign on and go to the mines.
> It is Filippe's opinion.
> What a good idea![78]

Filippe is Filippe Banguza, the paramount chief who was Gomukomu's patron. In the next song of the *ngodo*, however, a second call of the dancers, the joke suddenly turns out to be serious:

> Ha! We quarrel again! The same old trouble.
> The older girls must pay taxes.
> Natanele speak for me to the white man to let me be.[79]

The complaint is about the governor-general's Circular of October 1942, which, while tightening the labour laws, gave local administrators no guidance on how this was to be accomplished.[80] One consequence among many was an intensification of tax collection, including the taxation of women. Filippe Banguza's duties, as a state-recognised chief, included tax collection – and the joke about sending women to the mines (Natanele was a recruiter) suddenly becomes most bitter. This *ngodo* contains, as the song accompanying the dance itself, the complaint noted earlier of the woman whose husband has gone off to the Rand, leaving her with no clothes to wear, not even the 'black cloth' of mourning. The *ngodo* thus comments, from more than one angle, on the position of women in the labour system. Gomukomu's argument is then brought to a head in the *mzeno*. He complains that he himself has been forced to give up playing the *timbila* to go and work as a labourer on the Incomati banana plantations. He complains that the administrator is troubling everybody 'with his constant calling' to enlist. And he complains that the police are now beating people up indiscriminately – 'even the hands of chiefs' and 'even women' – as the new labour regulations are enforced. The Portuguese 'turn their backs' when any question of the people's welfare is raised:

> We got on the train and arrived at Sewe,
> And when we spoke about the matter of food,
> About the matter of food, they turned their backs.
> We overheard the Portuguese speaking about food,
> Speaking about food while their backs were turned.[81]

Significantly, the incident described here occurred during President Carmona's visit in 1939. Gomukomu, in effect, retorted to Katini's argument that the Portuguese have been seduced by the appeal of Chopi culture. Finally, Gomukomu explained that the whole problem of increased labour demands has to be understood in the context of 'the German war': 'The bloody fools of white men are fighting./Matijawo says they are like four-legged beasts.'[82] The presentation of the whole issue with a great deal of humour not represented in our selective quotations only reinforces the power of this *ngodo* and the quality of its analysis.[83]

This *ngodo* deals only with local effects of migrant labour. It does not mention what life was like on the Rand and makes no comment about whether working conditions there were very unpleasant or whether the pay was far too low. Instead, it focuses on the consequences for the local community of recruiting methods, of state-sponsored violence, and of the men's absence for such long periods of time.[84] The threat to the Chopi community is what matters.[85] We noted earlier that the world of these *migodo* is comparatively large. By contrast with the songs from Quelimane district, which show no awareness of a bigger world outside the immediate relations of power in which the people were trapped, the *migodo* seem to encompass great diversity and individuality. The impression, though, is misleading. Underlying the surface variety of opinion and of mood, the *migodo* express a fundamental concern for the health of the Chopi nation. Although the word itself appears only once in the forty-six songs (and then as a linguistic designation), Chopi identity, the good of the Chopi people, is the single central preoccupation to which all else relates.

The *migodo*, in short, express a feeling of ethnic nationalism that is completely absent from the Sena-Podzo or the Lomwe-Chuabo songs. It may, of course, be argued that the Chopi chiefs form a traditional elite, that the musicians are court praise-singers, and that the ethnic nationalism of the *migodo* is very much what we should expect of a group trying to perpetuate its position. But that nationalism has other and broader facets. In part, it reflects the historical conflict with the Gaza, who – as 'Shangaans' – are attacked in a number of songs. In part, too, it appears as a complex of reactions to the experience of working in the mines. For decades, the Chopi found themselves competing with ethnic groups from all over Southern Africa. On the Rand they were housed together in their own compounds, a system intended to exploit ethnic differences among workers as a labour control device. There the Chopi learned that their language and culture set them apart from others and that their musical prowess was seen as one of the most striking aspects of this culture.[86] Mining compounds contained areas for the competitive demonstration of such 'ethnic' skills as drumming and dancing, and the Chopi, like all migrant workers, were encouraged to consider themselves as 'tribesmen'. It was in this situation that they most clearly saw themselves as 'Chopi', regarding 'the efforts of other peoples with a great deal of disdain.'[87]

At the same time, the skill with which the Chopi had made migrant labour an integral part of their village life, so that as time passed a contract of recruitment came to be viewed as a 'second initiation' through which a young man had to pass before he was accepted as fully mature, only reinforced this sense of Chopi identity.[88] Most Chopi, then, appear to have identified with the concerns of their legitimate chiefs, who became the very symbols of the nation's history and integrity. In the *migodo* themselves, no distinctions are made between those who were inside and those outside the circle of power, as, for instance, are repeatedly made in early Zulu or Swazi or Sotho praise poetry.[89] There are no appeals to national unity and no attacks on individuals for failing to respect chiefly power, as again occur in later Zulu or Swazi or Sotho praise poetry. There is no straining of any sort after audience attention or audience agreement, and song after song proceeds on the assumption that the performance of the *migodo* is a community entertainment ('Come together with your wives' and the like).

Most important of all, however, is clearly the positive, anti-colonial aspect of Chopi nationalism. The targets for the most fierce attack in the *migodo* are consistently those that represent Portuguese authority, such as the court messenger – 'You Lekeni, you are as black as coal,/Son of Nyamandane, you are a terror!' – and the labour recruiters – 'Listen, they are off to their kraals as they are afraid they will be signed on' – and the state-appointed officials with their new demands for 'respect' –

> Just listen to the songs of Chigombe's village,
> To keep on saying "Good Day" is a nuisance.
> Makarite and Bubwane are in prison,
> Because they did not say 'Good Day.' . . .
> They had to go off to Quissico to say 'Good Day' there instead![90]

And, of course, the Portuguese themselves as colonial rulers were targets. The Chopi songs, in fact, attack all those targets that are differentiated in the Sena-Podzo and the Lomwe-Chuabo songs according to local circumstances, and they do so from a belief in a Chopi nation that is perfectly capable of managing its own affairs. Running through all the songs of the *migodo* are references to two contrasting systems of authority. On the one hand, representing the Chopi nation, are their genuine chiefs, with their councils of elders and their professional musicians. On the other are the usurpers of power – the administrators and *chefes do posto*, the court messengers, the police, the labour recruiters, the tax collectors, and the puppet chiefs and headmen. The fundamental problem is, 'They have taken the country, we know not how, and shared it out.'[91] In the total absence in the 1940s of any likely military or political solution to this problem, the *migodo* demonstrate confidence in Chopi institutions and revel in the special skills and vitality of Chopi culture. As was noted in the mid-1970s:

> Above all else it is the Chopi sense of identity which sets them apart from the other peoples with which they have contact. It is expressed in a feeling of pride in their 'Chopi-ness', an indefinable group identity. It is given focus in their language (chiChopi), patterns of marriage which link their various clans to each other, and most important in their music, which crystallizes their identity, and the lyrics which focus their in-group feelings against outsiders.[92]

As with the Sena-Podzo and the Lomwe-Chuabo, so with the Chopi, the perceptions and attitudes formed in response to colonial over-rule have proved tenacious.

This brings us back to the question of 'resistance' and 'collaboration'. These terms are, indeed, clumsy for describing both the psychology and the practice of African reactions to capitalism and colonialism. Of the groups described here, all of which belong to the same country at the same time and under the same system of colonial over-rule, only the Lomwe-Chuabo can be said to have 'resisted' in the sense that they refused utterly to come to terms with the regime: their resistance was expressed in large-scale migrations to Nyasaland. They, however, lived close to the border, and it is difficult to see how the Chopi could have adopted a similar course. For both the Chopi and the Sena-Podzo peoples, neither 'resistance' nor 'collaboration' adequately describes their perceptions or their behaviour. Nor do any other terms, like the various forms of 'consciousness', which are commonly applied, do justice to the degree to which what they thought and did made perfectly good sense in terms of their previous experience and of the realities of political, economic, and military power. A group's consciousness, especially in situations of only the most embryonic class divisions, must be carefully situated in the commonplace specificities of the historical context. What is clear, above all, from these songs is something of the quality of African life in Mozambique under Portuguese rule. A song like the following, from Gomukomu's *ngodo* of 1940, is as valuable as Bernardo Honwana's stories in providing us with a picture both of intolerable political pressure and of life going on nevertheless:

> Come together and make music for the new year!
> We fear only that our names will be written by the white men.
>
> Ngukyusa, my young brother, help me compose my music,
> I have no *mzeno* for my *timbila*.
>
> You said that you would care for me, my Gomukomu,
> You said that you would care for me, my Gomukomu.
> But now in my house I am left weaving alone.
>
> Manyina Mtumbu, you think you are beautiful because you are fair!
> Manyina Mtumbu, you think you are beautiful because you are fair!

> But you surely are sweet as the bees!
> Lekeni the messenger has come to call you.
> Filippe, our child, they will be the death of you with their calling.
>
> You, Lekeni, are sent on important affairs.
> You, Lekeni, are sent on important affairs.
> Yet you dally on the road, joking with the girls.
>   You people of Zandamela are called to the Court.
> Ngongondo, you fear the Court on account of your drinking.
>
> Come together and make music for the new year![93]

In this song, the complaints about labour recruitment, about the behaviour of court messengers like Lekeni, and about the system's lack of respect for Filippe Banguza, the paramount chief, provide the framework for the other comments – from Gomukomu's wife that she has been deserted while her husband composes his music, or about Manyina Mtumbu, whose light-coloured skin is making her vain, or about Ngongondo, the drunken chief of a neighbouring area. We get a sense of the texture of life in a Mozambican village under the *Estado Novo*, which is what social history should finally be all about.

NOTES

1 Earlier versions of this chapter were presented at seminars held under the auspices of the Centre for Southern African Studies, University of York, and the Southern African Research Program, Yale University. It was originally published in *American Historical Review*, vol. 88, no. 4 (1983), pp. 334–70. We are indebted to members of those seminars for suggestions, and especially to Patrick Harries, Neil Lazarus, and Jeanne Penvenne. Alice White and Manuel Sabão conducted our Sena-Podzo and Lomwe-Chuabo interviews and translated the songs.

2 Luis Bernardo Honwana, *We Killed Mangy-Dog and Other Stories* (London: Heinemann, 1969), pp. 47–8.

3 T. O. Ranger, *Revolt in Southern Rhodesia, 1896–7: A Study in African Resistance* (London: Heinemann, 1967), and 'Connexions between "primary resistance" movements and modern mass nationalism in East and Central Africa', *Journal of African History*, vol. 9, nos. 3 & 4 (1968), pp. 437–54, 631–42.

4 George Shepperson and Thomas Price, *Independent African* (Edinburgh: Edinburgh University Press, 1958), p. 402.

5 R. E. Robinson and J. Gallagher, 'The partition of Africa,' in F. H. Hinsley (ed.), *The New Cambridge Modern History*, Vol. 11 (Cambridge: Cambridge University Press, 1962), p. 620. Robinson and Gallagher further developed these notions in their classic *Africa and the Victorians: The Official Mind of Imperialism* (London: Macmillan, 1963).

6 Timothy C. Weiskel, 'Changing perspectives on African resistance movements and the case of the Baule peoples', in B. K. Swartz and Raymond E. Dumett (eds), *West African Cultural Dynamics: Archeological and Historical Perspectives* (The Hague: Mouton, 1980), pp. 550–51.

7 ibid.

8 Colin Bundy, 'The Emergence and Decline of a South African Peasantry', *African Affairs*, vol. 71, no. 285 (1972), pp. 369–88. idem, *The Rise and Fall of the South African Peasantry* (London: Heinemann, 1979); and Robin Palmer and Neil Parsons (eds), *The Roots of Rural Poverty in Central and Southern Africa* (London: Heinemann, 1977).

9 George Rudé, *Ideology and Popular Protest* (New York: Pantheon, 1980), p. 9.

10 Leroy Vail and Landeg White, 'Plantation protest: the history of a Mozambican song', *Journal of Southern African Studies*, vol. 5, no. 1 (1978), p. 23.

11 The songs from the 1940s and 1950s are not, in general, still sung, but songsters recalled them to mind at our request.

12 Andrew Tracey and Gei Zantziger filmed two *migodo* in the mid-1970s and published a companion volume to the two films: *A Companion to the Films 'Ngodo wa Mbanguzi' and 'Ngodo wa Mkandeni'* (n.p., 1976).

13 Hugh Tracey, *Chopi Musicians* (London: Oxford University Press, 1948), p. 10. Tracey provided both the Chopi versions of these songs and his own translations.

14 Tracey, *Chopi Musicians*, pp. 10–11.

15 ibid., p. 15.

16 For a detailed account of the history of this area from the early 1800s to 1975, see Leroy Vail and Landeg White, *Capitalism and Colonialism in Mozambique: A Study of Quelimane District* (London: Heinemann, 1980).

17 ibid., Chs. 4 and 8.

18 Group Interview, Muanivana Compound village, Luabo, Quelimane district, 26 August 1975. (All interviews are from Quelimane district unless otherwise specified.)

19 Vail and White, *Capitalism and Colonialism in Mozambique*, esp. pp. 308–14.

20 Sung in Chuabo by Janta Lakriman and Luis Prazo of Maganja da Costa, at Checanyama Compound village, Luabo, 30 August 1975. Luabo is one of the largest of the Sena Sugar Company's plantations. For the Chuabo version of this song, see Vail and White, *Capitalism and Colonialism in Mozambique*, p. 360.

21 Sung in Chuabo by Pedro Rovi and João Orfumane of Maganja da Costa, at Checanyama Compound village, Luabo, 30 August 1975. The italicised lines are those that are sung in chorus by the listeners. For the Chuabo version, see Vail and White, *Capitalism and Colonialism in Mozambique*, p. 361.

22 Sung in Chabo by Teresa Sabinu, at Macuse, 2 September 1977. The words in Chuabo are:

> Kokala miyo kino ovadiwa
> Kokala miyo kino ovadiwa
> Kokala miyo kino ovadiwa
> Ona, siyaya!
>
> Ndimangiwa, ndimangiwa
> Ndimangiwa, ndimangiwa
> Ndimangiwa, ndimangiwa
> Samasoa - ay!

For further discussion of these campaigns, see Leroy Vail and Landeg White, '"*Tawani, Machembero!*": forced rice and cotton cultivation on the Zambezi, 1935–1960', *Journal of African History*, vol. 19, no. 2, pp. 239–63; and Allen Isaacman *et al.*, '"Cotton is the mother of poverty": peasant resistance to forced cotton production in Mozambique, 1938–1961,' *International Journal of African Historical Studies*, vol. 13, no. 4 (1980), pp. 581–615.

23 Sung in Lomwe by Armena Muhinayula and Helena Souzinho of Mulevala, Ile, at Checanyama Compound village, Luabo, 30 August 1975. For the Lomwe version of this song, see Vail and White, *Capitalism and Colonialism in Mozambique*, p. 362.

24 Sung in Chuabo by Parose Kwiri of Baixo Licungo, at Juncua Compound village, Marromeu, Beira District, 7 September 1975. For the Chuabo version of this song, see Vail and White, *Capitalism and Colonialism in Mozambique*, p. 360.

25 Sung in Lomwe by Armando Francisco and Manuel Shipitela of Mulevala, Ile, at Muidi Compound village, Luabo, 9 August 1975. For the Lomwe version of this song, see Vail and White, *Capitalism and Colonialism in Mozambique*, p. 365.

26 Sung in Lomwe by Armando Francisco and Manuel Shipitela of Mulevala, Ile, at Muidi Compound village, Luabo, 9 August, 1975. For the Lomwe version of this song, see Vail and White, *Capitalism and Colonialism in Mozambique*, p. 365.

27   Sung in Chuabo by Familia Sayidana, at Macuse, 2 September 1977. The words in
     Chuabo are:

> Ay - ay - ay e - e Ay - ay - ay
>   e - e Ay - ay - ay
> Mundi velele
>   Ananga nomala
> Ananga nomala – labwa Kokora!

28   Sung in Lomwe by Armena Muhinayula, Helena Souzinho, and Casavera Fernando,
     from Mulevale, Ile, at Checanyama Compound village, Luabo, 30 August 1975. For
     the Lomwe version of this song, see Vail and White, *Capitalism and Colonialism in
     Mozambique*, pp. 352–3.
29   Vail and White, *Capitalism and Colonialism in Mozambique*, p. 316.
30   Public Record Office, Kew, FO 371/11989, G. Hornung to Pyke, 15 April 1927,
     enclosure in Pyke to the Foreign Office, 19 April 1927.
31   Sung in Chuabo, by Paterina Joao and Palmira Goodbye of Baixo Licungo, at Juncua
     Compound village, Marromeu, Beira District, 2 September 1975. For the Chuabo
     version of this song, see Vail and White, *Capitalism and Colonialism in Mozambique*,
     p. 354.
32   Group Interview, Muanavina village, Luabo, 26 August 1975.
33   For a full discussion, see Vail and White, *Capitalism and Colonialism in Mozambique*,
     Ch. 9.
34   ibid., pp. 272–8, 314–25.
35   Sung in Sena by Nkanda Brassa and the women of Madumo village, Luabo, 17 August
     1975. The words in Sena are:

> Varajin - ay
>   *Ay - ay*
> Varajin - ay
>   *Ay - ay*
> Varajin - ay
>   *Varajin aende kwawo Varajin anyanya.*

     For another song attacking Varajin, see Vail and White, *Capitalism and Colonialism in
     Mozambique*, pp. 321–2.
36   Sung in Sena by Julia Manico and the women of Mapangane village, Luabo, 10 August
     1975. The words in Sena are:

> Muripata
> Ndinalezera
>   *Ay - ay - ay, Ay - ay - ay*
> Ndinafuna kubala mwana, Muripata
>   *Ay - ay - ay, Ay - ay - ay.*
>
> Muripata
> Ndina gona
>   *Ay - ay - ay, Ay - ay - ay*
> Ndinafuna kubala mwana, Muripata
>   *Ay - ay - ay, Ay - ay - ay.*
>
> Muripata
> Sina lozoa
>   *Ay - ay - ay, Ay - ay - ay*
> Ndinafuna kubala mwana, Muripata
>   *Ay - ay - ay, Ay - ay - ay.*

37   Sung in Sena by Julia Manico and the women of Mapangane village, Luabo, 10 August
     1975. The words in Sena are:

> Dona Anna, Ay - ay
> *Ay - ay - ay*
> Ay - ay, Dona Anna
> *Ay - ay*
> *"Bwera uku!" ntekume*
> *"Bwera kunu!" ntekume*
>
> Dona Anna, Ay - ay
> *Ay - ay - ay*
> Ay - ay, ndikamangewa
> *Ay - ay*
> *"Bwera uku!" ntekume*
> *"Bwera kunu!" ntekume*
>
> Nakamangewa, Ay - ay
> *Ay - ay - ay*
> Ay - ay, mangewa na nkambala
> *Ay - ay*
> *"Bwera uku!" ntekume*
> *"Bwera kunu!" ntekume.*

Policeman: Lima, lima, lima, *ari!*
Menya, menya, menya muntu! *Ari*, menya menya!
Feda puta, mwanamba, feda puta.
Nkabilima tai polola tai
Nakugona lero, nakugona lero, nakugona lero.
Ari nkazi asafuna goniwa uyu. Ndakupata lero.

Woman: Nkabi nyo nyo nyo nyo pianga
Goniwa kwene mbwene.

Policeman: Feda puta.

> Dona Anna, baba,
> *Ay - ay - ay*
> Ay - ay, Tika mangewa, baba
> *Ay - ay*
> *"Bwera uku!" ntekume,*
> *"Bwera kunu!" ntekume.*

38 Sung in Podzo by Dose Chonze, Jiwa Todo, and the men of Madumo village, Luabo,
17 August 1975. The words in Podzo are:

> Oso Romeya - ay
> O - o
> *O*
> O - o
> *O*
> Oso Romeya - ay
> *Da nyinda kwiri ya mache damona.*

39 Sung in Podzo by Dose Chonze, Jiwa Todo, and the men of Madumo village, Luabo,
17 August 1975. The words in Podzo are:

> Weno,
> Ina kuta kupata
> Weno
> *O - ay - ay*
> Weno
> O - ay - ay
> Weno
> *Pa nyini pa mache.*

40   Sung in Sena by Jose Kashkinya, Madumo village, Luabo, 17 August 1975. For the Sena version of this song, see Vail and White, *Capitalism and Colonialism in Mozambique*, p. 363.
41   Group Interview, Madumo village, Luabo, 17 August 1975.
42   Performed in Sena by Vittoria Camacho and the women of Muanavina village, Luabo, 24 August 1975.
43   Sung in Sena by Fernando Nicolos and the women of Pirira village, Luabo, 5 August 1975. For a detailed discussion of this particular song and for the Sena version, see Vail and White, 'Plantation protest'. Also see Landeg White, 'Power and the praise poem', *Journal of Southern African Studies*, vol. 9, no. 1 (1982), pp. 8–32.
44   Group Interview, Mopeia, 20 September 1975.
45   See Vail and White, *Capitalism and Colonialism in Mozambique*, Ch. 9.
46   Tracey, *Chopi Musicians*, pp. 1–2.
47   ibid., p. 27. Also see, ibid., pp. 32, 75.
48   ibid., p. 43.
49   D. J. Webster, 'Kinship and cooperation: agnation, alternative structures, and the individual in Chopi society' (unpublished PhD dissertation, Rhodes University, Grahamstown, South Africa, 1975), p. 16.
50   Tracey, *Chopi Musicians*, p. 48.
51   ibid., p. 68.
52   For songs that imitate Katini's *ngodo* of 1940, see, for example, Tracey, *Chopi Musicians*, pp. 68, 75.
53   Tracey, *Chopi Musicians*, p. 53. The exact flavour of songs such as this one, as recorded by Tracey, must be considered elusive, for it appears that Tracey bowdlerised the songs he presented in his published work, removing obscenities. Compare, for example, Webster, 'Kinship and cooperation', p. 383, n.
54   Tracey, *Chopi Musicians*, p. 46. The black cloth refers to mourning garments.
55   ibid., p. 10.
56   ibid., p. 35.
57   ibid., pp. 32, 58.
58   ibid., pp. 4–7.
59   ibid., p. 14.
60   ibid., p. 16.
61   ibid., p. 20.
62   ibid., p. 21.
63   Webster, 'Kinship and cooperation', p. 14.
64   Tracey, *Chopi Musicians*, p. 23.
65   The proliferation of clandestine schools after 1939 to teach people arithmetic and, among other subjects, Portuguese also attests to the Chopi's sense of realism; Charles E. Fuller, 'An ethno-study of continuity and change in Gwambe culture' (unpublished PhD dissertation, Northwestern University, 1955), p. 242.
66   Tracey, *Chopi Musicians*, p. 25.
67   Interwoven with the theme of cultural victory is a second theme, the death of Katini's close friend and associate, Magengwe, who was also a claimant to the chieftainship of Wani Zavalo. He died in Lisbon, and his death in a foreign land at the moment of triumph coupled with Katini's suspicions that his death was not wholly unwelcome at court brings human folly and weakness back to the centre of the argument and turns the *mzeno* into a haunting dirge.
68   Tracey, *Chopi Musicians*, pp. 28–9.
69   Vail and White, 'Plantation protest', pp. 1–25 passim.
70   St. Vincent Erskine, 'Journey to Umzila's, South East Africa, in 1871–1872', *Journal of the Royal Geographical Society*, vol. 45 (1875), p. 53.
71   Frederick Elton, 'Journal of an exploration of the Limpopo River', *Journal of the Royal Geographical Society*, vol. 42 (1873), pp. 22–32.
72   Patrick Harries, 'Slavery, social incorporation, and surplus extraction: the nature of free and unfree labour in south-east Africa', *Journal of African History*, vol. 22, no. 3,

pp. 320–22. Also see Augusto Cabral, *Raças, usos, e costumes dos indigenas do districto de Inhambane* (Lourenzo Marques, 1910), p. 35.

73 Webster, 'Kinship and cooperation', Ch. 3; and Jeanne Penvenne to authors, 26 December 1981. Also see Daniel da Cruz, *Em Terros de Gaza* (Porto, 1910), pp. 154–74, 225–73, *passim*; Emma H. Haviland, *Under the Southern Cross; or, A Woman's Life Work for Africa* (Cincinnati: God's Bible School, 1928), pp. 306–12; Dora Earthy, *Valenge Women* (London: Oxford University Press, 1933), pp. 20–60; and Fuller, 'Continuity and change in Gwambe culture', pp. 150–51, 180.

74 Vail and White, *Capitalism and Colonialism in Mozambique*, pp. 200–16.

75 Brito Camacho, *Mozambique: Problemas Colonias* (Lisbon, 1926), pp. 19–33. Also see T. H. Henrikson, *Mozambique: A History* (London: Rex Collings, 1978), p. 121; Sherilynn Young, 'Fertility and famine: women's agricultural history in southern Mozambique', in Palmer and Parsons (eds), *Roots of Rural Poverty*, pp. 66–81; and Vail and White, *Capitalism and Colonialism in Mozambique*, pp. 205–11.

76 Tracey, *Chopi Musicians*, pp. 73, 80, 68. It should be noted, however, that Tracey's ties to the mines need to be explored in future work.

77 Tracey, *Chopi Musicians*, p. 30.

78 ibid., p. 41.

79 ibid., p. 43.

80 T. de Bettancourt, *Relatorio do Governador-General de Mocambique, 1940–42*, I (Lisbon, 1945); pp. 45–7.

81 Tracey, *Chopi Musicians*, p. 48.

82 ibid., p. 47.

83 For an example of the humour, see, for example, ibid., p. 48:

> There are women foolish enough to take grain from the bins
> to sell for beer!
> To take corn and waste it on drink.
> Listen, oh . . .!
> . . . The bottle is empty!!

84 For a discussion of the long-range effects of prolonged labour migrancy from Chopi country, see Fuller, 'Continuity and change in Gwambe culture', passim, and D. J. Webster, 'Migrant labour, social formations, and the proletarianisation of the Chopi of southern Mozambique', *African Perspectives*, vol. I (1978), pp. 154–74.

85 With the outbreak of the Second World War, the forced cotton regimen imposed in Quelimane district was also implemented in Chopi territory, and the initial results included brutal exploitation of the women and several severe famines. It is reasonable to assume that this topic was handled in later *Migodo*, but this research has not been done. Fuller, 'Continuity and change in Gwambe culture', pp. 152–3.

86 Fuller, 'Continuity and change in Gwambe culture, p. 242.

87 Webster, 'Kinship and cooperation', pp. 9, 79.

88 ibid., pp. 20, 81–3.

89 See White, 'Power and the praise poem', passim.

90 Tracey, *Chopi Musicians*, pp. 41, 59, 55.

91 ibid., p. 74.

92 Webster, 'Kinship and cooperation', pp. 8–9.

93 Tracey, *Chopi Musicians*, p. 41.

10 Fire on the mountain: resisting colonialism in Algeria

DAVID PROCHASKA

From the barbed wire-fenced backyard I could see the Edough mountains clearly. They rose abruptly from the flat plain surrounding the sprawling Annaba which had swallowed but not yet digested a huge glop of rural migrants. You could not see the small hamlet in the hills at almost 3000 feet, although it was little more than five miles away, partly because it was obscured by a dense stand of oak and pine, most of it cork oak. The Edough mountains extend clear the other side of this French hamlet which never quite became a British-style hill station, part of an on-again, off-again coastal range which stretches a good deal of the way to Algiers in the west and in the east to the Tunisian border and beyond.

This particular late summer evening, however, was different. Several spots of orange-red dotted the distant dark: the forest was on fire, and not in one but several places. It reminded me of nothing so much as the brush fires which sweep periodically through the chaparral of that other Mediterranean-like land, southern California. In California nearly annual fires in the dried-out late summer and early autumn are a part of life, and so too are the mudslides which follow just as inexorably after the winter rains on the newly-denuded hillsides.

I asked my Algerian landlord about these fires in the Edough. Scion and contemporary head of an old Annaba family which settled in town at the time the Spanish *reconquista* chased them out of the Iberian peninsula – or so they claim; with a background of family privilege and the personal manner that goes with it; patient, nonplussed, and always immaculately groomed – sweat and dirt are foreign bodies which have not yet spread to his part of the Mediterranean; a precise conversationalist, packaging his words neatly in his mouth prior to delivery, emphasising a point with a twist of the wrist, his fingers bent at the knuckles – no wonder with this combination of background and style he is a second-term member of the National Assembly.

He looked at me with a twinkle in his eyes and that engaging half-smile which I was never sure was the start of friendship or the beginning of a leer but probably a combination of both, and said, 'Some say it's bandits setting the fires.'

'Bandits? But why?'

'Perhaps they don't like the government,' he replied drily. I could never be sure how much of this he believed and how much he was simply passing on, as curious to see how the *étranger* would react as the stranger was to hear what he had to say.

Our story begins here, in the present, but also long before, before the French invaded Algeria in 1830. Annaba was the second city captured by the French after

Algiers, but Algerians had long been living in the Edough mountains as well as on the plains surrounding the city. Trees had once covered the plain as well as the mountains, but the Algerians had cut most of them down to graze their animals and grow their crops. One forested area which remained lay in the low hills south and a little east of town; it was called the Beni Salah forest after the Algerians who lived there. Of course, 'Algerians' is an anachronism: there were no Algerians in 1830. Those living in the Beni Salah forest called themselves the Beni Salah, and likewise in the Edough mountains we have the Beni Merouan and Beni M'Hamed, the Fedj Moussa and Ouichaoua, the Ouled Attia and Tréat, more then a dozen groups in all. Approximately three-quarters of these groups were Arab and a quarter were Berber.[1]

The number of Algerians in the Beni Salah forest was less than 2000 on the eve of the French conquest; those in the Edough mountains numbered some 3500.[2] Thus, the population density in the Edough was slightly more than fourteen persons per square kilometre, which is equivalent to one person every ten hectares, a very low density. The second major population consisted of animals. Cattle, sheep, goats, horses, and mules were prized possessions; their products ranged from food and woolen cloth to hides.

The Algerians in the Beni Salah and Edough had worked out a symbiotic relationship with nature; they existed in a precarious ecological equilibrium with the forests in which they lived. These Algerian forest dwellers both raised livestock and planted crops. They followed a four-year ecological cycle which they called *kçar(kusār)*. It began when towards the end of summer at the time the sirocco blew they set fire to the lower limbs of trees and underbrush in order to open up clearings in the forest. Thus, they complemented the fire ecology characteristic not only of the Annaba region, but of most of the Mediterranean where fires broke out routinely during the hot, dry days of late summer. By eliminating underbrush, fires permitted annuals to grow, which were crowded out otherwise by perennials. Moreover, the seed pods of certain perennials opened only as a result of fire. Lastly, the virtually fire-resistant bark of the cork and holm oaks was itself an adaptation to this fire ecology. These late summer fires combined with controlled burning produced, therefore, an abundant growth of tender, young plants the following spring which made for excellent grazing. Thus, it is not surprising that not only in the Annaba region but elsewhere in the Mediterranean people have complemented the fire ecology with controlled burning. Nor is it surprising that the fires which resulted have been a fact of life for centuries across southern France from Antibes to Provence, as well as in Corsica, Tunisia, and Morocco.[3]

After burning, the Algerians farmed intensively the newly opened-up areas for one year. They planted cereals and truck gardens of cantaloupes, water-melons, squashes, onions, and turnips. They rotated cereals with nitrogen-fixing legumes, including peas, lentils, and green beans to increase yields. They used the cinders produced in clearing as fertilizer. They irrigated the crops in a rudimentary fashion. And then after raising one year's crop, they let the fields lie fallow for three years, and as the new undergrowth came up, they grazed their herds of cows, sheep, and goats on it. After four years the entire cycle began over again.

To round out the basic *kçar* cycle, the Algerians tapped the forest for a variety of products which they used to fashion an interwoven and largely self-sufficient material culture. In the forest clearings they built their *gourbis*, or wood and thatch huts, from *diss* grass and tree leaves placed over poles made from tree limbs. Firewood provided fuel for both heating and cooking. They made charcoal and used it for fuel and to forge iron ploughshares. They obtained tannin from first Aleppo pines and later holm oaks to tan the hides of the animals they slaughtered. Their ploughs were wooden with at most an iron blade. They fabricated utensils, beehives, olive presses, and fencing for the fields from wood.

Finally, they supplemented their diet of *cous-cous*, or cracked wheat, meat, vegetables, unleavened bread, fruits, olives and goat's milk with acorns when the crops were insufficient or failed altogether. The effects of such a diet were graphically described in a series of sayings current among those living in the forest.

> I ate it [acorns] fresh,
> It gave me diarrhoea;
> I ate it with legumes,
> It gave me colic;
> I ate it with bouillon,
> It set my body on fire;
> I ate it straight,
> My guts swelled up;
> I ate it with sour milk,
> It burned my insides;
> I ate it with mallow leaves,
> My belly became bloated;
> I ate it with oil,
> I spent the night on my left side;
> I ate it with butter
> I stank like a dog.[4]

Thus, the goal of the Algerians in the Beni Salah forest and the Edough mountains was to maintain an ecological equilibrium between themselves, their animals, and the forest in which they lived. If the animals were numerous, the amount of land cultivated was reduced; if the cultivated land was extensive, the animal herds were reduced. But both cultivation and stock-raising were necessary, since neither alone could support a forest population. At the same time, too many people in too little forest put too great a strain on resources. The Algerians were forced then either to clear more land for cultivation and herding by fire, or to migrate elsewhere to a less populated area.

Moreover, the attitude of the Algerians towards the forest was congruent with their ecological relationship with it. If the forest belonged to the Turkish *bey* in theory, in practice it belonged to the Algerians who used it. After all, the forest was there, they had not created it, they had not planted the trees which grew there. Thus, the Algerians considered it there for their use but they did not claim it as their property; it was *'arsh*, the equivalent of communal land. But the areas cleared by *kçar* and cultivated intensively were different. These plots created out of the forest by dint of their own labour they considered theirs, they passed them down from father to son, generation after generation, and contested anyone who attempted to take them; these were *milk*, or the equivalent of private property.

Ecological balance between people, animals, and the forest was the ideal; the reality, however, needs to be nuanced in two important respects. In the first place, the delicate ecological web could be rent instantaneously and completely by any number of natural or human disasters: famine and disease, feuds and raids, forced migration. These forest-dwelling Algerians may have been spared the periodic economic depressions of industrial economies, but they faced instead the recurrent demographic crises characteristic of a subsistence-based economy. Crop failures combined with endemic and epidemic disease to produce periodic demographic crises which killed people off and kept population low. In addition to natural disasters, there were those caused by man. A Turkish *razzia*, or raid, could disperse a tribal grouping within a matter of hours temporarily or permanently. Such *razzias* were reprisals for failure to pay taxes, or banditry, or revolt. Furthermore, there were inter-tribal rivalries as well as intra-tribal conflicts.[5]

Secondly, momentary ecological equilibrium did not mean that the Algerians in the Beni Salah forest and Edough mountains were completely self-sufficient, cut off from the wider world. In fact, it was precisely the fragility of their ecological balance that led them to forge ties, primarily trading links, with the outside world which served among other things as a kind of primitive economic insurance. The Tréat extracted tannin from Aleppo pines and cork oaks and sold it. Hides were hauled to Annaba and other ports where they were sold to a French trading company, the *Compagnie d'Afrique*. The Ouichaoua let Moroccans chop firewood and make charcoal for a fee. The Beni Salah engaged in *achaba*, an arrangement by which they charged nomads from the Sahara for the right to pasture their flocks in the forest during the summer. Finally, the inhabitants of the Edough and Beni Salah exchanged these goods as well as an occasional handmade ploughshare or olive press, plus any surplus cereal crops at *sūqs*, or periodic markets, held in Annaba and elsewhere in the vicinity.[6]

In short, the Algerians of the Beni Salah and the Edough forests were preindustrial people practising a combination of subsistence agriculture and stock raising. They were largely self-sufficient but not totally so, oriented to the local Annaba region but with a number of links to the wider world.

And then the French came. First, they occupied the city of Annaba and changed the name to Bône. Second, they displaced the Algerians and peopled Bône with Europeans. Third, they grabbed the land on the plain outside Bône. Fourth, they mined the area for minerals. Fifth, they took over the forests in the Beni Salah and Edough mountains.

To glimpse what was going on in the forests, let us tag along with the English consul to Algeria on a trip he made by horseback to the Edough in 1868:

> Being obliged to visit Philippeville [a port west of Bône], I thought it right to make a tour through the province of Constantine to acquaint myself with the actual state of the country . . .
>
> 'La Safia' [in the Edough mountains] . . . three years ago was a magnificent forest of cork oak, interspersed with rich valleys more or less cleared for cultivation, and covering a superficies of 6500 acres [2600 ha].
>
> A concession of this forest was originally given to a French gentleman for a period of 40 years, subsequently extended to 90 years. In terms of this he was to enjoy the right of stripping the cork trees over the whole extent; and of cultivation and pasturage over about 225 acres [90 ha] of cleared land. From the latter portion the Arab occupants were ejected, much against their inclination, though not, I presume, without some sort of compensation. Subsequently, the right of pasturage over the whole forest was assumed by the *concessionnaire*, the claims of the Arabs were ignored, and they were only permitted to pasture their flocks within the forest on payment of rent to the *concessionnaire*.
>
> Such was the state of things when, in 1865, this concession was purchased by the London and Lisbon Cork Wood Company for the sum of 12,000 pounds.
>
> What I have described as having taken place at La Safia was going on all

around, everywhere the original *concessionnaires* were parting with the land which had been freely given to them as an inducement to colonization. Their successors were arrogating rights to which they had no legal claim, and a feeling of jealousy and distrust was engendered amongst the native population, which found an expression sometimes in open rebellion, but more frequently in acts of wanton destruction.[7]

In this report of his excursion, our English consul touches on all the salient features of French forest policy in Algeria. First, the French expropriated the forests from the Algerians in order to concede them to French businessmen. Second, the French applied to Algeria a forest code developed for France, which they enforced by fining the Algerians. Third, the Algerians resisted as best they could.

Like the Turks, the French state claimed legal ownership of the forests; unlike the Turks, the French conceded vast tracts of these state lands to private individuals to exploit commercially. Where the Algerians based their entire political economy on forest products and uses, the French wanted the forests primarily so they could use the cork oaks to make products such as bottle corks. Here is where the forests in the Bône region were so important, because the Edough and Beni Salah forests contained such a high proportion of cork oaks. The Bône region became, therefore, one of the leading centres in Algeria for the commercial production of cork oak, and Algeria became in turn one of the leading exporters of cork not only in the Mediterranean but in the world.[8] Who knows, perhaps Marcel Proust's cork-lined bedroom, in which he wrote *À la Recherche du Temps Perdu*, was made from Algerian cork, maybe even cork from the Edough.

The French Second Empire of Louis Napoléon conceded the prime cork oak forests of Algeria to wealthy individuals in the 1850s and 1860s. Businessmen, notables, and cronies of Napoléon all received vast tracts of forest. Charles de Lesseps, brother of the Suez Canal builder and member of the Emperor's entourage, obtained some 2800 ha in the Beni Salah forest. The original French *concessionnaire* of the La Safia forest described by our English consul was a leading *colon*, or settler of Algeria, by the name of Cès-Caupenne, whose father was a high Napoleonic official.[9]

What de Lesseps and Cès-Caupenne received, however, were no more than meagre plots compared to what certain other favourites obtained. The facts speak for themselves. The very first concession was granted in 1848 to a Parisian businessman named Lecoq, another person we will hear of again. Twenty-one

Table 1: Division of forest concessions in Constantine Province, 1862

Area of concession	Number of concessions	
less than 1000 ha	1 concession occupying	30 ha
1000–2000 ha	1 concession occupying	1635 ha
2000–5000 ha	16 concessions occupying	52465 ha
5000–10000 ha	6 concessions occupying	34343 ha
10000+ ha	4 concessions occupying	55320 ha
Total		143793 ha

Source: Nouschi, *Enquête*, p. 326.

additional lots were leased between 1852 and 1860, which added up to more than 200 000 ha. Of this total, nearly 150 000 ha were located in Constantine province;

of the thirty-four concessions in Algeria, twenty-eight were in Constantine province.[10] That these were large concessions is demonstrated by Table 1. Not only were three-quarters of all concessions situated in Constantine province, but half of those were located in the Bône region. Besides 10 000 ha of the Beni Salah forest, more than 50 000 ha in the Edough mountains had been parcelled out as detailed in Table 2.

Table 2: Major forest concessions in Edough mountains, 1862

Concessionaire	Area conceded (in ha)
Besson	17824 ha
Lucy, Falcon	11245 ha
Gary, Bure	6773 ha
Berthon-Lecoq	6654 ha
Martineau des Chenetz	5972 ha
Duprat	5418 ha
Cès-Caupenne	2656 ha
Total	56542 ha

Sources: AN: F 80 1785: Gouvernement Général de l'Algérie, *Mesures à prendre à l'occasion des incendies de forêts* (Algiers: Bastide, 1866); TEFA 1859–61, pp. 273, 275; TEFA 1862, pp. 311–15; Gouvernement Général de l'Algérie, *Statistique générale de l'Algérie* [hereafter SGA] 1867–72, p. 363; SGA 1879–81, p. 129.

In order to concede this land to a tiny number of French individuals it had been necessary to take it from a much larger number of Algerians – the next point made by the English consul. Now, it is not possible to say precisely how much forest land was taken from each group of Algerians and exactly how much was given to each concessionnaire.[11] We do know, however, how much forest was taken from seven of the dozen-odd groups of Algerians living in the Edough and Beni Salah forests (see Table 3). The Beni M'Hamed lost the least, 971 ha of 13 077 ha, or 7 per cent of their total holdings. At the other end of the scale, the Ouichaoua lost more than 80 per cent of their 29 183 ha, the single highest proportion, and the Reguegma segment of the Beni Salah lost a whopping 26 744 of 34 699 ha, the highest absolute amount.

After the Algerians were expropriated and the forests were conceded, the French forest code was applied. In state forests the code was administered directly by the forest service; in private forests, private forest guards were hired to enforce it. Developed originally for France and now applied to Algeria, this forest code symbolises the clash of two different political economies, for the French code of 1827 was based on the assumption that forested areas were uninhabited. Instead of French peasants who relied on the forest for pasture, firewood, and food, the introduction of capitalist property relations now meant that rural Frenchmen theoretically owned their own property and could purchase forest products on the market.

The main problem with the French forest code was that it did not fit the France of 1827. Perhaps it was applicable in north-eastern France where it was first adumbrated, but it already ran into severe difficulties when it was applied to the Midi and Corsica. And in the Ariège region of south-western France, the peasants waged between 1828 and 1872 the long drawn-out 'War of the *Demoiselles*', so

Table 3: Algerian land conceded to forest concessionaires in Bône region

Algerian group	Total holdings (in ha)	Private concessions (in ha)	State forests (in ha)	Comments
Beni Merouan	3472	2674	—	To London/Lisbon and Senhadja & Collo
Beni M'Hamed	13077	971	—	All to Besson
Beni Salah Ouled Serim	30654	8317	—	Includes both state and private forests
Beni Salah Reguegma	34699	26744	—	Includes both state and private forests
Fedj Moussa	11020	4335	3375	2093 ha to Besson, 2242 ha to Gary, Bure
Ouichaoua	29183	23749	2129	Divided between Berthon Lecoq, de Noireterre, and Besson
Ouled Attia	23052	15513	—	All to Besson
Tréat	13630	10992	—	4081 ha to Gary, Bure and 6910 ha to Besson

Sources: Nouschi, *Enquête*, p. 335; and references cited in note 1 above.

named because the men dressed as women to avoid recognition.[12] In short, such encroachments by a centralising French state have been treated by more than one historian as striking instances of internal colonialism.[13]

If the French forest code, based on liberal nineteenth-century principles of capitalist political economy, worked only imperfectly when applied in France, it takes little imagination to realise that it was virtually impossible to extend it to largely precapitalist Algeria. Moreover, this conflict was exacerbated in Algeria in a way that it was not in France because those enforcing the code and those having the code forced on them came from two radically different cultures. But that did not deter the French forest service; rather than meeting the Algerians halfway, they simply rode roughshod over them.

The forest code was applied to areas inhabited by Algerian groups, not to uninhabited areas. It forbade pasturing sheep and goats but allowed pigs, which were considered unclean by all devout Muslims. It banned construction of *gourbis*, or huts, less than one kilometre from the forest. It severely restricted the ability of Algerians to cultivate their plots in the forest, and to gather dead wood, cork or *diss* grass. The number of *procès-verbaux*, or violations of the forest code, rose dramatically. From an annual average of 3143 infractions in all Algerian state forests during 1874–8, the number of *procès-verbaux* increased to an average 6780 for Constantine province alone between 1886 and 1890. And, sure enough, the two main violations in 1890 were for pasturing livestock (50 per cent) and cutting firewood (25 per cent). Infractions entailed fines, often ludicrously high. At 2 fr per sheep and 4 fr per goat, the fines were often more than the value of the animals. A 50 fr fine for cultivating a plot in the forest was twice the purchase price of one hectare of forest.[14]

No wonder the Algerians prayed, 'Lord, save us from the forest guards!'[15] No wonder no less a colonialist than Jules Ferry complained angrily that in the eyes of the forest service 'the natives were always breaking the law.'[16] No wonder that the concessionaires counted the fines as part of their revenues, that 'Instead of exploiting the forest, they [the concessionaires] find it easier to exploit the Native.'[17]

The forest service clearly knew what it was doing. One official declared: 'the contemporary pastoral lifestyle of the Natives [is] incompatible with all civilisation. We do not want to leave, and we do not want to be assimilated. Trying to isolate us everywhere, they want to chase us out of Algeria or even better assimilate us.'[18] Eugène Étienne, one of the most important *colons* of Algeria and leader of the *parti colonial* in France, sounds like an expert on political economy: 'without eliminating user's rights and abolishing the [cultivated] enclaves, there will be no industrial exploitation of the forests.'[19]

As in all such situations, there is some question as to the extent to which the forest code was actually enforced. On the one hand, nominal French policy was to abolish the Algerians' traditional users' rights in exchange for 10 per cent of the forest taken. No doubt this is the 'sort of compensation' the English consul 'presumes' the French paid the Algerians. On the other hand, it is not at all clear when, where, and to what extent this policy was implemented. Traditional users' rights were maintained in theory in the Edough mountains at least as late as the 1890s. But in practice?[20]

Still, the forest guards first had to catch the Algerians. In the following passage an Algerian enumerates some but by no means all the methods he and his compatriots used to escape their attentions:

> What do you want me to do about the *procès-verbaux* of the guards? In the first place, I always know which work sites they're supposed to watch and which ones they're absent from – we have, we other natives, a kind of telegraph which although not similar to yours, is not any less rapid – I know right away what direction they're heading, so I have all the time necessary to get my herd out before their arrival. And then, if by chance I'm caught, I do what I can to make a good deal, if I can't succeed in proving that the herd isn't mine, that all the guards are wrong, that they have lied. As far as the fine? Oh well, I hold a big work party [*twîza*] on a nice day, all my friends and relatives come and work on my behalf at the site, knowing that if need be I would do the same for them, and my fine is paid without having cost me a *sou*.[21]

No doubt a 'good deal' could include a little *baksheesh*, or bribe, passing hands.[22]

How else did the Algerians respond to the combined assault on their traditional political economy represented by expropriation of their land and application of the forest code? In the 1860s our English consul referred to the Algerians' 'feelings of jealousy and distrust' at what the French had done. To say the least: listen to these Algerians still complaining loud and long in the 1890s:

> We come to you to set forth the grievances we have against the company of Lidon [perhaps l'Edough?] of Monsieur Barthon lakouk [sic].
> The members of this company have harmed us greatly by taking over the land that we have worked to support our family and the gardens planted with all kinds of fruit trees which we own.
> We no longer have access to these lands except as tenants of the Company.
> If we didn't want to rent them, they would plant cork oaks there and wouldn't miss an opportunity to cite us with *procès-verbaux*.
> They have fleeced us, to the point of throwing us into poverty.
> In short, we pay a double tax every year: a tax to the State, a tax to the

Company, and it's this latter one . . . that weighs most heavily on us.

It's no different with tree branches and *diss* grass, which we sell to earn our living, and which the Company makes us pay for at the rate of 1 franc 50 per month and per person.

We pay them as well 2 *sous* per limb, for the construction of our huts [*gourbis*].

It's been 15 years that we've been paying a tax to the Company named above, in fact a rent, etc.

We sent you a petition two months ago, which has gone unanswered.[23]

All of the Algerians' complaints against the concessionaires are contained in this petition. The concessionaires received communal land which the Algerians had used for years. One group of petitioners claimed, for example: 'We have a communal parcel in the commune of Ain Mokra (*'arsh* Ouichaoua) and the Compagnie Besson who is our neighbour claims possession of our land, we have owned these parcels for years and years, when this Company didn't even exist.[24]

For the government to have ceded the forests to private concessionaires was one thing. But for the government to have expropriated their forest enclaves as well – this violated the Algerian moral economy:

Will you tell us if you have only rented or sold the cork oak forests to the company, or if you have ceded also the uncultivated land and the plantations of fig trees and vineyards? . . . May the members of the company not try to fool you into thinking that they are independant of the forest, these uncultivated lands! For these parcels that we work come from our ancestors.[25]

Clearly, the concessionaires were acting abusively: selling olive trees to Italians who used them to make charcoal, expropriating land from the Algerians and renting it to their own forest guards, collecting rent from one Algerian for a *qūbba*, or shrine, built by another.[26] One response of the Algerians was to set their grievances down in formal petitions. Surely, this is evidence that at least some Algerians were beginning to learn the rules of the French colonial game. But if the Algerians we heard above did not even know the names of their tormentors ('Lidon' for London and Lisbon Cork, 'Barthon lakouk' for Berthon-Lecoq), how effective could their petitions have been? For equally clearly, the Algerians were getting no support from the government. The Algerians appealed to the Governor-General, the highest ranking colonial official in Algeria, as a distant but just lord. 'I am convinced in advance that you will honour a request from the father of a family suffering from hunger.'[27] However, the Governor-General did not respond, no matter what their relation to France, no matter what services they had rendered:

Berthon-Lecoq is my neighbour, he has expropriated me without any legal proceedings in spite of my protests . . . ·I am a poor old man, 85 years old, with no economic resources and father of [illegible] people . . . P.S. I have served France for 25 years as *shaykh* and so has my son . . .[28]

Playing cat-and-mouse with the forest guards and writing petitions to the Governor-General were two responses of the Algerians to the French. Yet another was to simply burn the forests down.

In September 1894 the following telegram was sent from Algiers to the Havas press agency in Paris:

Bône: Fires yesterday cause veritable disaster region all along Bône line. At Duvivier, several hundred places set on fire; to left and right road sixty-odd telegraph poles burned. Lines to Randon down. At Duvivier two farms set on fire. Vast brazier at Karezas. Lacombe property ravaged: vineyards; six thousand olive trees; 25 haystacks burned completely. Oil mill, farms, wine-cellars damaged; flames, smoke climb prodigious height covering top Edough and entire plain. At Randon, arabs cry, chant, indifferent to disaster. This morning, fires appear extinguished, but sky obscured by smoke; heat diminished.[29]

A news story appeared the same day with the headline, 'Monster fire in Algeria'.

Bône, 11 September.
  Since ten o'clock this morning, the sun has been obscured by intense smoke due to the forest fires. Five fires surround Bône: at Chaiba, at Ain-Chand, at the Lacombe farm, at Elmenia and behind the Edough mountains, from where immense flames can be seen.
  The heat is torrid.
  Tunis, 11 September.
  A strange phenomenon, which has not yet been explained, currently plunges the inhabitants of Tunis and the surrounding countryside into a profound astonishment.
  A band of smoke, appearing some 10 kilometers in width and a hundred in length, obscures the entire western part of the sky, producing a veritable eclipse of the sun, which appears no more than a fuliginous ball.
  The smoke, pushed by a violent sirocco wind, appears to come from an enormous fire to the west of Zagouan.
  The heat in town is suffocating, the thermometer has reached the highest mark of the season.

What the residents of Tunis saw, of course, was the cloud of smoke from the fires of Bône, 'But when one considers the distance which separates the two cities, one cannot think without horror about how large the conflagration must be.'[30] The distance between Bône and Tunis is over 250 km.

Further up the coast from Bône towards Algiers the writer Guy de Maupassant was touring the region when the fires of 1881 broke out:

During the night I thought I was in front of an immense, illuminated city. It was an entire burned mountain with all the underbrush already cooled, whereas the trunks of the oaks and olive trees remained incandescent, thousands of enormous upright coals . . . already no longer smoking, but like piles of colossal lights, all in a line or scattered about, resembling boulevards without end, squares, winding streets . . .[31]

Earlier, our English consul had invoked another artist to convey the scene he saw left by the fires of 1873 in the Edough mountains:

At about 20 kilos. from Bône the forest loses its verdure, and as far as the eye can reach, weird and blackened stumps, like a picture of Gustave Doré's, mark the effect of the fatal fires of 1873, which created such havoc here and in almost all the forests of Algeria.[32]

In an earlier report he had fleshed out his description of these same fires:

The recent fires, not only those wilfully caused by the insurgents in 1871, but those which took place so lately as August 1873, have been most disastrous. After an unusually severe sirocco, conflagrations burst forth in every part of the colony, but particularly in the province of Constantine, and hundreds of square miles of forest were consumed in a few days.
  At Bône the city was encircled by a belt of flame 50 kilometres in diameter,

which reached almost up to its gates, a great part of the magnificent forest of Edough was consumed, as were many others in all directions.[33]

And speaking of the English consul, whatever happened to the La Safia concession which we visited with him after the London and Lisbon Cork Company purchased it from Cès-Caupenne in 1865?

Three months after the English Company had taken possession:

> the forests were set on fire in twenty different places, and for two days the flames raged with incredible fury. Seventy miles [115 km] length of cork forests was burnt, property to the amount of 3,000,000 pounds was destroyed; and in the forest of La Safia alone, 600,000 noble oaks, four-fifths of the total quantity, were consumed.[34]

Fires had always broken out during the hot season in the Bône area as they had elsewhere in the Mediterranean due to the ecology of the region. But the French claimed that more and worse fires had occurred since they had arrived to colonialise the Algerians. Maybe, maybe not – there is no way of knowing. What we do know is that there were some fires every year, in some years there were more than others, and that the conflagrations of 1865, 1873 and 1894 described above plus those of 1859, 1863, 1870, 1876, 1881, and 1892 became literally engraved on the collective colonial memory, that is, they comprised an element of colonial political culture then in the process of formation.

As one of the most thickly forested areas of Algeria, the Bône region was hit frequently by large-scale fires. More fires struck Bône than elsewhere in Constantine province, and more occurred in Constantine province than the rest of Algeria. The government commissions which investigated the worst fires were based regularly in Bône. Of ten areas in Algeria hit heavily and repeatedly by fires, the two worst anywhere were the Beni Salah forest plus another stretch in Constantine province between Bône and the Tunisian border. Major fires broke out in the Beni Salah in 1872, 1877, 1887, 1894 and 1902. On the other hand, the extensive cork oaks of the Edough generally did not suffer fires commensurate with their number and extent, except for the fires of 1863 and 1865 and the conflagration of 1881 described above by Maupassant, which was perhaps the single worst fire in Algerian history.[35]

What were the causes of the fires? Let us listen to what people at the time had to say. The concessionaires blamed the Algerians pure and simple. De Lesseps, Besson and des Chenetz concluded their report on the 1865 fires: 'There is no disguising the fact that war by the arsonist's torch like that of the assassin's ambush or the rebel's gun originates from the same cause and aims at the same end: the ruin of the Colony.'[36] The settlers agreed. 'The Arab in Algeria, there is the enemy.'[37] And: '[s]etting fires is one of the particular forms of native banditry, a protest against France's possession of the forests, revenge for the trouble caused to their predatory habits [*habitudes de rapines*].'[38]

There was something to this claim, although not as much as the concessionaires and settlers thought. The clash between two opposed ways of life, between two political economies, forms the social-historical context in which the Algerians resisted the French in the forest. But the form this resistance took – arson – was itself the result of a tradition of Algerian resistance to French colonialism. Settlers and concessionaires at the time saw no difference between setting fire to the forests and the armed revolts mounted by the Algerians periodically against the French since 1830. The important revolt of 1871, mentioned above by the English consul,

Table 4: Forest fires in Algeria, 1853–1928

Year	Number	Area (in hectares)	Damage (in francs)
1853	n.a.	'very significant'	n.a.
1860	n.a.	10,000 (a)	n.a.
1863	n.a.	42,000 (b)	n.a.
1865	n.a.	163,954	2,212,676
1871	n.a.	'very significant'	n.a.
1873	177	75,313	n.a.
1874	n.a.	2,777	n.a.
1875	n.a.	n.a. (c)	n.a.
1876	120	55,172	441,881
1877	134	40,538 (d)	1,807,061
1878	164	8,156	617,324
1879	218	17,663	625,987
1880	137	20,881 (e)	533,245
1881	244	169,056 (f)	9,042,440
1882	130	4,018	188,751
1883	148	2,464	102,338
1884	147	3,232	205,185
1885	285	51,569	674,487
1886·	288	14,043	270,325
1887	395	53,714	1,560,920
1888	311	14,788	176,833
1889	309	17,807	522,389
1890	202	23,165	1,726,505
1891	393	45,924	2,906,459
1892	409	135,574	6,605,275
1893	398	47,757	5,303,752
1894	308	100,890	2,266,043
1895	250	32,907	324,661
1896	179	14,091	632,058
1897	396	79,203	2,468,062
1898	150	12,384	282,264
1899	272	16,099	547,766
1900	162	2,937	143,192
1901	135	9,687	259,111
1902	475	141,141	3,668,780
1903	388	94,398	5,329,047
1904	244	2,759	90,093
1905	255	7,676	274,084
1906	219	9,126	399,037
1907	211	4,457	92,809
1908	344	6,540	182,339
1909	278	9,751	653,050
1910	482	24,294	411,287
1911	322	16,309	339,148
1912	338	26,505	377,203
1913	696	138,191	3,086,138
1914	362	43,305	507,718
1915	237	9,350	151,149
1916	357	78,863	1,995,231
1917	359	95,453	4,265,095
1918	268	38,720	1,963,149

(continued)

Table 4 (continued)

Year	Number	Area (in hectares)	Damage (in francs)
1919	678	116,889	3,774,631
1920	846	83,986	3,980,288
1921	231	11,210	549,323
1922	601	89,473	3,891,153
1923	231	5,997	338,696
1924	457	62,360	5,726,597
1925	193	9,146	687,858
1926	526	81,985	41,515,062
1927	234	10,504	1,807,251
1928	177	13,339	3,025,681

Notes:
a. Constantine province only.
b. Constantine province only.
c. Minimal loss.
d. Of which, 36,750 ha in Constantine province.
e. Of which, 18,800 ha in Constantine province.
f. Of which, 155,000 ha in Constantine province.

Source: Annual government reports reprinted in Marc, *Notes sur les forêts*, pp. 364–5, 435.

had swept up the Algerians bordering Besson's concession. 'The course of the revolt was marked, each day, by the development of fires.'[39] However, the times were changing. Where the first decades of French rule from 1830 to about 1870 were characterised by armed revolts by the Algerians against the French, from 1870 onwards resistance became more muted, and it was not until the First World War, in 1916 in Batna, and after the Second World War, in 1945 in Sétif, that we see a recrudescence of revolts which lead in 1954 to the biggest armed rebellion of them all, the Algerian Revolution.

The region of Bône pretty much follows this pattern. One of the first and most intensively colonialised areas, it makes sense that revolts ended sooner here than elsewhere. In fact, the last armed revolt in the region was that of the Beni Salah in 1852. After attacking a group of military woodcutters and murdering a French army colonel, the Beni Salah sought refuge from French repression first in the forest, and then across the border in Tunisia. French repression was indeed harsh: they declared the Beni Salah collectively responsible, confiscated their land, and did not allow them to return from Tunisia until 1863. Just three years later in 1866 a major fire broke out in the Beni Salah forest, the first of several. After an important fire in 1877, the Beni Salah were again declared collectively responsible, the first time in Algeria this punishment had been meted out for arson. Clearly, Beni Salah resistance had not changed; what had changed was the form resistance took. They had resisted the Turks (the Beni Salah had acknowledged Turkish suzerainty only during the reign of the last *bey* of Constantine), they rebelled against French hegemony as long as they could, and then they resisted by setting fire to the forest. What changed, therefore, was less the Beni Salah tradition of resistance than the French ability to repress such resistance.[40]

The government concurred in part with the settlers and concessionaires who blamed forest fires on Algerian banditry. Thus, the official investigation of the

1881 fires 'reveals the natives promenading with lit torches in areas not yet on fire as well as in those where the fire had been put out,' and members of the commission came away with 'this conviction that the natives had followed a prearranged plan, obeyed a sort of signal'.[41] In cases of outright arson such as these, the government invoked the harsh penalty of collective responsibility, as we shall discuss in a minute. In the majority of cases, however, the government veered away from the views of both the concessionaires and settlers to argue that 'the majority of forest fires ... originate ... from setting fire to brush and undergrowth, operations which the natives' self interest ... leads them to take in order to increase and improve the pasture of their herds':[42] in other words, *kçar*, clearing by fire.

Consider Table 5, which presents causes of fires prepared by the government itself. Given the source, namely official forestry service reports, the data certainly cannot be said to be biased in favour of the Algerians, which makes the findings all the more interesting.

Table 5: Causes of fires in Algeria, 1886–1915 (percentages)

Accidental	8
Carelessness	32
Intentional	23
Unknown	37
Total	100

Source: Tabulation of probable causes based on official forestry reports in Marc, *Notes sur les forêts*, p. 396.

The single largest category is fires of unknown origin. More revealing is the fact that the causes of all fires listed in Table 5 result from 'appraisals which, for 4/5 of these cases are based only on probabilities.'[43] But let us continue. 'Accidental' appears to be a category of fires due to nature rather than to man. For example, it included those begun by lightning and flying sparks from locomotives. Thus, of 138 fires in the Bône region in 1902, another big year for fires, 49 were caused by lightning, not to mention another 14 ignited by sparks from locomotives and not included in the 138 investigated.[44] Carelessness caused fires such as the one which broke out in the Beni Salah forest when Italian workers neglected to clear the brush from around their cooking fire, which I will discuss below. It would seem, therefore, that fires caused by Europeans resulted from carelessness, since everyone knew that Europeans did not cause fires intentionally. By definition, intentional fires were those set by Algerians. In the view of the government and forest service, however, there were two sorts of intentional fires, namely, those resulting from *kçar* and those caused by arson pure and simple. Therefore, it emerges from the government's own analysis that less than one-quarter of all fires were caused by the Algerians, and moreover, that this one-quarter included fires due to *kçar* as well as arson.

Now, let us pursue this analysis one step further. Every year that a major conflagration occurred, a blue-ribbon commission was appointed to make a special report, and every commission concluded that, accidents and carelessness aside, *kçar* was the main cause of fires. But how then to prevent fires? Why, stop setting fires, stop practising *kçar*, of course! For example, the report on the 1873 fires first pinpointed *kçar* as the cause, then called for 'rigorous repression' of the Algerians as the solution, in fact, collective responsibility. Thus, the French placed the Algerians in a classic double-bind situation. They said in effect, 'You can live in the forest as long as you don't set fire to it,' to which the Algerians replied, 'How can we live in the forest without setting fire to it?'

Neither the concessionaires nor the settlers nor the government bothered generally to ask the Algerians what they thought, but when they did, the Algerians basically agreed with the government that the chief cause of fires was *kçar*.[45] But where the Algerians agreed on the fact of *kçar*, they disagreed totally with the European interpretation of *kçar*. While most Algerians most of the time denied Algerian arson, some Algerians admitted that some arson occurred some of the time. But arson was not due to *kçar*, it was due to European attacks on *kçar*, to attempts by the concessionaires and forest service to suppress the entire traditional Algerian way of life in the forest with *procès-verbaux*, individual and collective fines, and property seizures. 'Due to the citations [*procès-verbaux*], we've become so wretched that we are led to commit thefts and crimes,' avowed one group of Algerians. 'The settlers, the *compagnie des forêts du Fendeck* [Lucy, Falcon] who are our neighbours, ruin us with *procès-verbaux*.'[46] And the Algerian author of a pamphlet on the 1881 fires concluded, 'The fires in our canton must be attributed to the motive of revenge [against the forest companies which multiply the *procès-verbaux*].'[47]

Behind *kçar* the Algerians saw a French attack on their entire way of life; behind *kçar* the French discerned Algerian banditry powered by religious fanaticism. The French worried constantly about the danger of Sufi agitation among the Algerians, about the danger of pan-Islamic propaganda in mobilising the Algerians, about the rise of a *mahdi*, a chiliastic prophet who would drive the French from Algeria. The fires of 1881 came closest to fulfilling these French fears. The beginning of an Islamic century was always a propitious moment for an outbreak of Islamic millenarianism. The turn of the fourteenth Islamic century occurred in 1881–2, and was the year of the Sudanese Mahdi. Moreover, the government commission investigating the fires pointed to the influence of a revolt in the Algerian province of Oran and the French invasion of Tunisia in the same year as additional influences on the Algerians. The religious dimensions of popular resistance, especially in the case of 1881, undoubtedly merit closer scrutiny. Yet the French never demonstrated how events in the wider Islamic world impinged concretely on Algerians in the forest; they never uncovered actual conspiracies and arrested specific leaders, let alone a self-proclaimed *mahdi*.[48]

The question of Algerian arson merits one final comment. Liberal French historians writing in our own day have been more reluctant to ascribe at least some fires to arson than the Algerians themselves were at the time. Thus, Charles-Robert Ageron, the leading historian of colonial Algeria, is admittedly puzzled by the 1881 fires: 'It is very difficult for the historian to pronounce formally on the character of these fires.' The uncharacteristically high number of fires, the large area burned (over 150 000 ha in Constantine province alone), combined with the unwillingness of the Algerians to fight the fires all lead Ageron to 'the acceptance in some measure of the official version'.[49] Likewise, André Nouschi, author of the single best study of the effects of colonialism on Algerians in Constantine province, makes no attempt whatsoever to weigh causes or assess blame for forest fires.[50] Perhaps violent resistance such as arson, a variety of *la résistance chaude* as the Algerians say, ruffles liberal sensitivities, perhaps it runs against the liberal grain. If so, it is rather ironic that these same historians were staunch supporters of that other movement of resistance, the Algerian Revolution.

To conclude this section, we can say *kçar* led sometimes to uncontrolled fires, but more often it was the French intention to suppress *kçar* and the political economy of which it was a part that led Algerians to commit arson. Algerian banditry did not lead to forest fires; the French attack on the Algerians led some Algerians to respond by engaging in banditry, one form of which was setting the forest on fire. Thus, the Algerians committed more arson than historians think, but less than contemporary colonials thought.

How did the government and the settlers react to the Algerians allegedly burning down the forests? Consider the government first. The primary response of the government to forest fires was to hold the Algerians collectively responsible when it was felt arson had been committed. This was the other, grimmer face of the government's two-faced reaction to fires. Despite the fact that the government's own studies concluded that *kçar* was the main cause of fire, the government was not deterred from invoking the harsh penalty of collective responsibility.

Collective responsibility was a policy practised by the Turks, although there were instances of its application in metropolitan France as well. It was repugnant to liberal legal notions; it struck the liberal sensibility as unenlightened, absolutist, and retrograde. Collective responsibility took two forms: first, collective fines; and, second, wholesale confiscation of land, which could then be bought back usually at 40 per cent of the value of the Algerians' total possessions.

The first time the French applied collective responsibility in colonial Algeria for setting forest fires was 1877 when the Beni Salah had their lands confiscated. Of 7944 ha of cultivable land, 4199 ha were taken; they could buy back their land for 40 per cent of the value of all their holdings and belongings, which worked out to 280 000 fr in all. This sum was so high that a French official proposed that the Beni Salah be allowed to pay the money over six years at 50 000 fr per year for the first five years plus a final payment of 30 000 fr. '[T]he sum of 50 000 fr represents nearly the amount of taxes paid in 1877; it is all that the Beni Salah can pay over and above what they owe currently without being forced to sell off considerable numbers of their animals.'[51]

This was still far too heavy a price to pay, especially when combined with the land lost by the Beni Salah in 1876, which had gone to the creation of still more European villages in the plain of Bône. The Beni Salah complained to the Governor-General that the money they must pay 'will plunge them into a state of absolute misery and distress'.[52]

Most galling of all was the utter unfairness of the whole process as far as the Algerians were concerned. Neighbours of the Beni Salah, also hit with collective responsibility, 'request to leave the country in order to live elsewhere,' a French official reported:

> [f]or they are tired of being held responsible for fires in forests where they cannot live, and which are inhabited only by workers of all nationalities who come from all over, and who are careless about complying with the prescribed measures for preventing fires, knowing full well that in case of a conflagration, they are above suspicion and hence safe from punishment.[53]

There was no doubt something to their complaint. For example, the very next year in 1878 another fire broke out in the Beni Salah forest, and a colonial newspaper reported that:

[T]he fire started in a field kitchen adjoining a hut [*gourbi*] constructed by some Italian workers extracting tannin in the middle of the forest and which they had lived in a few days . . . As far as the cause of the fire is concerned, an examination of the area makes it possible to say that no underbrush had been cleared from around the *gourbi* in which the Italian workers lived . . .[54]

The reporter concluded by warning against blaming all fires on the Algerians. But tell it to the rest of the settler press.

In the end, even the government realised that the 1877 sanctions were simply too much for the Beni Salah to bear, and instead of confiscating more than 4000 ha, it took 2239 ha instead, but at the same time increased the total amount the Beni Salah had to pay. This was not the only time the Beni Salah were held collectively responsible for setting fire to the forest. Leaving aside the confiscation of their land as a result of the revolt of 1853, the Beni Salah were forced to pay a collective fine in 1894; and in 1902 an official commission concluded that they were guilty of having set fires willfully and had not fought the resulting fires sufficiently, both of which constituted 'insurrectional acts' and merited the application of collective responsibility. This same 1902 body recommended, furthermore, that since collective responsibility had not apparently deterred the Beni Salah in the past, this time they should be 'displaced and moved to unwooded areas,' and should be made to pay·a collective fine equal to ten times their annual tax. 'This fine may at first seem a little high, but it is not such that it may gravely impair the resources of the natives.'[55]

To the government the Beni Salah were clearly inveterate arsonists. Not so the Algerians living in the area of Jemmapes where some of the most destructive fires in the history of colonial Algeria occurred in 1881. Whether or not the Algerians set fire to the forests because they thought a *mahdi* was coming to chase the French out of Algeria, the French responded by applying collective responsibility in a wholesale manner. No fewer than thirty-seven groups had their land confiscated, and another twenty were assessed collective fines.[56]

As with the Beni Salah in 1877, the government came to realise over time that the fines were exorbitant, and lowered the amounts due. Still, however, the Algerians paid in the end 3 360 83 fr, or 336 fr per person. To put this in perspective, let us note simply that this sum is more than each Algerian was forced to pay who took part in the revolt of 1871, which was the single largest uprising in Algerian colonial history. Moreover, observers then and now have argued that in 1881 the government used collective responsibility as a ruse to obtain land for European colonisation.[57] No wonder the Algerians complained more than ten years after the fires that 'the sequestration is too heavy, it has ruined us: some among us have been dispossessed completely, reduced to utter poverty, and forced to become vagabonds.'[58] No wonder that to the invocation, 'Lord, deliver us from the forest guards!' the Algerians added, 'Lord, deliver us from the process-servers [for the confiscations].'[59] Worst of all, certain of the Algerian groups affected began literally to disintegrate. Tax revenues declined precipitously as the tax base eroded, a tell-tale sign that the Algerians were voting with their feet and leaving the countryside altogether.[60]

Yet another result of alleged Algerian arson was the settlers' invention of what I call 'the myth of Algerian banditry'. The gap between objective and subjective truth, that is between the number of fires actually set by the Algerians and the number of fires contemporaries alleged were set by Algerians, forms an ideological space which was filled by the fiction that Algerians were by definition bandits practising arson. This is not to say that Algerians did not commit arson; it is to say that the Algerians did not commit as much arson as the settlers said they did. This ideological gap emerged clearly in the following exchange concerning the 1881 fires which appeared in a settler newspaper:

> [F]from this day onward, we vow a just and implacable hatred for this brutish people, which our grandchildren will share, for we will be sure to tell them of their heinous crimes, which will become the legends of our evening gatherings during the winter . . . [Let us extinguish] in the blood of this vile race of slaves the ardent thirst of vengeance which devours us.[61]

To this racist outburst another settler responded in the same paper:

> Go among the Algerians after the forest fires and you will agree fully that not everything is rosy for them . . . When the Arab's self-interest is diametrically opposed to the accomplished act, be convinced that he is not guilty.[62]

Now, I would argue that at the root of the settler myth of Algerian banditry was a radical disagreement concerning the nature and practice of *kçar*. To put it simply, for the Algerians *kçar* entailed controlled burning to obtain pasture. For the French, however, *kçar* led to uncontrolled fires, and thus it was related to if not identical with arson, one form of Algerian outlaw behaviour. Thus, the commission investigating the 1902 fires asserted that:

> Thefts and murders are combined with forest fires . . . The fact of repeated acts of brigandage aimed at people and at the property of settlers . . . has demonstrated to it [the commission] that the question of grazing is not the sole cause of fires. One must also look for the cause in the antipathy of the Native for the European, in his habits of brigandage, in his desire to destroy which is so clearly demonstrated by where the fires break out and the choice of windy days.[63]

The settlers' interpretation of *kçar* led logically to the double-bind situation in which the French put the Algerians, for if *kçar* led to forest fires, then obviously the way to stop forest fires was to stop *kçar*. To take this a step further, I would argue that the French had no alternative but to equate *kçar* with banditry. For if the French had interpreted *kçar* as reasonable and rational rather than unreasonable and irrational, that is as a central component of the Algerian political economy, then the implication would have been either that the Algerians had just as much right to the forests as the French, or that the French had to move the Algerians out of the forest altogether. The French could not accept either implication. To admit even implicitly that the Algerians had a right to the forest was to undermine the entire colonialising mission of France in Algeria. And to accept the implication that to eliminate *kçar* the Algerians had to be eliminated from the forest was too extreme a measure even for the French to contemplate taking in colonial Algeria. On the one hand, therefore, the French waged a war of attrition against the Algerians and, on the other hand stigmatised both the practice of *kçar* and those who practised it by inventing the myth of Algerian banditry, the second reinforcing and rationalising the first.

In a similar manner the settlers selectively misinterpreted their experience to conform to their myth of Algerian banditry, as we can see from the following examples: 'when the natives don't want fires, there are no fires'; and 'the confiscation of lands imposed on the Beni Salah after the 1877 fires does not appear to have led them to mend their ways, since they have started to set fires again', so that '[A]ll in all, they are still better off than before'.[64]

For the settlers to legitimise their tendentious reading of Algerian experience, it was necessary to claim expert status. The government fell in with them. The 1902 commission declared that the 'people living in the countryside', that is, Europeans from whom it received depositions, were as a result of simply living in Algeria '*au courant* with the mentality of our Muslim subjects'.[65]

Together these elements comprised the settler myth of Algerian banditry. It was in turn a component of the larger settler colonial ideology shared to a greater or lesser extent by all Europeans. Government officials subscribed to it insofar as they correctly diagnosed the main cause of fires as *kçar* but incorrectly attempted to cure *kçar* by applying collective responsibility. Yet the myth of Algerian banditry was shared more by settlers and concessionaires than by government officials, which is why I have called it the settler myth of banditry. The following settler account makes this clear. With bitter sarcasm a Bône newspaper derided the official government report on the 1873 fires, which singled out *kçar* rather than banditry as the chief cause.

> It is . . . a document worthy of standing alongside the impudent theories of spontaneous combustion . . . Idiotic settlers, who accuse the Arabs of malevolence and who naively base your suspicions on what you have seen and heard, you are only ungrateful . . . Without fires, no harvests . . . it is the leader of the High Commission who informs you of that. Cry, if you want, for your fields, your haystacks, your woods, your olive trees, your farms; cry, if you want, for the victims of Bugeaud, . . . The settlers claim that the Arab sets fires to combat our domination. There is nothing to it: it is quite the opposite. If the Arab wanted to run us out, he would never set fires.[66]

Yet whenever the government applied collective responsibility as it did in 1881, then the settlers applauded the government, since both agreed in effect that the cause was Algerian bandits practising arson:

> The movement of colonisation which was stifled all along the littoral by the possession of the best land by the natives will now be strengthened and able to spread throughout the countryside . . . The mobile columns moving through the ravaged countryside will make it possible to smash all resistance, to terrify the bandits, and above all to prevent an explosion of trouble which had been feared at the announcement of the measure [of collective responsibility] imposed on the natives.[67]

There was no final victory, no clear-cut denouement to the campaign the French waged against the Algerians in the forest. The French did all they could to force the Algerians out of the forest short of physically displacing them. The Algerians yielded where they had to, relied on subterfuge when they could, and for the rest basically put up with things. After all, what were their alternatives?

If there were no clear-cut turning points, however, there were definite long-term consequences. First and foremost, the French squeezed the Algerians pitilessly. As discussed earlier, the Algerian political economy was based on a rough-and-ready equilibrium between land, people, and animals. Table 6 demonstrates the collapse of this equilibrium.

As French concessionaires and the forest service took over large tracts, the effective size of the forest shrank. Over time, the Algerians moved around less and settled down more; sedentarisation gradually replaced migration. There was no unexploited, uninhabited forest left to migrate to. At the same time, population

Table 6: Land, people, and animals in Bône region, 1856–1955

	Population	Cultivated land				Cows				Sheep		
		ha	ha/ person	% chg	No.	No./ person	% chg	No.	No./ person	% chg
Edough mts.										
1856	15729	9030	0.57		37226	2.37		46497	2.96	
1955	56154	18856	0.34	−40	32943	0.59	−75	51572	0.92	−69
Bône plain[a]										
1856	11139	6606	0.59		30006	2.69		36748	3.30	
1955	46237	12979	0.28	−52	15625	0.34	−87	20574	0.44	−87

Note a: Figures for Bône plain, which includes the Beni Salah, included here for purposes of comparison.

Source: Tomas, *Annaba*, p. 274.

grew, nearly quadrupling in the Edough mountains in a century. Thus, people were pinched. To feed more people, more forest was cut down and more land was cropped. Still, the amount of cultivated land per person declined by 40 per cent. More land cropped meant less land for grazing. The absolute number of animals remained constant, but the number of cows per person dropped by three-quarters and the number of sheep by two-thirds. These are the brute facts of the French assault on the Algerians in the forest.

How did the Algerians respond? Basically, they shifted to other work in the forest, or moved out of it altogether. The traditional political economy remained the basis of life of most Algerians in the forest, but they relied increasingly on the capitalist political economy introduced by the concessionaires. Noteworthy here is that in 1925 five out of every six workers in the forest were Algerians.[68] This was the result of a process which had been going on since the 1860s and 1870s. A concessionaire describes how it worked. 'The [Berber] Kabyles come, arrangements are made with their leaders for a certain amount of work, the work completed they are paid, they leave and the same ones don't return'. As a result, 'Already our Arabs are no longer content with a few days of work per month, they have gotten a taste of being well off, and as a result they are now changed people.'[69] To be sure, there were Algerians who clung to the old ways. Concessionaire Besson observed that the Algerians refused to fight fires saying, 'leave it, leave it! It's supposed to burn!' and that they 'prefer to nourish themselves with acorns and young sprouts of *diss* grass rather than emigrate' to earn money. But capitalism is catching on: 'there are always three who want work for every one who can be satisfied.' Moreover, Besson and de Lesseps 'don't have any other workers besides the natives . . . Thus, a prosperity of sorts has replaced . . . the misery that prevailed before Bône Europeans arrived.'[70]

By the 1890s, if not earlier, capitalist wage labour had become an essential part of the Algerian economy. The Ouichaoua still raised animals and grew cereals, although so much of their land had been taken that they did not have enough left to raise enough food to feed themselves. 'But the inhabitants find a considerable source of revenue in the important forest exploitations created by the Europeans on their land.' It is the same with the Tréat, who 'are poor due to so little land. However, the exploitation of their forests by the Europeans furnishes them with considerable income.'[71] Their main task? Stripping bark from the cork oaks three months of the year. All this is evidence that the Algerians were learning gradually the rules of the capitalist game. But if the Algerians were entering the capitalist economy, they were entering it at the bottom. Besson's operation was typical: a European foreman directing an Algerian work crew of day labourers paid 'at the going rate'.

Another rule connected with the capitalist game the Algerians were learning was how to litigate. The numerous petitions addressed to the Governor-General discussed earlier are a clear example of this. And there are others. At the time of the 1881 fires, Algerians wrote a pamphlet and sent a petition to the prefect outlining their grievances. A dispute over land in the Edough mountains lasted from the turn of the century well into the 1930s.[72] Are these examples of how popular protest modernised in a colonial context? Perhaps. Perhaps not. The efficacy of such litigation may be doubted, but why could it not be combined with arson as a means of resistance?

The second main response of the Algerians to the squeeze on their traditional way of life was to migrate from the forest to the city. The population in the Edough nearly quadrupled from the middle of the nineteenth century to the middle of the twentieth, but the population of Bône increased more than ten times to over 100 000. Some of this increase included Algerians from the Beni Salah and the Edough, and no doubt some ended up working in one of the four bottle-cork factories located in Bône in 1904. But if they had moved to the city, they had not moved up the socio-economic ladder. All four of the workshops were French-owned; both foremen were non-French Europeans (probably Italians, maybe Maltese); there were more European than Algerian skilled workers, more Algerian than European unskilled workers, and in every job category the Algerians received less pay for equal work.

Table 7: Bône bottle cork workers, 1904

	Skilled Workers			Unskilled workers		
	Number	Minimum Salary	Maximum Salary	Number	Minimum Salary	Maximum Salary
French	12	4 fr	4 fr 50	4	3 fr	4 fr
Other European	15	4 fr	4 fr 50	6	3 fr	3 fr 50
Algerian	11	3 fr 50	4 fr	25	2 fr 50	3 fr

Source: Archives de l'Assemblée Populaire Communale de Annaba, Statistique Industrielle. 1904.

For the Algerians, the capitalist labour market was a segmented one. Thus, capitalism came hand in hand with colonialism in Algeria. 'Modernisaton' came with the mailed fist of 'Westernization'. No wonder the Algerians said, 'The French, they're hard [*dur*].'

Today, the French have left and the Algerians have taken over, *l'Algérie française* has become *l'Algérie algérienne*. The name of Bône has been changed back to Annaba. The unrepresentative colonial state has been replaced by a one-party national state. When the national assembly is in session, my former landlord is put up in the swankiest hotel in downtown Algiers at government expense. When now-deceased President Boumedienne came to Annaba to whip up political enthusiasm at a rally staged in the soccer stadium, my maid Zaouia Dief, a poor, illiterate, self-effacing woman already old at 50, refused to go to the rally, saying she couldn't be bought for the loaf of bread every participant received at the gate.

Today, forest fires still occur in the Edough mountains. Are there bandits still setting the forests on fire? Have the Algerians internalised the French myth of

Algerian banditry? Whatever my landlord the member of parliament thinks, he isn't saying.

NOTES

1 Information on the Algerian groups in the Beni Salah forest and the Edough mountains can be found in Archives Nationales d'Outre-Mer d'Aix-en-Provence [hereafter AOM], M 90 (165) [Beni Merouan]; M 83bis (158) [Beni M'Hamed]; M 78 (275), 8 H 6 [Beni Salah]; M 78 (276), 21 KK 11, 8 H 6 [Djendel]; M 78 (282), 21 KK 12, 8 H 6 [Dramena]; M 83bis (55) [Fedj Moussa]; M 97 (259) [Ouichaoua]; M 84 (75) [Senhadja]; M 98 (267) [Tréat]; 21 KK 16, 8 H 6 [Talha]. On the proportion that were Arab and Berber respectively, see Ouanassa Siari-Tengour, 'Les populations rurales des communes mixtes de l'arrondissement de Bône (Annaba) de la fin du XIXe siècle à 1914' (unpublished thèse de 3e cycle, Université de Paris VII, Jussieu, 1981), annexes, Arabs and Berbers, cercle de Bône and cercle de l'Edough.

2 In 1845 the population of the Beni Salah was 1805, and the population living in the Edough mountains was 3148, which was divided into 132 four-person *gourbis*, or huts, 392 large five-person tents, and 220 small three-person tents. Cf. Ministère de la Guerre, *Tableau de la situation des établissements français dans l'Algérie* [hereafter TEFA] (Paris: Imprimerie Royale, 1838–68), vol. 1844–45, pp. 422–3; AOM, F 80 1785, Examen de la question forestière dans la subdivision de Bône, 1846. To arrive at an estimate of the population in 1830, these mid-1840s figures have been increased by roughly 10 per cent. On the problem of deriving 1830 totals from mid-1840s data, see André Nouschi, *Enquête sur le niveau de vie des populations rurales constantinoises de la conquête jusqu'en 1919* (Paris: Presses Universitaires de France, 1961), pp. 20–30.

3 British Parliamentary Papers [hereafter BPP], 1874, Vol. LXVII, Accounts and Papers – Commercial Reports [hereafter Comm Rep], Vol. 33, p. 1294; H. Marc, *Notes sur les forêts de l'Algérie* (Paris: Larose, 1930), p. 395; Charles-Robert Ageron, *Les Algériens musulmans et la France (1870–1919)* (Paris: Presses Universitaires de France, 1968), p. 111. For background on the Algerian forest economy, see the documents in AOM, F 80 1784 and 1785; and André Nouschi, 'Notes sur la vie traditionnelle des populations forestières algériennes', *Annales de Géographie*, vol. 68 (1959), pp. 525–35.

4 A. Hanoteau and A. Letourneau, *La Kabylie et les coutumes kabyles* Vol. 1 (Paris: 1872–1873), p. 510, quoted in Nouschi, *Enquête*, p. 42.

5 Adrien Berbrugger, 'Mémoire sur la peste en Algérie depuis 1522 jusqu'en 1819', *Exploration scientifique de l'Algérie pendant les années 1840, 1841, 1842* (Paris: Imprimerie Royale, 1844–67), séries II: Sciences médicales, vol. 2, pp. 214–15, 217, 219, 222–3, 232–3; M. Dureau de la Malle, *Province de Constantine* (Paris: Gide, 1837), pp. 142–3; Lucette Valensi, *Le Maghreb avant la prise d'Alger (1790–1830)* (Paris: Flammarion, 1969), pp. 37–9.

6 AOM, F 80 1785, Forêts du Kaidat de l'Edough, 1845; TEFA 1841, p. 281; Eugène Battistini, *Les forêts de chêne-liège de l'Algérie* (Paris, 1938), pp. 24–5; Ageron, *Les Algériens*, p. 127; Nouschi, 'Notes', p. 529.

7 BPP 1867–68, Vol. LXVIII, Comm Rep, Vol. 29, p. 203.

8 Report of concessionaire Cès-Caupenne to Governor-General, extracted in TEFA 1862, pp. 320–21; BPP 1893–94, Vol. XCIII, Comm Rep, Vol. 44, pp. 17–19; BPP 1895, Vol. XCVII, Comm Rep, Vol. 37, p. 57; AOM, 9 X 121, Rapport de la Commission d'Enquête, *Incendies de forêts en 1902 dans la région de Bône* (Algiers: Franceschi, 1903), p. 103.

9 Information on forest concessionaires is found in AOM, F 80 1783–1787; and Battistini, *Les forêts*.

10 Ageron, *Les Algériens*, p. 108; Nouschi, *Enquête*, pp. 326–7.

11 While the archival documents regarding Napoleon's concessions are very precise about the size of each concession, they are virtually silent when it comes to determining how much land was taken from which group of Algerians. To find the latter information, we must turn to another set of documents, which, however, were generated for a different reason. These data were produced from the application of the so-called *sénatus-consulte*, which classified and delimited the land of each group of Algerians. At the same time as the French took the measure of the Algerians, however, they used the knowledge obtained to exercise their colonial hegemony, for example, by converting communal *'arsh* to individual *milk* land to break up and sell landholdings more easily, to designate and set aside as yet unconceded forests as state forests, to divide Algerian groups into segments in order to rule them more effectively, or to attach them to *communes de pleine exercise*, full-fledged French communes run by local settlers. A second problem is that the *sénatus-consulte* was applied to some groups in the late 1860s but to others only in the 1890s (including many in the Edough). Yet another problem is that in the course of carrying out the *sénatus-consulte* the French frequently reorganised and renamed the Algerians. Despite these problems, they contain much rich information.

12 John M. Merriman, 'The *Demoiselles* of the Ariège, 1829–1831', in Merriman (ed.), *1830 in France* (New York: Franklin Watts, 1975), pp. 87–118.

13 See Eugen Weber, *Peasants Into Frenchmen* (Stanford: Stanford University Press, 1976), especially pp. 485–92 where he applies Frantz Fanon to nineteenth-century France. For a case study of cork workers who took over and ran briefly their own enterprises as producers' co-operatives during the Second French Republic of 1848, see Maurice Agulhon, *La République au village* (Paris: Plon, 1970), pp. 126–45, 305–60.

14 Nouschi, *Enquête*, p. 494.

15 L. Vignon, *La France en Algérie* (Paris: Hachette, 1893), p. 394, quoted in François Tomas, *Annaba et sa région* (Saint-Étienne: Guichard, 1977), p. 271.

16 Quoted in Ageron, *Les Algériens*, p. 128.

17 Conseiller general Mollet, quoted in ibid., p. 127.

18 Quoted in ibid., p. 123.

19 Quoted in ibid., p. 124.

20 In the course of carrying out the *sénatus-consulte*, the documents cited in note 1 above often state that user's rights are maintained – at the same time as the Algerians are petitioning to have the concessionaires observe their traditional user's rights.

21 AOM, F 80 1785, 'Les forêts algériennes', 20 November 1905.

22 In 1891 a prefect urged the surveillance of forest guards 'too often tempted in their solitude by the ease of illicit gains and a nearly certain immunity.' (Quoted in Nouschi, *Enquête*, p. 495).

23 AOM, M 97 (259) (Ouichaoua), Hadj Mahmoud ben Bou Abdallah, Ali ben Bou Abdallah ben Ammar and fourteen others to Governor-General, 2 August 1984.

24 ibid., Mohammed ben Cheick, Mohammed ben Cherif and others to Governor-General, 18 September 1893.

25 ibid., Hadj Mahmoud ben Bou Abdallah, Djebali Mohammed ben Ali and 'all of the inhabitants' of Ouichaoua to Governor-General, 23 November 1894.

26 ibid., Boudiaf Belkassem ben Mebrouk ben Abdelkader to Governor-General, 7 December 1894; Djaballi Mohamed ben Ali to Governor-General, 17 November 1894.

27 ibid., Djaballi Mohamed ben Ali to Governor-General, 17 November 1894.

28 ibid., El Hadj Mahmoud ben ben-Abdallah to Minister of the Interior, 19 September 1893.

29 AOM, F 80 1785.

30 AOM, F 80 1785, *Petit Journal*, 12 September 1894.

31 *Le Soleil*, quoted in Ageron, *Les Algériens*, p. 117.

32 BPP 1875, Vol. LXXVII, Comm Rep, Vol. 36, p. 1435.

33 BPP 1874, Vol. LXVII, Comm Rep, Vol. 33, p. 1294.

34 BPP 1867–68, Vol. LXVIII, Comm Rep, Vol. 29, p. 203.

35 Marc, *Notes sur les forêts*, pp. 390–91.

36 Quoted in Battistini, *Les forêts*, p. 51.
37 *L'Indépendant*, 16 April 1882, quoted in Nouschi, *Enquête*, p. 486.
38 Maudemain, president of syndicat des agriculteurs of Guelma, quoted in AOM, 9 X 121, *Incendies . . . en 1902*, p. 108.
39 AOM, F 80 1787, Besson, Note sur les forêts.
40 AOM, 8 H 6; TEFA 1850–52, pp. 13–14.
41 Quoted in Marc, *Notes sur les forêts*, p. 402.
42 ibid., pp. 397–8.
43 ibid, p. 396.
44 AOM, 9 X 121, *Incendies . . . en 1902*, p. 105.
45 Conclusions regarding Bône of commission investigating 1894 fires, quoted in Marc, *Notes sur les forêts*, p. 400.
46 Algerians of commune mixte Collo, quoted in Nouschi, *Enquête*, p. 494.
47 Ali ben Belqasem ben Mahoui, quoted in Ageron, *Les Algériens*, p. 118.
48 For a discussion of Islamic chiliasm in earlier Algerian revolts, see Peter von Sivers, 'The realm of justice: apocalyptic revolts in Algeria (1849–1879)', *Humaniora Islamica*, vol. 1 (1973), pp. 47–60.
49 Ageron, *Les Algériens*, p. 117. Similarly, '[P]erhaps the numerous fires in the cork oak forests of Jemmapes and the region of Philippeville in September 1870 were not all accidental.' (p. 110).
50 Nouschi, *Enquête*, pp. 334–5, 482–96 are the main passages where he discusses the fires.
51 General commanding Bône subdivision to general commanding Constantine division, 22 September 1878, quoted in Nouschi, *Enquête*, p. 483.
52 Petition of March 1879, quoted at loc. cit.
53 General commanding Bône subdivision to general commanding Constantine division, 9 June 1878, quoted in ibid., p. 484.
54 AOM, F 80 1785, *Correspondance Générale Algérienne*, 10 August 1878.
55 AOM, 9 X 121, *Incendies . . . en 1902*, pp. 121–2.
56 SGA 1879–81, p. 338.
57 Ageron, *Les Algériens*, p. 118.
58 Quoted in Nouschi, *Enquête*, p. 490.
59 Tomas, *Annaba*, p. 271.
60 Nouschi, *Enquête*, p. 491.
61 *L'Indépendant*, 29 August 1881, quoted in Ageron, *Les Algériens*, p. 117.
62 ibid., 30 September 1881.
63 AOM, 9 X 121, *Incendies . . . en 1902*, pp. 107, 118.
64 ibid., pp. 116–17, 119.
65 ibid., p. 11.
66 AOM, F 80 1783, *L'Est Algérien*, 26 February 1874. The 'victims of Bugeaud' refers to a group of seven European inhabitants of the hamlet at the crest of the Edough mountains who were burned to death in a fast-moving fire in September 1873.
67 *Indépendant*, 4 September 1881, quoted in Nouschi, *L'Enquête*, p. 485.
68 SGA 1925, p. 200.
69 AOM, F 80 1785, Gary, Bure, Enquête générale sur les incendies de forêts (1865).
70 AOM, F 80 1787, Besson, Note sur la mise en valeur des forêts de chêne-liège.
71 AOM, M 97 (259), Rapport (1901); M 98 (267), Rapport (1902).
72 AOM, M 97 (259) [Ouichaoua]; Ageron, *Les Algériens*, p. 118; Nouschi, *Enquête*, p. 486.

11 The war of degradation: black women's struggle against Orange Free State pass laws, 1913

JULIA C. WELLS

> At about nine o'clock this morning a long procession of about 600 women headed by a stout native Mrs Pankhurst, a flag wrapped around her voluminous shoulders, marched up Maitland Street, shouting and chattering towards the Magistrate's Court. The procession was a weirdly picturesque one, some of the women jumping about as though possessed, while others carried emblems of liberty in the shape of Union Jacks attached to walking sticks, knobkerries and broom handles.[1]

With this startling demonstration in May 1913, the townspeople of Bloemfontein finally became aware that the African and Coloured women of the town's locations were in a state of revolt. Although the events appeared 'weirdly picturesque' to white observers, they were in fact the opening salvo of an intensified public campaign to draw attention to black women's refusal to comply with local pass laws. At issue was the Free State's system of residential passes required of all African and Coloured urban dwellers, male and female, over the age of 16.[2] Passes were essentially identity documents signifying permission from white officials to live or work in one place. Specific details of pass requirements varied from town to town in the Orange Free State, but most municipalities tied eligibility for a residential pass to employment in the service of whites. In many cases all women (and men) over the age of 16, married or not, had to secure the signature of a white employer every month in order to reside legally within a town. This effectively eliminated the option to be a housewife in one's own home and prevented older children from attending schools. In addition to these onerous terms, the workers themselves had to pay a monthly fee of sixpence. Failure to comply with the pass laws carried a penalty of £5 (at a time when the average female wage in domestic service was 10s per month) or one month in prison. Most of all, blacks fiercely resented the frequent and harassing police spot checks for passes. Females suffered the additional hazard of assault and rape by unscrupulous policemen. Even the very act of waiting in line for the monthly passes subjected women to physical abuse from black policemen.

Although the women's objections to this system were clearly justified, it took a struggle of twenty years, capped by an unprecedented passive resistance campaign, to remedy the situation. Ultimately, however, they did succeed and theirs became one of the few cases in South African history in which the demands of black protesters were met. They key to this startling success lay in forcing the Union government to assert its authority over the provincial and municipal levels of government, in effect testing the terms of the Act of Union. Officials at the national level, as we shall see, were extremely reluctant to interfere with the local policies of one of the conquered Boer republics. But when a contest arose over how to deal with the civil disobedience, the Union government was forced to choose between respect for provincial autonomy and the danger of spreading black unrest. Ultimately it pressed for a uniform national policy consistent with the other three provinces, excluding women from pass requirements.

By securing the national government's acceptance of responsibility for the problems, the women placed the struggle in an economic context that was in fact conducive to a sympathetic settlement. When the Union of South Africa was formed in 1910, the 'native policy' it inherited was shaped primarily to meet the labour needs of the mining industry. Pass laws effectively channelled an adequate supply of male black labour into the mines. African women were expected to remain in rural homesteads, raising enough food to feed themselves and their children, enabling the mines to keep male wages low. Thus, women's traditional place in society suited the needs of mining, while the labour needs of other sectors of the economy were also met primarily by male migrant workers. Consequently, in the former British colonies of the Cape and Natal, African women were left unregulated to continue in their supportive roles. In the Transvaal only occasional local efforts to control black women occurred. This left the Free State as the only province in the Union to require passes of women.

The difference stemmed from a significantly lower black/white population ratio. Blacks outnumbered whites in the Free State only two to one, whereas the national average was close to four to one.[3] A condition of chronic labour shortages prevailed, placing black women in a unique position. Their labour, working directly for whites rather than subsidising migrants, was in high demand, drawing women into the wage-labour market in numbers unprecedented elsewhere in the nation. Domestic service was by far the largest sector of female employment. In the province as a whole, female servants outnumbered males more than two to one, while in the Transvaal, the province with the most similar social configuration, the numbers were nearly equal.[4] The uniqueness of the Free State's labour market was most evident in the urban areas. Bloemfontein employed female servants at over five times the rate for Johannesburg.[5] From the Free State point of view, women's passes were an integral part of a controlled labour market constructed to cope with a condition of labour scarcity.

The Republic of the Orange Free State had established its rigorous pass regulations prior to the Anglo-Boer War of 1899–1902. At the time protests from the urban black leaders to municipal authorities fell on deaf ears. Consequently many blacks believed the South African War heralded their liberation from the harsh Free State legislation. They expected the more liberal 'native policies' of the Cape to be extended to the former Boer republics. Voting rights, land ownership, educational benefits, relaxation of the general pass law and an end to all women's passes were eagerly anticipated. During this period two forums developed for articulating the political aspirations of both the coloured and African communities, treated alike under the law. The African Political Organisation (APO), under the leadership of Dr A. Abdurahman in Cape Town, represented the Coloured voice in the Free State, while the Native Vigilance Association, or Native Congress, as it was sometimes called, spoke for the Africans. The latter was one of the constituting bodies in the foundation of the South African Natives National Congress (SANNC) in 1912.

During the first decade after the war the legally constituted authority administering the Free State changed hands every few years. Initially ruled by the British as the Orange River Colony, it was then granted limited self-government in 1907. This was followed almost immediately by talks which led to the formation of the Union of South Africa in 1910. Prior to unification, delegations from both the APO and the Native Congress wandered from official to official, ranging from the mayor of Bloemfontein to the king of England, only to be told in each case that authority for changing these laws lay elsewhere. The one improvement offered by the British officials was a system of exemptions from pass laws for black males who had attained high levels of education or met certain property requirements. Since women in the rest of South Africa were not required to carry any passes, no provision was made for female exemptions in the Orange Free State.

When the Union of South Africa was formed in 1910, the black leadership of the Orange Free State was quick to call for a uniform national 'native policy'. After first being rebuffed by Bloemfontein officials, the Native Vigilance Association petitioned the new Minister of Native Affairs, Mr H. Burton, stressing the incidence of physical abuse meted out to women by arresting police and the financial hardship embodied in paying sixpence per month to purchase residential permits. Burton readily admitted 'the grievance is a substantial one,' and initiated several actions in search of an easy administrative solution to end passes for women.[6] First he wrote to the Bloemfontein municipality, asking for budget figures in an attempt to determine if the pass fees were in fact being used as a major source of revenue by the city. The reply from Bloemfontein denied this, although the records show that the income from residential passes had doubled over the previous five years, while other income and expenditure on black services remained the same. This brought the Town Council nearly £2000 in additional revenue.[7] Burton also wrote to the Justice Department, asking if the excesses of police methods could not be curbed. Extensive legal opinion was also sought in the hopes that the old Free State regulations might be ruled illegal. None of these efforts produced any substantial results.[8]

The Department of Native Affairs responded afresh to black political pressure on the issue in 1912. The founding of the South African Natives National Congress in Bloemfontein in January stimulated a new round of activity among the Free State women under the guidance of the experienced Cape politician and provincial representative for Tembuland, Walter Rubusana. The women circulated their own petition, collecting five thousand signatures from throughout the Free State, and sent a delegation of six women to deliver it to Burton in Cape Town. Once again, he assured them of his fullest sympathy on the question of women's passes and promised stronger action.[9]

Burton then wrote to the Provincial Administrator of the Free State, asking him to help find a way to get the passes removed. This rather lame initiative revealed the strength of the local opposition to altering the laws. The Administrator, no doubt sympathetic to the local basis of the female passes, took a survey of all the towns in the province. He asked if the municipalities wanted to have the women's pass laws changed and if they knew of any cases of physical abuse in their enforcement. The answers to both questions came back with virtually unanimous 'nos'.[10] Some towns indicated that if the laws were to be changed it should be in the direction of tighter controls, not less. Others admitted there had been some charges of rape in the past, but that since none of these resulted in conviction, it was not considered a problem.

The backdrop to these official inquiries was a worsening labour shortage in the towns of the Free State. During the war, Bloemfontein served as a major military base for the British, and consequently became a relatively prosperous safe haven from the ravages of the surrounding countryside. No doubt politically neutral rural blacks sought temporary refuge in greater numbers than their Boer counterparts. Once reconstruction re-established the rural economy, however, large numbers of Africans returned to their former homes. Following this brief post-war boom in black urbanisation, Bloemfontein lost 43 per cent of it black population between 1904 and 1911, compared to only a 5 per cent drop in whites. Similarly, Winburg lost 30 per cent of its black population while the number of whites remained constant.[11] Invariably the demand for domestic servants increased relative to the declining labour supply. The use of domestic servants in white households was the norm, even among poor whites. Census data suggests that by 1911 there was one black servant for every three white people in the Free State urban areas.[12] Apparently not satisfied with this ratio, whites in Bloemfontein complained that domestic help 'can't be got at any reasonable price'.[13] They first tried to establish a labour bureau in the hope that the forced registration and fingerprinting of all black residents would help them identify potential workers. When the black community totally refused to be fingerprinted the scheme was abandoned at the end of 1911.[14] During 1912 the Bloemfontein Town Council hired an official collector of arrears for the location in an attempt to force the residents to pay all overdue fees owed to the city for rent, water and sanitation services. When this also failed to coerce more workers into the labour market, frustrated whites called a public meeting in December 1912 to discuss 'the present acute native servant question'.[15] They were convinced that the location was full of women who could work but were unwilling. Believing that stricter enforcement of the existing regulations might solve the problem, the Town Council met with the police to discuss more rigorous tactics. A similar meeting took place in the town of Jagersfontein.[16]

The declining black population altered the economic position of the remaining black community in two important ways. The relatively higher demand for labour no doubt nudged wages up. The shortage of domestic servants also increased the women's bargaining power. Indeed, there is evidence of a power struggle between blacks and whites over terms of labour. Topping the list of white complaints about the servant problem was the desire to prevent workers from changing jobs, to compel servants to sleep-in on the employer's premises and 'to keep servants much more reserved'.[17] In other words, whites wanted limited job mobility, longer work hours and no complaints about work conditions. But if black women were in a position to command generally higher wages for their labour, many probably chose to do laundry work rather than enter domestic service. The advantage lay in being one's own boss, free to collect as many bundles of laundry from white homes to be washed, dried, ironed and returned as was manageable. Children could thus be better attended, or even incorporated into the work routine. In these strongly Christian communities, the Western ideal of the ever-present housewife held strong appeal.[18] The evidence suggests that laundry women indeed perceived themselves to be in a sufficiently strong position to demand improvements in their conditions of service. In Winburg they demanded the construction of a municipal washhouse to save them having to carry their loads five kilometres to the nearest river. The Jagersfontein laundry women petitioned their Town Council for exemption from laundry permit fees. Neither initiative, however, was successful.[19]

The second effect of the declining black population was less positive. While labour conditions may have benefited the average working woman, they did not bring similar relief to the urban black elite. By the turn of the twentieth century, a well-defined elite stratum of educated skilled individuals prospered in the Bloemfontein locations.[20] The brief post-war population boom enabled many to establish their own businesses, primarily in the building-related occupations of carpentry, plumbing, brickmaking and cartage. Others made a living from wagon-building, shoemaking or operating cafés. These entrepreneurs, along with teachers and ministers of religion, were among the few who were not employed directly by whites. Many of the men no doubt qualified for exemption from the pass laws or could afford to pay the fee for a monthly 'work-on-own-behalf' permit. Those with sufficient resources, who were official standholders, that is registered tenants of the city, also profited enormously from subletting rooms and second houses. At the height of the boom, city officials jealously condemned the 'rent-grabbing capitalists of the location,' citing the £9 per month they collected in rent.[21] With worker's wages averaging roughly £2 per month and domestic work paying about 10s to £1 10s, this lodgers' income secured the elite women's freedom from economic pressure to enter the wage labour market.

The population decline, however, coinciding with the Town Council's construction of additional locations, deprived many of this source of convenient revenue. The diminishing black population also sharply reduced the demand for the trades and services offered by the elite group. In addition, the decade prior to 1913 witnessed a steadily increasing attack on the black entrepreneur class from whites who resented the competition. Restrictions on black trading rights and limitations on licensing cut into profits.[22] All of these factors created adverse economic pressure on the elite stratum of society, no doubt forcing many of its women to look for wage employment for the first time. With few job options open to them, it seems likely that many would have turned to laundry work rather than domestic service, for the same reasons as the working class women. It offered to all the advantage of working one's own hours, the possibility of incorporating child labour and being free from the critical eye of a white mistress or master. Indeed, the public washhouse in Bloemfontein, complete with running hot water and steam irons, provided the ideal meeting place for the displaced elites and the aspiring housewives of the working class. Already thrown together in housing, in churches and now in the work place, the women were in an especially strong position to form a class alliance to fight the common enemy of all, the hated passes.

As white pressures to regain control of the black labour force entered a newly aggressive stage in 1913, the rage of the black women exploded. At last the white citizens of the Free State got the kind of tough pass-law enforcement for which they had been clamouring. The records do not indicate what triggered the new police aggression, but during May four times as many Bloemfontein women were arrested under pass laws as in the previous month.[23] The police method of waiting for women as they left the public washhouse struck at the nerve centre of the black community.

The women felt the time had come to take unprecedented action. On the afternoon of 28 May 1913, a female mass meeting took place in the Waaihoek location of Bloemfontein. The women concluded that a passive-resistance campaign was the only course left to them to get rid of the hated passes. They decided that 'no matter what happened or how they were treated they would refuse to carry passes.'[24] The local papers speculated that the women had been particularly inspired by reading in the press about the British suffragettes, whose militancy was then at its height.[25] The result was the march of two hundred infuriated women through the centre of town, demanding to see the mayor, an account of which was quoted at the beginning of this chapter.

Since the mayor was not in, a smaller deputation of the women's leaders returned the next day for an interview with him. At that time he told them that the women's pass law was the responsibility of the Union government and that there was nothing local officials could do to change the law. He did offer to ask the police to be more considerate in the way they went about making arrests, but warned the women that the laws must be obeyed. Not satisfied with his response, the women returned to the location, where that night they provoked the police into taking action. As the Bloemfontein *Friend* described it:

> Crowds of dusky Abigails crowded around the native police station. Here they tore up their passes, threw the pieces on the ground, defied the police in language more forcible than polite, conveyed the intelligence that rather than carry a pass they would suffer untold agonies and imprisonment.[26]

A total of eighty women representing fourteen different Coloured and African ethnic groups were arrested and ordered to appear before the magistrate the next day.

Among the arrested were the wives of the town's best-educated and skilled tradesmen: Martha Leshanie, wife of a brickmaker; Rutha Kabani, the plumber's wife; Catharina Simmonds, the wagon-maker's wife; Amelia Twayi and Martha Mapikela, carpenters' wives; Sarah Lichabu, the shoemaker's wife; Liza Japhtu, wife of a cab owner; Annie Erasmus, the trolley owner's wife; Emma Mamfrie, whose husband operated a Scotch cart; Jane Moroka, the schoolteacher's wife; and Katie Louw, the Anglican minister's daughter. The group included a dispro-portionately high number of Coloureds and averaged 32 years of age.[27] They appeared before the magistrate accompanied by six-hundred women from the location, all emphatic in their demand for the total removal of the pass laws. When the police tried to control the crowd, the women reacted with swift retaliation against their hated enemies. The APO newspaper boasted:

> On that day the Native women declared their womanhood. Six hundred daughters of Africa taught the arrogant whites a lesson that will never be forgotten . . . Sticks could be seen flourishing overhead and some came down with no gentle thwacks across the skulls of the police, who were bold enough to stem the onrush. 'We have done with pleading, we now demand', declared the women.[28]

During the course of this demonstration one of the women was reported to have shouted 'Votes for Women!' in apparent identification with the British suffra-gettes.

Before making their court appearance, another delegation of women again interviewed representatives of the city council, this time securing a promise to restrain the police from making any further arrests. Meanwhile, male members of the Native Vigilance Committee discussed the whole matter in some detail with the resident magistrate, Mr Ashburnham. Recognising that part of the difficulty was the failure of any public official to take responsibility for the pass laws and their possible revision, he promised to arrange a meeting between both city and provincial officials and black leaders to discuss the problem. Expressing great sympathy with the women's complaints, Ashburnham gave the crowd a warning that the laws had to be complied with, even while constitutional efforts were underway to change those laws. He then dismissed all the charges. After the other women returned to the location, the leader, Mrs Molisapoli remained behind still wrapped in her Union Jack, 'protesting her loyalty and willingness to die for the cause'.[29]

Within days the lead taken by the Bloemfontein women spread throughout the countryside. In the nearby town of Winburg on 2 June, a hymn-singing crowd consisting of 'all the women of the location' converged on city hall.[30] After informing the mayor that they intended not to carry passes any longer, they then quietly dispersed. The immediate official response was to do nothing, pending guidance from the municipal officials in Bloemfontein.[31]

At first it seemed that women's protests had produced the desired result as arrests for pass law infractions temporarily ceased. The calm, however, was short-lived. Scarcely two weeks after their initial protests, the women of Bloemfontein were again engaged in a physical battle with the police. The women believed that the mayor had promised no further arrests until the matter had been fully investigated. But on 16 June when a policeman tried to arrest one woman, three of her friends intervened and forcibly freed her. The police formed a special posse to arrest all four offending women and then took them to the charge office. Word of this apparent violation of the mayor's promise spread quickly throughout the locations and two to three hundred women soon gathered where the prisoners were being held. Katie Louw, one of the women's leaders, was called in to help mediate the tense situation. She arrived in time to see one of the white police officers strike her good friend and co-churchwoman, Greta Phillips, with a sjambok (short whip). Tensions exploded into a scuffle between women and police. The women threw rocks, bit and kicked the policemen, who in turn lashed out with their sjamboks and tried to wrestle off attacking women.[32] Eventually the police gained the upper hand and arrested thirty-four women on charges of public violence.

At the trial, the police insisted that the original arrest had not been for the usual breach of pass regulations, but a legitimate case of vagrancy. The women all testified that their gathering had been only to ascertain the reason for the arrests. They had been told that it was in fact for pass violations. This broken promise prompted their zealous actions, but the violence, they all claimed, was initiated by the police.[33] The thirty-four were all found guilty of public violence and since they refused to pay fines, were sent to prison for two months.

Either the women had read more into the mayor's promises than was intended or he blatantly lied to them about suspending pass arrests. A much less sympathetic version of the official response to the initial demonstrations was given to Winburg officials when they wrote to Bloemfontein for advice at this time. They were told that the Bloemfontein police were instructed to go ahead with prosecutions under pass laws as usual.[34] Following this example, the Winburg officials gave local women a warning that full prosecution under the pass regulations would begin in six days. Subsequently, on 1 July, Winburg began what was to become a fixed pattern for months to come, arresting six passless women at a time, thereby filling the local jail to capacity and then arresting six more when the sentences were served.

In Bloemfontein, when the meeting arranged by Magistrate Ashburnham between provincial and local officials took place, the matter of legal responsibility was used as an excuse for inaction. The mayor's patience was wearing thin and he told the women once again that the town council had no authority to change these laws in any way. The Provincial Administrator, Dr Ramsbottom, also disclaimed any responsibility for the regulations, pointing out that it was a concern of the Union government. Under the terms of the Letters Patent establishing the Union of South Africa in 1910, no local authority was allowed to change any laws touching on 'native policy'. This was left as a matter for national policy-making at the parliamentary level. Neither official admitted that the municipalities had the option simply not to enforce the law or to drop the monthly fees required to purchase a pass. When the women still insisted they would settle for 'nothing more or less than the total abolition of their passes,' they were told such a move would be 'quite impossible'.[35] In reporting this meeting to the town council and the press, the mayor further belittled their cause by declaring it 'not of a substantial nature' and a 'frivolous complaint'.[36]

While municipal officials were thus showing increasing irritation at the women's campaign, support from within the black community quickly blossomed. After being initially sceptical of the wisdom of the women's methods, black newspaper editors Solomon Plaatje, secretary of the SANNC, and Dr Abdurahman of the APO began championing the cause of their columns. Plaatje, under the banner headline 'War of Degradation', urged the women to fill the jails until their full liberty from pass laws was secured.[37] Abdurahman praised the women and advised them, 'when in prison not to do a stitch of work of any sort'.[38] Later he admonished that men 'might well hide our faces in shame and ponder in some secluded spot over the heroic stand made by Africa's daughters.'[39] The campaign also inspired many new poems, including a parody of Macaulay's 'Lays of Ancient Rome':

> Too long have they submitted
> To white malignity;
> No passes would they carry,
> But assert their dignity.
> They vowed no more to fawn or cringe,
> Nor creep to the tyrant's power;
> But to proclaim their womanhood,
> Their inherent, God-given dower.[40]

Faced with little sympathy from local officials, the women continued to refuse to buy their monthly passes, and suffered imprisonment during an exceptionally cold July winter. As their determination increased, they adopted blue ribbons as the sign of their struggle and wore them boldly to inform the public and the police of their stand.[41] On 8 July, the black women of Jagersfontein actively joined the campaign by presenting the town council with a petition from 334 women calling for exemptions from passes. But the response, as elsewhere, was simply that the town council had no jurisdiction to deal with these laws.[42] The mayor stressed the white community's needs, claiming that an end to pass laws would seriously affect local housewives; besides, in his opinion, it 'was no disgrace to be stopped and asked for a pass.'[43] In addition to the petition, a campaign of passive resistance started as older women refused to take out passes in solidarity with the younger ones who had been the targets of the police.[44] Using their July court appearance to articulate their complaints, they argued that passes should be valid for a full month regardless of the date of purchase of the previous pass. The town council was unsympathetic and simply passed a resolution stating that all passes expired on the eighth of each month.[45]

Similar efforts to negotiate a solution in Winburg produced no better results. In mid-July the protesting women sent a petition to the mayor asking if they could negotiate. He agreed, but the meeting resolved nothing, running into the same deadlock as all previous ones: town officials claimed they had no responsibility and all laws had to be obeyed until such time as someone else changed them.[46] At the end of July a special delegation of men from the mission churches and schools also approached the mayor, seeking relief from pass requirements which forced children over the age of sixteen to work for white employers rather than attend school.[47] The official response was consistently rigid.

As the stalemate continued through August and September two patterns of municipal reaction emerged. While Bloemfontein eventually chose to pursue a fairly liberal course, the smaller towns only aggravated tensions by taking a firm law-and-order approach. A new furore erupted in Winburg when, on 15 August, Ruth Pululu was found guilty of being without a pass. As the assistant school mistress and 23-year-old daughter of the Wesleyan minister, she was ineligible for a residential pass since she was neither married nor working for a white person. Her arrest at the school in front of her class provoked fresh outrage from the entire community.[48] With her commitment to the passive resistance campaign, going to jail for fourteen days meant a serious disruption of the school year. If she subsequently entered domestic service, as required by the law, the school would be forced to close.

With the legal assistance of Mr Baumann, a sympathetic local white, her case was postponed to give the location residents time to draw up yet another petition. When this failed to produce any results, the residents agreed to appeal her court case to test the validity of the law.[49] The situation took a more ominous turn when the location residents sent a petition of protest to the Minister of Native Affairs in Cape Town. In a strongly worded letter accompanying the petition, the Winburg Native Vigilance Association warned that the men were now prepared to embark on a full labour stoppage 'as they cannot stand by and see their wives and daughters innocently be prosecuted and imprisoned without being criminals.'[50] It closed with the observation that they did not expect any help from the Native Affairs Department, but felt it should stand warned of what might happen.

Alarmed by the prospect of a black general strike, Winburg's white residents called a public meeting for 25 September. In a last-minute effort to head off trouble the mayor called yet another meeting with the leaders of the protesting women early that morning, but they refused to abandon their campaign. That night, in an impressive show of white unity, farmers, professionals and working men joined together to demand a militant offensive against the black women. They proposed to abandon the six-person-at-a-time arrest pattern in favour of arresting and jailing every woman found in the town without a pass. They demanded the construction of new jails to accommodate all such women. One enthusiastic burger commented that he 'thought it manly of them by not giving in to the native women'.[51] Such a drastic step, however, would have seriously aggravated the situation, so the mayor wrote once again to the Attorney General and the Commissioner of Police in Bloemfontein seeking advice. The replies of both simply detailed the regulations in question and reviewed the normal penalties.[52]

When all these efforts failed to resolve the impasse, the white women of Winburg joined the fray. Although their concern reflected a certain degree of female solidarity, they also faced the profound impact of the proposed imprisonment of all their domestic servants. Touching off a heated round of anonymous correspondence in the Bloemfontein *Friend*, a mysterious writer calling herself 'Common Justice' defended the black women's stand in a biting letter. Expressing admiration for their courage, she claimed the town council was beaten and should admit it. She challenged other white women of Winburg not to allow the council to tyrannise them with threats to turn all the black women out of the location. Calling for a white women's protest demonstration, she claimed she would only reveal her identity 'when the hour comes'.[53]

This letter apparently did not reflect popular sentiment, as it sparked a barrage of abuse from other readers, mostly men, who defended a tough law-and-order stance. One writer referred to the proposed demonstration as 'a silly hoax' and warned:

> We citizens of Winburg are not prepared to allow such acts of folly to be perpetrated in our midst. I appeal to the citizens of Winburg to stand as a man in upholding their prestige and honour, and not to allow such loathsome nonsense to obtain any footing in their respected town of Winburg.[54]

Further letters asserted that the whole idea must surely be a joke and appealed to the white women to behave responsibly. 'There are bounds even to chivalry', another declared.[55] One writer suggested a simple solution. If the white women refused to do the housework left undone by their imprisoned servants, soon enough their men would relent and change the whole system.[56]

In the end, the white women's demonstration was called off because the black women 'threatened to take uninvited so overwhelming a part in the procession as to utterly confuse its character.'[57] 'Common Justice' then tried to resolve the mystery of her identity by offering to pay two guineas to the local orphanage for anyone who had written a nasty letter and was also willing to reveal his or her name. By the end of October, the furore died down, to the chagrin of at least one reader of the *Friend* who thanked 'Common Justice' for having provided considerable relief from the boredom of the usual reading in the newspaper. In a long and colourful epic poem dedicated to 'Common Justice', the writer, Tante (aunt) from Rouxville, captured the white women's dilemma, 'If our responsibilities we shirk, dear sister, think of it, we'll have to work!'[58]

Tensions also exploded in Jagersfontein by the end of September. Sixty women had been charged that month for being passless and they used their court appearance once again to dramatise their grievances. Before entering the court house, they lined themselves in a row across the street, clapping, singing and wearing blue ribbons. Inside, their leader, a 'jet-black Mozambique lady' named Aploon Vorster, insisted simply, 'we are a peaceful people; we refuse to carry passes.'[59] She was sentenced to thirty-days' hard labour, while most of the other protesters received seven-day sentences. That afternoon the campaign once again lost its 'passive' quality when crowds of blue-ribboned women attacked the jail cart transporting the convicted women to the Fauresmith prison. The police could only disperse the crowd by sjamboking women, using the town's new fire hose and patrolling the streets on horseback.[60]

On other fronts, quieter negotiations were more earnestly pursued. The Department of Native Affairs finally became directly involved in the struggle in response to the tense situation in Winburg. First, after investigating the legality of the Winburg pass regulations, the Department ascertained their full legitimacy. Relief could not come merely through a reinterpretation of the law.[61] In addition to the letters and petitions sent by the black residents, the magistrate from Winburg wrote to the Department officials, suggesting that they could fully appreciate the volatile nature of black opinion only after an on-the-spot investigation. In compliance with this request, the Department of Native Affairs sent under-secretary Edward Dower as its official representative on 16 September. His mission was twofold: to consult with Free State blacks about the workings of the newly passed Natives Land Act and to look into the question of passes for women.[62] Stopping first at Thaba Nchu, Dower met with various black leaders including Katie Louw from Bloemfontein and Aploon Vorster from Jagersfon-tein.[63] His plea for the women to stop their civil disobedience fell on deaf ears and Katie Louw advised him to 'leave the women alone in their determination as they preferred to await the outcome of their struggle in jail.'[64] As a result of this meeting, Dower met with both municipal authorities and the Commissioner of Police, urging them to take extra-legal action to reduce drastically the rigour of their law enforcement. Details of this meeting are not available, but Bloemfontein officials did respond according to the wishes of the Department of Native Affairs.

Indeed, Bloemfontein's reaction to the continuing conflict was far more sophisticated than Winburg's. The town council's immediate response to the women's crisis was the establishment in June of a special native affairs committee within its own ranks.[65] Later, when faced with the strength of the resisters' determination, it sought ways of tackling the labour shortage problem on other fronts, minimising the need for women's passes. In early September, the city won a court case establishing its right to prosecute criminally anyone owing back-rents and fees.[66] This significantly altered the dynamics of the struggle, as it gave the authorities a powerful new tool through which to apply economic pressure on the black residents. At the same time, members of the elite of the location, eager for a voice in local government, sent a petition to the mayor suggesting the establishment of a black advisory board. As natural leaders of the black community, they asserted, they could control the location, helping with both the collection of arrears and the placement of female servants.[67] If both problems were solved, they hoped, the rigorous and harassing enforcement of pass laws which affected their own wives at the washhouse, could be abandoned. Since they were mostly self-employed and therefore not the targets of labour-coercive measures, they did not challenge the municipality's motives. The town council saw this as a way out of its dilemma and an opportunity to divide the black community along class lines. If the aspirations of the elite were appeased separately, they could be used as agents of control over the masses.

To prove his good intentions, Mayor Ivan Haarburger consulted with the location residents in a public meeting on 5 September. After conceding that representatives could be elected by the location residents rather than appointed by the town council, he agreed to the establishment of such an advisory board. Haarburger took this to mean that the black representatives would see to it that arrears were paid up and the women's civil disobedience would end. Although he still insisted that the municipality could do nothing to change the pass laws, he admitted for the first time that other steps could be taken, if so desired by local officials. He told his audience, 'if only they work harmoniously with the Council, the Pass Laws can be carried out in such a manner that no hardships need occur.'[68] He concluded by stating that he would now inform the Administrator that the blacks were willing to co-operate.

Haarburger obviously misread the situation by assuming that the elite men controlled the women's campaign. The women maintained their female solidarity and refused to be divided along class lines. Since women of even the elite stratum did not receive the same exemptions from pass laws as their own husbands, they could not set themselves apart from the masses. In this and subsequent meetings the men consistently claimed that they had advised against passive resistance and could do nothing about it, lamenting, 'We have now given up this matter into the hands of the Almighty to work it up in his own way . . . what is impossible with man, is possible with Him.'[69] The women continued to express loyalty to their struggle and the resistance continued long after this initial mayor's meeting, although precise figures on the numbers arrested are not available.

The mayor was in fact the only one to make any concessions, but he proceeded as if the impasse had been overcome through mutual compromise. By the time of the next mayor's meeting in October, regulations for the advisory board had been drawn up and seven men elected to serve. In the November meeting, Joseph Twayi expressed special appreciation for the sympathy shown by the mayor and commissioner of police regarding women's passes, stating 'there now seemed to be more satisfaction among the people in the location.'[70] Indeed, the Orange Free State commissioner of police admitted that in the past there had been times when the pass laws were enforced with 'relentless vigour,' but that now it was being done 'quite tactfully'.[71] Mayor Haarburger also carried the matter one step closer to resolution by taking the initiative to contact Senator W. P. Schreiner, a known supporter of black rights, on the question of introducing legislation in parliament. By this time the legal advice given to the local, provincial and Union government officials was unanimous in claiming there was no other way to change the laws.[72]

Much of the pressure on municipal officials to capitulate came from the sheer persistence of the campaigning women. Throughout the course of the various negotiations, the resisters' commitment was bolstered by increasing support from the black community nationwide as well as part of the white public. From the beginning, both *APO* and *Tsala ea Batho* gave thorough coverage to the campaign. By August both newspapers regularly urged their readers to send contributions to the newly formed Orange Free State Native and Coloured Women's Association.[73] The money went for the care of families of imprisoned women and to cover medical costs incurred by the wretched prison conditions. In the Cape, APO women held special fund-raising events, while donations poured in from throughout the Union. During July both Plaatje and a liberal British journalist, Dewdney Drew, visited the resisting women imprisoned in Kroonstad. What they saw gave them fresh ammunition to take their case to the general white reading public outside of the Free State. Articles appeared in the popular mission journal, The *Christian Express* and the *Cape Argus*, detailing the very severe treatment endured by the women in prison, as well as the primary reasons for the campaign.[74]

Perhaps the greatest public outcry on the issue came when Dr Adburahman gave a militant speech of condemnation at the September annual meeting of the APO in Kimberley. Claiming that the Orange Free State pass laws were in fact a modern-day slavery, he called for all blacks throughout the Union to unite in a massive labour strike to demand full justice.[75] Katie Louw and Thomas Mapikela from Bloemfontein were on hand to read out reports on the course of the women's struggle. The entire gathering gave hearty endorsement to the resisters, urging other women to follow suit.[76] The excitement generated from this meeting only spurred on the resisters.[77] Bitter rebuttals to Abdurahman's extreme militancy appeared in the Bloemfontein *Friend* and *Cape Times*, while the commissioner of police for the Orange Free State accused him of 'gross libel' for his allegations.[78]

But Addurahman refused to retract any of his statements and used his replies to further expound on Free State brutality.

When the passive resistance wound down in 1914 the results were still inconclusive. Intervention by the Native Affairs Department had successfully persuaded authorities to ease up considerably on the rate of female arrests. Local tension over this semi-solution, however, surfaced on occasion when heavy-handed police tactics reminded black residents that the laws were still on the books and could be rigorously enforced at any time. In November 1913, Bloemfontein's black police constables retaliated against location residents for exposing cases of police brutality. When one white and one black constable were convicted on charges of indecent assault, the police went on a binge of entering people's homes in search of passes and waiting for the women at water taps and as they left church. Vehement complaints to the mayor put a stop to such behaviour and even won the sympathy of the *Friend*.[79]

A similar situation occurred in Jagersfontein in March 1914. Armed police surrounded the location and then made a door-to-door search for passes, demanding them at gunpoint. After making a record high number of arrests from this sweep, prosecutions for female passes were dropped altogether.[80]

The campaign, however, had achieved absolute clarification of statutory responsibility for women's passes. Although the laws were local and unique to the Free State, only the Union government had the power to change them. Ironically, by the end of 1913, the women, the Bloemfontein municipality and the Department of Native Affairs all pressed for a legislative solution, although each no doubt envisaged a different type of bill. The women continued not to buy passes and turned their attention to formal lobbying. Early in 1914 the Orange Free State Native and Coloured Women's Association sent yet another petition to the Governor-General of South Africa, urging him to use his influence to secure the necessary legislation. It also circulated appeals to various white voters and legislators, seeking their assistance.[81] In March 1914, a special select committee was appointed to investigate the pass laws throughout South Africa. It produced a Draft Bill to Amend Pass Laws, published in June, with provision for the Governor-General to initiate action by which Orange Free State municipalities could repeal their regulations requiring passes for women.[82] No changes could be made without the full co-operation of the municipalities. In essence the bill represented an attempt by the Union government to wash its hands of the responsibility for deciding whether or not women should carry passes.

The outbreak of the First World War, however, pre-empted any debate over, or passage of the bill. When the SANNC resolved to abstain from any political confrontation during the war, the women gave up their passive resistance. Throughout this period of uneasy truce, both sides continued to press for a more satisfactory solution. Both the SANNC and the Wesleyan Methodist Church kept the issue alive through formal appeals and resolutions.[83] The Department of Native Affairs sought a solution that would divide women along class lines and discriminate against the unmarried. First, in 1914, it suggested that the existing laws might be interpreted to exempt all married women. Then in 1916 it pressed for the automatic exemption of the wives and children of exempted men. Both ideas were flatly rejected by the Provincial Administrator of the Orange Free State.[84]

However, pressure from a new Transvaal APO leader, Talbot Williams, finally pushed the Department of Native Affairs into taking more assertive action. In a series of meetings in Pretoria, Williams convinced Secretary of Native Affairs, A. L. Barrett, of the gravity of the stalemate.[85] In 1917 the Justice Department, at the prompting of the Department of Native Affairs, issued a ministerial order directing the police in the Free State not to arrest women for passes unless first receiving a warrant from the municipality.[86] If complied with, this would eliminate the random and harassing aspect of pass-law enforcement.

This order from the Justice and Native Affairs Departments marked a major policy shift at the Union government level as it set a precedent of intervention. After so many years of fruitless attempts to arrive at an administrative solution, the Native Affairs Department ultimately secured full relief for the women in 1923 when the Native Urban Areas Act was passed, specifying that residential passes applied to males only. Although the women's pass crisis originated from labour shortages in the Orange Free State, it was finally resolved by Union government officials whose primary concern was keeping the public peace. When the Department of Native Affairs did accept full responsibility, the very definition of the problem changed. The question was taken out of the local context and was placed within the national social configuration. Basing their judgment on the norms of the other three provinces, Native Affairs officials viewed female passes as undesirable, primarily a product of excesses in Free State policy and a threat to national security.

The women's victory, however, was two-edged. While women enjoyed their freedom from the harassment of pass laws, their communities continued to face the economic pressures placed on them by municipal authorities. The coercive designs of the female pass laws were simply replaced by new more sophisticated means of control including rigorous collection of fees, internal monitoring of location life by the advisory board, stricter enforcement of other regulations such as the night curfew and occasional police blitzes. Nevertheless women's freedom from pass laws and extra-economic coercion was secured nationwide for all classes of women for nearly fifty years to come. In so doing the women established an unmatched political precedent for the black community.

NOTES

1 This account was carried by the Bloemfontein *Post*, 30 May 1913, an English language daily newspaper and in *Tsala ea Batho*, 14 June 1913, a newspaper for black readers edited by Solomon Plaatje in Kimberley. The latter version named the leader, Mrs Molisapoli.

2 The term 'Coloured' designates people of mixed-race descent according to South African historical usage, and although unpopular today, is maintained here for clarity. Law 8 of 1893 made the earliest provision for residential passes in urban areas in the Republic of the Orange Free State.

3 *Union of South Africa Census 1911* (hereafter 1911 *Census*), Vol. I, *Populations of the People*, Table 1 – Union of South Africa.

4 1911 *Census*, Vol. V, *Occupations of the People*, Table XXI – Occupations of the People: Census Districts According to Classes.

5 1911 *Census*, Vol. V, Table XX – Occupations of the People: Urban Areas, Summary According to Classes.

6 South African State Archives, Pretoria, Department of Native Affairs (hereafter DNA),

Pass Laws Memorandum, 1919, p. 4.

7 Orange Free State Archives, Bloemfontein (hereafter OFSA), Orange Free State Provincial Secretary (hereafter OFS PS), Letter to Under Secretary of the Interior, Bloemfontein, from Acting Secretary of Native Affairs, 5 September 1910; and Bloemfontein Mayor's Minutes (hereafter BMM), 1906 and 1910.

8 See DNA, Pass Laws Memorandum, 1919, pp. 4–12.

9 Burton claimed that the matter had already been discussed at cabinet level. Cape Town *APO*, 6 April 1912.

10 OFSA, Provincial Administrator, Orange Free State (hereafter PA OFS), Letter to Minister of Native Affairs, 26 September 1912.

11 1911 *Census*, Vol. I, Table XVII – Return of Population: Urban Centres; and *Orange River Colony Census 1904*, Population of Chief Towns, p. vi.

12 1911 *Census*, Table XVI.5 – Occupations of the People – Orange Free State and Table XVII, Return of the Population: Urban Centres.

13 OFSA, Bloemfontein Town Council Minute Book (hereafter BTCMB), 1 December 1910.

14 BTCMB, 16 June 1910 and 7 December 1911.

15 BTCMB, 12 December 1912.

16 The Bloemfontein meeting was called to discuss 'the present acute native servant question'. Bloemfontein *Friend*, 6 December 1912; the Jagersfontein municipality ordered the police to enforce the pass laws more rigorously according to the *APO*, 16 November 1912.

17 These complaints were made at the 1910 meeting to discuss the servant question. BTCMB, 1 December 1910.

18 According to the 1911 *Census*, 84 per cent of Bloemfontein's black population claimed church membership; in Winburg the figure was 94 per cent. Vol. VI, Religion of the People, Table VII, Religions of the People in Detail: Urban Areas.

19 DNA, Pass Laws Memo, p. 15.

20 The elites were noted for wearing silks and satins and living in houses with lace curtains and filled with photographs and knickknacks. A special location was built for the wealthier Coloureds which included a cricket pitch, lawn tennis, football and croquet grounds and a cycle track. See South African Native Affairs Commission Report (hereafter SANAC), Cape Town, 1904–1905, Vol. 1, par. 380; BMM, 1906, Report of Native Locations, 68; and *The Mission Field*, February 1915, publication of the Society for the Propagation of the Gospel in Foreign Parts, London.

21 SANAC, pars. 326 and 301; and OFSA, *Bloemfontein Corporation Year Book 1910*, 'The Native Locations and the Native Public Health,' by D. M. Tomory, Medical Officer of Health, p. 78.

22 See resolutions of Bloemfontein Town Council, 1 September 1904; 6 September 1906; 18 June 1908; 16 June 1910 and 2 July 1910.

23 OFSA, Bloemfontein Criminal Record Book (hereafter BCRB), 1913. Perhaps new commanding officers took over when the old South African Constabulary was replaced by the South African Police in April.

24 *Post*, 29 May 1913.

25 ibid. The Bloemfontein Criminal Records show that Georgina Taaibosch was arrested during the May swoop and then twice more during the passive resistance campaign. Her daughter claims she was an avid reader who made her children tiptoe through the room if she was reading the newspaper. She in turn shared her information with the community to the extent that she was described as 'the community's newspaper'. Since the local papers gave generous coverage to the British suffragettes, it is indeed likely that someone like Georgina Taaibosch popularised their tactics. Personal interview, Bloemfontein, November 1979.

26 Bloemfontein *Friend*, an English-language daily newspaper, 31 May 1913.

27 This list is compiled from a number of records and from information gained through interviews; however much is left to conjecture as the similarity in family names does not

necessarily mean that two people are spouses. What appears here is an educated guess. OFSA, Bloemfontein Town Clerk Correspondence (hereafter BTCC), 29 October 1913, List of Africans holding shoemaking licenses in Bloemfontein Location; BTCC, 13 November 1913, List of Licensed tradesmen in Bloemfontein; (BCRB), 1913. Personal interviews, Bloemfontein, November 1979. Of the arrested women, 34 per cent were Coloured, although Coloureds made up only 17 per cent of the black female population. 1911 *Census*, Table VII.

28 *APO*, 14 June 1913.
29 *Post*, 30 May 1913.
30 *Tsala*, 14 June 1913.
31 OFSA, Winburg Municipal Minute Book (hereafter WMMB), 6 June and 20 June 1913.
32 *Friend*, 17 June 1913; and OFSA, Bloemfontein Court Record, *The King vs Johanna Botha and 33 Others*, 24–28 June 1913.
33 *The King vs Johanna Botha.*
34 WMMB, 20 June 1913.
35 *Friend*, 4 July 1913; BTCMB, 4 July 1913.
36 ibid.
37 *Tsala*, 21 June 1913.
38 *APO*, 14 June 1913.
39 *APO*, 28 June 1913.
40 ibid.
41 *Friend*, 26 August 1913.
42 OFSA, Jagersfontein Town Council Minute Book (hereafter JTCMB), 22 July 1913. The Jagersfontein women delayed their entry into the campaign because during June many of the Coloured leaders were building new houses as the result of forced removals from previously integrated neighbourhoods. *APO*, 9 August 1913.
43 *APO*, 9 August 1913.
44 The average age of women arrested for pass offences prior to the resistance was 20. Once the campaign started, however, the average age rose to 27. OFSA, Jagersfontein Criminal Record Book (hereafter JCRB), 1912–14.
45 JTCMB, 22 July 1913.
46 WMMB, 14 July 1913.
47 ibid.
48 *Tsala*, 23 August 1913.
49 South African State Archives, Pretoria, Governor General's Office (hereafter GG), Letter to Secretary of Native Affairs from Winburg Native Vigilance Association, 16 August 1913. This case was ultimately lost on appeal to the Attorney General on 16 February 1914. OFSA, Winburg Criminal Record Book (hereafter WCRB), 1912–13.
50 GG, letter.
51 WMMB, 25 September 1913.
52 WMMB, 23 October 1913.
53 *Friend*, 11 October 1913.
54 *Friend*, 16 October 1913.
55 ibid.
56 *Friend*, 18 October 1913.
57 *Friend*, 22 October 1913.
58 ibid., 31 October 1913. It is unlikely that Common Justice's true identity was unknown in the small town of Winburg. This historian, writing seventy years later, speculates that it was Marie Baumann, wife of the lawyer who defended the black cases in court, daughter of a liberal Dutch missionary and enthusiastic patron of the local orphanage. For information on her family background, see *Orange Days, Memoirs of Eighty Years Ago in the Old Orange Free State*, written by her sister, Caroline Van Heyningen (Pietermaritzburg, 1965).
59 *Friend*, 29 September 1913; and Solomon Plaatje, *Native Life in South Africa* (London,

1919), p. 94. Aploon Vorster, or Susan, as she was listed in the court records, appears to have been the 45-year-old head of a rather militant family. Other Vorster women, all listed in the records as Mozambicans, arrested for having no passes in 1913 and again in early 1914, were Elizabeth, 19; Rose, 20; and Maria, 17. In addition to these three, Sophia, 19; and Emma, 21, were among the victims of stepped up pass raids in November 1912. JCRB, 1912–14.

60 *Friend*, 29 September 1913.

61 DNA, Pass Laws Memo, pp. 16–17.

62 The Natives Land Act of 1913 is notorious for depriving African people of the right to own land outside of restricted areas in South Africa. Since it was passed only after the beginning of the passive resistance, it seems it had little direct bearing on the women's struggle. Over time it may have contributed to black urbanisation, thus alleviating labour shortages in towns; however, in 1913 its primary significance was the political furore it created in the black community.

63 Representative women from Winburg did not attend this meeting because Dower had scheduled a separate visit to Winburg. *Tsala*, 4 October 1913.

64 ibid.

65 BTCMB, 6 June 1913.

66 *Friend*, 6 September 1913.

67 BTCC, Minutes of Interview between His Worship the Mayor, certain Ministers of Religion, and Representatives of the Natives held at Waaihook, Friday, 5 September 1913 at 8 p.m.

68 ibid.

69 ibid.

70 BTCC, Minutes of Public Meeting held at Waaihoek on 2nd December 1913.

71 *Post*, 1 October 1913.

72 BTCC, Legal Opinion submitted by H. F. Blaine, 14 October 1913; and DNA, Pass Laws Memo, pp. 4–11.

73 See *APO*, 9 August, 23 August, 20 September, 8 November, 1913; 24 January 1914; and *Tsala*, 10 August, 27 September, 4 October 1913.

74 *Lovedale Christian Express*, 1 November 1913; Cape Town Cape *Argus*, 20 October 1913.

75 *Tsala*, 4 October 1913.

76 *APO*, 4 October 1913.

77 At a report-back meeting following the APO conference in Kimberley, Bloemfontein's resisters vowed to carry on despite the hardships before a cheering crowd. *APO*, 25 October 1913.

78 *Post*, 1 October; *Friend*, 30 September, 4 October; *APO*, 4 October 1913.

79 BTCC, Minutes of Public Meeting held at Waaihoek on 2 December, 1913; and *Friend*, 4 December 1913.

80 *APO*, 21 March 1914; and JCRB, 1912–14. In ·March eighty-three women were arrested for being without passes, then no more female arrests occurred for the rest of the year.

81 GG, Petition of the OFS Native and Coloured Women to Viscount Gladstone, 27 January 1914; and Circular letter from OFS Native and Coloured Women's Association, 18 February 1914.

82 Union of South Africa, *First and Second Report of Select Committee on Native Affairs June 1914.*

83 OFSA, Minutes of the Municipal Association of the Orange Free State, 28–29 March 1916; South African State Archives, Pretoria, Prime Minister's Office, Reply to Secretary of the Synod, Wesleyan Methodist Church of South Africa, 21 February 1916.

84 OFSA, Orange Free State Provincial Secretary, Letter from Under Secretary of Native Affairs, 12 February 1914; Letter to Under Secretary of Native Affairs, 5 March 1914; Letter from Department of Native Affairs, 8 February 1916; Letter to Department of Native Affairs, February 1916.

85 South African State Archives, Pretoria, Department of Native Affairs, Intradepart-

mental communication, 16 March 1917.
86 Union of South Africa, *Report of Interdepartmental Committee on the Native Pass Laws*, 3 August 1919, p. 25. U.G. 41–1922. 25.

12 Property, protest and politics in Bugisu, Uganda

STEPHEN G. BUNKER

Many recent analyses of East African politics argue that national states there have increased their political control and their economic exploitation of the peasants who make up much of their population. With a few exceptions[1] they treat national peasantries as passive victims.[2] I argue in this chapter that freeholding peasants who produce valuable export crops can sustain effective challenges to central state power.

These challenges, however, are often stimulated by local leaders in pursuit of their own political goals. The use and manipulation of peasant protest against the state raises the question of who benefits and who loses in these struggles. Peasant protest often provides crucial leverage in struggles to open political space for locally based power groups. These groups are usually more sensitive to peasant demands than are power holders in the central state. On the other hand, their stimulation of peasant protest may seriously disrupt local economies and politics in ways which impose heavy costs on the peasants.

I examine these questions in a study of how coffee-growing peasants in Bugisu, Uganda, protested against crop prices and marketing conditions and how local aspirants to representative positions in government and in the local crop-marketing co-operative manipulated these protests in their struggles to achieve autonomy from the central state and in their competitions with each other.

Peasants have seldom been seen as an effective political force. Marx's analysis of peasants mobilised behind a revolution which ultimately consumed them in *The Eighteenth Brumaire* sets the tone for other studies.[3] Observers have described peasants as the backbone but also the final victims of Latin American independence movements and of the Mexican Revolution.[4] They imply that peasants react to and advance new economic and political arrangements in which they only partially participate. In this view, peasants may be essential to the establishment of new regimes, but they finally succumb to new forms of exploitation under the regimes they help establish.[5] Analyses of peasant rebellion and resistance explain that peasants are too bound to the soil and to its seasonal routines, live and work in excessive isolation, and are too much subject to manipulation by 'Kulak' members of their own class to sustain their own political impetus.[6]

These notions about the peasantry's liminal political status are replicated in discussions of the peasantries' economic position. This 'awkward class',[7] caught between modes of production but having none of its own, is seen by many as producing values and as reproducing labour essential to a capitalist mode of production which increasingly smothers it but finally maintains it in order to exploit it.[8]

These same themes appear in the long and complex debate about the articulation of modes of production in Africa. Ideas which started with Luxemburg's[9] and Lenin's[10] rather different notions about capitalism's need for primitive accumulation and new markets at its own fringes continue with Meillassoux's[11] statements that non-capitalist forms survive because capitalism needs them and Bettelheim's[12] notions that capitalism partially dissolves, but finally conserves, elements of the other modes of production which it encounters. By attributing the existence and the persistence of African peasantries to capitalism, these authors commit major errors of reifying and attributing intentionality to capitalism itself, and ignoring the autonomous role of peasant societies in reproducing themselves.[13]

Without denying the tremendous impact that penetration by capitalist regimes of exchange, labour recruitment, and production has had on the development of peasantries in Africa, I insist that Africans actively participated in the shaping of their own economic and social organisation. Africans resisted the pressures imposed by colonial and capitalist agents, and they took advantage of new opportunities which those agents provided. Their responses to these pressures and opportunities engendered new forms of power based on new alliances and new conflicts within their societies.[14] These new groupings struggled against the colonial state and against each other for control of key political and economic resources. I believe that it is in these struggles, rather than in the abstract notion of capitalism, that we must search for the dynamics which moulded colonial and post-colonial society in Africa. In this sense, we must see peasantries as actively participating in the formation, reproduction, and conservation of their societies.

I will show in this study that the Bagisu acted collectively to resist exploitation by the state and to retain control over their own labour. They used the state's dependence on their crops to increase their own influence on local economic and political processes and institutions. Each resolution of their protests or resistance against the state changed the economic and political environment in which they acted and thus established the conditions in which subsequent protest occurred.

New ascendant groups and political institutions emerged from this struggle. They were directly accessible to only a privileged few, but because the economy depended on peasant production, and because political change was driven by peasant protest, the Bagisu peasants achieved and maintained considerable political power.[15] This power, however, was not of the type which a conventional analysis of legislative and executive institutions could illuminate. Peasants do not appear clearly in the available documentary evidence. Rather, the power of the peasantry, and the political centrality of their protest, can only be deduced from the political actions and statements of the holders of official positions and writers of reports and memoranda whose names and words fill the archives.

The Bagisu peasants and the Ugandan state

Bugisu's fertile soils and mountainous elevations provide a disproportionate share of the value of exportable coffee in Uganda, a country whose national state depends on coffee and cotton raised by smallholding peasants for approximately 90 per cent of its foreign revenues.[16] The state's dependence on coffee exports combined with a regime of freehold tenure to provide the Bagisu peasantry a bargaining threat which was denied to the state itself, namely a withdrawal from commercial agriculture. The fact that the Bagisu could partially withdraw from the market into subsistence, while the state clearly could not, to some extent equalised a balance of forces which in most other respects greatly favoured the national state. Because the state derived revenues necessary to its own existence from a freeholding peasantry, it was caught in a very tight bind. The implementation of its own political and economic projects required that the state 'capture' the peasantry[17] in order to increase its rate of appropriation from it, but if it attempted to increase the rate of surplus appropriation the peasantry could threaten to withdraw from commercial production, from which the state could easily appropriate surplus, into subsistence production, which the state could neither sell nor tax.

The Ugandan state, like many others in Africa,[18] attempted to resolve this dilemma through direct control of crop markets. Market control in a freeholding peasant economy such as Bugisu's, however, offers the major avenue to wealth and power at the local level. Local power groups have therefore persistently challenged the state's control of crop marketing by mobilising local peasants behind demands for higher crop prices and more local control of and participation in crop purchasing, processing, and sale. To the extent that these groups can convince the state that peasants will reduce their participation in cash-cropping, they can force the state to delegate market control to them.

Coffee farmers who threaten to destroy their trees are threatening to withdraw from production for at least four years, the time that new plantings take to reach production, and to require considerable capital to re-establish their trees. The destruction of coffee trees, therefore, is a potent threat, both against the state and against neighbours who continue to collaborate with the state by selling coffee. In Bugisu, the threat of crop reduction or withdrawal from coffee cultivation has been a constant factor in the state's tolerance of local demands for control. Even though actual destruction of coffee trees has occurred infrequently, the state has usually attempted to accommodate local claims to power.[19] I will describe incidents of property destruction and withdrawal from the market. These incidents, however, were only a particularly violent manifestation of a continuing struggle between the Bagisu peasants, the state, and the local leaders who mediated between the state and the peasantry. The threat that the Bagisu would withdraw from the market has been manipulated only symbolically through most of this struggle. The Bagisu have in fact come to depend on crop markets, and even partial withdrawal is costly to them. The state, however, has succumbed rapidly when they actually did withdraw from production or sale, and these rare incidents have maintained the credibility of this threat.

The issue of who wins and who loses when peasants protest is a difficult one, in Bugisu as elsewhere. Local political leaders, themselves peasants or of peasant origin, directed and orchestrated protest. They used peasant protests in their own struggles for power against the national state and against each other. Leaders, as a class, have benefited more from these struggles than have other Bagisu. At the same time, though, their ability to expand their own power depends on their capacity to mobilise the Bagisu. In order to maintain popular support, individual politicians have been obliged to struggle for local rights against the state.[20] Successful assertion of these rights benefits the Bagisu as a people. It also reduces the likelihood that the state will allow the particular politicians who achieve these rights to remain in office. The farmers consistently want higher prices: the government wants lower prices. The government demands higher coffee quality, while most peasants oppose quality-control measures because they impose additional labour requirements. To stay in power, the leaders must demand higher prices and resist policies which increase labour demands. The leaders protect the peasants against the state. At the same time, however, their manipulation of peasant protest may reduce crops and lead to destruction and violence which is very costly to the peasants.

The favourable environment of Bugisu District and the social organisation of the Bagisu have strengthened their ability to make demands on the state. Bugisu District covers 3030 sq km of eastern Uganda and includes a segment of the eastern and southern slopes of Mount Elgon, a volcanic formation on the Kenya border reaching 4300 m from the 1200 m-high plain surrounding it. The Bagisu compose the bulk of the district's half-million inhabitants and nearly all of its coffee growers. Mount Elgon provides both the rich soil and the elevation required to grow *arabica* coffee. Government regulations to protect coffee quality prohibit cultivation of the less valuable *robusta* coffee anywhere in the district, effectively limiting coffee cultivation to the mountain slopes. Cotton is the major cash crop grown on the plain, although a considerable amount of produce, especially bananas, is sold in local markets. Hoes and machetes are the basic tools of agricultural production; only transport and processing – neither of which are directly controlled by the peasants – are mechanised.

The Bagisu were almost exclusively smallholders, with freehold, individualised tenure over their own land. Technological and cultural restraints – i.e., the delicate, labour-intensive requirements of harvesting and pulping coffee, equal division of land among sons of the same womb, and lineage controls over the alienation of land[21] – restricted the possibilities for concentrated land tenure and meant that the labour available within households established the boundaries of coffee-plantation size. For most peasant households, withdrawal to subsistence was a real, though undesirable, option. Coffee cultivation maintained a prosperous, relatively homogenous peasantry. Except for the highly urbanised Mengo District, in which Kampala is located, Bugisu District had the highest per capita income in Uganda.[22]

The dilemma which the state faced in its attempts to increase coffee production and improve coffee quality in Bugisu emerged from the special nature of the Bagisu as a peasantry in the Uganda economy. Throughout the 1960s, coffee consistently provided over 50 per cent of Uganda's foreign revenues. Even though Bugisu *arabica* coffee accounted for only 7.2 per cent of total land in coffee and 5.2 per cent of total coffee volume, its superiority over the *robusta* coffee which predominated in the rest of Uganda brought so much higher prices that it accounted for 9.97 per cent of total value to farmers in 1961.[23]

I have described elsewhere how the Bagisu response to indirect rule and its use by the British to promote and administer cash cropping gave rise to successive ascendant groups – chiefs, civil servants, and politicians – and how these groups used the threat of peasant withdrawal from export agriculture to gain greater participation and representation in local government and increasing control of crop marketing.[24] In 1954, after a long campaign which involved intense mobilisation of the Bagisu peasants, a coalition of civil servants and political leaders with support from some of the chiefs was permitted to establish a co-operative (The Bugisu Co-operative Union, BCU), which took over all the assets and the jurisdiction of the state agency which had controlled coffee sales since 1933. The BCU was composed of smaller units, Growers Co-operative Societies (GCS) which purchased coffee from their own members and sold it to the BCU.

The GCSs were organised according to, and bounded by, traditional lineage divisions. This arrangement incorporated, and assured representation of, traditionally cohesive and territorially defined groups. Representatives from each GCS constituted the Annual General Meeting which was formally the paramount authority in the union. These representatives elected the BCU committee, which was empowered to direct the policy of the union. The BCU committee in turn hired a secretary-manager to run the union and direct its permanent staff. In principle, the committee set general policy and served as guardian and representative of the growers' interests, while the secretary-manager ran the 'day-to-day' business of the union.

The 1955 ordinance which empowered the BCU, however, also created the Bugisu Coffee Board, which was given broad powers over the union. The BCU sent representatives to the board, but they did not constitute the majority there. All of the union's activities and assets were subject to regulation by the board according to the board's discretion.[25] The BCU's leaders were very much aware and resentful of the board's powers. They particularly objected to the board's control over coffee prices and over the reserve funds which had accumulated during the period of direct colonial control.[26]

The BCU's leaders had mobilised popular support for a Bagisu-controlled co-operative by showing how this would improve the general welfare. From 1940 until 1954, the leaders had stressed to the Bagisu that they could receive much more money for their crops if the trade was in Bagisu hands. Not only would the middlemen be eliminated, they said, but the theft by the British of large portions of what should belong to the Bagisu would end. The leaders had achieved power through these promises, and in order to maintain their power they had to provide at least some of what they had offered. The power of the board threatened their ability to satisfy peasant demands.

The immediate issues against the board involved the powers of Bagisu leaders, but these leaders were constrained to press for peasant demands in order to keep their positions. They could use peasant protest against the state, but they had to beware of peasant discontent against themselves. Higher prices and lowered quality controls for coffee constituted immediate and tangible evidence that these leaders could keep their promises. Neither they nor the peasant members were much concerned with the establishment or the continuation of a solid financial position for the BCU, as they saw the union as already tremendously wealthy and hardly in need of solicitous attention. The leaders of the BCU were much more interested in questioning the colonial state's co-operative officers about the disposition of the coffee reserve funds than they were in listening to arguments that the coffee prices they were demanding for the farmer would cost the union a loss.[27] Their objections to the farmers that the board was holding down prices found sympathetic ears.

They also wanted to control BCU personnel appointments, because they needed crucially placed allies on the staff to pursue their struggle against the board. They were especially eager to take over the responsible and high-paying jobs in the industry which had until then been held by Europeans and Asians. Technically they had the right to change these employees, or at least to argue very strongly to the board that they be changed. The Department of Co-operative Development

(DCD), however, hinted very directly that if the union did not accept and retain the 'expert and experienced professionals' who were in the department's view essential to the industry's smooth and profitable running, the union would not be registered.[28] The BCU committee officers acquiesced, but, as with their acceptance of the board, did so with a bad will, and their resentment against the European staff caused considerable friction and problems. The Bagisu leaders attempted to direct peasant discontent against the European staff as well as against European domination of the board.

The BCU committee did insist on the appointment of the politically powerful Paulo Mugoya as assistant coffee manager. Mugoya was a son of the first Mugisu appointed as a county chief, nephew of a very powerful sub-county chief, and cousin of the first Mugisu to serve as Bugisu District's secretary-general. Members of this family in Mugoya's generation had received more schooling than most of the other groups in Bugisu, and therefore supplied a large number of chiefs, government employees, and union workers. Mugoya, because of his education and family connections, had held a series of agricultural jobs in government. He had worked closely with the organisers of the union. He was a strong force in, and later chairman of, the District Council. He was closely allied to the committee, most of whose members were also on the District Council.

There was direct and open hostility between him and the European manager of the union staff, Roland G. Woods. Woods had opposed his appointment. Mugoya obviously had much greater access to and influence over the committee than would be normal in conventional staff–committee relations in a co-operative, and his privileged position enabled him to circumvent many of Woods' directives and any sanctions Woods might have wished to apply to him. Mugoya was able to use his District Council position to mobilise both popular and official protest against Woods and against the board, but he also used his powers in ways which embarrassed the BCU committee and disrupted staff management. He carried a long-standing feud between his lineage, the Bamuddu, and a neighbouring lineage, the Babeza, directly into the BCU's internal affairs.

There had been repeated disputes over the boundaries and the division of political position between these two lineages, many of them resulting in fights and deaths. The Bamuddu had occupied numerous high-level chiefships, and the Babeza claimed these chiefs had discriminated against them politically. The Babeza had compensated for their loss of political power by sending many of their children on to higher education. The Babeza consequently held a disproportionate number of official administrative positions in the district, so the conflict between the two lineages had extensive repercussions.

Gimugu, the BCU's chief book-keeper and later chief accountant, and Wakiro, the staff secretary, were both Babeza. Mugoya attempted to reduce their authority by refusing to acknowledge their functions and by criticising and obstructing their work. The two Babeza allied themselves with Woods and against Mugoya. Mugoya and Gimugu became involved in a long and bitter struggle over position in the BCU. This lineage-based rivalry, as well as resentment against the continued employment of Europeans, increasingly impeded discipline and efficiency throughout the union staff. The feud between the Babeza and the Bamuddu, and Mugoya's willingness to manipulate kin- and lineage-based divisions in the pursuit of his own political goals, became a major factor in the violent protest movement against the BCU which emerged in 1960.

Woods resigned in 1957, claiming that these conflicts made it impossible for him to run the Union.[29] Mugoya was appointed coffee manager. He was now in a position to exacerbate his feud with the two Babeza, which he did through more and more open criticism of Gimugu and a rather petulant anger at receiving instructions through Wakiro, who as secretary was responsible for transmitting all committee decisions to the staff.[30]

The committee expressed concern that the feud between Babeza and Bamuddu was impeding the management of the Union, but its dependence on Mugoya and on the rest of the District Council in its fight against the board constrained it from taking effective actions to discipline the staff. Several other members of the BCU staff were also serving on the District Council. These included the chief accountant, the cashier, and the assistant manager. While these men probably spent less time on council business than Mugoya did, there were reports and official complaints about staff members not being present during working hours because of council duties.[31] The committee passed a resolution against staff participation in politics, but it was never implemented.

The line between staff and committee was thus obscured by important connections based on politics and friendship between individual committee members and individual staff members. Because the mutual support of committee and staff members was so crucial to each, and because the extension of the common interests of each outside of the union made their alliances, especially against the state and the board which protected its interests, very useful, neither side was eager to exercise its function of controlling the other. The pursuit of BCU autonomy, the use of the union as a vehicle by which the Bagisu could come into their proper rights as a strong and respected people and the political ambitions of individual committee and staff members all predominated over worries about organisational efficiency.

The committee directly involved the various GCSs to develop consensus and support for the BCU and protest against the board. It met frequently and called many general meetings, that is, meetings to which each GCS sent delegates. Unless there was some particularly urgent problem which required such attendance, general meetings were supposed in the by-laws to be called only once a year; but between September 1954 and September 1958, eighteen general meetings were called. Attendance of GCS representatives at these meetings ranged from 80 to 139. The meetings were long; one lasted two days, another three, a third seven, and the final one lasted thirteen days. Representatives received food and travel allowance, and overnight expenses if meetings lasted more than a day, so this form of decisionmaking became fairly expensive. The paid trips to town, however, served to reward influential members of the various GSCs. Through them, the BCU committee could propagate the reasons for a strong union and generate support for itself.

Kitutu, the committee president, was in the main office almost every day, as he insisted on being informed on what was going on there and on giving advice and orders. He and various members of his committee also spent a great deal of time travelling to various GCSs, giving talks there, settling disputes and solving problems, urging people to join the co-operative and follow its rules, and speaking against the board's continued control.

In 1957, a protest movement against the BCU monopoly strengthened the BCU's position against the board. Committee members were able to attribute the movement to low coffee prices, and eventually achieved the increases they had been demanding.

The state strikes back

B. B. N. Mafabi, a former sub-county chief who had provided important support to the campaign to establish the BCU, and who had served a term on the committee, started this movement. He claimed that if the committee were not so extravagant it could pay its members higher prices. He founded a separate marketing organisation, and applied for a licence to buy and export coffee. Two visiting MPs from England recommended that the group be granted a licence. This reinforced Mafabi's claims and strengthened the determination of his followers. At Mafabi's urging, they withheld their coffee in order to sell it through the new organisation. Mafabi was not granted a licence, however, even though he continued his appeals for over a year. This coffee was thus temporarily lost, and in danger of spoiling.[32]

The BCU leaders were able to use this manifestation of peasant discontent as a demonstration that they should be allowed to raise coffee prices. Under the threat that peasants would withhold more coffee, the committee convinced the board to authorise higher prices for the 1957/58 season. As a result, the union suffered its first loss. The committee's success in gaining higher prices, moreover, led soon after to state intervention and an abrupt loss of the autonomy it sought.

The ambiguities of the union's being perceived and used simultaneously as a vehicle of ethnic self-assertion and as a business operation created a constant tension between the political preoccupations of both staff and committee and the need to continue conducting daily business. Increasingly it was the latter which was neglected. It was clearly the state's purpose to ensure that the BCU made a profit and maintained the commercial viability of the crop, that certain laws and principles of business were maintained by the committee, and that the crop continue as a source of income for the protectorate. The committee's ability to mobilise the peasantry, however, meant that direct state opposition to or intervention in the union posed the danger of provoking disruption of coffee deliveries. Thus, even though government agents were empowered and encouraged to supervise the union's activities, they were generally reluctant to oppose Bagisu wishes even when they were in disagreement with them. In one particularly revealing case, the Commissioner for Co-operative Development (CCD) strongly urged the BCU committee to file a suit for embezzlement in which two of its members were directly involved, but neither he nor the board took any independent action when the committee refused. In another case, the CCD appealed directly to the protectorate governor, Sir Andrew Cohen, when the board, with a majority of the BCU representatives concurring, accepted a higher construction bid from a lower-rated company. Cohen responded that 'the people should choose whoever they want as their builders'.[33]

The delicate balance between the Protectorate government's tolerance of Bagisu demands for autonomy and its concern for the continuation of coffee crop revenues finally collapsed over the coffee prices paid to the farmers. Officially the board controlled prices, and the board was responsible for setting realistic ones. When BCU leaders used the threat of protest and crop withholding to force coffee prices to 2.3 East African shillings per pound, instead of the 1.5 shillings which the

European board members had recommended for the 1957/58 crop, the governor appointed two more Europeans to the board. The Bagisu objected, plausibly enough, that the governor had acted to guarantee a clear European majority. They resolved to boycott all future meetings.[34] This meant in effect that there was to be no further communication between the governing or controlling body and the implementing one, a clearly impossible situation. Shortly thereafter, the governor appointed a commission of inquiry to consider the state of the BCU and to make recommendations for its improvement.

The commission conducted interviews and received evidence in Bugisu for a month in April and May 1958. It faulted the board's handling of relations with the Bagisu, but most of its criticism was directed at the union itself, primarily for bad business management as seen in lack of staff control, badly kept accounts and records, waste, and extravagance.[35] It attributed this to a highly inefficient accounting system but also blamed extensive staff involvement in district politics. It sharply criticised the large amounts of money spent on transportation for committee and general meetings, and said that Kitutu, the committee president, was far too dominant in the union. It claimed that the farmers were disaffected and that they felt very much in the union's power, rather than feeling that the union was their organisation and could be controlled by them.

The commission backed its assertion of 'the growers' feeling of helplessness, however, with only a single quotation; 'One of them actually said "we feel like prisoners, unable to make any choice",' and with a reference to 'a few members' who wished to return to individual bargaining with private traders.[36] The paucity of documentation for this claim contrasts strongly with the detailed evidence it presented of mismanagement and inefficiencies. The claim that the growers felt helpless in their relations with the BCU also appears incongruent with the commission's own statement that all of the GCSs and their members supported the BCU's campaign against the board and its boycott of board meetings.

The commission also warned that the rapid deterioration in the quality of Bugisu coffee during the preceding year and a half had driven many of the BCU's international buyers to seek other sources. It recommended reinstating some form of quality distinctions with bonus incentives in order to re-establish Bugisu's international reputation.

Finally, the commission recommended that the Uganda co-operative laws be revised to allow the Commissioner for Co-operative Development, if he was convinced that a society was not being run satisfactorily, to appoint a suitable person to run it with full control over its liabilities and assets and with 'the powers, rights, and privileges of the duly constituted committee'.[37]

It was this last recommendation which had the greatest effect on the union. The law was enacted as proposed, and the commissioner of the DCD appointed W. E. Neal, a British co-operative officer and a trained accountant, as supervising manager of the union. Most of the BCU committee were replaced at the next general meeting.

Bagisu leaders now turned the anti-government suspicions and hostilities which they had aroused against the board, against Neal, the supervising manager, and the DCD for which he worked. The committee urged the GCSs to oppose and obstruct the union while it was run by the supervising manager. District Annual Reports from 1959 and 1960[38] indicate that the District Council and the GCSs supported the BCU committee's campaign against the supervising manager. Many GCSs submitted complaints to the BCU and to the Department of

Co-operative Development against the union's operating procedures. The supervising manager did improve the financial condition of the union and of some of the GCSs dramatically in his two years, but both the BCU committee and numerous GCS officers complained that he and his assistants rode roughshod over them. The 1961 District Annual Report claimed that improved payments for coffee and larger bonuses had led to greater co-operation between the GCSs and the supervising manager. The petitions to government for his withdrawal continued, however, and the most significant protest movement in the BCU's history was gaining momentum at exactly this time.

BCU committee members consistently objected to Neal's decisions and actions. They initiated numerous resolutions to various branches of government to have him removed and sent occasional delegations to Kampala to petition the government and the Commissioner for Co-operative Development for his dismissal.

National independence, politics, and conflict within the BCU (1958–64)

The imposition of direct government control over the BCU coincided with the first stages of the constitutional conventions and parliamentary elections prior to internal self-government and final independence. The increased emphasis on African nationalism and the intensifying struggle between African parties and politicians for control over the independent state promoted and sharpened divisions within the union and resistance to the colonial state.[39] The state could not have chosen a less opportune time to restrict BCU autonomy.

The close nexus between the struggles for power at the national, district, and local levels on the one hand, and the mobilisation of peasant protest on the other, are most dramatically seen during the turbulent years prior to independence. The new political positions in an increasingly representative District Council and powerful new roles such as Member of Parliament and District or County Party Chairman strengthened the base from which the Bagisu could demand local control. The intense fights to control these roles and the rich rewards they promised led to the development of powerful factions both outside and inside the union.

The BCU was a central element in the division of power in the district, so political struggles enhanced the widespread opposition to the government-imposed supervising manager and the totality of his power. New forms of peasant protest emerged from outside the BCU. As before, Bagisu leaders used these protests, first against the state and then against each other.

Local protest against the state and against Neal was intensified by a second separatist movement which not only threatened the integrity of the BCU's monopoly, but eventually threatened severe crop reduction. This movement, led by a northern politician and later Member of Parliament, Stephen Muduku, tried to establish a rival to the BCU, the Bugisu Coffee Marketing Association (BCMA). Members of at least twenty GCSs joined the BCMA, and three of these GCSs ultimately left the BCU altogether.[40] The BCMA was strongest in Buwalasi, where Mugoya's lineage predominated, and was alternately encouraged and manipulated by Mugoya in his own campaigns for power within the BCU.[41] Like the earlier separatist group, the BCMA was organised around the promise of higher coffee prices, the relaxation of quality controls, and the provision of marketing arrangements more convenient to the growers. Even though the BCMA ultimately failed, the BCU was pressured to accede to popular demands on all these issues. The BCMA bought coffee from 1960 to 1963, and was appealing for a licence during all of that time. Its existence led the government to propose splitting the union up, a move which the committee resisted strongly. Government vacillated on its position about giving the BCMA a marketing licence, so the issue was one of continuing concern to the union, which was very eager to maintain the monopoly.[42] Thus, even while the BCU was struggling to affirm its autonomy from government, both in its resistance to the supervising manager and to the board, it was attempting to maintain the exclusive monopoly which it derived from the government.

The BCMA did, however, provide the BCU leaders with a platform from which to attack Neal and the DCD. They invoked the peasant discontent which the BCMA represented to underscore their claims that Neal and his appointees did not respect and, in fact, discriminated against the Bagisu. They were able to blame Neal directly as well. The BCMA's officers claimed that they had founded the organisation when one of them, a GCS representative, had been beaten by one of Neal's mill employees, a European. When the GCS committee complained to Neal he dismissed them, telling them to try to sell their coffee elsewhere.[43] They decided to follow his suggestion. This episode heightened the feelings that Neal was abrupt and insensitive to Bagisu dignity.

Thus, even though the BCU committee was deeply opposed to the BCMA, its existence was convenient to them in their fight against Neal. The conflict between the BCU and the BCMA was intense. It heightened enmities between opposed factions at the district level and between lineages at the sub-county level. Within some GCSs, the fights between adherents of the BCU and followers of the BCMA led to killings, beatings, and house-burnings. Yonosani Mudebo, the BCMA secretary, was Mugoya's brother, and there were persistent reports that Mugoya was encouraging the BCMA and giving it confidential information and advice. This intensified the BCU's internal conflicts. To the extent that the union's leaders could impute these problems to Neal, they could further justify resistance to him.[44]

The BCMA strategy was to urge farmers to withhold their coffee for sale in the BCU-affiliated GCSs, and to work toward withdrawing the GCSs from the BCU. The co-operative officer for Bugisu had relayed word from the DCD that any GCS with a 75 per cent vote in favour could withdraw from the BCU. The BCMA had not been granted a licence, but the DCD did not appear likely to prevent BCMA sales.[45] In any event, Muduku was free to make other arrangements for sale.

Although Muduku was from a lineage further up the northern valley, BCMA strength was centred in Buwalasi, the sub-county which included the feuding Babeza and Bamuddu, and in neighbouring sub-counties, where political and lineage tensions had always run high. Tensions and divisions between lineages in Buwalasi were exacerbated by the BCMA, and determined in part which GCSs joined the BCMA. The Bamuddu, Mugoya's lineage, were particularly influential in the BCMA. The Babeza tended to remain loyal to the BCU. The feud between these two lineages, in turn, sharpened the conflict between the BCU and the BCMA. The campaign to urge farmers to withhold coffee and to vote to leave the BCU became violent quite early. Farmers who favoured the BCMA started to slash the coffee trees of farmers who continued to sell their coffee to the BCU and who refused to vote for withdrawal from the BCU. BCU supporters retaliated in kind. There were a number of violent confrontations between the opposing sides, and several murders and house-burnings were attributed to revenge for coffee slashing. I was unable to find any estimate of the number of trees destroyed, but the 1961 District Annual Report referred to 'widespread intimidation and tree-slashing' in Buwalasi and Buyobo.[46] People I interviewed in 1970 reported that there had been extensive damage to coffee trees there, and some further up the valley. There was also some disruption of coffee collection due to threats to truckdrivers' lives. The same annual report estimates that 15 per cent of the growers supported the BCMA. In eight of the twenty GCSs affected, BCMA support was considered to be very strong; three finally did withdraw from the union.

The BCMA was the single most important form of peasant protest which the BCU committee used against Neal, but they also manipulated popular concern with coffee prices and opposition to quality controls in their campaign against him. Neal was especially concerned with the decline in coffee quality. Mugoya and various committee members exploited popular resistance to quality controls by publicly questioning Neal's assessment of the coffee crop and by criticising his proposals to initiate more stringent grading procedures. Mugoya consistently opposed this idea, and was increasingly successful in turning the committee's attention to other possibilities. In 1960 he claimed that in fact Bugisu coffee had not declined in quality, but that the BCU was being discriminated against at the Nairobi auctions by their own broker in complicity with Neal.[47]

Price differentials had never been popular among the majority of the farmers, although a small number of GCSs strongly supported them; and any argument against them was likely to find favour. It was in the short-term interests of committee members to look outside the union for the reasons of the falling prices for Bugisu coffee. They supported Mugoya's claim, allowing him to discredit Neal further and gain support among the rest of the farmers. By doing so, they could bring another complaint against Neal and not weaken their own position by proposing burdensome and unpopular solutions.

The campaign against Neal and the board was exacerbated by the increasing power in the political roles open to Bagisu and by the fact that many of the committee members either occupied these roles or were competing for them. Again, BCU leaders used peasant discontent and violence to give more force to their arguments against the state. Mutenio, the committee president; Gunigina, the vice-president; and J. N. K. Wakholi, the treasurer, were the main leaders of an anti-government faction in the District Council which was trying to change the district's tax base by reducing crop taxes and poll taxes and increasing taxes on salaries. The debates over this issue and the general feeling among growers that they were taxed unfairly and arbitrarily sparked off serious rioting in Bugisu and neighbouring Bukedi districts in 1960.[48] The District Council invoked popular resentment and violence to reduce taxes radically, overriding Bugisu Native Administration objections. It also cut the number of chiefs to eliminate the resulting deficit.

Faced with increasing levels of violence and generalised opposition to most aspects of colonial administration, the DCD withdrew Neal in late 1961, six months ahead of schedule. In March 1962, the Board was abolished. After more than eight years of protest, the BCU formally gained full autonomy at the local level, though it was still subject to statutory supervision and audit by the DCD.

The BCMA and power struggles within the BCU

While Neal remained as supervising manager, he posed a common threat which tended to submerge the major dissensions within the union. The BCU committee used the BCMA and peasant protests about prices and quality control against him. Even though the committee members were concerned about the threat which the BCMA posed to the BCU monopoly, it had also served them as a demonstration of the harm done by Neal's and his mill manager's lack of sensitivity to Bagisu aspirations and dignity. After Neal left, the dissensions within the union re-emerged, sharpened by competition for the newly available political opportunities in the district. Bagisu competitors for power now turned the BCMA and other peasant protests into issues which they used against each other. As before, this competition was played out through appeals to the Bagisu farmers for support and, once again, these appeals centred on questions of coffee prices, collection and grading procedures, and bonus payments. The new BCU committee, and especially Mutenio, the successful businessman elected as president, were very concerned to avoid the kinds of problems which had provoked the DCD intervention of 1958. For the first time, the committee concerned itself with keeping coffee prices low enough to guarantee the BCU an operating surplus and to improve quality sufficiently to guarantee a favourable market. These concerns were not popular with the growers, but Mutenio felt they were essential to avoid another DCD intervention.[49] As these were exactly the same issues that the leaders of the BCMA were using to mobilise support, there was inevitably common cause between the BCMA leaders and dissident factions within the union. As a result, the committee started to denounce pressure from within the union for higher prices. Statements to the farmers promising higher prices were now seen as 'treacherous' support for the BCMA.[50]

While the DCD was not particularly interested in protecting the BCU monopoly, it was concerned to maintain coffee production. Worries about the destruction of coffee trees, crop withholding, and disrupted deliveries associated with the BCMA were clearly an important factor in Neal's early withdrawal and in the DCD's accession to changes in the staff arrangements which Neal had instituted.[51] Neither Neal's withdrawal, nor the reappointment of Mugoya as head of the BCU staff, however, ended the problems the BCU had with the BCMA and with Mugoya's apparent support of it. Nor did it silence Muduku's accusations, or make him willing to negotiate with the BCU. Mugoya refused to subordinate himself to the committee and continued to use the BCMA and its demands as a platform to speak against the committee and to expand his own power within the union. Once Neal had left, the BCU committee was the sole object of the BCMA's criticisms, with Mugoya on the one hand maintaining that prices were too low and that quality had not deteriorated, and with Muduku on the other maintaining that he could negotiate more favourable brokerage arrangements in Nairobi.

Mutenio, the BCU committee president, tried hard and consistently to keep politics out of union considerations, but the intensifying struggle betwen the two national parties, the Democratic Party (DP) and the Uganda People's Congress (UPC), the political positions of the committee members, and the political activities of the BCMA all combined to bring party politics into the BCU.[52] Mutenio and Gunigina, the BCU vice-president, were both founding members of the DP in Bugisu, and held high positions in it. Most of the rest of the committee were UPC members. Mugoya had stood for the Legislative Council in 1959 as an independent, but had lost to Muduku, the DP candidate. It became increasingly evident that Mugoya planned to run again.

These conflicts were extended out into the general membership as a result of Muduku's attempt to identify the BCMA as a DP organisation, a ploy he needed to gain favour with the government under which he was trying to get a coffee selling and export licence. This use of party connections made Mutenio's position as head of the DP in the district especially problematic within the union itself. No matter how much he tried to keep the committee from division along political lines, he himself was too vulnerable a target there. There were increasing rumours that Mugoya was working with the BCMA and using it to gain a stronger position in the DP. Yonosani Mudebo, secretary and founding member of the BCMA, and Mugoya's brother, told me that Mugoya was never a member of BCMA, but that he gave him, Mudebo, a great deal of advice on how to run it. Mugoya himself, though he denied in a 1961 general meeting that he was helping the BCMA, did admit that he had told some of the farmers in the north at a public meeting that the BCU was not paying them fair prices. Mutenio consistently disallowed any political discussion during meetings, but several times it was claimed that the DP was going to ruin the union. The BCMA, Mugoya's activities, and the attacks on Mutenio all served increasingly to divide the committee on party lines.

By 1962, Mugoya was openly defying the committee's authority. At the beginning of the 1962/63 season, he promised the growers that they would receive two shillings a pound for their coffee, a price which was 55 cents over what the BCU committee estimated it could pay given that year's world coffee market situation.[53] The committee felt itself in a very tight situation; the BCMA seemed to have a very good chance of getting its licence under the DP government which had just won the national elections. The committee would risk another state intervention if it raised prices so high that the union lost money. The committee had heard that a Nairobi company had offered Muduku a price which would enable him to pay two shillings a pound in order to get the BCMA established.[34] There was a great concern on the committee that the BCU would lose large numbers of its members to the BCMA if it could not give them the price Mugoya had promised. Mutenio, however, threatened to resign if the higher price was paid, arguing that it would hurt the union badly and that it went against all of the principles of co-operation and education of the members. He did accede, however, to a resolution to petition the Bugisu Coffee Trust Fund for the amount necessary to make up the difference in price to the farmers.[55]

Mugoya was obviously becoming a political and financial liability to the union. He consistently opted for actions and policies which would enhance his popularity with the farmers. He was increasingly careless with confidential committee and official information, and claimed authority beyond the powers of his own position. The committee had twice reprimanded him, Gimugu, and Wakiro for the problems their conflict was causing within the staff.[56] There were also complaints against him on the staff for neglecting parts of his work and for spending large periods of time away from his office. The major sources of complaints against him, however, were in accusations of complicity with the BCMA. These ranged from claims that he was actively supporting the BCMA to complaints that he had told the growers that Muduku was correct in his claims that the BCU could pay higher prices to the farmers. Finally, the committee appointed three of its own members, headed by George Waisi, a young and energetic northerner who had worked for the Bugisu Coffee Board and for the BCU before being elected to the committee, to investigate and submit a report on staff problems. The DP government did recognise the BCMA and gave it a marketing licence, which the BCU appealed against unsuccessfully. A second election, however, brought a coalition of the UPC and the Kabaka Yekka (the Buganda royalist party) to power and the trading licence was revoked.[57]

Shortly before the BCMA licence was revoked, in January 1963, Waisi and four other committee members submitted the report of their investigations. The four had spent a full month hearing evidence and complaints from farmers in GCSs throughout the coffee-growing area. They reported that they had met attempted obstruction by politicians and some union employees thought to be connected to the BCMA.

They claimed that their report reflected what they were told by the farmers, and its acceptance by the annual general meeting tends to corroborate their claim.[58] Nonetheless, it was not an unbiased account. Waisi was by now a likely candidate for the BCU committee presidency, and so was in direct competition for power with Mugoya. Whether or not the report he submitted was an exact picture of what he heard in the GCSs, it reflected heavily the same political directions which he was taking; and it ultimately worked very much to his advantage.

There is no doubt that there was great discontent among the coffee growers. Waisi's report emphasised the number of times farmers said that both prices and procedures had been much better under both Woods and Neal. According to the report, the farmers were also very unhappy with what seemed to be inordinate staff privileges and compensation. They complained about excessive pay scales, inordinate and unbalanced pay raises, and the number of employees who were allowed to use union vehicles and other equipment, often for their own purposes.

The report mentioned the various problems caused by the feud between the Bamaddu and the Babeza on the staff, but said that the membership resented the domination by the Bawalasi, the clan which included the two lineages, of so many of the top staff positions. Other complaints about the staff included charges of 'clanism' and favouritism in hiring generally, of redundant posts, and of staff favouritism to their own areas and kinsmen.

By far the greatest part of the report, and the sharpest of its criticisms, were directed against Mugoya and against Dombodo, his assistant. The report claimed that Mugoya had been directly involved in establishing the BCMA, that he had repeatedly given it financial and organisational advice, and that he had made the brokerage arrangements for it in Nairobi. It accused him of being most responsible for the disenchantment of the farmers with the union, for having promised them impossibly high prices, and for alienating many of the GCS members by bad management. It also took Mugoya to task for being primarily responsible for the continuation of the feud with the Babeza staff members.

While Waisi was making his report, Mugoya had mastered enough support on the committee to fire Gimugu. He managed this, however, just as his own power was collapsing. The union had received enough money from the Bugisu Coffee Fund to make up a part of the difference between their estimate and Mugoya's promised prices to the farmers, but the committee was angry at having to use this recourse, and the farmers were disillusioned with Mugoya because they had received less than he had promised them. The growing evidence of Mugoya's political use of both the BCU and the BCMA cost him the rest of the support he had previously had on the committee. Finally, the end of the BCMA as an effective threat left the BCU much less susceptible to the kind of pressure which Mugoya had been exerting. The next annual general meeting, which met in March 1963, was highly indignant at Gimugu's firing. The meeting dismissed Mugoya, thus eliminating what had been one of the major influences in the union and incidentally ending whatever chances Mugoya might have had to win political office.[59]

In September 1963, the BCU annual general meeting elected Waisi as its president. With Mugoya and Gimugu both fired, Waisi was by far the most influential man in the BCU. Mugoya's old post, that of the general manager, and that of his assistant were both abolished. Wakiro was appointed as secretary manager, the newly established senior clerical position. Unlike either Gimugu or Mugoya, Wakiro had never been in any political or leadership positions. He had no other basis of power than his own job, and showed no disposition to challenge the dominance which Waisi was asserting.

Waisi had used peasant protest and complaints to eliminate the other major foci of power in the BCU, on the one hand by defeating them entirely, as in the case of Mugoya, and on the other hand by dividing the base of power in such a way as to leave himself without competition, as in the staff reorganisation.[60] In contrast with Kitutu, the first BCU president, who also was without important competition within the union itself, Waisi was in charge of an organisation free of the Bugisu Coffee Board's control, and so had much more freedom to implement his own policies. He had achieved this power precisely through interactions with and promises to the farmers, offering them the benefits of a stronger union and the solutions to their discontents. His ability to do so, however, depended ultimately on the autonomy which the political manipulation of peasant protest had earlier won for the union. The eventual defeat of the BCMA, and Waisi's election as president, was effectively a UPC triumph, as it brought a committee with close political ties to that party into power. It also left the dissidents, the breakaway societies which had joined the BCMA, vulnerable to politically based reprisals when they tried to rejoin the BCU. As late as 1971, some of these societies had not been allowed to reincorporate, and their members were forced to sell through GCSs controlled by other lineages and located in other areas.

Conclusion

The BCMA's direct threat to the BCU and to the DCD ended in 1963, but its effects continued long afterward. The BCMA was a major factor in the BCU's fight to regain its own autonomy. Mugoya's attempts to use the BCMA to challenge the committee resulted in his own eventual dismissal and the emergence of the BCU committee president as the predominant force within the BCU. The threat of withdrawal from coffee production and sale, which underlay the entire campaign to establish the BCU and later to resist government intervention, was greatly strengthened by the destruction and violence of the BCMA's challenge to the BCU. BCU leaders continued to manipulate this threat in their struggles against state control. In all of these ways, the BCMA contributed to, and was exploited in, the political strategies of Bagisu leaders.

The Bagisu peasants who lost lives, coffee trees, and houses, and the greater numbers who had trouble selling their coffee later on, all paid dearly for the BCMA separatist movement. These costs, however, were concentrated in Buwalasi and Buyobo. For the rest of the coffee growers in Bugisu, the effects of the

BCMA may have been much more beneficial. While it is true that the BCMA and other protest movements were stimulated, orchestrated, and manipulated by Bagisu leaders in their struggles against the state and against each other, it is also true that these leaders could only maintain support by agitating for higher prices and for less labour-absorbing quality requirements. Indirectly, therefore, these leaders' political strategies focused peasant demands and communicated them to the state. In all of these struggles, peasant protest, and the predominance of peasant production which made their protest economically and politically significant provided the basic levers for the threats and bargaining between the state and local claimants to power. Clearly, the conflict in the union and over its control were costly to the efficiency of the union as a commercial organisation. The particular issues behind which it was possible to mobilise the popular support which the Bagisu leaders needed in their struggles for power were not those which would have made the union itself a more profitable operation. In this sense they can be seen as prejudicial to the Bagisu themselves if we assume that a cost-efficient union would have directly benefited them. What these struggles did achieve, however, was precisely that neither the union nor the government had clear or direct control over the coffee crop. This moderated the extent to which the Bagisu themselves depended on commercial crops. The prices available through protest and freedom from the extra work involved in carrying cherry[61] to a central pulping station, and in maintaining quality control, left the Bagisu closer to the subsistence economy which was finally the basis for any political resistance that they could organise against government or the commercial cropping system represented by the BCU.

NOTES

1 Goran Hyden, *Beyond Ujamaa in Tanzania: Underdevelopment and an Uncaptured Peasantry* (Berkeley: University of California Press, 1980). F. Holmquist, 'Defending peasant political space in independent Africa', *Canadian Journal of African Studies*, vol. 14, no. 1 (1980), pp. 157–68. E. A. Brett, *Colonialism and Underdevelopment in East Africa* (New York: Nok, 1973).

2 M. Mamdani, *Politics and Class Formation in Uganda* (New York: Monthly Review Press, 1976). I. Shivji, *Class Struggles in Tanzania* (New York: Monthly Review Press, 1976).

3 Karl Marx, *The Eighteenth Brumaire of Louis Bonaparte* (New York: International Publishers, 1963).

4 See, e.g., B. R. Hamnett, *Revolucion y Contrarrevolución en México y el Peru* (México: Fondo de Cultura Economómica, 1978).

5 B. Anderson and J. D. Cockcroft, 'Control and cooptation in Mexican politics', *International Journal of Comparative Sociology*, vol. 7, no. 1 (1966), pp. 11–28.

6 E. Wolf, *Peasant Wars of the Twentieth Century* (New York: Harper and Row, 1969).

7 cf. T. Shanin, *The Awkward Class* (London: Oxford University Press, 1972).

8 A. de Janvry, *The Agrarian Question and Reformism in Latin America* (Baltimore: Johns Hopkins University Press, 1981).

9 R. Luxemburg, *The Accumulation of Capital* (Routledge & Kegan Paul, 1951).

10 V. I. Lenin, *Imperialism, the Highest Form of Capitalism* (New York: International Publishers, 1939).

11 C. Meillassoux, 'The social organization of the peasantry; the economic basis of kinship', *Journal of Peasant Studies*, vol. 1 (1973), p. 89.

12 C. Bettelheim, 'Theoretical comments' in A. Emmanuel, *Unequal Exchange* (New York: Monthly Review Press, 1972), Appendix I, pp. 271–322.
13 cf. Hyden, *Beyond Ujamaa.*
14 See Robin Palmer and Neil Parsons (eds), *The Roots of Rural Poverty in East and Central Africa* (Berkeley: University of California Press, 1977) for applications of these ideas in different African case studies.
15 S. Bunker, 'Center–local struggles for bureaucratic control in Bugisu, Uganda', *American Ethnologist*, vol. 10, no. 4 (1983), pp. 749–69.
16 C. Young, 'Agricultural policy in Uganda: capability and choice', in M. Lofchie (ed.), *The State of the Nations* (Berkeley: University of California Press, 1971), pp. 141–64.
17 Hyden, *Beyond Ujamaa.*
18 R. Bates, *Markets and States in Tropical Africa. The Political Basis of Agricultural Policies* (Berkeley: University of California Press, 1981).
19 Bunker, 'Center–local struggles'.
20 S. Bunker, 'Dependency, inequality, and development policy: a case from Bugisu, Uganda', *British Journal of Sociology*, vol. 34, no. 2 (1983), pp. 182–207.
21 Members need lineage approval for all land sales outside the lineage. In all other regards, Bagisu had individualised and free ownership of their own land; i.e., they did not pay rent, and were thus 'freeholding'. See also Bunker, 'Dependency'.
22 Uganda Protectorate, *Statistical Abstracts* (Entebbe: Government Printer, 1958).
23 Republic of Uganda, *Statistical Abstracts* (Entebbe: Government Printer, 1969).
24 Bunker, 'Center–local struggles'; *idem*, 'Dependency'.
25 Uganda Protectorate, *The Bugisu Coffee Ordinance* (Entebbe: Government Printer, 1955).
26 Bugisu Co-operative Union (hereafter BCU) Committee Minutes, 1954–8, also various letters in BCU and Department of Co-operative Development files, Mbale, and my own interview with former members of the BCU committee and of the Bugisu Coffee Board, 1969–71.
27 BCU Committee, Minutes, 1954–8, BCU, Mbale.
28 BCU Committee Minutes, 1954.
29 Interview in Nairobi, 1970.
30 BCU staff memos, 1957–8, in BCU files, Mbale and interviews with former committee members and staff, 1969–71.
31 BCU Committee Minutes, 1956–8.
32 Interviews with Mafabi and with former BCU staff and committee, 1970–71.
33 Letter from Cohen to Senior Cooperative Officer, Bugisu/Sebei, 17 October 1956; Department of Co-operative Development (hereafter DCD) correspondence, 1956–7, in DCD files, Mbale; interviews with former Bugisu Co-operative Board members, 1970–71.
34 BCU Committee Minutes, February–March 1957.
35 Uganda Protectorate, *Report of the Commission of Inquiry into the Affairs of the Bugisu Cooperative Union, Ltd. Sessional Paper No. 14* (Entebbe: Government Printer, 1958).
36 ibid., p. 18.
37 ibid., p. 34.
38 Bugisu Native Administration, files in Bugisu District Administration, Mbale.
39 C. Young, N. P. Sherman, and T. Rose, *Cooperatives and Development in Ghana and Uganda* (Madison, University of Wisconsin Press, 1981), pp. 42–6.
40 Bugisu Native Administration, District Annual Report, 1961, filed in Bugisu District Administration, Mbale.
41 Interviews with former BCMA and BCU officers; BCU correspondence and committee minutes, 1960–63; DCD correspondence filed in Mbale, 1960–63.
42 The information about the BCMA presented here has been pieced together from interviews with Muduku, Mudebo, and other leaders and members of the movement, and with BCU staff and committee members, from BCU, District Council, District Administration, and DCD correspondence, memos, and reports, and from BCU Committee and Annual General Meeting Minutes. All documents were consulted in

the files and archives of these agencies in Mbale. All individuals were interviewed at least once as part of a systematic survey of all BCU leaders' and high ranking staff members' career histories; further interviews to clarify events were conducted with Muduku, Mugoya, Mudebo, Waisi, Mutenio, Gimugu, and Gunigina. In addition, stories of the BCMA came up frequently in conversations and interviews with coffee growers in Buwalasi and Buyobo.

43 An undated letter from the committee of Buhasa, GCS, Budadiri, to Senior Co-operative Officer, Bugisu/Sebei in DCD correspondence for 1960, makes this complaint against the mill manager.

44 BCU Committee Minutes and correspondence and interviews conducted in Buwalasi and Buyobo, 1970–71.

45 DCD correspondence, Mbale, 1960–62.

46 Bugisu Native Administration, filed in Bugisu District Administration, Mbale.

47 BCU Committee Minutes and correspondence, 1960–61.

48 Uganda Protectorate, *Report of the Commission of Inquiry into the Affairs of the Bugisu District Council* (Entebbe: Government Printer, 1960).

49 Interview, Mbale, June 1970.

50 BCU Committee Minutes, 1962.

51 DCD correspondence, 1951–62, filed in Mbale offices.

52 Mutenio's statements in BCU Committee Minutes and in Annual General Meetings, 1961–62, BCU files in Mbale, and interviews with Mutenio and other members of his committee, Mbale, 1970–71.

53 Interviews with members of the Mutenio committee and with former officers of the BCMA.

54 BCU Committee Minutes, 1962.

55 BCU Committee Minutes and interview with Mutenio.

56 BCU Committee Minutes, 1962.

57 Young et al., *Cooperatives*, pp. 43, 78–9.

58 I found the discussions of the report in various committee minutes, but the copy I actually read was lent to me by Waisi.

59 Annual General Meeting Minutes, 1963, BCU, Mbale.

60 Waisi was also instrumental in eliminating the cotton section from the BCU. This move removed competition from other leaders who grew that crop. See Bunker, 'Center–local struggles'.

61 'Cherry' is the full, un-dried coffee bean still surrounded by a plum-like meat and skin: it weighs abour four times as much as dried coffee bean.

SECTION IV
Rebellion

13 'Black snake, white snake': Bāhta Hagos and his revolt against Italian overrule in Eritrea, 1894

RICHARD CAULK

Schoolchildren in Akkala Guzāy remember the refrain of a song which boasts of the boldness of the Italian client whose revolt in December 1894 precipitated war: 'Bāhta, Bāhta, Bāhta of Sāgānayt: Master of the Italians; mastered Sanguinetti. Are women born of [such] a lion?'[1] Bāhta's revolt led to the Battle of Adwā, and within a generation subjects of Eritrea regarded Bāhta as a martyr for the motherland, Ethiopia.[2] In this guise he has been officially praised since the incorporation of Eritrea into Ethiopia.[3] Yet for all its drama, his brief revolt is most notable for the completeness of its failure. Nonetheless, the whole of his career is of interest because as a commoner he rose to high position in northern Ethiopia where the provincial nobility and gentry monopolised all secular avenues for advancement.

Farmer's son and an outlaw

Neither side of Bāhta's family was linked by elders to title-holders. He was born to wealth, they told a researcher in 1970, and spent his childhood herding some of the numerous cattle his father acquired after settling in Sāgānayti, the chief place of his native district.[4] Bāhta married a woman of the village of no greater consequence than himself apparently having waited until he was in his early thirties. This was just before he became an outlaw in 1875.[5] By then his father, Hagos Andu, had been dead for ten years. Hagos had been the victim of a judicial murder in the struggle by his neighbours to preserve the semi-republican traditions which set Akkala Guzāy off from the rest of the trans-Marab ('Marab-mellāsh' to highlanders) and from most of the other Tegreññā-speaking areas as well.

The communes of Akkala Guzāy were grouped into seven houses or clans. From the reign of Tēwodros II (1855–68), the headship (*shum-gult*) of each became undisguisedly hereditary, with the emperor rather than an assembly of heads of communes confirming succession to the civil title of *kantibā*. All adult males in a locale continued to meet to choose the head of their commune (*shum-addi*), although first-born sons ordinarily succeeded their fathers in this office also. In the hope of subverting the communal assembly of Sāgānayti, Hagos Andu allied himself with Tēwodros's governor in Hāmāsēn and Sarāyē, *Dajāzmāch* Hāylu of Sa'āzagā, who helped a friend of Hagos obtain the title of *kantibā* and autocratic

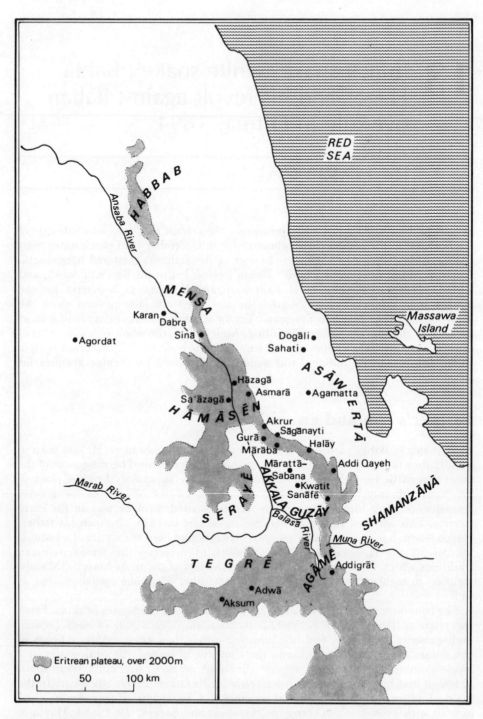

Map 4: *Ethiopia north of the middle Marab: Marab-mellāsh*

powers in the Kārnasham district of neighbouring Hāmāsēn.[6] This probably took place in early 1861. Traditions say that four years later Hāylu's great rival in his struggles for primacy, Walda Mikā'ēl, from the adjacent village of Hāzagā, rebelled, and went to his wife's kinsmen in Akkala Guzāy. Fearing Hāylu as an ally of the overweaning Hagos Andu, the office-holding families there threw in their lot with Walda Mikā'ēl, and with him suffered a terrible defeat. Alone Walda Mikā'ēl fled south of the Marab to Adwā where he put the future Yohannes IV (1872–89) in his debt by joining his attacks upon partisans of the Emperor Tēwodros.[7] Meanwhile, Hagos Andu's neighbours had convened an assembly to denounce his machinations with Hāylu, and killed him. The reputation of being the son of an over-ambitious upstart was remembered once Bāhta became a favoured Italian *protegé.* Tradition also records that the people of Akkala Guzāy later rose and killed a local *protegé* of Walda Mikā'ēl, whom he tried to place over all their seven houses.[8]

Almost nothing is known about Bāhta's life when he farmed at Sāgānayti in his father's stead. However, at some point he did convert from the Ethiopian Orthodox Church to the Roman Catholic Church. Informants say that he did this as an adult after the death of Bishop Justin de Jacobis, who died in 1860.[9] The clergy and many of the inhabitants of Sāgānayti transferred their allegiance, and their buildings and tithes, in June 1866. The Lazarist missionaries themselves report the hardships from which converts were escaping: locusts, drought and pillage by factions among the gentry of the Marab-Mellāsh as Tēwodros's authority decayed in the last years of his reign. The people rose under their *kantibā* against the soldiery, but quarrels among the villagers paralysed defence.[10] Many joined in the rapine.[11] Others sought aid from the French missionaries.[12] As a result, the many new Catholics of Sāgānayti, some 900 by the 1870s,[13] were subjected to more than the usual oppression of soldiers and officials. On the eve of his coronation as emperor, the soon-to-be Yohannes tried to restore by force the authority of his Orthodox bishop. A new school at Sāgānayti was burned down in November 1871 and the newly harvested crops carried off. Many in eastern Akkala Guzāy recanted. Those who hid rather than abjure allegiance to the Catholic bishop were left 'not a calf or a goat', an Ethiopian Catholic eyewitness later recalled.[14] What happened to the herds inherited from Hagos Andu is not recorded. But Bāhta, who later in life was an intensely devout Catholic, probably did not lapse.

Bāhta Hagos became an outlaw (*sheftā*) on the eve of the Ethio-Egyptian War of 1875–6 because of a blood feud with the family of Emperor Yohannes's uncle, Arāyā the Elder. Yohannes had appointed Arāyā as overlord (*shum-nagārit*) of Akkala Guzāy. In the first week of October 1875, Arāyā's 18-year-old son, Embāyē, visited Sāgānayti to demand money, having milked the part of the province assigned to him. He shoved aside the commander of the troops his father had quartered there, wounded the Catholic priest when he tried to intervene, and killed a brother of Hagos Andu who grabbed hold of him. This provoked fury. Embāyē's retainers fired, but the enraged villagers allowed them no time to reload their single-shot muskets, before driving them out of Sāgānayti with staves and spears. The villagers killed two dozen. Half a dozen men of Sāgānayti also died, including other kinsmen of Bāhta. Bāhta now took vengeance for his family, killing the prince with a spear thrust.[15] With the other young men of the village Bāhta and his two younger brothers, Kāssu and Sangal, fled into the broken country below the escarpment with some of Embāyē's firearms.[16]

Most of the fugitives soon returned to their farms. However, in the latter part of 1876 Alulā, trusted lieutenant of Yohannes, burned Bāhta's house and those of his relatives and confiscated their cattle and other property, following the emperor's triumph over the Egyptians. Alulā had brought his army from Tegrē through Akkala Guzāy in pursuit of Walda Mikā'ēl of Hāzagā, Walda Mikā'ēl having welcomed the invaders.[17] Armed by the Egyptians, Walda Mikā'ēl killed Hāylu of Sa'āzagā and ravaged the countryside.

From Alulā's first passage through the highlands of the Marab-mellāsh, Bāhta and his brothers made a permanent camp at Agamatta (or Agamada) among the Muslim Asāwertā, a branch of the pastoral Afār, midway down to Massawa and overlooking the caravan route between Akkala Guzāy and the coast.[18] The brothers added to their stock of firearms and ammunition by waylaying the escorts Aiāyā and Alulā sent with traders. They also despoiled the caravans. Until then they had simply been peasants according to Bāhta's confessor, *Abbā* Kefla Māryām.[19] Now they were *sheftā*. And a violent life it was; at about this time Bāhta executed his brother, Kāssu, in a quarrel which involved both disposition of plunder and Bāhta's authority as elder to Kāssu. Four years after Bāhta's fixing his camp at Agamatta, in 1880 Alulā's deputy drove Bāhta and Sangal to take refuge near the Egyptian forts at Karan in the Ansabā valley north of Asmarā.

The martial skills which the brothers had learned since 1875 stood them in good stead in their new haunt. The Egyptian commander at Karan recruited them after the rains of 1880 to hamper. Alulā's sweeps through the Ansabā for livestock and other taxes. On behalf of the emperor, in 1882 Alulā imposed tithes upon the pastoral Muslim Habbab for the monastery of Dabra Sinā. Bāhta and Sangal were sent to organise the Habbab resistance, and with them looted Dabra Sinā and killed some of the monks.[21] They found many other opportunities as freelances. The Habbab and their sedentary neighbours were accomplished cattle raiders and 'lived in enmity and war'.[22] The Habbab, Mensā and other Tegrē-speaking peoples of the Ansabā gloried in exploits of daring and theft.[23] However, they did not know well the use of firearms.[24]

Political events soon broke Bāhta's ties with the Habbab. Egypt's African empire collapsed. In September 1885 Alulā and his allies in the valleys of the Ansabā and the Gāsh defeated adherents of the Sudanese Mahdi. The Habbab leaders now went down to Massawa to sign a treaty of protection with the Italians who had landed in February replacing the bankrupt khedive. In reprisal Alulā and his allies scourged the Habbab lands both before and after the rains of 1886.[25] Outlawry had become most hazardous.

Collaborator

The first Italo-Ethiopian war broke out in January 1887 when Alulā surprised a relief column at Dogāli in the coastal plain. The war brought malcontents from the gentry of the Marab-mellāsh into Italian service petitioning for aid against rivals, and pursuing the possibilities which colonial service opened for those without customary claims to fiefs and appointments. In March 1888, Yohannes besieged the forts of the Italian expeditionary corps at Sahati near Dogāli, but the siege failed. The expeditionary corps was then repatriated because of the fierce summer heat and the cost, and Italian officers took in hand the irregulars (*bāndā*, Ital. sing.). Before the end of 1888, an Akkala Guzāy *bāndā* had been created out of refugees on the coast. At the beginning of 1889, the Massawa command made its first agreements with Bāhta who had come down to the coast like many others.[26] They had him station himself at his old hideout in Agamatta to keep the Asāwertā from the influence of *Rās* Arāyā's son Dabbab.[27]

The emperor's cousin had become a bandit chief among the Asāwertā in 1882, being peeved with the mediocrity of his title *fitāwrāri*. But hardly had Yohannes mobilised his vast army of February 1888 than Dabbab slipped away to the emperor's camp, taking several hundred rifles the Italians had given him to arm recruits. This won him the title *dajāzmāch* and, on the imperial army's withdrawal before the rains, the government of Akkala Guzāy. 'I wished to be the only Chief without any other Chiefs on my side', he wrote explaining his defection. 'They made other Chiefs, that made me angry, and I decided to desert them.'[28] This is the familiar pattern of noble rebellion.

In August 1888, an attempt by 800 Asāwertā and other *bāndā* to surprise Dabbab at Sāgānayti ended in the deaths of more than 300 of the attackers and all five of their Italian officers. The attackers had outnumbered Dabbab's retainers, but the villagers were alerted by the Asāwertā firing on some shepherds and the whole population rose, women as well as men, in the defence of the village.[29] Reportedly Dabbab promised outlaws the emperor's pardon in scraping together reinforcements.[30] Yet even when it was learned that Yohannes and Arāyā the Elder had been killed fighting the Sudanese in March 1889, the Italian government dared further intervention in the highlands only through stipendiaries. Thus they helped Walda Mikā'ēl's son-in-law, Kefla Yasus, oust Alulā's lieutenant from Karan. And once more they armed Dabbab when he returned to Tegrē before the rains of 1889 to challenge Yohannes's appointed heir, his bastard son, *Rās* Mangashā. In return Dabbab had invited the Italians at Massawa to occupy Akkala Guzāy themselves. The Massawa command was still inclined to rely upon his legitimacy there when, in July 1889, Bāhta wrote with evident satisfaction that Dabbab had walked into a trap laid for him by Alulā and had been arrested. Akkala Guzāy lay open to the parvenu.

Until he was killed in September 1891, Dabbab Arāyā had many followers in Akkala Guzāy. Bāhta could muster only some 300 riflemen when protecting the flank of the Italian advance to occupy Asmarā on 2–3 August 1889. He had the further task of bringing over to the Italians the many leaders of his home province who remained undecided. A sudden mobilisation in mid-August to check Alulā's belated attempt to repossess himself of his command in the Marab-mellāsh gave Bāhta and other leaders of *bāndā* in Akkala Guzāy a chance of proving anew their usefulness. The usefulness of the *bāndā* was coming into question as the Italians began regrouping recruits under Italian officers and subjecting them to European-style drill and discipline as uniformed regulars called *askāri*. On 1 January 1890, Massawa and its dependencies were renamed Eritrea. For a few days at the end of that same month the governor of Eritrea occupied Adwā to vindicate the honour of Italian arms on the anniversary of the 'massacre' of Dogali. Less than half of his force of 5100 were Italians or *askāri*. The majority were *bāndā*. Bāhta's following had grown to 700 riflemen.

At the beginning of May 1889, a bargain struck between Menilek and the Italians at Wechālē in Wallo gave Menilek peace and closed Massawa to the arms trade to all but him, while Menilek ceded Asmarā with most of Hāmāsēn, the Ansabā valley, and much of the coast to Italy. Akkala Guzāy south of Halāy and Sāgānayti was left to Ethiopia. Early in November 1889, however, soldiers sent into the upper Munā by Menilek's governor of Agāmē, *Shum-Agāmē* Sebhat Aragāwi, ran into Italian irregulars and Italian officers. The Italian patrol forced the Agāmē to retire from Shamanzānā, a place long contested between the governors of Akkala Guzāy and of Agāmē. With the withdrawal of the governor of Eritrea and regular troops to Asmarā and Karan in February 1890, all Akkala Guzāy, including those parts which had not been assigned to Italy by the Treaty of Wechālē, were recognised by the Italians as dependent upon Bāhta whom they installed at Sāgānayti with the title *dajāzmāch*. Already by the end of December 1889, he had recruited *bāndā* to garrison Gurā for him and obtained vows of loyalty from notables commanding armed men in other districts where the *kantibā* of Sāgānayti had never exercised authority.

While respecting the propriety of candidates being from the office-holding families as Tēwodros and Yohannes had, the officer attached to Bāhta completed the disenfranchisement of assemblies by appointing heirs to the headship of the communes as well as investing heads of clans as Tēwodros and Yohannes had done. The tax rights of suzerainty also expanded further by denial of all earlier exemptions. Informants explain that custom limited the power of the heads of communes to presiding over the adult males of the village whenever an important matter of general concern was brought to the attention of his spokesman. This agent, whom the *shum addi* appointed, convened the assembly. In day-to-day affairs, village elders acted without involving the *shum addi*'s agent. In nothing of importance could a *shum addi* decide matters on his own. Bāhti tampered with this essential. To the admiration of the Italians, he settled debate with a word. In his presence litigants assumed the same obsequiousness as in the aristocratic traditions of Hāmāsēn, Tegrē and the Amharā lands.

Baronial powers could be pursued in Akkala Guzāy only with outside support. Bāhta's sponsors gave him ammunition, cash and stores of rifles, for which men readily volunteered as retainers throughout Ethiopia. Italy's clients also received supplies of grain. This was of great consequence in 1889–90. Livestock sent from India for the Italian expedition in 1887–8 had introduced cattle disease. This disease soon spread to the highlands on both sides of the Marab. Plough oxen and herds as far south as the Blue Nile bend and Shawā died. Waves of destitute men, women and children migrated from the Tegreññā-speaking lands towards the Italian garrisons, begging succour from the stocks of grain they imported by sea. The ranks of the *askāri* and *bāndā* swelled with volunteers driven by hunger and the respect which bearing arms enjoyed as a manly occupation. At the end of 1889, Bāhta appealed to the Massawa command for protection against Sebhat of Agāmē. Famine in Agāmē forced this mighty neighbour to ask for a truce, biographical traditions of Sebhat's descendants recall, so that he could safely send guards to buy grain when a surplus suddenly became available in southern Akkala Guzāy.[41]

In quick succession, when they had served their purpose, the Italians disarmed, detained, and, in some cases, executed Kefla Yasus and other collaborating notables from Hāmāsēn. Those forewarned went into exile in Tegrē or retired to the borderlands. Some made their way to the court of the new emperor in Addis Ababā. Arrests continued into 1891–2. But Bāhta's fortunes met no such reverse. In the early 1890s, he had a stock of 2000 rifles and had issued many more of them than the 500 *bāndā* he was authorised to raise. The Italians accounted him 'the most powerful leader in the colony'.[42] Yet his people were wretched. Just before the rains of 1891, an Italian parliamentary commission saw signs of great misery in Akkala Guzāy. Bāhta told the visitors that much land had gone out of cultivation for want of oxen. Women had adopted the digging stick and planted a few crops. But many still could not afford seed grain two years after the beginning of the Great Famine.[43] Bāhta himself prospered. He pocketed the stipend paid by the. Italians for the upkeep of the *bāndā* and paid his soldiers very little. Nonetheless, all the able-bodied young men flocked to be enrolled for he feasted them twice daily in the Ethiopian fashion of a great lord.[44] His part of the grain levy, labour services to work his fields, and cash from court and market fees made him rich. It was remarked, however, that he was more feared than loved.[45]

Politically Bāhta was not successful in making good his pretensions to lordship. *Kantibā* Metalkā of Mārattā-Sabana south of Sāgānayti scorned his pretensions to govern the whole of the Seven Houses: 'I do not know any Bāhta here . . . I do not know any person above me here'.[46] Metalkā's retainers held off Bāhta's soldiers when they arrived to arrest him. The quarrel was settled only formally by the marriage of the *kantibā*'s daughter to Bāhta's eldest son whom Bāhta made a *grāzmāch* and put in charge of part of the *bāndā*. His eldest daughter, Tamnit, was married to the son of *Kantibā* Asmerom of Mārabā just west of Sāgānayti, another neighbour proud of his autonomy under Italian lordship. The marriage of Bāhta's second daughter, Lamlam, to Menilek Sebhatu, the son of another Italian-made *dajāzmāch*, Sebhatu Bakit of Kārnasham, strengthened Bāhta's ties with the descendants of his father's hated Hāmāsēn ally.[47] Both of these sons-in-law were required to become Catholics which suggests that the grooms' families valued the alliance. However Asmerom was biding his time. As his brother's second-in-command, with the title *azmāch*, Sangal Hagos was responsible for all the southern parts of Akkala Guẓāy to the Balasā and the Munā. By treaty these remained Ethiopian. In fact people there were loyal to local men and ignored Sangal's authority while they betrayed no knowledge of Menilek's. When visiting Asmarā, Bāhta cut a fine figure parading with the pomp of a great Ethiopian lord, whatever the reality of his political powers. Trumpeters and a shield-bearer preceded him with men-at-arms. A hundred riflemen escorted him as he rode a white mule with the silver trappings reserved in Ethiopia to the nobility. He lacked only the kettle drums (*ngārit*, Eth. sing.) of a genuine *dajāzmāch* of highest rank in the empire.

Before returning home for leave in June 1893, General Oreste Baratieri, the governor of Eritrea, disbanded a number of irregulars including the largest unit, that of *Dajāzmāch* Sebhatu Bakit of Kārnasham who was perhaps Bāhta's closest friend outside of his family. Only the Mahdist threat to the colony during the governor's absence, and the adventurism of careerist officers, interrupted replacing irregulars led by clients with *askāri* and a militia reserve. Bāhta campaigned to Agordat in late December 1893 and to Kassala after Baratieri's return in July 1894. The colony's frontiers had been pushed to the middle Marab, the Balasā and Munā rivers in the south despite Menilek's bitter objections. The occupation of Kassala so enlarged them that the Italian command dared not disband any of the 10 000 soldiers who were available in Eritrea in 1894. Besides, of all the freebooters who had given the Italians such easy access to the highlands in 1889, Bāhta was thought the least likely to betray them.[49]

The revolt

Informants single out Sebhatu's fall as the signal which set Bāhta to search for allies below the Marab and in Addis Ababa.[50] In September 1894, an Italian agent at the capital discovered that Menilek, crowned emperor in November 1889, had been in correspondence with Bāhta Hagos since 1893. Italy's favourite client had petitioned Yohannes's imperial successor for pardon and laid claim to both Akkala Guzāy and Hāmāsēn. Bāhta also made overtures to Mangashā Yohannes for an alliance against the Italians but had received no encouragement before Mangashā, *Rās* Alulā, and other commanders from Tegrē went to pay homage to Menilek in May–June 1894.[51]

News of Mangashā's visit made many of the notables of the Marab-mellāsh restive. The display of concord between the emperor and Mangashā, Mangashā's equally public attempts to get Italian aid, was expected to presage some great event in the north.[52] Bāhta may also have become anxious because of the welcome at the imperial court of the leader of the House of Sa'āzagā, Hāylu's grandson Abarrā. Abarrā had eluded arrest in Hāmāsēn when the Italians sought to cast him off and fled to Tegrē. There, before accompanying Mangashā to Addis Ababā, he was reconciled with Walda Mikā'ēl of Hāzagā, the latter remaining in forced retirement at Aksum until his death in 1906. In this way the feud which had torn the Marab-mellāsh apart for over a generation before the arrival of the Italians was formally ended, and Abarrā of Sa'āzagā became an attractive candidate should Hāmāsēn be restored to the empire.[53] In the *rās*'s absence, Bāhta pressed Mangashā's deputy in Tegrē for a response to his request for an alliance and got courtesies in answer. During the rainy season following the visit, Bāhta sent a messenger to Menilek announcing that he could no longer tolerate foreigners in the lands of his fathers. Before a non-commital reply could be dispatched, the more heated message arrived: 'Will Your Majesty constrain me to remain a slave to the Italians forever?'[54]

Those Tegreans who remained at the court argued that if nothing was done the whole north would fall into Italian hands as so many young men had crossed the Marab to enrol as *askāri*. According to gossip, the Shawāns suspected the northerners of wanting to embroil them in a border war in order to weaken Menilek's hold over the empire. Mangashā wrote from Tegrē early in the rainy season of 1894 asking to be charged with liberating the trans-Marab and was told that Menilek wanted no fighting. 'What is to be done, as we discussed,' Menilek replied in August, 'is for you to send repeatedly messages from *Maskaram* [the month in which the rains end] and as the country is evacuated, occupy it.'[55] Bāhta's messengers were sent back after the rains with strict warnings that he should not rebel if he wished to be counted among the emperor's friends. Menilek explained that he wanted nothing to jeopardise negotiations with Rome for a peaceful return to the frontiers laid down in the Treaty of Wechālē and its annexes.[56]

By early December 1894, a conspiracy existed for Bāhta to raise a general rebellion to greet a Tegrean advance into the colony. Some have claimed that this was a Catholic plot. The Italian, General Baratieri, civil governor of Eritrea and defeated commander at the Battle of Adwā, first put forward this self-serving claim, alleging that the French Lazarists had engineered the conspiracy because he was having them replaced by Italian Capuchins. An Eritrean long in Italian employ, Gabra Mikā'ēl Germu, in his study of Ethiopian relations with Italy, accepts this claim. But correspondence sequestered by the general does not bear this out. Nor did the Capuchins, who landed at Massawa on 2 December, believe it. The arrangements, such as they were, were the work of Mangashā's right-hand man, Tadlā of Aybā. Menilek had not been informed. And no time had been fixed for Bāhta to act, although the Tegreans were mobilising and had agreed to assemble near the Balasā.[57]

In the first days of December 1894, disquieted by the spy's report from Addis Ababā, Baratieri suddenly warned the new resident at Sāgānayti, Lieutenant Giovanni Sanguinetti, that he could no longer count upon Bāhta's good faith.[58] Baratieri was at Karan supervising dispositions against the Mahdists and could not disband Bāhta's irregulars because he expected an attack upon Kassala. At Sāgānayti on the evening of 14 December, Bāhta, his brother Sangal, and Bāhta's son, Gabra Madhen, went as usual to have coffee with Sanguinetti. They threw the resident to the ground, bound him and arrested the two Italian telegraphists and Sanguinetti's secretary, Gabra Egzi'abehēr. They also cut the telegraph wires north to Asmarā. 'God will punish you; Italy is great', Sanguinetti supposedly warned. 'Ethiopia is greater yet', reportedly was Bāhta's retort.[59] Gabra Mikā'ēl Germu's informants seem to have given a more colourful account, for he says that, sitting on the resident's chest, Bāhta taunted him: 'You with eyes of a cat, hair of a monkey and with white lips; you donkey of the sea who have come having crossed seven rivers, do you not know yet that the Ethiopian empire is superior to Italy, that the Ethiopian people are twice as numerous and twice as brave as the Italians, . . . You donkey!'[60]

The sources put forward a variety of explanations for Bāhta's sudden action. The French missionary at Akrur near Sāgānayti heard that on arriving at the resident's house Bāhta had seen Sanguinetti strike a notable from Degsā with the butt of a rifle. 'I will hit you too', Sanguinetti warned when Bāhta interfered, thus provoking the trio to act.[61] According to his confessor, *Abbā* Kefla Māryām, and others who knew him, Bāhta was bitterly at odds with the new resident, because Sanguinetti had taken a prized rifle from Bāhta. He had also taken the wife of one of Bāhta's retainers as his concubine. When Bāhta supported the aggrieved man, he was told: 'Let it be understood I can take any woman who pleases me, even your wife if I want.'[62] This may be a folk explanation equating sexual licence with political excess. In Akrur and elsewhere in eastern Akkala Guzāy the older generation knows the story of Bāhta's wife objecting to his carrying out orders to supply handsome young boys and girls for a fair in Rome. His youngest daughter, they say, set the example by refusing to go. They claim that carrying women off and sexually abusing them caused the insurrection.[63] A variant recounted by Bāhta's granddaughter has his wife urge him to obey orders. He stilled her by asking, 'How would you like to us to have to send our own children?'[64] None of these circumstances is mentioned by the resident's secretary in his eyewitness account. He confirms Baratieri's grudging admission that Bāhta's disaffection was part of the rise of nationalist sentiment. The brash jingoism of the Italian officers in the colony had offended collaborators. To this extent Baratieri was right to say that a warlike people proud of their antecedents could not be untouched by living in proximity since 1885 with such civilised notions as having a fatherland.[65]

In the course of the uprising, Bāhta left no doubt that this was no mere twist in the career of an opportunist. He made himself the spokesman for wider discontents than those of Italy's disgruntled client governors. Failing to win over Sanguinetti's secretary, Gabra Egzi'abehēr, by offers of fitting him out as commander and providing his blind mother in Asmarā with a pension, Bāhta angrily admonished his misplaced loyalty: 'The Italians curse us, seize our land; I want to free you . . . let us drive the Italians out and be our own masters.'[66] That very night, 14

December 1894, in calling the people of Akkala Guzāy to mobilise, Bāhta had a similar sounding edict published on the market place:[67]

> I have freed you from that government which has come from overseas to despoil you, to take your lands, to prevent you from farming your *rest* without paying taxes, which prohibits you from cutting firewood in the forests.

Eritreans early in this century told the story of how when Sangal hesitated to betray a lordship to which both had sworn, Bāhta dismissed his scruples: 'Oh my brother, Sangal, don't be silly! Once a white snake has bitten you, you will find no cure for it.'[68] Baratieri recalled that on both sides of the colonial border the saying circulated: 'You can recover from the bite of a black snake, but you will never get better from the white snake's bite.'[69]

The setting aside of land for Italian settlement from 1892 had raised a popular clamour. 'The *rās* raid our crops and stock whenever they can,' one group complained, 'but they leave us our fields which we can plant anew . . . the Italians take even the land . . .'[70]

The Italian dream of redirecting the flow of emigrants from the New World to Eritrea soon proved illusory. However, Baratieri had rushed to claim as state property lands abandoned by cultivators since the cattle disease of 1888. In 1893 some 19 000 hectares had been reserved for European use. In 1894 almost fifteen times that amount was set aside. This last action went far beyond anything the civilian commissioner's modest schemes of colonisation called for, and was done precisely because in 1894 the Eritreans were reported to be taking a renewed interest in the lands vacated when their plough oxen had died.[71] With the renewal of prosperity and of the population after 1892, Eritreans were reoccupying their farms when the first half dozen Italian families arrived in November–December 1894 to settle in the highlands. Local informants told the Capuchin missionaries that Bāhta was truly afraid that Italy would deprive the people of all their inherited rights in land (*rest*).[72]

'My helpers are all Ethiopians', Bāhta bragged on 14–15 December.[73] He told Gabra Egzi'abehēr that the emperor and *Rās* Mangashā Yohannes were behind the revolt. The Muslim Asāwertā as far as Massawa would join him he predicted, and all the *askāri* in Italy's pay. Even the Mahdists were ready, he said.[74] By mid-December, the Tegreans were assembling and expected Bāhta to arrest Sanguinetti and cut the telephone line. Bāhta called upon these allies only after carrying out his part of the plan.[75] But the revolt at Sāgānayti neither united nor inflamed the Eritreans despite its appeal to pent up resentments of both high and low. It was over quickly. This is not the only reason it did not become a general insurrection.

On hearing at Karan on 15 December 1894 that the telegraph south of Asmara had been cut, Baratieri telegraphed Asmarā ordering Pietro Toselli to march his *askāri* into Akkala Guzāy. These regulars loyally followed their commander whom they much admired. With a swiftness which European troops could never match, the *askāri* arrived within sight of Sāgānayti the very next day, 16 December. *Kantibā* Asmerom made no effort to delay them as they passed through his residence. For two days he supplied Toselli's men while they stalled the rebels with talk.[76] Toselli estimated that Bāhta had as many as 2000 riflemen with him.[77] At dawn on 18 December, Toselli dropped the charade of negotiating and advanced into Sāgānayti. It was empty. During the night, Bāhta had eluded the *askāri* watch and gone to attack Halāy.

Eritreans explain that Bāhta decided to move closer to the frontier when no word came from his Tegrean allies on the fourth day of the rebellion. First, he wanted to take the arms and other munitions stored at the small fort at Halāy.[78] Only 250 *askāri* were on duty there. But Bāhta delayed an assault. He sent his two Tegrean Catholic priests, *Abbā* Kefla Māryām and *Abbā* Takla Hāymānot, who had negotiated with Toselli, to parley again. They offered safe-conduct to Asmarā if the Halāy *askāri* would lay down their arms and surrender the fort's magazine intact.[79] Bāhta seems to have ignored the lesson from Yohannes's unsuccessful siege in 1888 of Dogāli in the coastal plain, this siege having shown the advantage European-led troops had on the defensive in fortified positions. Perhaps Bāhta was just overly confident of the effect on morale of his having arrived with a force overwhelmingly more numerous than the defenders.

The Italian commander at Halāy drew out the *pourparlers* in order to give Toselli time to overtake the rebels. Consequently, not until the beginning of the afternoon did Bāhta attack. For nearly three hours the *askāri* held off Bāhta's force of some 1500 men. When the besieged were at last on the point of succumbing, the attackers found themselves caught in crossfire. Encumbered with four guns, some 1200 *askāri* and *Kantibā* Asmerom's retainers had marched from Sāgānayti while Bāhta talked. Mist hindered the relief. The rebels were put to rout nonetheless. Eleven *askāri* died; twenty-two were wounded.[80] The *bāndā* of Akkala Guzāy and their allies dispersed pell-mell, leaving their leader behind among the dead.

Some said that Bāhta had fallen in flight. Others insisted that, once he saw his army abandoning the field, he stood his ground with determination in order to set an heroic example. Elders in 1970–71 quoted his despairing curse: 'Ah! Akkala Guzāy! May you bear only women . . .!'[81] Bāhta's confessor and the other priest had stayed with him, 'burning with love for their country', Gabra Mikā'ēl Germu contends.[82] However, the defeat at Halāy and the loss of their commander disheartened Bāhta's soldiers. Reaching Sāgānayti too late to be used by the Italian command in delaying battle, the new Capuchin bishop apostolic Michele da Carbonara obtained the surrender of 1200 rifles in exchange for an amnesty for those who returned peaceably to their villages and disarmed.[83] Of the *bāndā* of Akkala Guzāy only 300 had acquitted themselves with sufficient reliability towards the Italians for Baratieri to let them retain their arms.[84] Nonetheless Bāhta's son, Gabra Madhen, tried without success to organise guerrilla forces in the neighbourhood of Sāgānayti to which he returned. Discontent with colonial rule grew but led to migration of the Habbab *en masse* into Mahdist territory in early 1895, or uncoordinated acts of violence.[85]

Informants recall that while Baratieri was disarming Akkala Guzāy, Bāhta's son swooped down upon Sāgānayti to burn the house of one of his father's lieutenants, *Qañāzmāch* Mikā'ēl. Mikā'ēl maintained his own *bāndā* with a small stipend which the Italians paid him and informed on Bāhta whose position he coveted. He remained secretly loyal to the Italians on 14–15 December making it possible for the resident's secretary to preserve his master's personal possessions and cash box against the rebels. The Italians rewarded Mikā'ēl by appointing him over Sāgānayti with the rank of *dajāzmāch*, although he too was a fervent Catholic. He soon fought for them in the war with Menilek.[86] Bāhta's son also burnt the whole village where *Kantibā* Asmerom lived. Two other places which had helped Toselli suffered a like fate. Some of Bāhta's followers remained in hiding until after the Battle of Adwā (1 March 1896), when a Catholic priest from Addigrāt arranged for an Italian pardon. The basis for a guerrilla movement did not exist. Gabra

Madhen Bāhta had to go to live as a *sheftā* among the Asāwertā, later slipping away to the Tegrē in time to join Menilek's armies which arrived there in December 1895–January 1896.[87]

The indifference of people in Akkala Guzāy to Bāhta's revolt is documented by Gabra Egzi'abehēr's account of his peregrinations with Sanguinetti after 14 December. Almost at once the hostages had been sent towards the Balasā under escort of Bāhta's brother-in-law. *Bāshā* Tasfā Māryām. While the Battle of Halāy was raging and afterwards, but before the news of Bāhta's death became known, the column of prisoners passed through areas in southern Akkala Guzāy which were little excited by the insurrection. When it did become known that Bāhta had been killed, Tasfā Māryām began to bargain with his prisoners for pardons for himself and the *bāndā* of the escort. While Gabra Madhen was nearby trying to rally partisans, peasants in one locality tried to chain the escort in order to claim a reward from Baratieri. Even before they learned that the *askāri* were burning villages suspected of sympathy with the rebellion, Tasfā Māryām's party had begun to circle back towards Sāgānayti. Only at this point did the *bāndā* of the escort argue among themselves about whether they ought not turn the hostages over to Gabra Madhen, Bāhta's son. Tasfā Māryām and the second-in-command, *Bāshā* Taklē, chose to deal with the Italians. Thus before Gabra Madhen's incendiary descent upon Asmerom's residence at Marabā, Sanguinetti, the two Italian telegraphists, Gabra Egzi'abehēr and Sanguinetti's servants arrived there led by Bāhta's two lieutenants with the escort in tow.[88]

The hostages were warmly welcomed by *Kantibā* Asmerom. Bāhta's brother-in-law presented himself to the Italians as a candidate for Bāhta's position in Akkala Guzāy. They rejected Tasfā Māryām according to Akkala Guzāy traditions with the rebuke: 'Those who betray their own brothers cannot serve us.'[89] This says more about the moral which informants wished to draw, than it does about Italian practice. The rivalry of near kin was ingrained from competition in feudal service. Similarly, the loyalty expected of personal retainers made it natural that Sanguinetti's secretary and his household servants should have remained with him. Gabra Egzi'abehēr had been under-secretary at the age of 13 to Dabbab Arāyā and, in 1889, briefly shared his imprisonment in Tegrē before escaping to Asmarā. He regarded as a second father the officer who first took him into service at a time when not to be in Italian employ was to risk starvation. Eventually he did defect, after Adwā, but it is not surprising that he was deaf to Bāhta's appeals in December 1894, and continued to serve Sanguinetti until their release.[90] Given the ties between a commander and his men in Ethiopia, there is nothing far-fetched in the Italians' reporting that *askāri* on leave in Akkala Guzāy ignored Bāhta's edict and reminded him that the Italian officers they served were their leaders, not he.[91] The failure of others to join the rebellion had less obvious causes.

The Lazarist priest at Akrur had supposed at the start of the revolt that the Asāwertā would aid Bāhta.[92] No such thing happened. One section came to fight for Toselli and their leader, Adām, claimed the governorship of Akkala Guzāy as a reward.[93] A foe of Bāhta's was readily found to replace the *kantibā* of Halāy, who, along with five others implicated in the revolt, was tried and executed in February 1895. The new appointee, *Qañāzmāch* Mahāri, remained unswervingly at the Italians' disposal, hoping for Bāhta's position in the weeks following the Battle of Adwā, when the colony was all but defenceless except for such *bāndā*.[94] Underlying Bāhta's failure to arouse greater sympathy is something more fundamental than the petty ambitions of an overly numerous gentry. By 1893, he had many enemies:

among the office-holding families because he menaced their local independence; among the commonality because they regretted the growth of autocratic power. The Italians observed that his paramountcy rested upon military domination.[95]

As was natural, Bāhta recruited many of his kinsmen into the *bāndā*. His rule came to be seen as favouring his father's home district, Sana Daglē, of which Sāgānayti is the centre. He turned a blind eye when the people of this district, many of them related to him, got the better of the Asāwertā in fracas which had been part of competition over pasturage along the escarpment for many years. The Asāwertā found their once-flourishing herds greatly reduced and themselves without redress. Sanā Daglē alone was excused from time to time from taxes, informants say, and Bāhta allowed the Catholics there to apply what they owed to the building of churches. He built others himself for the Catholics in this favoured part of eastern Akkala Guzāy. He drove out Orthodox clergy who dared challenge his high-handedness in the district and transferred parishes and their revenues to the Catholic clergy there. Thus there was dissension even in the one district where he is well remembered today. 'We destroyed you when our heart was cruel,' a lament ascribed to Bāhta at Halāy runs: 'Today we are wanting you when we are in trouble.'[96]

After Bāhta's death, most of his die-hard followers had fled south with Sangal and *Fitāwrāri* Tasfu of Kwatit (Coatit), who had joined in the attack upon Halāy. With Sangal, he crossed the Balasā to Mangashā's camp between Addigrāt and Adwā. They brought the Tegreans a remnant of some 500 men.[97] Tasfu re-crossed the border less than a month later, on 12 January 1895, in an army of 12 000 riflemen and 7000 spearmen which Mangashā Yohannes and his lords had raised. This imposing army marched to Kwatit intending to snap the lines of communication of General Baratieri who in late December 1894 had occupied Adwā once more.

No general insurrection welcomed the Tegreans and their allies from the trans-Marab. However, the governor of Eritrea had only some 3200 *askāri* and less than 650 trustworthy *bāndā* for the defence of the colony's southern frontiers. He reached Kwatit ahead of them. The Italo-Eritrean force was much outnumbered; but it was more plentifully supplied with ammunition, and alone had artillery. And the *askāri* fought tenaciously for paymasters whose generalship justified, in this part of the second Italo-Ethiopian War, the attachment which their courage inspired. In a second battle, at Sanāfē in southern Akkala Guzāy later in January 1895, Mangashā's great army fell apart. Hardly firing a shot, it fled back into Tegrē. Far from liberating the trans-Marab, the events of December 1894–January 1895 exposed the whole north to partition. It was to halt this that Menilek marched to Adwā.

For his service in the second Italo-Ethiopian War of 1895–6, Menilek made Sangal Hagos a *dajāzmāch*. In 1900, he obtained a pardon from the Italians who recognised this title and nominated him a notable on his return to Sāgānayti where he died in 1924. *Kantibā* Asmerom's son, Tasammā, already in 1894 a deputy for him, rose steadily. Under the British occupation after the war of 1935–41 he became a *rās*, a pre-eminence he lent to the Unionist Party, the principal Eritrean organisation of the later 1940s and early 1950s which actively supported a political link with Ethiopia.[98]

Bāhta Hagos acted out a career typical of many northern Ethiopian leaders.[99] Exceptional in his humble origins, he nonetheless started his independent activities in 1875 with an act of rebellion, which was provoked by the defence of his family's honour in a blood feud. He took to the bush and for over ten years lived as a *sheftā* or brigand, preying on caravans, and allying himself with pastoral lowlanders. The creation of the Italian colony of Eritrea in 1890 provided this enterprising commoner with a chance for political advancement. Bāhta led irregular troops, known as *bāndā* which helped the Italians establish their colony, and was rewarded with the governorship of his native province of Akkala Guzāy. His skilful deployment of the resources which the Italians made available to him in the form of firearms and grain enabled him to build a sizeable following. However, the same years in which Bāhta was building his own power were also years of growing tension between Italian Eritrea and the Ethiopia of Emperor Menilek. Menilek, having come to the throne in 1890 in succession to the Tegrean prince Yohannes IV, steadily built his power in Tegrē province, just to the south of Akkala Guzāy, and began to challenge the Italians for the allegiance and loyalty of the Eritrean notables. That loyalty was the more easily won as the Italians began to offend local sentiment in Eritrea, and then struck a heavy blow at the basis of peasant society through alienating land for European settlement. Bāhta rebelled in the name of the traditional rights of the people of Akkala Guzāy against a government 'which prohibits you from cutting firewood in the forest'. Bahta chose to ally himself with the Tegrean appointees of Emperor Menilek to the south, thereby choosing the black snake over the white. Yet, in spite of the popularity and justice of the causes for which he rose, his rebellion drew little support. His rivals distrusted his ambition; his people hated him as oppressive; and potential allies had been alienated. In the short run, at least, the white snake prevailed.

NOTES

1 Tselote Heskias, History Department senior, Addis Ababa University, 1971, collected this while teaching at Sanāfē. Such a song was current by the Third Italo-Ethiopian War, that of 1935–41, according to informants from Eritrea interviewed at a funeral in Addis Ababa by Rezene Tekle, 3rd-year History major, Addis Ababa University, 1970–71.

2 Institute of Ethiopian Studies, Addis Ababa University (hereafter IES), *Azmāch* Gabra Mikā'ēl Germu, 'Ityoppyānnā Itālyā', Amharic MS, MS No. 324, fol. 194. The late Gabra Mikā'ēl worked in the Eritrean colonial administration and explains that in this history, dated 30 *Miyāzyā* 1911 Eth. Cal. (= 8 May 1919 AD), 'I am presenting to readers things which I have collected from books and old men' (fol. 137); the books were clearly Italian.

3 *The Ethiopian Herald*, 4 September 1976 in a summary of an article the day before on Eritrea in the Amharic daily, *Addis Zaman*.

4 Mohammed Nur Abdu, 'Political history of Akala Guzay: 1885–1895' (unpublished BA thesis, Addis Ababa University, May 1972), p. 15; he suggests *c.* 1851 as his birthdate.

5 Rosalia Pianavia-Vivaldi, *Tre anni in Eritrea* (Milan: Cogliati, 1901), p. 237; Ruffillo Perini (ed.), *Di qua dal Marèb: Marèb-mellàsc'* (Florence: Tipografia cooperativa, 1905), p. 250. Perini makes him in his fifties in 1894, and therefore born in the early 1840s.

6 Perini, *Di qua dal Marèb*, pp. 247–51.

7 Johannes Kolmodin (ed. and trans., with Bahta Tesfa Yohannes), 'Traditions de Tsazzega et Hazzega: traduction française', *Archives d'Etudes Orientales*, vol. 5, no. 2 (Upsala, 1915), Paras. 209, 213–14, pp. 146, 148–51.

8 Mohammad Nur Abdu, 'Political history', p. 37.

9 ibid., p. 15.

10 Archives of the Vicar Apostolic, Asmara (hereafter AVA), Mgr. Louis Bel, diary., 4:6, fol. 88. Luigi Fusella (trans.), 'Le lettere del *Dabtara* Asseggakhāñ', *Rassegna di Studi Etiopici*, vol. 12 (1953), pp. 88–90.

11 Fusella, 'Lettere del *Dabtara* Asseggakhāñ'.

12 Louis Bel diary, fol. 101.

13 Aleme Eshete, *La Mission Catholique Lazariste en Ethiopie*, Institut d'Histoire des Pays d'Outre Mer, Etudes et Documents No. 2 (Aix-en-Provence, 1971), p. 52.

14 Bibliothèque Nationale (Paris), Manuscrits éthiopiens 192/Collection Mondon Vidail-het 104, fol. 17, Anonymous Amharic history.

15 Francesco da Offeio, *I Cappuccini nella Colonia Eritrea: ricordi* (Rome, 1910), pp. 25–6. Duflos to De Sarzec, Sanā Daglē, 18 October 1875, Jean de Coursac (ed.), *Une page de l'histoire d'Ethiopie; le règne de Yohannes depuis son avènement jusqu'à ses victoires de 1875 sur l'armée égyptienne* (Romans, 1926), pp. 160–61.

16 Da Offeio, *I Cappuccini*, p. 27.

17 Alulā's rise from peasant to *rās* is a famous exception to the norm of feudal service in the north: Haggai Erlich, 'Alula, "The Son of Qubi"; a "King's Man" in Ethiopia, 1875–1897', *Journal of African History*, vol. 15, no. 2 (1974), pp. 261–74.

18 Perini, *Di qua dal Marèb*, p. 248; Pianavia-Vivaldi, *Tre anni*, p. 238; Carlo Conti Rossini, *Italia ed Etiopia dal Trattato d'Uccialli alla Battaglia di Adua* (Rome: Istituto per l'Oriente, 1935), p. 110.

19 Da Offeio, *I Cappuccini*, p. 27.

20 Haggai Erlich, *Ethiopia and Eritrea During the Scramble for Africa: A Political Biography of Ras Alula, 1875–1897* (E. Lansing: Michigan State University, African Studies Center/Tel Aviv: Shiloah Centre for Middle Eastern and African Studies, 1982), p. 33.

21 ibid., p. 35.

22 Enno Littmann (ed. and trans., with Naffa wad Etman), *Publications of the Princeton Expedition to Abyssinia*, Vol. 2, *Tales, Customs, Names, and Dirges of the Tigre Tribes: English Translations* (Leyden: Brill, 1910), p. 42.

23 ibid., p. 202.

24 Ferndinando Martini, *Nell'Affrica Italiana. Impressioni e ricordi* (Milan: Fratelli Treves, 1895, 5th edn), pp. 151–2.

25 Erlich, *Ethiopia and Eritrea*, pp. 93–4, 100.

26 Alessandro Sapelli, *Memorie d'Africa: 1883–1906* (Bologna: Zanichelli, 1935), p. 29; Pianavia-Vivaldi, *Tre anni*, p. 239.

27 Conti Rossini, *Italia ed Etiopia*, p. 110.

28 Public Record Office, Kew, FO 403/123, No. 50, Enclosure 2, Dabbab to Hogg, Adwā, 23 December 1888, English trans. See chapter by D. Crummey (Chapter 6, above) for other, similar acts of rebellion by noblemen.

29 Archives of the French Foreign Ministry, Paris, Correspondence politique des consuls, Massaouah 5, fol. 272, Mercinier to Affaires Etrangères, Massawa, 18 August 1888.

30 Baldissera to Guerra, Massawa, 10 September 1888, Carlo Giglio (ed.), *Etiopia-Mar Rosso*, VII. *Documenti* (Rome, 1972), Italian Foreign Ministry, Comitato per la Documentazione dell'Opera dell'Italia in Africa, No. 101 Allegato.
It is regretted that Richard Caulk, before his death, did not include reference numbers in the text for footnotes 31 to 40 and 48.

31 ibid., noṡ. 200, 252, 254. Bāhta to Baldissera, n.p., cited in Erlich, *Ethiopia and Eritrea*, p. 157, n. 65.

32 Erlich, *Ethiopia and Eritrea*, p. 165.

33 Pianavia-Vivaldi, *Tre anni*, p. 239.

34 Massimo Vitale, *L'Opera dell'Esercito*, II, *Avvenimenti militari e impiego*, Parte Prima, *Africa orientale 1868–1934* (Rome, 1962), Italian Foreign Ministry, Comitato per la Documentazione dell'Opera dell'Italia in Africa, pp. 50, 54.

35 Conti Rossini, *Italia ed Etiopia*, p. 110.

36 Giglio (ed.), *Documenti*, VII, Nos. 332 and 371. Mohammed Nur Abdu, 'Political history', p. 37 ff.
37 Perini, *Di qua dal Marèb*, p. 251.
38 Mohammed Nur Abdu, 'Political history', p. 36.
39 Martini, *Nell'Africa*, p. 151.
40 Giglio (ed.), *Documenti*, VII, No. 371. Vitale, *Avvenimenti militari*, p. 52.
41 *Grāzmāch* Assefa Zewde, 'YeKebur *Rās* Sebhat yaheywat Tārik'. I am indebted to Assefa's grandson, Taddesse Gebre Egziabeher, History graduate, Addis Ababa University, 1971, for transcribing and translating this family history.
42 Pianavia-Vivaldi, *Tre anni*, p. 239.
43 Martini, *Nell'Affrica*, p. 160.
44 Cosimo Caruso, 'Ricordi africani (1889–1896)', *Politica Internazionale*, vol. 42 (1938–9), p. 450.
45 Pianavia-Vivaldi, *Tre anni*, p. 239. Perini, *Di qua dal Marèb*, p. 250.
46 Mohammed Nur Abdu, 'Political history', p. 38. cf. Perini, *Di qua dal Marèb*.
47 Perini, *Di qua dal Marèb*, pp. 249–50.
48 Pianavia-Vivaldi, *Tre anni*, pp. 240–41.
49 Oreste Baratieri, *Mémoires d'Afrique: 1892–1896* (Paris: Delagrove, 1898), pp. 104–5.
50 Informants from Akkala Guzāy interviewed, Addis Ababā, 1971, by Rezene Tekle, 3rd year History major.
51 Carlo Zaghi (ed.), 'L'Italia ed l'Etiopia alla Vigilia di Adua nei Dispacci segreti di Luigi Capucci: contributo alla biografia di un grande pionere', *Gli Annali dell'Africa Italiana*, vol. 4, no. 2 (1941), no. 55, p. 552. Conti Rossini, *Italia ed Etiopia*, p. 112.
52 Gabra Mikā'ēl Germu, 'Ityoppyānnā Itālyā', fol. 186.
53 Zaghi, 'Dispacci segreti', pp. 531, 549. Kolmodin, 'Traditions', Paras. 280, 283, 288, pp. 198–200, 203–4.
54 Conti Rossini, *Italia ed Etiopia*, p. 115.
55 ibid., p. 109 quoting Italian trans.
56 ibid., p. 113. Zaghi, 'Dispacci segreti', Nos. XV and XXII, pp. 530, 533.
57 Conti Rossini, *Italia ed Etiopia*, p. 115.
58 ibid., p. 114. Baratieri to Affari Esteri, Asmara, 22 December 1894; Italy, *Atti Parlamentari, XIX Legislatura, prima session 1895–1896 (Rudini-Caetini-Ricotti-Blanc)*, communicati il 27 aprile 1896, No. XIII bis Avvenimenti d'Africa: Gen. 1895–Marzo 1896 (Rome, 1896), Doc. 11.
59 Vico Mantegazza, *La Guerra in Africa*, Vol. 1 (Florence: Successori Le Monnier, 2 vols, 1896, p. 256. Also Pianavia-Vivaldi, *Tre anni*, p. 242.
60 Gabra Mikā'ēl Germu, 'Ityoppyānnā Itālyā', fols. 187 and 187 *verso*.
61 Archives of the Italian Foreign Ministry (Rome), Documenti Diplomatici Italiani Serie Confidenziale, XCIV, No. 531, Annesso 14, p. 218, Castan to Barthez, Akrur, 16 December 1894 (?).
62 Da Offeio, *I Cappuccini*, p. 30.
63 Dr. Iyob Takle from Akrur, interviewed by Habtom Gebre Mikael, History major, Addis Ababa University, June 1971.
64 Mohammed Nur Abdu, 'Political history', p. 27.
65 Baratieri, *Mémoires*, p. 75.
66 Pianavia-Vivaldi, *Tre anni*, p. 258.
67 Italian trans. quoted in Roberto Battaglia, *La Prima Guerra d'Africa* (Turin, 1958), p. 594. Gabra Mika'el Germu, 'Ityoppyānnā Itālyā', fol. 187 *verso* explains the occasion but does not give the text.
68 Kolmodin, 'Traditions', Para. 285, p. 201.
69 Baratieri, *Mémoires*, p. 75.
70 Quoted in Angelo Del Boca, *Gli Italiani in Africa Orientale* (Rome: Laterza 2 vols., 1976, 1979), Vol. 1, *Dell'Unita alla Marcia su Roma* (1976), p. 526.
71 ibid., p. 519.
72 Da Offeio, *I Cappuccini*, p. 29. The connection between expropriation and the revolt has

been noted in Richard Pankhurst, *Economic History of Ethiopia 1800–1935* (Addis Ababa: Haile Sellassie I University Press, 1968), p 173.

73   Gabra Mikā'ēl Germu, 'Ityoppyānnā Itālyā', fol. 187 *verso*.
74   Pianavia-Vivaldi, *Tre anni*, p. 258.
75   Conti Rossini, *Italia ed Etiopia*, p. 115.
76   Mohammed Nur Abdu, 'Political history', p. 38.
77   Del Boca, *Gli Italiani in Africa*, Vol. 1, p. 528.
78   Gabra Mikā'ēl Germu, 'Ityoppyānnā Itālyā', fol. 188.
79   Da Offeio, *I Cappuccini*, p. 31. Kolmodin, 'Traditions', Para. 285, p. 201.
80   Baratieri, *Mémoires*, p. 109.
81   Mohammed Nur Abdu, 'Political history', p. 30.
82   'Ityoppyānnā Itālyā', fol. 188.
83   Da Offeio, *I Cappuccini*, pp. 31–3.
84   Baratieri, *Mémoires*, p. 110.
85   Del Boca, *Gli Italiani in Africa*, Vol. 1, p. 539.
86   Caruso, 'Ricordi', p. 451, from a visit to Sāgānayti in 1892. Pianavia-Vivaldi, *Tre anni*, p. 261. Mohammed Nur Abdu, 'Political history', p. 31.
87   Mohammed Nur Abdu, 'Political history', p. 31. cf. Del Boca, *Gli Italiani in Africa*, I, p. 529.
88   Pianavia-Vivaldi, *Tre anni*, pp. 261–5.
89   Mohammed Nur Abdu, 'Political history', p. 31.
90   Pianavia-Vivaldi, *Tre anni*, pp. 257–8, 266–8.
91   Archives of the French Foreign Ministry (Paris), Mémoires et Documents, Afrique/ 138: Abyssinie 5, fol. 498, French Military Attaché to Ministry of War, Rome, 10 August 1895.
92   Archives of the Italian Foreign Ministry (Rome), Documenti Diplomatici Italiani Serie Confidenziale, XCIV, No. 531, Annesso 14, p. 218. Castan to Barthez, 16 December 1984 (?).
93   Mohammed Nur Abdu, 'Political History', p. 42.
94   Sapelli, *Memorie*, p. 130. On the half dozen executions, see Del Boca, *Gli Italiani in Africa*, I, pp. 528–9.
95   Perini, *Di qua dal Mareb*, p. 248.
96   Mohammed Nur Abdu, 'Political history', p. 41, trans. note 22, p. 65; on his divisive religious policy, p. 40, see also J. Theodore Bent, *The Sacred City of the Ethiopians: Travel and Research in Abyssinia in 1893* (London: Longmans Green, 1896), p. 213.
97   Baratieri to Affari Esteri, *via* Massawa, 22 December 1894, *Avvenimenti d'Africa: Gen. 1895–Marzo 1896*, Doc. 10.
98   Giuseppe Puglisi, *Chi e? dell'Eritrea 1952. Dizionario biografico* (Asmara: Agenzia Regina, 1952), pp. 271, 281.
99   See my 'Bad men of the borders: *Shum* and *shefta* in northern Ethiopia in the 19th century', *International Journal of African Historical Studies*, vol. 17, no. 2 (1984), pp. 201–27.

14 'The drum is greater than the shout': the 1912 rebellion in northern Rwanda[1]

ALISON L. DES FORGES

Rebellions, particularly those in Africa during the colonial period, used to be described as simple movements of basically similar peoples directed to the single end of abolishing foreign rule. But we are now beginning to appreciate how diverse was the membership and how flexible the goals of some of these movements. The rebellion in northern Rwanda in 1912, though short-lived, exhibited much of the complexity of composition, aims and leadership found in more stable political formations. It drew support from every rank of society, pulling in some adherents as individuals, others as blocs of kinsmen. It attacked two sets of oppressors, one Rwandan, one foreign: the relative importance of each target depended on the given location and moment. Although inspired by the heroism of a local resister, the movement finally took its orders from outsiders, a paradox in view of the apparently parochial attitudes of the participants. Its leaders successfully exploited both the older legitimacy linked to the Rwandan kingship and newer charismatic sources of power.

Most of the people of northern Rwanda lived largely from agriculture, although they also raised goats, sheep and, less often, cattle. These cultivators were known as Abahutu (Hutu) and were concentrated particularly in the well-watered, fertile highlands of north-central and north-western Rwanda. A far smaller number, fewer than 10 per cent, were pastoralists called Abatuutsi (Tutsi). Some of them had occupied scattered holdings interspersed among the fields of the cultivators, but most had settled in the north-east where the land fell off to a hotter, drier plains area. The third and smallest group of people, the Abatwa (Twa), were descended from the original population of hunters and gatherers. They continued to live by 'milking' the forests and swamps, either in the old way or in the more recent adaptation of preying on passers-by. While all three groups had long exchanged goods and services and had come to share a common language as well as other cultural traits, each remained a distinctive unit. The Abahutu and Abatuutsi treated each other with respect but both groups scorned the Abatwa.

In the seventeenth and eighteenth centuries, the rulers (*abaami*) of Rwanda had begun asserting their right to govern this northern region. They soon learned that their troops could not count on easy victories in the area. Though their warriors were better trained and better organised than the local people, they were fewer in number, lacked intimate knowledge of the terrain and had no clearly superior weapons. With no overwhelming military advantage, the rulers frequently sought to extend their influence through alliances with the leaders of the small states and large kin groups which effectively dominated the region. Some northerners refused such arrangements outright. But others, especially those struggling for power within or between these various units, accepted royal support willingly and in return provided the rulers with produce, cattle or other goods and services from their communities. Over time these arrangements took on the force of custom and both sides felt bound to each other in a relationship of reciprocal trust. Northerners never lost sight of their own interests however. When they found rulers failing to protect them adequately against dangers of human or supernatural origin – enemy attack, disease, drought, famine – or asking too high a price for such protection, they withheld their contributions, took to arms or fled beyond the limits of central control.[2]

During the nineteenth century, the Rwandan state grew stronger and its rulers more ambitious. 'The drum is greater than the shout' became an accepted proverb, meaning the power of the state exceeds that of the people. Towards the end of the century, a great ruler named Rwabugiri conquered much new territory and reformed the military and administrative systems. He was the first to establish royal residences on the periphery of the kingdom, including several in the north. From them he brought many more northerners under royal authority and collected greatly increased amounts of goods and services.[3]

The expansion of the state caused changes in the social structure of the north. At the top formed a new, small stratum of notables from the central kingdom, most of them ordered by Rwabugiri to settle in the region. They disliked living in the cool, damp mountains and made no secret of their distaste for the 'land of the Abahutu'.[4] They were not necessarily more powerful than the strongest local leaders, but their closer connection with the court seemed to them sufficient cause to claim higher status than any northerner. Because they were few in number and usually lived mostly in the vicinity of the royal residences, they constituted an annoyance rather than a serious threat to the home-grown leaders. These prominent local men were in the process of transforming their own positions in response to the new opportunities offered by the expansion of central control. Under the guise of representing the ruler, some were raiding their neighbours for produce or cattle which would be delivered only in part, if at all, to the royal residence. Claiming to be able to summon the royal troops, they were forcing other kin groups to acknowledge their authority or to surrender some of their land. Naturally these agressive leaders shared their gains with their kinsmen, but in distributing the booty they allocated a disproportionate part to themselves. They attracted clients from weaker or less enterprising kin groups and settled them on lands confiscated from others. Formerly first among equals within the kin group, these leaders became strong enough to start exercising unquestioned authority over their own kin as well. One Rwandan remembered the head of his lineage exploiting his relatives by collecting two cows for every one he passed on to the ruler.[5]

In expanding into the north, Rwandan rulers came into touch with another set of leaders, mediums for the spirit Nyabingi. Once a woman powerful at the court of Ndorwa before that state had been defeated by Rwanda in the eighteenth century, Nyabingi had been transformed after her death into a spirit able to influence daily events. During the nineteenth century a number of mediums 'chosen' by Nyabingi began interceding with her for those troubled by illness, poverty or infertility. From mid-century on, several mediums, women as well as men, began gathering their adherents into recognisable political units, some of which persisted for several decades. In addition to exercising political and judicial authority, they collected considerable offerings. While all the mediums claimed to speak for Nyabingi, each had his or her own manifestation of the spirit. Occasionally mediums co-operated with one another, but they were more often rivals and sometimes even bitter enemies. Conscious as each was of his or her own interests, the medium adopted different policies in dealing with Rwandan expansion. Many remained within their autonomous spheres of interest and avoided contact with the Rwandan ruler, while a few decided to ally with him. Those who resisted Rwabugiri's advances, however, became best known. In a number of battles they caused serious losses to royal troops. Rwabugiri killed several leading mediums but could not destroy the spirit. It simply moved on to another host, infusing him or her with all the authority held by those fallen. Thus legitimacy grew rather than diminished as it was passed down the line of resisters.[6]

In 1895 Rwabugiri died suddenly and the expansion in the north ended just as abruptly. As the news of his death spread, many of the notables fled back to the central kingdom. Those who delayed were more often than not chased off by the local people. Rwabugiri's first successor, weak and ill, had no time to be concerned with the north for he was soon killed in a coup by Kanjogera, the most powerful of Rwabugiri's wives, and her brothers. They put in power Musiinga, the son of Rwabugiri and Kanjogera, but since he was still a child, they themselves became the rulers. In the first years of the reign, Kanjogera and her assistants did little more than quell uprisings, both within the central kingdom and on the periphery. One of the most threatening focused on another of Rwabugiri's wives, a woman named Muserekande, who had supposedly also given birth to a son eligible to rule. She was supported for some months by the people of the east and north-east but then was defeated by troops from the court. Kanjogera claimed that this rival and her son had been killed, but Muserekande's supporters continued to believe that they had only fled beyond the north-eastern frontier. The court sent forces also into north-central and north-western Rwanda to punish the local people for having expelled the notables. It was unable, however, to reimpose these agents of central control or to reinstate payments on the scale of those demanded by Rwabugiri.[7]

As the court forces succeeded in restoring at least a semblance of order, opposition to Kanjogera and her brothers increasingly took the form of intrigues among the elites of the central kingdom. Complicating these factional disputes was the arrival of the Europeans. Quick to realise both the opportunities and the threats posed by this new source of power, the queen-mother and her aides consented to a German protectorate over Rwanda in return for support against rivals. At first the Germans established only two small military outposts on Lake Kivu and entrusted overall supervision of Rwanda to an administrator resident in neighbouring Burundi. He had no way to govern Rwanda, then with a population of about one million, except through the existing structure. Though concerned to disturb the system as little as possible, he did insist that both missionaries and traders be permitted to carry on their occupations within Rwanda. These foreigners, sometimes through ignorance, sometimes deliberately, caused disruptions which required frequent attention from the court.[8]

In the decade after Rwabugiri's death, the court was most worried about controlling notables and foreigners within the central kingdom. But both groups were posing equally important challenges in the north. Around 1900, notables who had discovered the wealth of the region during Rwabugiri's time began taking control of holdings, either by using their own warriors or by making arrangements with local leaders. While some acted under the pretence of representing the court, others made not even a nod in the direction of royal authority. The foreigners constituted an even greater threat. In the early years of the century, Germany, Belgium and Great Britain were disputing boundaries which would finally be settled only in 1911 between Rwanda, Uganda and the Congo. Military expeditions and parties of surveyors from all three colonial powers regularly paraded back and forth through the region, making, negating and reasserting claims to various pieces of territory. Traders, particularly those seeking ivory, were beating a path from East Africa across northern Rwanda into the Congo. They often requisitioned supplies or porters along the way with little or no recompense and ignored all protests, whether from the people, the court or the colonial administration. Missionaries, Roman Catholic White Fathers, had settled at Rwaza in north-central Rwanda and at Nyundo in the north-west. Rich and well-armed, they had set about imposing their own version of order and making demands on their neighbours for various goods and services.[9]

Local leaders too were exercising authority over growing numbers of people. Some were drawing on resources offered by notables or foreigners as they had once drawn exclusively on support offered by the court. Others had profited from more basic social and economic changes. Significant numbers of people had migrated into the north, either from central Rwanda or from the other side of Lake Kivu, where there had been warfare for years. At the same time, some local groups had suffered loss of lives and wealth in a series of famines and epidemics, both human and bovine, during the 1890s. Those who had escaped disaster through good fortune or recent arrival, were displacing those who had been weakened. The unfortunates were forced to flee to kin elsewhere or to become clients of local strong men.[10]

One of the most outstanding of these local leaders was Rukara, who dominated much of the wealthy region of Mulera. He was the third generation of leaders of the Abarashi kin group to profit from ties with powerful outsiders. His grandfather had been one of the first in the area to co-operate with Rwandan rulers. His father had been counted one of Rwabugiri's most loyal supporters and had supplied him with warriors and with ivory taken in the forests beyond Lake Kivu. Among the rewards given him by the court had been an aristocratic wife from the central kingdom who had given birth to Rukara himself. Always loyal to the court, Rukara's father had died fighting soldiers from the Congo who had intruded into Rwandan territory. Rukara was then an adolescent but he had assumed leadership of the kin group anyway. Within a few years he had firm command over seven or eight thousand men and exercised influence over several thousand more. He had a herd of fifteen hundred cattle, making him the richest as well as the strongest local leader in the area. He had attracted many clients from outside the kin group whom he had settled on land at his disposal. He formed them into his own guard and used them to ensure the obedience of kin and stranger alike. The killing of his father had supposedly given him an abiding hatred of Europeans and all those in their employ, but he managed to develop productive ties with both the missionaries at Rwaza, near his home, and with transient colonial agents. Secure in his own power and confident of such backing from outsiders, Rukara began showing less respect to the court. At one point he went so far as to insult Kanjogera herself when she returned a judgment against him in a judicial case.[11]

More worrisome to the court than men like Rukara were the local leaders who were building their power in opposition to the court. Basebya, one of the Abatwa of the northern forest, had been recruited with some of his fellows into the service of

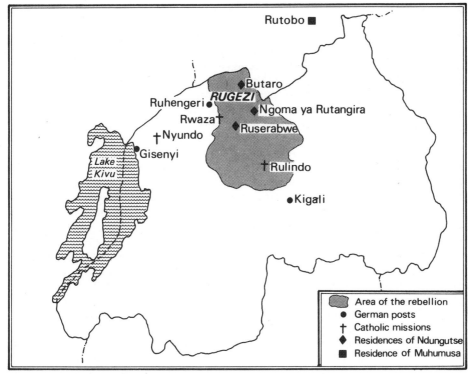

Map 5: *The 1912 rebellion in Rwanda*

Rwabugiri. Skilful and courageous archers, they had formed the personal guard of the ruler and had lived very comfortably. After his death, they had served his first successor but had then opposed Musiinga's installation. Disgusted by the factionalism of the new regime, they had severed all ties with the rulers and had withdrawn to the vast Rugezi swamp in north-central Rwanda to hunt and to prey on nearby cultivators. After a famine in 1905 weakened many cultivators in the vicinity, the Abatwa multiplied their raids and drove them away. Many of the fields of the once productive region returned to deserted bush. Those who remained either paid tribute to the bandits to protect themselves or actively joined their band. One observer remarked, 'Any Umuhutu who wanted his fill of meat, who wanted his fill of beans, joined Basebya's following and they came ten thousand strong to pillage the hills.'[12] Along with Basebya and his men, a growing number of Nyabingi mediums refused to acknowledge the authority of the Rwandan ruler. They claimed that their exercise of power, based on a tie with the spirit, was as legitimate as that of any holder of the Rwandan kingship. Muhumusa, who was becoming one of the most respected of these religious leaders, had been extending her authority over people around Rutobo in the north-east since the beginning of the century. The Europeans would eventually assign this region to Uganda, but the Rwandan court regarded the territory as its own and was disturbed by Muhumusa's increasing dominance there.[13]

By 1905 the court had disposed of problems in the central kingdom sufficiently to be able to turn its energies more to the north. Musiinga, who was now an adult and playing a larger role in government, was anxious to resume the expansion pursued so vigorously by his father. But expansion in this reign was to differ in two important ways from that of Rwabugiri's era. First, neither Musiinga nor Kanjogera took up residence in the north but instead relied on notables to rule for them. Some they sent for that purpose; others they co-opted from among those already established on the spot. These agents were far more numerous than those named by Rwabugiri and they operated with far less royal supervision. Second, the court and its agents called on the guns of the Europeans and so acquired a decisive advantage over the local people. The German administrator regarded the northerners as 'wild and inclined to disobedience'.[14] He wanted them more intensively ruled both to facilitate his own work and to ensure that neither British nor Belgian agents could claim that the area was not actually controlled by Rwanda. He usually had only several dozen soldiers at his disposal and he did not always send them where and when the court and its notables most wanted them. But he dispatched them often enough to permit the notables to give the impression that the soldiers were at their command. One northerner who had witnessed attacks called down upon his community by agents of the court remembered the image of a notable standing on a hilltop pointing out with his spear the houses which he wanted the soldiers to destroy.[15] Missionaries, who were accepted as part of the larger colonial establishment, also helped the notables by permitting their employees to go along on rounds to collect taxes or give orders. As one angry group of northerners remarked to an agent of the court who had come on such an errand with the backing of some mission employees, 'If you had not had the of "Bwana" with you, you would have gotten nothing here but sticks to chase you away.'[16]

The proliferation of notables necessarily made central rule more costly: now all these intermediaries had to be fed as well as the court itself. Not only were there more mouths to feed, but appetites had grown as notables found it easier to collect taxes. The court, too, expected larger returns because it had more agents with greater power in the field. Northerners resented the increased demands in and of themselves, but found them all the more offensive because they were enforced with the help of foreign power and because they were adding more to the personal wealth of the notables than to the public treasury of ruler and state. In multiplying the number of notables, the court sometimes left vague the limits of their commands, either through ignorance of local geography or from a desire to stimulate conflict between rival agents. Some holdings were allocated territorially, others in terms of lineage; this also led to confusion over lines of authority. While northerners could sometimes profit from these situations by playing off one notable against another, often they suffered from the conflicts and ended up either having to provide men to fight in skirmishes between the rivals or having to pay taxes to more than one agent.[17]

Rule by the notables cost the northerners in terms of status as well as of wealth. Instead of having frequent direct contact with the ruler, northerners now ordinarily had to pass through notables to communicate with the court. Since Musiinga and Kanjogera did not travel in the region, anyone who wanted to approach them had to make the trek south. Local leaders who had been accustomed to delivering their contributions personally as willing acknowledgment of royal sovereignty were now obliged to hand them over to notables who then took credit for gathering them so successfully. Rulers of small states like Busigi in the north-east and Bushiru in the north-west found themselves placed under notables more eager to exploit them than to uphold their customary privileges. Ritual specialists of Bumbogo and Buberuka, whose regions had been free of taxation in return for their participation in ceremonies at court, discovered that their long-standing immunity had been suddenly cut back to their own hills of residence and that their dependents elsewhere had all come under direct rule by notables.[18] When agents of the court had needed the co-operation of the northerners, they had treated local leaders as respected allies. But as they grew stronger and more numerous, they became increasingly conscious of themselves as an elite. Exemplars of what they believed to be the superior culture of the central kingdom, they grew more open in showing contempt for the people of the hills who, as one refined notable put it, 'could not even piss without baring their arses'.[19] Since the notables owned many cattle, the term Abatuutsi or pastoralists, took on the meaning of 'the elite', while the word for cultivators, Abahutu, was used increasingly in the sense of 'the common people'.

Neither the court nor the notables saw much profit in occupying the relatively poor north-east so they imposed their demands on this region from a distance. Court agents came more often than in the past and required more when they came, but they still returned to the central kingdom after each foray. In the richer north-central and north-western regions, however, court and notables favoured moving more aggressively. There the agents arrived with their own groups of warriors and clients to settle permanently on the most inviting pieces of land. They took plots which were not extraordinarily large, but in taking them they revealed a most significant assumption: that their rights superseded those of the cultivators who had cleared the land. While actual loss of fields affected only a few, this assumption threatened many. The northerners knew that whoever controlled the land controlled the people who lived on it. Dispossession was a far more powerful sanction than raiding and burning homes and crops, the usual methods of punishment in the past. As the notables took up residence in the north, they had greater opportunity as well as greater means for interfering with local life. And so they did. Soon they were deciding conflicts which had always before been handled between or within kin groups. Next they were dictating the choice of successors to deceased heads of lineages.[20] The boldest went further: they challenged the right of the local people to control their own labour. Notables of the Abatsobe and Abaskyete lineages particularly favoured by Kanjogera, for example, required people in their vicinity to come to cultivate their fields once or twice a week. The burden in and of itself was not overwhelming since the service was required by household, usually of five or six adults, rather than by individual. But, as with the appropriation of land, the assumption behind the demand and the eventual development which it forecast represented the real threat. Except for reciprocal sharing of labour between neighbours or kinsmen, the only people who cultivated for others were clients who laboured in return for the use of land. To cultivate for the notables meant to acquiesce in their claims to control the land; it meant exchanging the status of *abagabo* – free men in charge of their own fields and responsible for their own destiny – for that of clients, subservient to the will of the patron.[21]

At the same time as the notables were imposing new demands, Europeans also were requiring goods and services from the northerners. In 1907 the Germans designated a Resident for Rwanda. The Resident, of course, required a European-style capital and soon began building a new town at Kigali. He looked to the northern forests for the wood and to the northerners for the thousands of days of labour needed to transport it. He also gave orders for constructing a road into the north, another project requiring hundreds of labourers every day. The missionaries too were determined that their churches and homes should reflect the scale and solidity of their civilisation. When building the church at Rwaza, for example, they needed 800 labourers daily. Missionaries and administrators paid salaries to their regular workers but they usually paid nothing to these masses of unskilled labourers whom they requisitioned through the notables. They counted the notables' co-operation in recruitment as fair return for aid they had given them in the past, although they often thanked them with gifts as well.[22] Other Europeans travelling through the area paid ridiculously low prices or nothing at all for the food and firewood needed by their caravans, which could number a hundred men or more. Employees of the Europeans and notables in charge of assembling the provisions usually increased the amount collected sufficiently to provide a handsome profit for themselves. One contemporary observer commented: 'It is likely that the [A]Bahutu have not yet seen the end of their miseries . . . The new burdens will not erase the old ones but will be added to them and the [A]Batutsi will not fail to find a new source of profit in them.'[23]

The northerners responded to the demands by notables and foreigners in different ways. Some turned to the court, invoking the protection which the ruler was expected to provide. Their spokesmen recalled how faithfully they had served rulers past and present, then protested that the notables 'are making a mockery of our loyalty'.[24] Even when the court professed sympathy for their grievances, it did little or nothing to redress them; it was afraid of internal political repercussions and of displeasing the Germans. When one northern leader named Nzaramba sought out Musiinga to complain about exploitation by the notable Biganda, he received nothing but empty consolation. Ordered to return home, he inquired bitterly: 'Shall I return to see my kin whom Biganda has slain with his spears? Shall I return to drink the milk which Biganda has stolen? Shall I return to the home which no longer exists?'[25] With appeals to the court so unproductive, some northerners sought protection from the new sources of power within the region, one or another of the notables, the Germans or the missionaries. But such alliances were costly both in real goods and in self-respect and most lasted only a short time.

Northerners who chose active opposition – and most did at one time or another – robbed the outsiders, burned their houses, disrupted their travel and communications. They ordinarily organised their attacks within the kin group and neighbourhood, but they did share information and offer sanctuary to resisters from outside their own communities.[26] Some local people preferred not to fight and withdrew to clear new lands in the forest, beyond the reach of agents of the court. But a larger number chose a more potent form of withdrawal; they switched their allegiance to leaders opposed to the court. Many in north-central Rwanda joined Basebya. He and his group proclaimed no political goals and attacked all who had not paid for their protection, elite and ordinary people alike. But their life outside the law made them by definition enemies of the ruler; hence all who chose to share their existence proclaimed their opposition to Musiinga. People further to the north and east rallied around the medium Muhumusa. At first she had attracted adherents solely through her connection with the spirit Nyabingi, but sometime after 1905 she decided to tap the legitimacy provided by the Rwandan royal tradition as well. She had learned of Muserekande, the wife of Rwabugiri and rival of Kanjogera, who had led an uprising to put her son in power a decade before. Having heard that some Rwandans believed this rebel leader to be still alive, Muhumusa claimed her identity. By 1909 Musiinga and Kanjogera began to suspect that both Basebya and Muhumusa had contacts with important figures at court. Basebya was said to be passing a share of his booty to high-ranking officials who in turn informed him of the movements of troops sent to catch him. Muhumusa had apparently drawn the attention of legitimists at court whose hopes of ousting Musiinga exceeded their doubts about her claims. At the request of the court, the Germans sent their soldiers against both Basebya and Muhumusa. The expedition against the bandit, complete with Maxim guns, captured some of his cattle but not Basebya. He and most of his men escaped to sanctuary within the swamp. That against Muhumusa had more success and brought the medium back to Kigali where she was kept in detention for a few months. When her popularity continued to grow, the Germans yielded to Musiinga's entreaties and exiled her to Bukoba, a town well outside Rwandan boundaries.[27]

In the early years of the century, Rukara had been one of the local leaders most skilful in exploiting ties with the court and the Europeans to build his own power. As pressures from the central kingdom increased, he did not hesitate to voice his displeasure. His independent behaviour provoked the rulers so much that they ordered him killed in 1907. But Rukara saved himself by winning the protection of a passing German. For the next year he remained at home, avoiding contact with the court and its agents. In 1908 he and his men attacked the caravan of a European who had supported Rukara's opponent in a local dispute. At German request, the court summoned him to account for his act. He refused to go. The court responded by placing him under the command of a notable. Weakened by this blow from above, Rukara was then attacked from below. Many of his kinsmen had tired of obeying his orders and had even taken to calling his 'Umutuutsi' in criticism of his authoritarianism. One traveller present when Rukara assigned tasks to his kinsmen witnessed one of them refuse an order with the indignant comment, 'Am I not also an Umurashi?'[28] The dissatisfied Abarashi had coalesced behind Rukara's cousin Sebuyange who promised to conduct lineage affairs in a more customary fashion. As Rukara was becoming convinced of the disadvantages of relying on the help of outsiders, Sebuyange was actively soliciting it. Afraid that the missionaries were going to back his disgruntled kinsman, Rukara gave up all pretence of maintaining good relations with them. Asked to send them supplies, he provided rotten eggs. Kept waiting one day when he had called to see Father Loupias, the superior at the Rwaza mission, he finally sat down in Loupias' own chair. His followers, knowing that in Rwanda the stool of the head of the house is reserved for his sole use, urged him to move quickly. Rukara supposedly answered in a voice meant to carry, 'Why? Isn't my arse as good as his?'[29] By early 1910 he and Loupias had become so hostile to one another that the hot-tempered missionary slapped Rukara in the face during one of their disputes.

Sebuyange had appealed to the court to install him instead of Rukara as head of the Abarashi. An envoy arrived in the north to proclaim the decision of the court. Afraid to announce it without support, he enlisted Loupias and several employees of the mission to accompany him to Rukara's residence. There Loupias supervised the announcement of the decision: the Abarashi were to divide, each to follow the leader of his choice. The decision was a severe blow to Rukara, but it was immediately challenged by another messenger who claimed to be a later envoy from the court. Loupias ordered the matter postponed pending further clarification and then agreed to hear a case in which several of Rukara's men were accused of cattle theft. Loupias decided that the Abarashi had to return the cattle in question to the plaintiff, a notable whom Rukara hated. Still smarting from the pronouncement of the first envoy, Rukara declared that his kinsman would not comply with the missionary's decision. Loupias, a massive man, grabbed Rukara with one hand, his rifle with the other. Rukara called his kinsmen to his aid and two of them killed Loupias with their spears.[30]

Rukara had acted to defend himself, not to strike a blow against colonial oppression. Even one of Loupias' fellow missionaries virtually excused his behaviour, given his cause to fear the impetuous Loupias. The northerners, however, interpreted the incident differently. Although Rukara had not raised his own hand against Loupias, they made him into a hero of the resistance, the one who dared to kill a powerful European. Rukara fled to Belgian-ruled territory where he took shelter with his wife's kin. His people also found refuge with distant kin or hid in the extensive network of caves in the neighbouring volcanoes. The Germans devastated their homes and fields, killed a number of Abarashi and took prisoner those whom they had smoked out of the caves. Of course they most wanted Rukara himself and offered the enormous number of one hundred cattle for his capture. But they could find no one who would agree to gather information about him, far less try to lay hands on him. After a short time, Rukara returned to his home region. He rarely stayed long in one place but circulated throughout the area. Twice he attacked other northerners whom he took to be enemies, but for the most part he lived well-protected and well-provided with contributions from the people of the area. As the months passed and he continued to evade capture, he began encouraging others to attack the Europeans also. He argued that if they had not been able to avenge one of their dead, they would not be able to avenge two. A leader in Bugoye eighty kilometres to the west sent to request one of Rukara's spears, hoping it would give him special strength in opposing the foreigners. Rukara became so important as a symbol of resistance that both Rwandan parties opposed to him before the killing ended by giving him support. His rival kinsman Sebuyange helped him elude the Europeans several times. And the court, perhaps honouring a course it wished it could follow, sent him gifts of cattle, the supreme accolade to the brave warrior.[31]

Although Rukara inspired a few individuals to attack notables or Europeans, he gathered no mass movement behind him. He had a vision of restoring autonomy to the north; he had courage and ability to command; he had contacts with all the people of influence in his area and through them access to important resources. But he could claim no legitimacy to elicit the ultimate commitment. Of course he could not pretend to be legitimate in the royalist tradition since everyone knew he was not the son of a past ruler. Nor had he experienced the kind of spiritual crisis which would have allowed him to claim contact with Nyabingi and thus permitted him to draw on that more recently developed tradition of legitimacy. In addition, Rukara had risen to power by vigorously asserting the interests of the Abarashi against all rivals. Many other northerners whom he had harassed respected him enough as hero not to betray him, but they were not about to ignore the experience of the past and put themselves at his command.

For more than a year after Loupias' killing no leader emerged to raise the rebellion which many were anticipating. Then in July 1911 Muhumusa escaped from detention in Bukoba and returned to her home at Rutobo. This news generated tremendous excitement. She had come, many people said, to expel the Europeans. Others believed she would oust Musiinga and put a son of her own in power. But it seemed at first that such hopes would be disappointed. Although she had returned with the special aura of one who has outwitted the oppressor, she seemed more interested in rebuilding her own autonomous centre of power, based on Nyabingi, than on fighting either the Europeans or the Rwandan court. As Muhumusa tried to extend her control to the west, however, she was opposed by several local leaders who had been using alliances with rival mediums and with the British to establish their own areas of control. When she attacked them, they called on the protection of their own manifestations of the spirit and on the foreigners. The British agent knew from her reputation that Muhumusa was a woman to be reckoned with and he felt obliged to protect those of her enemies who had chosen the British side. But he hung back from decisive action because he was unsure whether she belonged under British or German jurisdiction. Thus on the first two occasions when British troops confronted Muhumusa, they withdrew, giving her easy victories. Not knowing why the British hesitated, her people assumed they were afraid and jeered, 'You chicken-eaters, how would you dare fight Muhumuza?'[32] Even after such dramatic success, Muhumusa was unable? to unite the people of this area. Some still preferred to stay under British protection and retain their attachment to their own particular Nyabingi mediums. In the meantime, people in the region twenty or thirty miles further south learned of the victories and took them as confirmation of Muhumusa's legitimacy. Some revived the story of her being Muserekande, the wife of Rwabugiri who had disappeared. Others continued to place their faith in the powers granted her by Nyabingi. Basebya, long familiar with the spirit, had supported other mediums near his own home base, but now he arrived to put his men at her service and to offer her links to other influential leaders further south. Perhaps it was he who persuaded her that people nearer the central kingdom were ready to rise up. She apparently decided to take on the role of resistance leader there and began travelling about with a child said to be the rightful heir to the Rwandan kingship. But in September 1911, before she could move south, she was attacked by a joint British-German force well supplied with a cannon and sixty-five rifles. Her followers, who had believed that enemy bullets would turn to water, were armed with spears. Forty of them were killed and she herself was captured with a minor injury to her foot. Soon after the British relegated her to Kampala because they had heard that a party of Rwandans was preparing to liberate her once more.[33]

As the news of Muhumusa's capture spread through northern Rwanda, people kept expecting to hear of a second miraculous escape. Thinking the uprising was about to begin, a number of Rwandans attacked Europeans and killed some of their supporters in October, November and December. At about the same time, people began transferring their hopes to a new leader, a young man named Ndungutse. Said to have been one of Muhumusa's close associates, he claimed to have received her power to overcome the oppressors. In the same way that mediums could pass their legitimacy to followers, so Muhumusa had handed on to him the right to lead the resistance against the exploiters. That the drive to resist might contain within itself moral force sufficient to legitimate the exercise of power was a new idea. It had been hinted at two years earlier when admirers had sought to draw on the special force of the spear of Rukara. Now it gathered strength and identity as it outdistanced the Nyabingi tradition which had first promoted it. Endowed with this more general and secular moral force Ndungutse could appeal to any Rwandan who opposed Europeans or the court, not just to the more limited number of Nyabingi adherents clustered near the north-eastern frontier.[34]

At the start of January 1912, Ndungutse established a residence at Butaro, just east of the Rugezi swamp. As he began planning to attack the Rwandan heartland, he decided to broaden his appeal by claiming to be the legitimate ruler of Rwanda.

Much as Muhumusa had mixed royalist and Nyabingi claims, Ndungutse now added royal parentage to the earlier justification of his right to rule. A stranger to the region, no one knew his real parents. This made it easier for people to accept whichever of the versions of his royal ancestry reached their ears: some said he was the son of Rwabugiri, others said the grandson; still others believed he had come not to rule himself but to prepare the way for a younger brother named Biregeya. While people differed on why they believed Ndungutse legitimate, they generally agreed on why Musiinga was illegitimate. Everyone knew he had usurped power fifteen years before. Despite active propaganda by the court, people still called the regime 'Cyiimyamaboko': 'It is force that rules'.[35] Musiinga's inability to defend the kingdom was simply further proof – indeed the natural consequence – of his lack of legitimacy. He had failed to bring the Abatwa bandits under control. He had not been able to prevent the Europeans from multiplying and from increasing their demands year by year. Nor had he saved his people from exploitation by the notables: appropriation of their goods, forced labour, threats to their land, interference in community life, disruption of their relations with the court itself. Unable or unwilling to protect his people, Musiinga had forfeited their loyalty.

As Ndungutse increasingly stressed his legitimacy as ruler, he also shifted the goals of his movement. When still closely linked with Muhumusa, he had exhorted people to expel Europeans from the region by local action. But as he became more firmly rooted in the north, he talked less about ending foreign exploitation and more about driving the notables back to the central kingdom. He warned, however, that oppression could be halted for good only by unseating Musiinga. No longer could local interests be assured by local action; only by restoring order at the centre could protection be guaranteed on the periphery. Thus he moved his followers from resistance to rebellion[36]

With a small band of followers, Ndungutse began making good on his promises in January. His first attacks against the notables succeeded and within a week or two people flocked to his support by the thousands. The rebels took one hill after another, driving the notables from the region, burning their hated residences, capturing their cattle and other goods. After five or six weeks, they controlled a solid triangle of territory pointed down at the heart of the kingdom: all of Buberuka, Kibali and Bumbogo and sections of adjacent Mulera, Bukonya and Buriza were in their hands. They were within a few hours march of the German capital at Kigali by the end of February or the beginning of March. As word of the rebellion spread, notables from fifty or sixty kilometres away deserted their posts to seek safety to the south. Ndungutse began to establish the framework of an administration and set up three residences in different parts of the liberated area.[37]

Basebya and his men were Ndungutse's earliest and most effective supporters. They had been with him when Muhumusa was captured, had formed his escort as he moved south, and had launched the first raids which established his reputation. Basebya, with his past experience at court, adjusted easily to Ndungutse's royalist pretensions. He shaved his head in the style of the aristocrats and wore cotton cloth, then usually afforded only by the rich. He grouped his men into two military formations modelled on those at court, each with its own name and insignia. One was made up of Abatwa, the other of cultivators and pastoralists. The Abatwa had been raiding for more than a decade so it is hard to judge how much their participation in the rebellion was simply a continuation of their old strategy for survival and how much it represented real commitment to political change. They knew, of course, that attacks on notables would net rich booty. And they sometimes plundered even supporters of Ndungutse, seeming to indicate that their greed outpaced their political awareness. But they certainly opposed the court and Europeans who had killed a number of their group in previous years and probably relished hopes of damaging them at the same time as enriching themselves. For the pastoralists and cultivators, the political significance of their participation was clearer. To have left their homes and usual occupations to follow an Umutwa bandit was an extraordinary step. It reflected their political alienation as much as their economic hardships. Rather than immediate profit, they sought the restoration of order which would permit them to return to their normal ways of living.[38]

Beyond the services of his men, Basebya offered Ndungutse access to the network of alliances he had created in building his own power. He spoke for him to influential lineage heads and to the Nyabingi mediums. He arranged for Abatwa of the western forest to cut communications between the German posts at Ruhengeri and Gisenyi and between the missions of Rwaza and Nyundo. He had other contacts who could supply him with guns.[39]

Once the Abatwa made the first moves, the local leaders rallied to the rebellion, bringing their kinsmen with them. These men who had prided themselves on their ties to the court responded to Ndungutse's claim of royal descent. Those who had shared no common interests in the past, and even those who had been enemies, accepted equally his promises of protection; as an outsider, he had no established favourites or opponents in the region. Many leaders came spontaneously to give him gifts and acknowledge his authority. Following the pattern usual in the central kingdom, they left their sons at Ndungutse's side when they returned home themselves. Ndungutse summoned to his residence those few leaders who had failed to present themselves on their own. They came too. Most of Ndungutse's followers were cultivators just as were most of the people of the north. But some pastoralists joined him too, either singly or by groups of two or three. Owners of small herds and resident in the area a long time, they had suffered along with the cultivators from the expansion of central control and European demands.[40]

Of all the local leaders, the one most important to the rebellion was Rukara, the man who could call out the largest lineage in the north-west, the ultimate hero of resistance. He came to Ndungutse in January when the uprising was just taking hold in the north-central region. At their first meeting, Rukara made clear that he was following Ndungutse as a resistance leader, not as a legitimate candidate for the kingship. He is said to have told him outright: 'You claim to be the son of Rwabugiri? You do not deserve to be called his son. You are not even fit to be his Umutwa. As for me [I know because] I have lived there at court. If you agree, let us unite, but do not talk to me any more about being the son of Rwabugiri.'[41] Realising the necessity of Rukara's support, Ndungutse took this insulting speech in his stride and refused to allow his indignant followers to punish Rukara. Rukara apparently still felt some loyalty to Musiinga, from whom he had received gifts of cattle, and wanted to direct the rebellion more against the foreigners than against the court. When he protested the decline in Ndungutse's threats against the Europeans, the rebel leader supposedly assured him that he would eventually render the Europeans harmless: 'I will be the master of the Europeans. You have nothing to fear from them. If I put goat milk on a stool, they will lick it off just like dogs.'[42] The two agreed to put aside their differences in emphasis and to continue working together.

Even while pursuing links with the influential leaders, Ndungutse was also developing his popularity with the masses. He went out regularly to visit the hills, to greet his people and accept their gifts. As he was carried about in his litter – a form of transportation used by kings – people flocked to admire him. Tall and handsome, he looked like an aristocrat. He wore his hair in long strands interwoven with many beads. One of his loyal followers told a sceptical missionary: 'We love Ndungutse and he loves us. If you could only see how fine he is! You cannot look at him without feeling tears come to your eyes.'[43] More eager to appear ruler than rebel, Ndungutse assembled the trappings of royalty. He had a drum created for him by one of the drum-makers of the court. He ordered one of the northern lineages which provided musicians for Musiinga to send him some also. He established contact with the most famous rain-maker of the kingdom whose ancestors had served the court for some ten generations. Perhaps most convincing to the people was hearing that Ndungutse had begun performing religious ceremonies to honour the royal ancestors. Surely, they believed, if he dared to do that without actually being of royal blood, he would have been rapidly and drastically punished by the spirits.[44]

As Ndungutse was gathering support in the north, he attracted the attention of notables in the central kingdom who were anxious for a change in regime. A dozen or so who had been most harassed by Kanjogera discreetly made contact with the rebel leader. Although they were all even better placed than Rukara to know that Ndungutse's claims to royal birth were pure fiction, they were willing to co-operate with him to oust their enemies. These allies from the centre may have provided Ndungutse with some information but contributed little else in concrete aid. Still, their involvement helped change the movement from a regional disturbance by unruly hill people into a pressing threat to Musiinga's rule. It encouraged Ndungutse to believe that he might actually win the whole country.[45]

Musiinga and Kanjogera watched the rebels multiply and grow stronger and they began to suspect the collaboration of some of the notables at court. They implored the Germans to destroy the movement or to permit royal troops to attack it. The regular head of the German administration was in Europe at the time and his replacement hesitated to act. He was uninformed enough to believe that Ndungutse was really the legitimate heir to the kingship and he even entertained the notion of putting him in power. While he wired the Governor in Dar es Salaam for instructions, he refused the pleas of the court for an immediate attack. But he did set up a cordon of four posts along the road from Kigali to the north-western post of Ruhengeri to try to contain the spread of the rebellion. When the court sent its own warriors to the north, the administrator refused to let them attack either. He allowed some to join the cordon but insisted that the rest return to the central kingdom.[46]

The rebels naturally took heart from this relative inaction. As Ndungutse came closer to his goal of actually ruling Rwanda, he decided to pursue negotiations with the Europeans. In January he had already prohibited his men from attacking the property of the missionaries. In February and March, thinking the Germans' hesitancy indicated the possibility of real support, he began to address fine rhetoric and gifts of cattle to both missionaries and administrators. The Fathers at Rwaza declined his cattle three times and showed no enthusiasm for being named his honorary 'maternal uncles'. The missionaries at the recently founded station of Rulindo, however, accepted his gifts, as did a representative of the Germans. When an important lineage head and his men killed several soldiers and Rwandan

Christians, Ndungutse offered to help capture them. Ironically, the assassins had probably been spurred to act by the rebellion itself. The Europeans ignored Ndungutse's offer so the flexible rebel then extended sanctuary to the killers.[47]

Even as Ndungutse was doing his best to establish good relations with the missionaries and Germans, he was still seen by many as a saviour from all foreigners. Particularly in areas where European demands had caused great suffering, such as in Bugoye along Lake Kivu, he was hailed as the fierce opponent of the colonialists.[48] Then, in early April, Ndungutse decided to make the ultimate concession to the colonial administration: he would give them Rukara. At this very time the administrator had received orders to quell the rebellion and was beginning to organise his expedition against it. Ndungutse may have known this and may have hoped to persuade him to change his mind by handing over Rukara. Or he may have known nothing of these plans and simply decided to take the step to impress the Europeans with his good intentions. When he had tried previously to establish relations with the missionaries at Rwaza, they had replied that they would not deal with anyone who harboured the killer of their colleague.[49] In deciding that European support was more important to winning the kingship than the continued loyalty of the Abarashi, Ndungutse may have been right. But in thinking that he stood a serious chance of winning that support and in imagining that anger at his betrayal would be limited to Rukara's kin, he had erred grievously. Perhaps his judgment was affected by a desire to rid himself of an arrogant and potentially dangerous rival. When Ndungutse's men took Rukara to lead him away, he understood full well what was happening. He is said to have told Ndungutse in parting, 'I have the coals; you bring the tobacco,' meaning that where I go, you will soon follow.[50]

Indeed it was only four days later that a combined force of German troops and three thousand warriors from the court took Ndungutse's residence by surprise at dawn. They killed about fifty men in a brief battle. The Abatwa, including Basebya, vanished through the grass walls of the houses and enclosure. A spy who had been planted among the rebels by the court pointed out Ndungutse to the German administrator. As Ndungutse tried to leap the enclosure at the back, the administrator shot him dead. For many, Ndungutse had actually been lost four days before when he had surrendered Rukara and shattered his own identity as protector.[51]

A week later Rukara was tried and found guilty of having killed Loupias. After hearing that he was to be hanged, Rukara remarked, 'When a man has a great name, he must be prepared to die for it.'[52] While walking to the gallows, his hands bound together Rukara managed to seize the bayonet of the soldier preceding him and to kill one more of the enemy before others shot him dead.[53]

Shortly after, Basebya allowed himself to be deceived into a meeting with a notable that turned out to be an ambush. It is hard to imagine the wily bandit trusting any notable so far unless he had been one of his protectors for a long time. Soldiers disguised as servants of the notable took Basebya prisoner. The Germans then shot him too, making him the last of the three leading rebels to die by betrayal from a fellow Rwandan.[54]

Throughout April and into May, the German soldiers and court warriors swept first west, then back east across northern Rwanda. Their orders were to punish 'the insubordinate peoples and chiefs by causing the greatest possible damage until they are completely submitted.'[55] A missionary who witnessed the sad spectacle wrote: 'The war continues; the [A]Batutsi massacre, are without mercy, half of the population . . . will be destroyed. Groups of women are led away and will become the booty of the great chiefs.'[56] Having lost their leaders, the people submitted for the time being. The notables, so recently humiliated by the rebels, returned in triumph to rebuild – with local labour – even grander residences. Instead of just encouraging more intensive occupation, the Germans now insisted on it. They saw to it that several of the tougher notables were given charge of the most independent districts. More central control – increased taxation, forced labour and interference in community life – was the immediate result of the rebellion, but it was achieved at the cost of ever-increasing dependence on European firepower. When the First World War forced the Germans to flee a Belgian-British advance, the notables too had to withdraw to the central kingdom. Five years later the new colonial administration, the Belgians, helped agents from the centre once again begin taking control of the north. But the northerners never accepted their domination and beginning in 1959 they helped lead the revolution against it. In the republic which followed, they would participate fully in power at the centre.

The basic issue of the 1912 rebellion was the equitable exchange of protection for loyalty. Over the years northerners had worked out varying arrangements with the court in which they had given their loyalty – and certain concrete manifestations of it – in return for royal protection. Beginning in 1905, they were subject to demands which they judged unacceptable, because of the amounts involved, because of the methods used to enforce them, and because of the proportion which benefited notables or foreigners rather than the ruler responsible for their safety. Objectionable in and of themselves, the demands also demonstrated Musiinga's failure to protect his people. The northerners cared little whether the failure resulted from lack of the power or of the will to protect, they saw it as violating their arrangements. They withdrew their loyalty from Musiinga and gave it to Ndungutse who promised to restore order not just in the region, but in the kingdom as a whole. In defending customary rights and in seeking to preserve the system by reforming it, they resembled rebels in many similar movements in Europe and Asia.[57]

While many of the rebels aimed first to remove Musiinga from power, expecting all needed changes to follow after that, not all shared this goal. Rukara particularly stood out as one who thought more limited local action against the notables could recall the ruler to his obligations. Others, also focusing more on the region than the country as a whole, wanted to direct their attacks primarily at Europeans and their agents. As leader Ndungutse doubtless knew how to exploit ambiguities and leave contradictions among these goals unresolved. But the people themselves were adept at seeing the movement as they wanted it to be. Because they focused selectively on just those elements which met their needs, people of all social groups and many diverse interests – from Abatwa bandits through common cultivators, ordinary pastoralists, local leaders to notables of the highest circles at court – were able to commit themselves to this uprising.

In discussing peasant wars, Eric Wolf has pointed out the importance of 'marginal' people, 'rootless intellectuals', as catalysts to action.[58] In the 1912 rebellion, both Ndungutse and Muhumusa seem to have filled this role. They were 'intellectuals' who put people into contact with a wider world of ideas and goals. But perhaps even more important, they were truly 'rootless'. In a society where individuals were identified in the context of their whole lineages, these two were strangers whose parents were unknown. Among a people of high historical consciousness, these two were outsiders with reputations but no known past. In contrast to them, Rukara, the well-known local hero, was hindered by his very rootedness in the community from ever becoming leader of a mass movement. It was easier for outsiders to make new alliances than for local leaders to end old vendettas. Basebya, although a northerner, was in a sense also a 'marginal man'. Having lived at court, he could offer his followers a taste of ideas and habits unfamiliar in the region. But he too was set apart as much by his origins as by his past experience. As one of a group scorned by the vast majority of Rwandans, he would not ordinarily have been acceptable as a colleague, far less as a leader. In the first decade of the century, however, circumstances were sufficiently threatening and his skills sufficiently needed for hundreds to accept his commands.[59]

Ndungutse and Muhumusa both had to recognise that the Rwandan royalist tradition offered the most widely accepted basis for legitimacy in the area and both adopted it as they set their sights on moving into the Rwandan heartland. But Muhumusa owed her first appeal to her power as a medium for Nyabingi and she never gave up this legitimacy even as she added a royal identity to it. Ndungutse inherited a charismatic power from her which enabled him to attract those northerners who rejected his claim to royalty. Not specifically linked to Nyabingi and useful only in confronting oppression, this more amorphous spirit justified resistance and gave it the hope of success. Once Ndungutse had been killed, this spirit passed to others. In 1913, several times during the First World War, in 1928 and again in 1935, strangers claiming to be Ndungutse inspired groups of northerners to take to arms against their exploiters. One European commented in 1935 that many middle-aged men had said that they had been hearing of Ndungutse all their lives and 'that if he still exists he must be a very old man'. The European concluded that Ndungutse was 'at present merely a name'.[60] He was wrong. Ndungutse had become a spirit, the incarnation of resistance to oppression. By taking on this identity resisters too could acquire legitimacy. While Ndungutse as ruler dominated the 1912 rebellion, Ndungutse as rebel assumed the more important role in later years, foreshadowing the time when the shout would prove greater than the drum.

NOTES

1 This paper is based on field work supported by the Foreign Area Fellowship Program in 1969 and the American Council of Learned Societies-Social Sciences Research Council in 1980–81. I acknowledge gratefully their assistance. I wish to thank the Institut National de Recherche Scientifique, Butare, and its Director, Cyprien Rugamba, under whose auspices this research was undertaken. François Ntaganira, Jean-Baptiste Barugahare, Antoine Rutayisire, and Jean-Baptiste Munyandamutsa helped me to understand Ikin'yarwanda and the Rwandan way of doing things.

2 Nkomeyeho; Ngorore; Bamurakura; Rutamu; Gumira; Ngaboyisonga; Bitenderi; Gasimba; Ndenzago; Bazirake; Rusabagira. Rwandan informants are identified more fully in the Appendix to this chapter. See also Jan Czekanowski, *Forschungen im Nil-Kongo-Zwischengebiet*, Vol. 1 (Leipzig: Klinkhardt, 1917), pp. 247–8, 264, 269. For a study of a similar process of expansion in south-western Rwanda, see M. Catharine Atterbury Newbury, 'The cohesion of oppression: a century of clientship in Kinyaga, Rwanda' (unpublished PhD thesis, University of Wisconsin-Madison, 1975).

3   Ngaboyisonga; Ndenzago; Simbagaya. For a detailed narrative of Rwabugiri's reign, see Alexis Kagame, *Un Abrégé de l'Histoire du Rwanda de 1853 à 1972* (Butare: Editions Universitaires du Rwanda, 1975), pp. 26–103.

4   Rutamu; Gumira.

5   Bitendiri; Gasimba; Ndenzago; Rusabagira. Czekanowski, *Forschungen*, pp. 247, 264, 269.

6   Rushaki; Bilihanze; Serutongi; Kanyamugenge; Bitenderi; Ngurube; Kimonyo; Gasekuru. Anon., *Historique et Chronologie du Rwanda* (n.p., n.d.), p. 153.

7   Alison Des Forges, 'Defeat is the only bad news: Rwanda under Musiinga' (unpublished PhD thesis, Yale University, 1972), pp. 21–102.

8   ibid.; Nyirakabuga.

9   Makeri; Rutabagisha; Ngerageze; Kamere; Mutabazi; Busuhuko; Mutarambirwa. R. P. Martin, 'Notice historique sur le Bumbogo' and [Robert] Schmidt, 'Notes sur l'histoire récente de la province du Bumbogo', both in the J. M. Derscheid Collection of Documents on Rwanda and Burundi, consulted courtesy of Professor Rene Lemarchand. William Roger Louis, *Ruanda-Urundi, 1884–1919* (Oxford: Clarendon Press, 1963), pp. 124–5. Diaire de Rwaza, February through October 1904, 24–30 December 1904, 1–29 January and 3–7 March, 1905, consulted at the parish of Rwaza, but now also available at the Archives des Missionnaires de Notre Dame de l'Afrique (Pères Blancs), Via Aurelia, Rome, Italy.

10 F. Geraud, 'The settlement of the Bakiga', in Donald Denoon (ed.), *A History of Kigezi* (Kampala: The National Trust, n.d.), pp. 50–52.

11 Kamere; Semarora. Czekanowski, *Forschungen*, pp. 246–8. Diaire de Rwaza, 4 April 1908. Felix Dufays, *Pages d'epopée africaine: jours troublés* (Ixelles: Librairie Coloniale, 1928), pp. 65–7.

12 Rusabagira. The number is not meant to be taken literally. Dufays, *Pages d'epopée africaine*, pp. 58–62.

13 Birahita; Kambali; Mikubanyo; Serutongi. Marcel Pauwels, 'Le Culte de Nyabingi', *Anthropos*, vol. 46, no. 3–4 (1951), pp. 337–57; Elizabeth Hopkins, 'The Nyabingi cult of southwestern Uganda', in Robert I. Rotberg and Ali A. Mazrui (eds), *Protest and Power in Black Africa* (New York, Oxford University Press, 1970), pp. 258–336. J. E. T. Philipps, 'The Nabingi – Anti-European Secret Society in Africa', *Congo*, 1928, pp. 319–21; M. J. Bessell, 'Nyabingi', *Uganda Journal*, vol. 6 (1938), pp. 73–86.

14 [German Resident Richard] Kandt to the Fathers at the Mission of Kabgayi, J. N. 1356, 3 January 1911, Correspondance Officielle, Archives de l'Archéveché de Kigali.

15 Rusabagira.

16 Diaire de Rwaza, 17 March 1905.

17 Karyabgite; Gumira; Colline. Martin and Schmidt, Derscheid Collection.

18 Sebagenda; Mitima; Bichunchu; Semarora; Ndenzago. Martin; and Anon., 'Situation Politique du Bumbogo, 12/31/39', Derscheid Collection. *Historique et Chronologie*, pp. 153, 157–8.

19 Nturo.

20 Karyabgite; Miruho; Mutarambirwa; Kazeyire; Kanyandekwe; Munyampeta; Sabini. Diaire de Rwaza, February through April 1905. I. Reisdorff, 'Enquêtes Fonciers au Rwanda', mimeographed, 1952, p. 18.

21 Rusabagira; Karyabgite; Makeri. Diaire de Rwaza, 16 August 1905, 10 January 1909. Cahier de Conseil, Mission de Rulindo, 18 April 1909, Archives des Missionnaires de Notre Dame de l'Afrique, Rome. Martin and 'Situation Politique du Bumbogo, 12/31/29', Derscheid Collection. H. Willems, 'Administration du Territoire du Point de Vue Indigene, Residence du Ruanda, Territoire du Ruhengeri', Prefectural Archives, Ruhengeri.

22 Diaire de Rwaza, 4, 28 April, 1 July 1904, 1 December 1911; Diaire de Nyundo, 3 May, 15 and 29 July, 8 August 1911, Archives des Missionnaires de Notre Dame de l'Afrique, Rome; Dufays, *Pages d'épopée africaines*, pp. 30–31. Czekanowski, *Forschungen*, p. 271.

23 Diaire de Nyundo, 25 August 1911.
24 Toringabo. Also Bahinbano; Makeri; Rutabagisha.
25 Gasimba.
26 Gumira; Guriro; Nkuriye; Mitima; Nyamuhinda; Mutabazi; Toringabo; Gashyekero; Sebitende; Habyarinka; Gahakwa. Diaire de Rwaza, 12 January 1904, 12 November 1907, 5 October 1909, 3 May 1910. Diaire de Nyundo, 18 January, 19 February 1908. Willems, 'Administration du . . . Ruhengeri'. Czekanowski, *Forschungen*, p. 274.
27 Karyabgite; Nyirakabuga. Diaire de Rwaza, 30 March, 23 August, 1909. Diaire de Kabgayi, 10 July 1909, Diaire de Nzaza, 24 June 1909, Archives des Missionnaires de Notre Dame de l'Afrique, Rome. [Resident] Von Grawert [to Governor], J. no. 18 [1907], Archives Africaines, Dossier Filme No. 1727 (1205), Brussels. Louis, *Ruanda-Urundi*, pp. 145–8.
28 Czekanowski, *Forschungen*, p. 247.
29 Habyarinka; Ndenzago; Rugirankana; Semarora; Mitima.
30 Semarora; Rugirankana; Mutabazi; Ndenzago. Diaire de Rwaza, 5–9 April, 12–26 April 1910; Diaire de Kabgayi, 9 April 1910. Kandt to Governor, H. No. 1200, 16 November 1910, Dossier Filme, no. 1727, Archives Africaines, Brussels. Father Léon Classe to White Fathers, 21 October 1910, Correspondance Religieuse, Mission de Rwaza; Father Delmas to Lieutenant Falkenstein, 13 October 1910, Correspondance Officielle, Mission de Rwaza, consulted at the parish of Rwaza.
31 Diaire de Rwaza, 5 May, 11 June, 10–11 October 1910. Father Delmas to Mgr. Hirth, 25 July 1910 and 20 November 1910, Correspondance Religieuse, Archives des Missionnaires de Notre Dame de l'Afrique, Rome.
32 Yowana Ssebalijja, 'Memories of Rukiba and other places', in Denoon, *History of Kigezi*, p. 182. Karyabgite; Rwatangabo; Nyangabo; Mburanumwe. Diaire de Nyundo, 12 November 1911; Diaire de Rwaza, November 1911. Kandt to White Fathers, 7 November 1911 and Delmas to Resident 3 November 1911, Rwaza Correspondance Officielle. P. Ngologoza, *Kigezi and Its People* (Dar es Salaam, East African Literature Bureau, 1969), pp. 51–4. Hopkins, 'Nyabingi cult', pp. 271–3.
33 Ngologoza, *Kigezi*, p. 54. Ssebalijja, 'Memories of Rukiba', pp. 183–4; Hopkins, 'Nyabingi cult', pp. 273–4. F. Bamunuka-Rukara, 'Bakiga resistance and adaptation to British rule', Denoon, *History of Kigezi*, p. 270. D. G. Schloback, 'Die Vermarkung des deutsch-englischen Ruanda-Grenze, 1911', *Deutsches Kolonialblatt*, 1912, pp. 1041–6. E. M. Jack, *On the Congo Frontier* (London: T. Fisher Unwin, 1914), pp. 239–40.
34 Diaire de Nyundo, 12 November 1911; Diaire de Rwaza, November 1911. Delmas to Resident, 3 November 1911, Correspondance Officielle de Rwaza.
35 Semusaza; Byahene; Gumira.
36 Karyabgite; Nyirakabuga; Nturo; Ndenzago. Diaire de Rwaza, 8, 11 February 1912. Diaire de Kansi, 9 February 1912, Archives des Missionnaires de Notre Dame de l'Afrique, Rome. Louis, *Ruanda-Urundi*, p. 154. Jean-Pierre Chrétien, 'La revolte de Ndungutse (1912), Forces traditionnelles et pression coloniale au Rwanda allemand', *Revue Française d'Histoire d'Outre-Mer*, vol. 59, no. 217 (1972), pp. 645–80.
37 Rwatangabo; Nyangabo; Mburanumwe; Karyabgite. Diaire de Rwaza, 24 February, 26 February, 7 April 1912. Chrétien, 'Revolte de Ndungutse', pp. 648–9.
38 Karyabgite; Mugabontazi; Muhama; Habyarinka; Kanyamudari. Diaire de Rwaza, 3 February 1906; Diaire de Kabgayi, 7 February 1906; Dufays, *Pages d'épopée africaines*, p. 61.
39 Kambali; Birahita; Mikubanyo; Serutongi; Kanyamugenge. Diaire de Nyundo, 2 February 1912.
40 Mikubanyo. Diaire de Rwaza, 7 April 1912. 'Jahresbericht, Ruanda, 1911 by Resident A. I. Gudowius', Archives Africaines, Dossier Filme 1729 (1205). Dufaye, *Pages d'épopée africaines*, pp. 74–5.
41 Karyabgite. Kagame, *Abrégé*, p. 163.
42 Karyabgite.
43 Diaire de Rwaza, 22 February 1912. Gumira; Mugabontazi; Karyabgite; Nyirakabuga.

44 Gudowius, 'Jahresbericht 1911'. Karyabgite; Bazatoha; Semarora. Diaire de Rwaza, 4 March, 7 April 1912. Chrétien, 'Révolte de Ndungutse', p. 657. Kagame, *Abrégé*, p. 161.
45 Karyabgite. Godowius, 'Jahresberight 1911'.
46 idem. Semarora; Bazatoha. Diaire de Rwaza, 15, 21, 24–26 February 1912; Diaire de Kansi, 9, 26 February 1912.
47 Diaire de Rwaza, 5 January, 6, 10, 29 February, 7 April 1912. Father Delmas to Resident 26 July 1912, Delmas to Musiinga 11 September 1912, Rwaza Correspondance Officielle.
48 Diaire de Nyundo, 2 February 1912. Chrétien, 'Revolte de Ndungutse', pp. 667–8.
   Diaire de Rwaza, 5 January, 6 February, 7–8 April 1912. Chrétien, 'Revolte de Ndungutse', pp. 674–5.
50 Karyabgite.
51 Karyabgite; Rwatangabo; Semarora; Bazatoha; Habyarinka; Kanyamudari; Ngaboyisonga. Diaire de Rwaza, 11, 13 April 1912; Diaire de Kansi, 21 April 1912. *Historique et Chronologie*, pp. 143–4. Chrétien, 'Revolte de Ndungutse', pp. 650–51. Kagame, *Abrégé*, pp. 163–6.
52 Busuhuko; Ngerageze.
53 Semarora; Mitima; Rugiranakana; Habyarinka; Kanyamudari; Ngaboyisonga. Diaire de Rwaza, 18 April 1912. Dufays, *Pages d'épopée*, p. 76. *Historique et Chronologie*, p. 142.
54 Nyirakabuga; Karyabgite; Mutarimbirwa; Toringabo; Muhama; Mugabontazi; Habyarinka. Willems, 'Administration du … Ruhengeri'. *Historique et Chronologie*, pp. 144–5. Kagame, *Abrégé*, pp. 167–9.
55 Louis, *Ruanda-Urundi*, p. 156.
56 Diaire de Rwaza, 3 May 1912.
57 See Roland Mousnier, *Peasant Uprisings in Seventeenth Century France, Russia and China* (New York: Harper and Row, 1970); Barrington Moore, Jr., *Social Origins of Dictatorship and Democracy* (Boston: Beacon Press, 1966); and James C. Scott, *The Moral Economy of the Peasant* (New Haven: Yale University Press, 1976).
58 Eric R. Wolf, *Peasant Wars of the Twentieth Century* (New York: Harper & Row, 1969), pp. 287–9.
59 Some Rwandans tried to make his status accord better with his power by saying his mother was actually one of the Abatuutsi elite, not one of the scorned Abatwa.
60 Hopkins, 'Nyabingi cult', p. 322.

Appendix I: Rwandan Informants

Name	Father's Name	Clan	Location of Interview
Bahinbano	Makambira	Abagesera	Rushaki, Byumba
Bamurakura	Bigumire	Abazigaba	Kiyombe, Byumba
Bazirake	Mishura	Abagesera	Bugarura, Ruhengeri
Bichunchu	Mpakaniye	Ababanda	Gatonde, Ruhengeri
Bilihanze	Ngwabinje	Abazigaba	Kiyombe, Byumba
Birahita	Butozo	Abasiinga	Butaro, Byumba
Bitenderi	Ndinda	Abasiinga	Muvumba, Byumba
Busuhuko	Mihiko	Abasiindi	Bugarura, Ruhengeri
Byahene	Ruhunga	Abanyiginya	Nyakibungo, Butare
Colline	Segatwa	Abatsobe	Nyarugenge, Kigali
Gahakwa	Nyamuco	Abungura	Muremure, Ruhengeri
Gasekuru	Biruga	Abasiinga	Butaro, Byumba
Gashyekero	Mvuyekure	Abagesera	Karago, Byumba

Gasimba	Magumirwa	Abazigaba	Butoozo, Byumba
Gumira	Rugamvu	Abanyiginya	Mayaga, Butare
Guriro, Isaie	—	Abagesera	Karago, Byumba
Habyarinka	Rwamagege	Abasiinga	Bugarura, Ruhengeri
Kambali	Muhuruzi	Abasiinga	Butaro, Byumba
Kamere	Nshaka	Abasiinga	Rwaza, Ruhengeri
Kanyamudari	Kimonyo	Abazigaba	Bugarura, Ruhengeri
Kanyamugenge	Gacinya	Abacyaba	Butaro, Byumba
Kanyandekwe	Ntabandwa	Abasiindi	Muvumba, Byumba
Karyabgite	Sendashonga	Abashambo	Tumba, Byumba
Kazeyire	Bwankazi	Abasiinga	Muvumba, Byumba
Kimonyo	Rwamihingo	Abazigaba	Muvumba, Byumba
Makeri	Rwamagaju	Abasiindi	Tumba, Byumba
Mburanumwe	Miburo	Abazigaba	Butaro, Byumba
Mikubanyo	Nsekuye	Abacyaba	Butaro, Byumba
Miruho	Gateba	Abasiindi	Gatonde, Ruhengeri
Mitima	Bigaruka	Abazigaba	Buhanda, Ruhengeri
Mugabontazi	Kinyana	Abatsobe	Gisozi, Kigali
Muhama	Ntabana	Abanyiginya	Kabuye, Kigali
Munyampeta	Rwakarengwa	Abashambo	Muvumba, Byumba
Mutabazi	Bishaka	Abagesera	Rwaza, Ruhengeri
Mutarambirwa	Muhirwa	Abagesera	Gatonde, Ruhengeri
Ndenzago, Ignace	—	Abagesera	Rwaza, Ruhengeri
Ngaboyisonga	Muheto	Abasiinga	Rwaza, Ruhengeri
Ngerageze	Mushokambere	Ababanda	Bugarura, Ruhengeri
Ngorore	Bwansheja	Abazigaba	Kiyomba, Byumba
Ngurube	Semirasano	Abakono	Muvumba, Byumba
Nkomeyeho	Muzerwa	Abagesera	Kiyombe, Byumba
Nkuriye	Kamonyo	Abagesera	Bushiru, Ruhengeri
Nturo	Sebinagana	Abazigaba	Ngoma, Butare
Nyamuhinda	Rugiri	Abasiinga	Gaseke, Ruhengeri
Nyangabo	Sebutama	Abazigaba	Butaro, Byumba
Nyirakabuga	Cyigenza	Abega	Nzaza, Kibungo
Rugirankana, Joseph	—	Abasiinga	Rwaza, Ruhengeri
Rusabagira	Nshunguyinka	Ababanda	Gatonde, Ruhengeri
Rushaki	Banyoya	Abazigaba	Kiyombe, Byumba
Rutabagisha	Nzarubara	Abasiindi	Tumba, Byumba
Rutamu	Kabera	Abanyiginya	Nyakizu, Butare
Rwatangabo	Ntiruhuga	Abazigaba	Butaro, Byumba
Sabini	Sebiruri	Abasiinga	Gaseke, Ruhengeri
Sebagenda	Bizuru	Abagesera	Giciye, Ruhengeri
Sebitenge	Rutungura	Abasiinga	Bushiru, Ruhengeri
Semarora	Birara	Abasiinga	Bumara, Ruhengeri
Semusaza	Muyoboke	Abasiindi	Murama, Gikongoro
Serutongi	Ndalizengana	Abazigaba	Butaro, Byumba
Simbagaya	Mirimo	Abazigaba	Rwaza, Ruhengeri
Toringabo	Mihigo	Abungura	Giciye, Ruhengeri

15 Mekatalili and the role of women in Giriama resistance

CYNTHIA BRANTLEY

In our past studies of African resistance to colonial rule, particularly agrarian protest, we have assumed that men alone were resisting. We have taken little opportunity to explore the role of women. The problem with studies of Giriama resistance against the British, including my own book, was that one necessarily had to focus on the actual fighting which took place from mid-August to October 1914.[1] But there are two especially critical components of this history. One is that, despite the degree to which all Giriama, including government headmen, opposed the tax demands and the call for their young men to become labourers, the men lacked the means to respond effectively to British demands. The second, more telling aspect is that an old woman, Mekatalili (Manyazi wa Menza), called women together to deal with the threat and the potential disruption that British demands were causing. Mekatalili directed a successful effort, beginning in June 1913, to stop labour recruiting, re-establish the traditional government, and reclaim 'Giriama for the Wa Giriama [sic]'.[2] Mekatalili's campaign offers a rare opportunity to learn which issues the women were raising and why they became so vitally involved in resistance. An examination of several sources – contemporary statements made by participants, documentation recorded by British officials at the time, and oral interviews conducted in 1970 – reveal the problems viewed from the Giriama perspective.[3]

Mekatalili, specifically, was critical in giving Giriama protest its initial form. The main questions are why and how the women took the initiative in a situation where the British were usurping local Giriama government with the main aim, from the Giriama perspective, of obtaining labourers: locally, as porters or workers on roads and buildings, but mainly as contract workers, either for the government in Mombasa or for burgeoning European plantations along the coast. This paper will argue that, since the Giriama traditional political system lacked the means to respond effectively, Mekatalili acted like a charismatic politician, coalescing the interests of Giriama women, powerless young men, and ageing elders to meet the challenge by reviving old customs and opposing all Giriama who were assisting the British.

Before we examine when Mekatalili did, we need to know the background which places this resistance in perspective. Nineteenth-century Giriama history was marked by several developments crucial to understanding the position of the Giriama when the British arrived to administer them in 1912. From mid-century, there had been a steady dispersal of population around and beyond an initial, fortified settlement, called a *kaya*, in the forest west of Mombasa. They moved northward in the hinterland and, after passing Kilifi, they moved into lands nearer

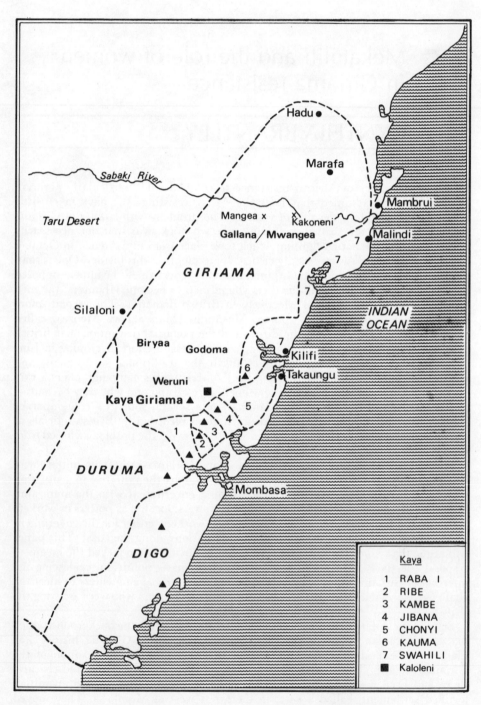

Map 6: *Nineteenth century Giriama expansion in relation to other Mijikenda*[74]

the coast. They eventually dominated the hinterland of Malindi and by 1890 they had extended north beyond the Sabaki River. They preferred the interior in order to graze cattle as well as grow crops, but by the end of the century they had lost almost all their cherished cattle to Maasai raids and to disease. They then became almost exclusively agricultural producers. Out of this migration grew a logical modification of their governmental system based on a single council of elders in the *kaya* into one based on numerous local councils of elders. With these migrations, generational conflicts emerged as young men became less dependent on their elders. Rituals became less critical. However medicines, controlled mostly by senior elders, continued to be used for many problems and for ultimate arbitration.

The main challenge to Giriama independence came from Arabs who had developed coastal maize plantations, using slave labourers, after the mid-nineteenth century. The Arabs tried a few times to enslave numerous Giriama and fought with them occasionally, but the Giriama always won on their own terrain. Some Giriama became slaves, some became Muslims, and a few became slave-owning producers in their own right. By the third quarter of the century, however, the Arabs and the Giriama had made an accommodation and had established good trade relations, so that Giriama-produced maize provided a significant proportion of the exports from Malindi.[4]

However, British domination of the coast resulted in the abolition of the slave trade in the 1870s which led to the failure of the Arab plantations. When the British came into conflict with the Mazrui Arabs of Takaungu in 1890, the Giriama initially gave refuge to their trading allies, until pressure and circumstances made it clear that the British had the upper hand. Giriama were convinced of the need to remain neutral and some were forced to take an oath as a measure of their good faith. The British-identified spokesman, Ngonyo wa Mwavuo, who had become a large maize producer with the help of slave and client labourers and had added to his riches as an ivory trader, was given permission to live at Marafa, north of the Sabaki River, as a reward for his negotiations. Except for some erratic tax collections and an occasional visit from the British District Commissioner at Malindi, Giriama were rarely exposed to British government from 1895, when the Protectorate was declared, until 1912, when the British sent their first officer into the hinterland to activate administration and obtain Giriama labourers for European-owned coastal plantations. These efforts were begun with hopes to recover (from cotton, rubber, and sisal) the former agricultural successes of the Arabs. British–Giriama contacts were minimal. Besides erratic tax collection, there were occasional requests for porters for administrative safaris and workers for the plantations. Giriama would usually agree to work, probably to be polite, but rarely showed up unless they were hungry and then worked for only a short time. There were a few mission efforts made – by the Church Missionary Society at Jilore, Vitengeni, and Kaloleni and by the Catholics at Mwabayanyundo. Giriama reception of Christianity was cool, at best, and people were mostly attracted to the fertile land these missions made available and, in a few cases, the chance to attend school through mission auspices. But the missions had only a few adherents. By 1912, only Mwabayanyundo and Kaloleni retained European missionaries, Vitengeni was served by an itinerant white preacher, and Jilore had been left to the local Giriama to run.[5]

It was in a special context that the British colonial administration established its first interior station among the Giriama and sent Arthur Champion there as the first assistant district commissioner. The coast and its hinterland were areas that needed to be drawn into the larger colonial administration of British East Africa; there was a serious labour shortage which the Giriama were earmarked to meet; and the London syndicates owning coastal plantations were putting pressure on the newly appointed provincial commissioner, Charles Hobley, to provide labourers to make their efforts successful.[6]

The Giriama met the demands for taxes and labourers with little co-operation. The headmen who had initially been designated by Giriama to collect taxes in 1901 when the British required them to, had done their job reluctantly and sporadically, without drawing much attention from either Giriama or British authorities. Now they were given the challenge of being in charge of councils, collecting taxes, clearing roads, trying criminals, and obtaining labourers. The elevation of headmen to greater authority seemed innocuous enough to the British, but it strained at the very place where the modification of Giriama government had become the weakest. Modifications of Giriama government had facilitated their economic expansion but had not prepared them for a unified, organised response to the kind of external threat that they experienced from British colonialism. Traditionally, men were initiated as councillors and a ruling generation was installed approximately every forty-five years. The presidency rotated among the senior spokesmen of the six clans. Participation in either the initiation or installation rituals had been easy when all Giriama lived in or near the *kaya*. Once they were dispersed over one-hundred-and-fifty kilometres, the rituals were at first modified and eventually abandoned. The last ruling generation to be installed was the one called Kavuta, which came to rule in the 1870s.

In 1912, there were four ranks of Giriama elders. Those of highest rank had been initiated as the senior members of the Kavuta ruling generation. Only three remained alive, including Pembe wa Bemebere, who resided in the *kaya*; Bogosho wa Menza, who lived near the *kaya* in Bamba; and Wanje wa Mwadorikola, who resided north of the Sabaki River. Next were the representatives of the six Giriama clans. This included the three Kavuta elders plus three others, slightly younger. Traditionally, all of these men were supposed to reside in the *kaya* or at least go there on call when needed, but distance and changing times had made both practices fall away. They lived in various locales and the practice of their authority had diminished, except locally. However, these men controlled the most powerful Giriama medicines, and were the senior members of the four secret societies: Fisi, Habasi, Gohu, and Kinyenzi.

At the third level were junior elders of the Kavuta ruling generation who, by the accident of the timing of their birth, would always remain junior elders, theoretically to be replaced by younger men who would become the senior elders of the next ruling generation. But installation procedures had been so delayed, that the chances were slim that the remaining Kavuta elders had the knowledge or inclination to initiate the next group. The Giriama had basically moved beyond this traditional system by their own design. Although British representatives did not realise it, their problem was that the Giriama had designated government headmen in 1901 from this junior group precisely *because* they were regarded as less significant elders.

The fourth group comprised the men called *nyere*. The term once referred to young men prior to their initiation as councillors. Now the distinction was blurred; men became councillors in various ways, but anyone still designated as *nyere* remained in the weakest position of all.

It is no wonder that the British could not comprehend the ranking of elders because, even among the Giriama, there was confusion and conflict. This is best represented by a single person, Ngonyo wa Mwavuo, the man who had become wealthy and influential as a result of his production and trade. His father had come from the nearby Digo tribe; he became a Giriama by joining a clan, but he had never been initiated into rank. Since Ngonyo was given permission to move north of the Sabaki River following the Mazrui rebellion in 1895, he was considered by Giriama and British alike to be a leader and spokesman. And, because he could not be ignored, he was designated an elder by the Giriama and a headman by the British. Wanje, the Kavuta in the north, resided far west of Ngonyo and was totally unknown to the British before 1913.

Whereas rank had once been rigidly determined by age there was now less certainty. This was especially true for the able-bodied 'young men', whom the British wanted as labourers. They had, because of changes and conflict within the Giriama system, no hopes of being initiated into the ruling generation of the future, although they were of the appropriate rank and age to serve in that capacity. Their main access to power was through individual achievement as producers.

This breakdown in the political system, more than any other factor, helps to explain why Mekatalili was in an opportune place to raise the issues confronting the Giriama. We will see how she incorporated the young men's wishes, the influence of the Kavuta elders, and Ngonyo's prestige to rally the Giriama against those people – primarily the headmen – who were assisting the British.

Why did Mekatilili come to the front and how did she do it? Giriama women had little ostensible power in Giriama society. They did not have a parallel council to the men's. Women became ranked individually, as councillors through the ranking of their husbands, more for prestige than function. The women had one secret society, Kifudu, which was best known for ecstatic dances performed at funerals, but they also had an oath, called *mwanza*, 'on which the most binding oaths for women [were] sworn.'[7] Their greatest influence, however, probably came from their powers of persuasion, as daughters, sisters, wives and mothers.

Mekatalili's first concern seemed to come from her own observations. She witnessed the pressure of the British officials who were active near her home. The first intensified labour demands in 1912 were made among the northern Giriama close to Mekatalili's home. Champion established his first administrative centre at a place called Mangea, which was just south of the Sabaki River and west of Jilore. Though Champion was following direct orders, he tried to do too much, too soon. He attempted to determine a census, collect taxes, confirm headmen and their salaries, set up the native councils, and obtain labourers. Moreover, he was to stop the trade in palm wine to alleviate drunkenness, and, because the British had made the ivory trade illegal, he confiscated tusks at every opportunity, thus cutting severely into two sources of Giriama wealth. Because the Giriama traded grain for rupees, they had money to pay their taxes, however reluctantly, when caught by Champion. Even the government headmen were unco-operative in providing labourers, particularly since their authority over the young men was ineffective. The Giriama disliked Champion's forcefulness and were also 'angry that Champion's porters were raping the women'[8] though Champion was probably unaware of this initially. Despite his vigorous work, in his first report in May 1913, Champion argued that the British should drop their labour policy and give assistance to Giriama agricultural production.[9]

The Giriama saw no signs of Champion's recommendations being implemented. Instead, demands intensified. The pressures for labourers from Protectorate headquarters and from London investors were too critical. Hobley, the provincial commissioner, held fast to the original plan and in June 1913 toured personally, as a demonstration of the government's power. He reiterated Giriama responsibility for supplying labour. In addition, he offered a rather jolting interpretation of the role of 'government' headmen and 'government' councils and suggested that Giriama rights to some of their land might be in jeopardy.[10]

Mekatalili was among the Giriama who heard Hobley speak, but she was unconvinced and unafraid. He 'spoke to them on every possible occasion on the question of supplying labour'.[11] The Giriama responded that they were 'afraid that the Europeans would ship them off to some distant land'. (He wrote in his report later that 'this might have been alleged of the Arabs some years ago but I was astonished at the idea being coupled with Europeans and asked if it had occurred, of course no cases were forthcoming'.[12] He thus dismissed the entire objection as irrational and disruptive.) Hobley's argument that the Giriama were losing money by not supplying labourers was unconvincing to agricultural producers who had the rupees to pay their taxes. He wrote in his report that 'The usual panacea suggested to get labourers is to increase taxation but I doubt this would do anything except increase the burden on the already overworked women.'[13] The Giriama did not need the money from contract labour, and the issue of slavery was a real one to them. They showed suspicion of British intentions. The young men argued in their talks with Champion that labourers had gone to the coast and never returned, and that the call for labour was a government bluff: 'If the government had wanted labourers, then government would not have bought off Arab slaves and removed all obligations to work against their will.'[14]

But labour was not the only issue of disagreement. Giriama councils involved all eligible elders, and no one person had authority over the others. Yet Hobley argued that the 'Native Councils' were no longer simply Giriama councils, but were to serve government in executive and judicial functions. Believing that fear of witchcraft was the main deterrent to the headmen's compliance, he urged the people to defy the medicine men. Although he was trying to assure the Giriama that British power was stronger than the medicines, he was threatening the power of the senior elders who controlled the medicines by disallowing their source of income, their realm of authority, and the basis of their ultimate power. He told the Giriama that if they failed to obey, he would send a military patrol to ensure obedience, and that he proposed to 'acknowledge their tenure of the country north of the Sabaki River only as tenants at will of the government'.[15] This clear suggestion that the Giriama were on the verge of losing their most fertile trans-Sabaki lands, where over a quarter of the population resided, was a serious threat to Giriama. All the people in northern Giriama became extremely agitated, and Mekatalili decided to do something about it. She began a campaign calling on the women and drew upon a tradition of a Giriama prophetess, Mepoho.

Giriama tradition holds that Mepoho prophesied the disruption of Giriama society by the coming of white men. Mepoho, who probably lived sometime in the nineteenth century, argued that just as Arabs had vessels which travelled on the water, white men would come and travel just as fast on land, and even in the sky.

The Giriama would no longer bear big children and the land would go bad. Youth would disobey and disrespect their elders. Boys would take snuff and chew tobacco, girls would marry young, and elders would no longer be able to exercise their power. The Giriama believe that Mepoho did not die a normal death. After her prophecy, she was swallowed by the earth.[16]

Mekatalili found focus here. She argued that the land was being spoiled and the Giriama were losing their customs. She saw the Europeans as the disruptive threat. She called upon the Giriama to save their children – sons and daughters – and to end the conflict between elders and the youth. The men had been unsuccessful at retaining Giriama culture; the women tried to revive it. In my 1970 interviews, a few Giriama confused Mekatalili and Mephoh, assuming they were the same person.[17]

Mekatalili had no special influence or position within the traditional Giriama system. Evidence indicates that she was a widow, and that she had no high rank as councillor. She did, however, live in Wanje wa Mwadorikola's village and was an extremely charismatic, popular woman. She stirred up the the people's interest in the situation as she saw it.[18] After Hobley's visit, she called the women together. In her statement made when she was first caught, she said:

> The rains last year failed and in Mangea and Goshi [south of the Sabaki] there was no food. I heard that the rains had failed on account of the introduction of cents and rupees into the country. I went to Ngonyo to ask if this was true. He told me it was not so. Ngonyo said that the departed spirits of the kaya Giriama are destroying the country. I called together the women and each woman brought maize which was exchanged for two male goats, which were then sacrificed. Ngonyo examined the entrails and said the evidence showed that the spirits were to blame and not the Europeans.[19]

Champion's version was that she argued that the headmen would 'sell young men to the Europeans, that Europeans would send them over the sea and they would be sold as slaves and never see their native land again.'[20] Champion argued that she 'collected a following under the guise of making harmless enquiries into the question of the rains (always a draw,)'[21] but when the district commissioner had first sent Champion into the hinterland he told him it would be difficult to collect for taxes because the rains had been short and there had been drought that year.[22] In November, after she was accused, Mekatalili said:

> There was a meeting at Makambweni where a large number of women were present. I called it . . . Any woman could have gathered the meeting together. No special power is required. I am not a mganaga [witch doctor] and have no medicine. We told Ngonyo that the Europeans in Mangea wanted labour to work at the monthly wage on the grounds that government paid salaries to headmen.[23]

One simple solution, then, would be to remove headmen from their positions and salaries.

A statement made by Mkowa, who was the headman of Garashi, indicated that the elders in his district were unwilling to offer their sons for work until they could consult with Wanje: 'I said this is nothing to do with Wanje. It is the shauri [problem] of the government.'[24] He said he was called to Ngonyo's village where Mekatalili had a large following of women and there was a crowd of elders outside.

> Katalili [sic] told me to give the bags of rupees which government had given me to induce me to send out the young men to work. I replied that I had not

received any rupees. They said if you will not give up the rupees, we shall go and make trouble against you of [sic] Malindi.[25]

After they left Ngonyo's the women went to discuss with Wanje in Masendeni. Wanje said, 'She did not consult me before she called this meeting. She is not a witch. The strength of that woman lies in her energy and powers of oratory.'[26] Mekatalili gained Wanje's support and they both encouraged many people from the northern areas to travel to the *kaya* to consult with the elders, propitiate the ancestors, air their grievances, and restore Giriama government. The women levied fines of five rupees for elders and three for warriors, presumably to have money to pay the ritual costs. Most people responded positively.[27] People who lived in the location of Mangea South heard about the call but refused to go because there had been an incident there in 1908 where young men had attacked their elders and had been severely punished by the British district commissioner.[28] Elders from Kaloleni, a location near the *kaya* but also near the railway (which would allow troops to reach them easily) said later that they did not attend. 'It was a very bad business. We did not approve of it. Our aim is to stay quiet and prosperous under the government and we do not like these secret doings.'[29] It is obvious that those who knew the power of the British government held back. Ngonyo himself, who remembered British actions from the time of the Mazrui rebellion, absolved the Europeans by encouraging the Giriama to go to the *kaya*.[30]

It is impossible to know how many people actually came to the *kaya*, but it was considered to be a big gathering both by the Giriama and the British. They were not all there for the entire time and the first ones probably arrived in early July. Some were still there by mid-August. 'Women came of themselves. Some ran away, others were brought by their husbands.'[31] Young men came along with many elders, three major Kavuta elders, but no headmen. The gathering was dominated by people from the north who wanted to spread the word of the problems to the southern Giriama. The crowd eventually included representatives from all areas of Giriamaland.[32]

All regions of Giriama provided a sheep to sacrifice to the ancestors, and they called in a rain doctor to pray for rain. Mekatalili led the discussions and she complained first about labour demands and then about the jurisdiction of the traditional elders being undermined. She indicated that vital customs were being spoiled and that wages which headmen received gave the government the belief that they could demand labour. She said 'Every elder must pay it back. And if they dare not send it themselves [I myself will] take it to "Bibi Queen", the mother of the Europeans.' [This is a reference to Florence Deed the missionary of Kaloleni who was called Bibi Didi.] 'We are not to fear the Europeans', she said.[33] As a response to the argument that headmen were betraying Giriama, she suggested that headmen should refuse their salaries as a solution to the labour problem.

Having had their discussion, sacrifices and prayers, they then evoked the authority of the ancestors, the elders, and the women by swearing powerful oaths.[34] Mekatalili took no part in the casting of any of these oaths, but she was significant in raising the issues. Oaths were powerful agents among the Giriama. Often the mere threat of an oath was sufficient to change a person's behaviour. But the three oaths – Fisi, *mwanza* of Kifudu, and Mukushekushe – sworn at this gathering were particularly severe because one carried the penalty of death and another was a mother's curse. The Fisi (Hyaena) oath, controlled by a few of the most senior elders, was mostly used as a last appeal in judicial cases with the council of elders and also, as Champion writes, 'used by the tribal elders when ordinary means have failed to check any innovation or departure from ancient custom which they may consider detrimental'.[35] The women's Kifudu oath usually concerned matters of women's behaviour and the fertility of the soil. The Mukushekushe oath was the mother's curse. Male offenders became deaf or dumb and women offenders became infertile. According to Kadidi wa Bembere, one of the women of Biryaa who 'laid' the Mukushekushe: 'Mekatalili's business was to gather the people to checkmate the government's request for labour. She was not however at the oath, nor is she one of our chief women. She was in the kaya, but her grade is too low to permit of her taking part in the oath.'[36]

There is some confusion over how many times the oaths were sworn and the particular proscriptions attached to each by those who took part, probably a deliberate action to elicit fear of death among all Giriama for disobeying any items of the oaths.[37] Those Giriama present used the occasion to air a large number of grievances. The terms of the oaths indicate the degree to which there was dissatisfaction among elders, women and young men. Each oath was sworn, using medicines, sacrifices and ceremony, cursing anyone who did not obey its proscriptions to be 'caught' by its penalty. It was probably not difficult to convince the Kavuta elders to restore their traditional authority and to elevate their councils over those of the headmen, who were now getting most of the arbitration fees. Their oath was thus aimed at the headmen, sworn against Giriama who refused to pay council fees to the elders instead. Included in this oath was anyone who showed the police or hut-counters where people lived, any who attend or witnessed at the headmen's courts, any who counted more than one wife for tax purposes, anyone who helped construct government roads and buildings and anyone who provided information about ivory. They swore against men appearing in shirts and trousers, calling them by the contemptuous term 'korobi' or 'call boys'.[38] Since the women had convinced the gathering that headmen had received money to obtain labourers, the elders swore against government headmen who refused either to give up their salaries or their own sons in response to the demands.

The primary aims of the Kifudu oath sworn by the women were to restore traditions, regulate women's behaviour, and prevent any undermining of the economy. They swore against the bewitching of goats and crops and against those parents who allowed their daughters to marry strangers, especially the Duruma to the south. The Duruma, as matrilineal cattle-keepers, valued Giriama wives who offered a means for Duruma sons to inherit from their fathers.

The women swore against those who washed with soap, stipulating the use of *mukone* (castor oil) sap for bodies, and ash from *mukulu* for clothes. They swore against people who washed clothes at places where people obtained household water. They also swore against the acceptance of low fees for sexual intercourse. This apparently related to the one means women had to gain independent income. The Kifudu society had a ritual in which the women could have a night of sexual intercourse with whomever they wished. They could charge a fee and then use that money for their own celebrations and ceremony. Nobody could be accused of adultery for this action.[39] It seems possible that by asking for increased fees, the women could obtain money to pay the taxes so that they could keep their sons at home. They also swore the Mukushekushe oath, which was a family oath whereby a woman washed her body, including her genitals, threw the water on the ground and disowned any or all of her children who acted against the interests of the family. Thus by bringing together the power of the Fisi oath of the men, the *mwanza* oath of the women's society, and the family oath of Mukushekushe, the

Giriama at the *kaya* on this occasion were acting to recapture Giriama unity and, in that way, oppose the British. There was never at any time any discussion of or preparation for war against the British.[40]

Fisi and Mukushekushe oaths had so long been used to regulate proper behaviour that a rumour that they had been sworn was generally sufficient to cause their proscriptions to be followed. In any case, most of the Giriama were predisposed to opposing the British taxes, labour demands, and judicial system, and to supporting a revival of traditional Giriama government. Only the Fisi oath was known to British officers, who regarded it as the most powerful Giriama oath. This probably reflected the respect that Giriama themselves demonstrated for it. The Fisi had been used periodically throughout Giriamaland to forbid washing in vital water holes during drought, and to cleanse areas of witches. When the medicine was buried, the entire water supply was considered to carry the medicine. Anyone who broke the terms of the oath and drank water was expected to die. During the famine of Magunia in 1899/1900, all the waterholes of Giriamaland were put under the protection of the Fisi oath. In the Mazrui rebellion of 1894, the British officer in charge insisted that the Giriama swear this oath to support the government, but it is a proscriptive, not an assertive, oath. Some Giriama, under Ngonyo's urging, performed a ceremony, but it was administered improperly without attendance by Fisi medicine men. After the oaths were taken at the *kaya* in August 1913, the water used in the oaths was carried throughout Giriamaland by the women. They sprinkled it in water holes as the oath was repeated. In addition to the oaths, the Giriama at the *kaya* made a major decision to reconstitute their traditional councils. Wanje was designated the chief judge of the north, and Pembe and Bogosho were to share duties in the south. They wanted to restore the power of the *kaya*, and encourage many elders to return there to live.[41]

This oathing ceremony at the *kaya* gained momentum from another incident that occurred simultaneously in the north. While people were still gathered at the *kaya*, some young men from Chakama became much bolder in their opposition to recruiting, especially against Champion. Their actions resulted in additional Giriama anger against the British. On 10 August, Champion detained nine young men who had 'displayed a most defiant attitude' when he suggested they go to work. He 'exercised pressure' by having them carry stones for the station at Mangea. On 13 August, some thirty armed men appeared. They attacked the camp, and a nervous police officer fired without orders, killing one man.[42] The attack had been made to free the prisoners, but it was also the culmination of growing hostility. Although it was a Giriama who had been killed, Hobley wanted Giriama punished. Champion held that the loss of a life was sufficient penalty and that the matter should be dropped. Giriama tradition held that when a man was killed by another, the offending clan had to pay blood money. But Hobley saw a chance to control the Giriama. He demanded either the 'ringleaders' for punishment, or for the people to hand over one hundred men for three months paid labour in Mombasa. If they refused, he would send a forty-man patrol.[43]

The matter was further complicated by the fact that the dead man's nearest relative was Ngonyo. Champion offered to pay blood money, but Ngonyo refused, arguing that if the British wanted to institute European law, then the dead man's killer should be punished. Eventually both sides backed down. The British arrested several young men and collected a fine of sixty goats from extremely disgruntled Giriama. Ngonyo, who had subsequently refused to pay his taxes in protest, finally sent some of his people to carry stones as tax payment, probably because he did not want the growing friction to reach a point of explosion. But Giriama remained angry that 'men had been put to crush corn, work they considered women's labour',[44] and Ngonyo must have known that Giriama everywhere would resent the fact that the Europeans had not paid blood money.[45] Acting on their own as warriors, the young men were unable to resolve the labour question, but the incident stirred up Giriama in all regions, including those meeting in the *kaya*. They were angry that the government had not only failed to pay blood money, but that it had also done nothing about two Giriama who had died working on the coastal estate of the Baluchi.[46] The Giriama labour situation was explosive and confronted British officers directly.

Mekatalili's energy and powers of oratory had provided a focus for the general grievances but she had no control over the oaths that were sworn. It would therefore be wrong to regard the meeting at the *kaya* and the proscriptions that emerged from the oaths sworn there as a well-planned and tightly organised campaign. Mekatalili almost certainly did not originate all the action that eventually transpired at the *kaya* nor did the three Kavuta elders apparently work in conjunction and try to control the outcome of the movement. The women were behaving independently. People brought grievances to the *kaya* which were then heard by the elders and supported by oaths. The main aim, apparently, was to restore Giriama unity.[47]

The result of this oathing was immediate; British administration simply ground to a halt. The oaths were secret, anyone who told of them faced death. Champion could not get porters; headmen's councils did not meet; no one paid taxes. Even loyal headmen were afraid. One, in Garashi, was told: 'Even you, Mkowa, we can shoot you with arrows because of this business of the Wazungu [Europeans].'[48]

One man, who had earlier asked to replace his uncle as headman, refused Champion's offer of the office because of the oath: 'I am a Nyika [Giriama], and I shall be killed by it.'[49] The headmen were caught in the middle. It was they whom the British officers blamed for lack of co-operation from the Giriama. And it was they, along with clerks and mission converts, who had most reason to fear the oaths. The headmen knew that the medicine men would not hesitate to poison them if they did not meet the terms of the oath.

Mekatalili's campaign was working: headmen were unable to perform their duties. When Champion eventually learned what had happened to halt government activities, he was furious. He pursued Mekatalili until finally, on 17 October, with the help of Mkowa and his deputy, he arrested her near Garashi. She signed a statement in Champion's presence, but admitted no wrongdoing. Later that same day Champion arrested Wanje (probably at his nearby home in Masendeni). In his October report, Champion conceded that Mekatalili's campaign had been effective: 'Every Giriama is much more afraid of the Kiraho [oath] than of the government; the WaGiriama [sic] boast openly that the government are afraid to fight them and for that very reason have never done so.'[50]

Hobley was apparently even more furious than Champion. He seemed determined to punish the Giriama fully enough to achieve total control and end this troublesome opposition.[51] On 11 November, he made an expedition to the *kaya*. He was accompanied by quite an entourage: a police captain and twenty-four policemen; four Giriama from the *kaya* region who had been pressured to inform Hobley of the oathing; and Champion and another assistant district commissioner who together brought eight headmen and their retinue of forty-three men from the north. Hobley first went to Bogosho's house in Biryaa and then led the group to the *kaya*. Hobley confronted the elders with accusations three different times. Each time all of the elders, including those he had brought, went into private council and each time they returned and admitted a little bit more. They finally pinpointed Mekatalili and the women as the initiators of the campaign.

In retrospect, Hobley's punishment seems severe. He blamed 'agitators' and took none of the Giriama grievances or arguments seriously. He forced the men and women to remove their oaths. Mekatalili and Wanje remained imprisoned and did not participate in the deliberations. They were later deported for five years and sent a thousand kilometres away to highland Kisii, each with a blanket and a small sum for food. Hobley levied a fine of 1500 rupees to pay for his expedition, and he made the Giriama promise to swear a pro-government oath as soon as the crops had been harvested. (There was no assertive oath available, but he was interpreting the oath of 1894 as such.) He forced them to agree, very reluctantly, to move the *kaya* to a more centralised location, and he evacuated and closed the old *kaya*. Although it was never discussed, he later added to the list of punishments the evacuation of the trans-Sabaki.

Presumably Hobley expected that the reprisals could reverse the situation in Giriama so that government headmen would thereafter hold their positions of power. He had discredited the last of the traditional elders: Bogosho through humiliation, Pembe through forced evacuation of the *kaya* and loss of his primary place among the Giriama, and Wanje through deportation. And he had rid the countryside of Mekatalili to whom he referred as the 'witch' who had stirred up the trouble.[52] Had he been correct, then the Giriama should have settled down and begun to provide support for the government councils, collect the fine, provide labourers as requested and, since he had made it one of the requirements, evacuate the trans-Sabaki without any difficulty. This was not the case, however. They paid the fine reluctantly, and it was mostly pro-British Giriama who made the payments. Pressured to perform, Giriama made a mockery of the pro-government oath that was sworn the following March. And even though individuals agreed to the move south of the Sabaki, they began to store extra grain, make additional bows and arrows, and send their cattle up to the Tana River in preparation for the possibility that they might be forced to move.[53]

The British-Giriama war broke out almost a year after the original oathing. When the fighting began in August 1914, most Giriama viewed it as a defensive measure against massive labour demands and the evacuation threat. The British sought desperately needed porters to assist in the campaigns of the First World War in nearby German East Africa. British–Giriama fighting was precipitated by the rape of a Giriama woman by one of Champion's policemen recruiting porters at Vitengeni, and by the dynamiting of the *kaya* as punishment because the Giriama had not built a new one. An outbreak of fighting was guaranteed by the appearance of a company of Kings' African Rifles who had arrived to deal with Giriama unrest after Champion was 'besieged' by Giriama warriors in his camp for a night following the rape. The British used the available troops to attempt evacuation of the trans-Sabaki, to give the Giriama a 'sharp lesson', and to force Giriama compliance. They burned huts and fired at all Giriama who failed to demonstrate immediate support. Giriama fighting lasted over a month. More than a hundred-and-fifty Giriama were killed. The British lost only one policeman.[54]

Mekatalili's role in the 1914 fight is uncertain, but was apparently minor. She and Wanje had escaped from their prison at Kisii in April and, despite their old age, had made their way back to the coast. They conducted meetings and levied fees, but they were caught and sent back to Kisii before fighting began. After three months of troop expeditions, Giriamaland was in total disarray, and the next year was spent with mopping-up forays, completing the evacuation of the trans-Sabaki, obtaining a thousand labourers, and collecting a 100 000-shilling fine. The entire administrative system was disrupted. Famine followed and the British brought in relief food. A 1916 official inquiry into the rights of ex-slaves to coastal lands naturally brought into question Giriama rights to the trans-Sabaki. The number of people evacuated was determined to have been far greater than the government had originally indicated, and the Giriama were allowed to return to their land in 1918.[55]

Meanwhile the native administration still barely functioned and all European officers had returned to the coast. Because of this difficulty, the British became receptive to Giriama requests to revive the traditional council system of the *kaya*. In an ironic twist, this idea received added stimulation when Mekatalili and Wanje were pardoned and returned from exile. They were greeted by a feeling of enthusiasm rare among the Giriama.[56] Even though enough essential customary knowledge had been lost which prevented the *kaya* ever being fully restored, Wanje and Mekatalili went to live in the *kaya* with British support. Wanje became the official head of the *kaya* council and Mekatalili became leader of the women's council, a position which had heretofore been non-existent. They held no official place in the administration, but they became *de facto* leaders of the Giriama in the same *kaya* that had been closed and dynamited as a 'hotbed' of their 'sedition'.[57] They stayed there until they each died, several years later. The men's council, in modified form, survives to this day. The women's council lasted only through Mekatalili's lifetime. In essential ways, although the Giriama had lost the fighting, they won their resistance. British administration remained skeletal; few Giriama subsequently became labourers; Giriama recovered their trans-Sabaki lands; they revived the traditional council as much as was possible, and they restored the *kaya*.

Why did the women get involved? Giriama women were dependent upon their sons for their ultimate position in society, as in many African societies. Their claim to the land that they worked and on which they would live out their lives was ultimately determined through their sons. Women lacked economic independence. Security was first assured through their fathers and then through their husbands, but final security came via their sons. Care in their old age made women dependent on their sons. It is likely, then, that the women were afraid of losing their rights to land and being destitute in their old age if their sons were taken as workers. Their concerns that their men would go off to be labourers made them fearful that they might lose them to slavery. It had happened in the past. Men had gone to the coast and become Muslims and not come back to be part of Giriama society. Some had, in fact, been shipped off to Arabia as slaves after they had been convinced to go work at the coast. A few were captured. Though slavery was not a risk with the British, Giriama fears were based on past reality.

Moreover, even the British recognised that women in Giriama society were the agricultural labourers.[58] Women planted, cultivated, and harvested the grain and produced most of the maize the Giriama sold through the Arab networks to Arabia. It was, as well, the Giriama women who carried the loads of maize from the hinterland out to Malindi for export. As a result, British officers viewed the men as idle, which was why they assumed they would want contract work. But especially since the Giriama had not been in the trans-Sabaki area for a long time,

Giriama women needed their men for first clearings of land and to build houses. Women also needed their sons to be the first settlers into new lands which were no longer controlled by the elders. This included land further north of the trans-Sabaki and land in what was called the coastal strip: ten miles which had been previously claimed by the Arabs through the Sultan of Zanzibar. If they lost their sons and husbands, it was clear that women's own agricultural work would be increased. Elders needed the young men, too, since they were dependent on their wives and sons to provide them with food and beer and to work the land they claimed. A large source of the elders' income came from fees presented to them by younger men for rituals and for resolving conflicts. In addition, the women and all of Giriama society compared men who worked as agricultural labourers with the slaves who had worked as labourers for the Arabs on the coastal plantations. Thus, for a man to be a worker in the fields for somebody else was equated with being a slave, and there was intense opposition to this role of indignity for young Giriama men.[59]

Another critical factor is the concern Giriama women had for maintaining the family. Much testimony indicates the degree to which the Giriama felt that their children were going to be lost to them. The women would say to the men, 'You do not feel the pain about the young men.'[60] 'You do not know the difficulties of labour pains and such.'[61] 'With our sons working on the coast in the old days, mothers used to cry.'[62] One said, 'The Giriama feel only their daughters can be given away, but not their sons. Anyway, Champion was not offering cowries [money] for our sons.'[63] 'He [Champion] was given our children because we had been defeated, but we didn't like it.'[64] In the 1970s, the symbol with which most Giriama defined the Giriama war was the story about the chicks:

And the Giriama said, who will give away his sons to go and be killed? You try taking the chicks. And Champion took a chick and the hen flapped and attacked him. And the Giriama said, You see what this hen has done? If you take our sons, we will do the same.[65]

Given this specific history of the Giriama and the Giriama relationship to their sons in an expanding economy, the Giriama were preconditioned to continue developing their economy and were unreceptive to demands for contract, wage labour. Young Giriama did not themselves want to work, and they might have relied on their fathers to protect their position, but the weakening of a unified council of elders for all Giriama over the years and a tradition of generational conflict between elders and youth left Giriama with few options to refuse British requests in 1912.

The women were concerned primarily with the overall threat to their culture, manifested mostly in the fear of loss of their children. Sons would be lost to contract labour, which portended losing them altogether as in the past, and daughters in marriages to Swahili and Duruma, which would take them beyond participation in Giriama life. They were trying to prevent Mepoho's prophesy from coming true. Given that the ruling elders had not responded effectively, and given that they were losing prestige and income to 'headmen', Mekatalili wanted to restore the elders' power, stop customs which were an infringement on Giriama stability, undermine the effectiveness of the headmen, and keep Giriama children at home. The youth had no way to prevent this change on their own. Nor did the women directly. But by bringing them all together at the *kaya* and by swearing oaths that would stop assistance to the headmen, protect the crops, and protect their children and prevent labour recruitment, Mekatalili offered a unified response. She *was* the ringleader, but she did not cause the problems. Most Giriama, including some headmen, did not want to provide labourers either. She gave Giriama the means to oppose the demands. She was neither a prophetess nor a witch. She held no rank and used no medicines herself.

It is odd, in a society where government resided exclusively in the hands of men, that this campaign should have been led by a woman, and one who was not a leader in any traditional sense. It might have been expected that Mekatalili would have drawn her authority among the Giriama from either the women's council or the Kifudu secret society, but it was, instead, her anguish over the growing disintegration of Giriama society that led her to convince others to act. Her legitimacy emerged primarily from her charisma. She was an effective and emotional public speaker, and as she began to publicise the injustices she felt, she found many Giriama who agreed with her. Although her legitimacy was non-traditional, her plea appealed to tradition.

Mekatalili was successful for several reasons. She was far removed from both the traditional and the British co-opted administrative networks. She sought no power for herself. Because she was a woman, British officers initially dismissed reports of her actions as 'old wive's tales'.[66] It worked to her advantage to be outside the formal structure of Giriama government while at the same time receiving support from some of its members. She was an astute politician. She consulted Ngonyo and Wanje before she acted. They represented the modern and traditional elders, respectively. Because of generational conflict, the Giriama probably would not have been receptive to any campaign launched by the Kavuta elders to restore their own power. Men such as Ngonyo who had become wealthy would have opposed any loss of their independence. Mekatalili gave the traditional elders the opportunity, however. But, most critically, she convinced the women that something needed to be done.

The exact relationship between Mekatalili and Wanje remains unclear. In some reports he was called her 'male helper', in others, her 'son-in-law'.[67] These two people were unrelated by marriage or blood. Long before her campaign, Mekatalili had married a man whose home was Masendeni, so she and Wanje had been neighbours. Later, after they were imprisoned together up-country, they cared for each as man and wife. The use of the relationship terms is a reflection of this period as well as their experience after their return to Giriamaland, when they lived in the *kaya* as head of the newly established men's and women's councils.[68]

When the Giriama talk of Mekatalili today, they say that it was the '*Christians* who thought *she* had spoiled things and that is why she was accused'.[69] But they refer mostly to her power of persuasion. 'She went out, I don't know why. She was a woman, but she was like a man.'[70] 'Mekatalili was not afraid to speak out. And that is how she became famous.'[71] 'When she talked, in a crowd of people, they listened.'[72] 'People would follow her because they knew she was speaking the truth.'[73]

NOTES

1 Cynthia Brantley, *The Giriama and Colonial Resistance in Kenya, 1800–1920* (Berkeley: University of California Press, 1981). The Giriama, who are the largest of the Mijikenda group, occupy the woodland plateau of Kenya's immediate coastal hinterland between the towns of Mombasa in the south and Mambrui in the north. Numbering approximately 150 000 in 1969 and occupying almost 6500 sq km, they lived in scattered patrilineal, patrilocal homesteads.

2 Kenya National Archives (hereafter KNA):CP 5/336-IV, Charles Dundas, 'Report on the Giriama Rising', 25 October 1914.

3 KNA: CP 5/336, Files on the Giriama Rising, Volumes I–V. Giriama Historical Texts (hereafter GHT), compiled by Cynthia Brantley [Smith], 1970–71, on file in the Archives, History Department, University of Nairobi, Nairobi, Kenya, and in possession of the author.

4 See Brantley, *Giriama Resistance*, Chs 2, 3, 4; and Cynthia Brantley, 'Gerontocratic government: age-sets in pre-colonial Giriama', *Africa*, vol. 48, no. 3 (1978), pp. 248–64.

5 See Brantley, *Giriama Resistance*, Chs 4 and 5.

6 'East African Protectorate, Native Labour Commission, 1912–1913, Report and Evidence' (Nairobi, 1913). KNA:CP 9/205, East African Coast Planters Association, 'Report of the Proceedings, London, 25 June 1912.' D. A. Low, 'British East Africa: the establishment of British rule, 1895–1912' in Vincent Harlow and E. M. Chilver (eds), *History of East Africa*, Vol. 2 (Oxford: Clarendon Press, 1965), p. 47. The following discussion is based on Brantley, *Giriama Resistance*, Chs 2–5 and Brantley, 'Gerontocratic government'.

7 Arthur M. Champion, *The Agiryama of Kenya*, John Middleton (ed.), Occasional Paper No. 25 (London: Royal Anthropological Institute of Great Britain and Ireland, 1967), p. 23.

8 KNA:CP 5/336-I, 'Interview Report' Ralph Skene, District Commissioner, Malindi, 30 March 1913.

9 KNA:CP 8/157, Arthur M. Champion, 'Report on the WaNyika covering a period from October 1912 to May 1913 by A. D. C. Giriama', 30 May 1913.

10 KNA:DC/MAL/2/1, Charles Hobley, '1913 Tour', 29 July 1913.

11 ibid.

12 ibid.

13 ibid.

14 KNA:CP 9/925, P.C. Mombasa to Chief Secretary, Nairobi, 5 May 1913.

15 Hobley, '1913 Tour'.

16 GHT: general and especially Paul Mitsanze (Mwabayanyundo), 29 December 1970; Kayafungo Elders (Kayafungo), 31 December 1970; Daniel Ngumbo (Jilore), 15 December 1970.

17 ibid.

18 The general material about Mekatalili comes from GHT, especially interviews 5, 7, 9, 11, 16, 19, 37, 40, and 55.

19 KNA:CP 5/336-II, 'Mekatalili's Statement', Garashi, 17 October 1913.

20 KNA:CP 5/336-I, Arthur Champion, 'October Report on the Present Condition of the WaGiriama', October 1913.

21 ibid.

22 KNA:CP 6/425, 'Instructions to New ADC Giryama', 24 December 1912.

23 KNA:CP 9/403, 'Mekatalili's Statement', 14 November 1913.

24 KNA:CP 9/403, 'Mkowa wa Gobwe's Statement', 15 November 1913.

25 ibid.

26 KNA:CP 9/403, 'Wanje wa Mwadorikola's Statement', 14 November 1913.

27 KNA:CP 5/336-I, 'Women's Statement', 14 October 1913.

28 KNA:CP 5/336-I, 'Ziro wa Luganje's Statement', 4 October 1913.

29 KNA:CP 9/403, 'Kitu wa Siria's Statement', 14 November 1913.

30 KNA:CP 9/403, 'Mekatalili's Statement', 14 November 1913.
31 KNA:CP 5/336-I, Charles Hobley, 'Notes from Baraza Outside Kaya Giriama', 14 November 1913.
32 This general discussion is based on ibid., Giriama 'Statements', and GHT.
33 KNA:CP 9/403, 'Nziji wa Yaa's Statement', 14 November 1913.
34 See Brantley, *Giriama Resistance*, Chs 2–5, and Cynthia Brantley, 'An historical perspective of the Giriama and witchcraft control', *Africa*, vol. 49, no. 3 (1979), pp. 112–33.
35 Champion, *The Agiryama of Kenya*, p. 35.
36 KNA:CP 5/336-I, 'Kadidi wa Bembere's Statement', 12 November 1913.
37 See 'Baraza Notes' and Giriama 'Statements'.
38 GHT: Luganje wa Masha, Vitengeni, 23 December 1970.
39 Champion, *The Agiryama of Kenya*, p. 24.
40 KNA:CP 5/336-I, Ziro wa Luganje, 'Second Statement', 15 November 1913; 'Kombi wa Yeri's Statement', 7 October 1913. GHT: Katoi wa Kiti (Kajiweni) 2 April 1971; Mwinga wa Gunga (Kinarani) 6 April 1971 and 14 June 1971; Hawe Charo (Takaungu) 2 April 1971.
41 Hobley, 'Baraza Notes', Champion, 'October Report'. GHT: Mulanda wa Wanje (Msabaha) 5 May 1971; Kobogo wa Masha (Kajiweni) 2 April 1971. Brantley, *Giriama Resistance*, Chs 2–5.
42 KNA:CP 5/336-I, Champion to Skene, 17 August 1913; Skene to Hobley, 19 August 1913. GHT: Kadzumbi wa Ngari (Chakama), 15 December 1970.
43 KNA:CP 5/336-I, Hobley to Chief Secretary, Nairobi, 26 August 1913.
44 KNA:CP 5/33-I, Skene's Report, 29 October 1913.
45 KNA:CP 5/336-I, Telegram, Hobley to Chief Secretary, Nairobi, 28 August 1913; Skene to Hobley, 31 October 1913 and 30 November 1913. GHT: Mwavuo wa Menza (Marafa) 16 December 1970.
46 KNA:CP 5/336-I, Logan, ADC Takaungu to Provincial Commissioner, 4 November 1913.
47 See Giriama 'Statements'.
48 KNA:CP 5/336-I, 'Mkowa wa Gobwe's Statement', 23 September 1913.
49 KNA:CP 5/336-I, 'Kombi wa Yeri's Statement', 7 October 1913.
50 'Mekatalili's Statement', 17 October 1913; KNA:CP 5/336-I, Champion, 'October Report'.
51 This information comes from 'Baraza Notes', Giriama 'Statements', and KNA:CP 5/336-I, Hobley, 'Giriama District Report on Political Situation and Evidence', Mombasa, 19 November 1913.
52 Hobley, 'Political Situation'. In his memoirs, Hobley refers to her as a 'half-mad woman': Charles Hobley, *From Chartered Company to Crown Colony* (London: Frank Cass, 1970 reprint [1st edn, 1929]), p. 165.
53 Brantley, *Giriama Resistance*, Ch. 6.
54 KNA:CP 5/336-IV, Acting Chief Secretary, Nairobi to ADC Mombasa, telegram, 7 September 1914. KNA:CP 16/38, Trail, 'Annual Report, Nyika District', 16 June 1915. Brantley, *Giriama Resistance*, Ch. 7.
55 Brantley, *Giriama Resistance*, Ch. 7.
56 Annual Report, 1818–19. KNA:CP 16/49, R. W. Lambert, ADC Rabai, 27 June 1919. Champion, 'October Report'.
57 Champion, 'October Report'.
58 Hobley, '1913 Tour'. KNA:KFI 72/6/13, Champion, 'Memorandum on Labour Supply and the Wagiriama'. n.d.
59 GHT: Luganje wa Masha (Vitengeni), 23 December 1970; Mwavuo wa Menza (Marafa), 16 December 1970.
60 GHT: Bambare wa Charo (Garashi), 18 December 1970.
61 GHT: Mwavuo wa Menza (Marafa), 16 December 1970.
62 GHT: Matsunga wa Maita (Bungale), 17 December 1970.

63 GHT: Pembe wa Bembere (Kayafungo), 31 December 1970.
64 GHT: Gona wa Nguma (Jilore), 14 December 1970.
65 GHT: Hawe Karisa Nyevu Makarye and Ishamel Toya (Jilore), 15 December 1970.
66 Champion, 'October Report'.
67 KNA:CP 5/336, Files on the Giriama Rising.
68 GHT: Gona wa Nguma (Jilore), 14 December 1970.
69 GHT: Luganje wa Mahsa (Vitengeni), 23 December 1970.
70 GHT: Nyundo wa Mwamure (Sekoke), 22 December 1970.
71 GHT: Bambare wa Charo (Garashi), 18 December 1970.
72 GHT: Shadrack Kambi (Jilore), 15 December 1970.
73 GHT: Nzaro wa Chai (Garashi), 18 December 1970.
74 Map 6 based on Cynthia Brantley-Smith, 'The Giriama Rising, 1914: focus for political development in the Kenyan hinterland 1860–1963', (unpublished PhD dissertation, University of California, Los Angeles, 1973) and Thomas Spear, *The Kaya Complex: A History of the Mijikenda Peoples of the Kenya Coast to 1900* (Nairobi, 1978).

16 Class, ideology and the Bambatha rebellion

SHULA MARKS

The 1906 Poll Tax or Bambatha uprising in Natal was the last armed resistance to proletarianisation by Africans, and a crucial moment in the consolidation and restructuring of colonial domination and settler accumulation in twentieth-century South Africa. Over the past ten or fifteen years, the rebellion has been incorporated in a contradictory rhetoric of revolution and warning. For African nationalists, now dedicated to guerrilla action, the uprising is 'the last heroic form of primary resistance to conquest', and Bambatha has joined the pantheon of pan-South African heroes,[1] while further to the left it has been appropriated to reveal 'the revolutionary masses', the rural proletariat, poor peasants and migrant workers, as the leading force in the class struggle. Unlike the 'richer peasants' who sided with the repression or 'the Christian landed petty bourgoisie' who opposed 'armed resistance in favour of "modern" constitutional petitioning, wringing its hands when armed struggle was at its height', the masses took on the might of a state which was attempting to shift the burden of taxation onto their shoulders.[2] The lessons seem clear enough.

In Natal, the legacy of the Poll Tax uprising has always been more ambiguous. There, the experience of the Bambatha rebellion has been deeply etched in the consciousness of every political leader since 1906; in 1927, Solomon ka Dinuzulu, son and heir to the last Zulu king, warned his people against the activities of the Industrial and Commercial Workers' Union. 'We shall have,' he argued, 'a. repetition of the Bambata affair if the authorities do not watch this movement. I warn you . . . do not enter this dirty business, as there may be bloodshed'.[3]

As late as April 1976, on the eve of the children's uprising in Soweto, Chief Gatsha Buthelezi remarked that what he feared most in South Africa was 'another Bambatha rebellion'. On many occasions he has justified his non-violent stance with allusions to the events of 1906, vaguely threatening the state with the possibilities of uncontrollable African violence if government policies of *apartheid* remain unchanged, while restraining his followers by warning of the dangers of a similarly unequal and state-provoked confrontation.[4] Yet the lessons of 1906 are both more significant and less clear cut for contemporary struggles than the conflicting rhetoric would allow. The revolutionary potential of different strata within the African populace does not fall so readily to order, nor is popular consciousness – the complex interaction of the material realities of the present with past experience crystallised in culture – as unambiguous or uncontradictory as the polemic purports.

In 1906, the small British colony of Natal was the scene of a major uprising against colonial rule.[5] Ever since the promulgation at the end of 1905 of a new Poll

Tax, rumours had abounded that Africans were about to rise against its imposition, and against their colonial masters. In many parts of the colony Africans killed off their white animals and destroyed tools of European manufacture. Acts of defiance against magistrates attempting to collect the tax before it was legally enforceable culminated on 8 February in the killing of a white policeman. A wave of hysteria swept Natal, martial law was declared throughout the colony, and for the next six months troops put down the 'rebellion', burning crops and homesteads, fining and flogging alleged dissidents (according to the *Times of Natal*, by 1907 some 5000 men had been officially flogged, 700 of them having had their backs 'lashed to ribbons' during the uprising)[6] and deposing chiefs suspected of disloyalty.

Although the phase of overt military conflict came to an end in August 1906, with the lifting of martial law, unrest continued throughout 1907. Following the end of martial law, wanted 'rebels' continued to roam Zululand, armed and apparently protected by the local populace from discovery. They appeared to have been behind the murders in Zululand in 1906 and 1907 of three chiefs who had remained loyal to the government, an unpopular magistrate and a couple of so-called witchdoctors or medicine men. The continued upheavals in Zululand and the neighbouring districts, and the total failure of the state to find the perpetrators, reinforced the constant rumours which linked the name of Dinuzulu ka Cetshwayo, the son and successor of the last independent king, both to the uncaptured rebels, the murders, and indeed to the Bambatha uprising itself. At the end of 1907, Dinuzulu was finally arrested and brought to court on twenty-three counts of high treason. Despite a massive and lengthy state trial, he was found guilty on only two-and-a-half of these counts. Sentenced to four years' imprisonment and a £100 fine, he was released from prison by the Prime Minister, General Louis Botha, as an act of clemency following the unification of the South African colonies in 1910, but remained in exile in Middelburg in the Transvaal until his death in 1913.

It is estimated that between 3000 and 4000 Africans lost their lives during the rebellion, while almost 7000 were gaoled. In one region alone, about thirty-thousand people were rendered homeless. By way of contrast, only some two-dozen whites lost their lives in the entire course of the disturbances, including six civilians, and one soldier who accidentally shot himself. When, at the end of the uprising, the Governor of Natal, Sir Henry McCallum, wrote to the Colonial Office in London requesting that imperial war medals be struck for the men 'who distinguished themselves in the recent battle', the great disparity between the number of fatalities, the brutal nature of the struggle, including the decapitation of the leading 'rebel', Bambatha, and the conflict between the colonial and imperial governments over the conduct of the campaign led Winston Churchill, then Under Secretary of State at the Colonial Office to minute caustically:

> There were, I think, nearly a dozen casualties among these devoted men in the course of their prolonged operations and more than four or even five are dead on the field of honour. In these circumstances it is evident that some special consideration should be shown to the survivors. But I should hesitate to press upon them an Imperial medal in view of the distaste which this colony has evinced for outside interference of all kinds. A copper medal bearing Bambatha's head seems to be the most appropriate memento of their sacrifices and their triumphs.[7]

This disparity in numbers, together with other evidence, led me to conclude in *Reluctant Rebellion* that despite the hostility against white rule manifested in the white-animal killing and the rumours of an imminent uprising, there was no co-ordinated or premeditated plan amongst Africans to overthrow colonial rule in 1906. Contrary to the contentions of the Natal government in justifying its declaration of martial law after a relatively minor incident, it was the declaration of martial law and the troop movements which followed which provoked African resistance. Faced by the troop presence, many decided that they 'might as well die fighting'. Thus, by acting on their conviction that the African populace was about to rise, the Natal settlers provoked the very rebellion they feared, as a kind of self-fulfilling prophecy.[8] In this sense the 'rebellion' was indeed 'reluctant'.

Archdeacon Charles Johnson, an Anglican missionary at St Augustine's in Nqutu district, at the heart of the conflict, was aware of this at the time. As he wrote to the Society for the Propagation of the Gospel headquarters in London:

> Many thinking people have been asking themselves, 'What are we to do with this teeming native population?' Some strong-handed men have thought that the time was ripe for the solving of the great question. They knew that there was a general widespread spirit of dissatisfaction amongst the natives of Natal, the Free State and Transvaal, but especially in Natal, and they commenced the suppression in the fierce hope that the spirit of rebellion might so spread through the land and engender a war of practical extermination. I fully believe that they were imbued with the notion that this was the only safe way of dealing with the native question and they are greatly disappointed that the spirit of rebellion was not strong enough to bring more than a moiety of the native people under the influence of the rifle. Over and over again it was said, 'They are only sitting on the fence, it shall be our endeavour to push them over'; and again, speaking of the big chiefs, 'We must endeavour to bring him in if possible'. Yes, they have been honest and outspoken enough – the wish being father to the thought, they prophesied the rebellion would spread throughout South Africa; had they been true prophets the necessity of solving the native question would have been solved for this generation at least.[9]

While Johnson was probably correct to stress the way in which 'the measures adopted' to deal with 'the first indication of dissatisfaction . . . were nothing more than an invitation to people who were irritated to go on to worse',[10] his interpretation of the underlying motives for the military action in Natal 'to engender a war of practical extermination' exaggerates the desire of the military and the colonists to actually 'exterminate' the African populace. Whatever the savagery of the military and indeed of the inflamed colonists in 1906, the nature of the problem for early twentieth-century Natal to solve was not so much *what* 'to do with this teeming native population', but *how* to get this teeming African population to work for the white man and woman – and preferably for the white man and woman in Natal – at the lowest possible cost, for the longest possible time. The objective was not extermination but the creation of a suitably subordinate and disciplined workforce. From this point of view, the suppression of the rebellion was a crucial moment in the consolidation and restructuring of settler capitalism in Natal.

In the last decades of the nineteenth century, the political economy of Natal was undergoing far-reaching changes, spurred by the mineral revolution on the Witwatersrand. As Henry Slater has pointed out:

Natal lay abreast one of the most important transport arteries between the coastal ports and the mining areas. A vast internal market opened up for its coal, timber, sugar, maize, cattle and dairy produce. Intensive branch-line construction served to link the rural areas to the main transport routes. A boost was given to the mercantile activities in Natal's towns, and to the beginnings of secondary industry. These developments brought about a substantial increase in the price of land, and a marked increase in the demand for labour on behalf of white employers, both local and beyond Natal's borders[11]

The most important demand for African land and labour came from the dramatically expanding capitalist farming sector in Natal: from 1891 to 1908–9, the area cultivated by white farmers increased from 85 000 acres to 541 000 (a 530 per cent increase), while agricultural output rose from 129 925 tons in 1893 to over 850 000 tons in 1904.[12] In fact it was not so much the 'teeming' African population which is striking in this period, but the rapid increase of the white population, which more than doubled in the thirteen years between 1891 and 1904.[13]

The pressures being felt by Africans from what was now a fully self-governing colonial state went way beyond anything experienced in the earlier years of colonisation. Whatever the form of their tenure, the encroachments of the state and of commercial farming were becoming inescapable. From all sides there was a demand for African land and African labour. At the same time, a series of natural disasters followed by the devastating rinderpest epizootic of 1896–7 and East Coast Fever from 1904 onwards destroyed the main source of African wealth, their cattle, and further undermined the ability of peasants to remain outside the colonial labour market. In area after area, rural social relationships were disrupted as land companies sold off their land to capitalist farmers intent on securing labour rather than rent from African tenants.[14] Magistrates wrote of the eviction of 'squatters' from lands which had been theirs for generations, or their transformation into labour tenants, while population pressures on the reserve lands closed these as an outlet for the dissatisfied and dispossessed.[15] Each year more and more young men were forced to make their way to town, not simply to earn money for taxation, but to provide food for their families.

The process was not confined to Natal proper. The annexation of Zululand by Natal in 1897 also opened up that territory to the demands of settler accumulation. There, in 1903, a joint imperial and colonial delimitation commission set aside land for settler occupation; their activities were only made possible by the ravages of the rinderpest epizootic on Zulu cattle in the last years of the nineteenth century, which had temporarily emptied the land of its herds. This expropriation was on top of the loss of valuable grazing lands in the districts of Utrecht, Vryheid and Paulpietersberg, which the Boers had taken during the Zulu civil wars of the 1880s. Annexed by the Transvaal as the New Republic in 1884, it was taken over by Natal in 1902 after the Anglo-South African war.

If the demand for African labour even before the war had been insatiable, the South African war escalated both the costs and the scarcity of labour. All over South Africa, the demands from the army and the boom which followed the war raised African wages, to the consternation of both capital and the colonial state. In Natal, the militancy of dockworkers during and immediately after the war, when their services were indispensable, had forced up rates of pay for *togt* labour (daily paid labour), while on the coal mines, too, these years saw a considerable leap in wages.[16] The capitalist response to this was swift and sharp: at a special conference called by the governor, Sir Henry McCallum, at the beginning of 1902 to discuss labour problems it was decided by all employers that the average wage of Africans should be reduced to 30s a month. McCallum regarded the reduction of wages as so essential that he offered to use his powers as Supreme Chief (the peculiar designation which the Governor of Natal took unto himself in relation to the colony's African inhabitants) to call out any additional labour required by the armed forces, the largest single employers of labour towards the end of the war. He also indicated his readiness to break up any strike action on the part of disaffected Africans under martial law or by making use of the very extensive powers conferred on him by the Natal Code, as Supreme Chief.[17] In the same year, the British Commanding Officer in Natal, Major-General Fetherstonhaugh, remarked that 'the native labour question' was 'the only matter for some time to come to cause any disturbance sufficiently serious to call for the presence of troops.'[18]

The undercurrent of hostility which whites were nervously picking up through innumerable spy reports and rumours was related to these unprecedented pressures on African land and labour. And the counteraction, the crushing of the Poll Tax 'rebellion', whether seen as an attempt to nip African resistance in the bud or forcibly to divorce African peasants from the means of production, undoubtedly had the effect of accelerating their proletarianisation. In the proclaimed labour areas of the Transvaal, the number of workers from Natal-Zululand increased by more than half between 1906 and 1909, while in the latter year, according to one estimate, 80 per cent of the adult males in Zululand were away at work. Although the numbers subsequently fell considerably, this was clearly crucial in the making of Zululand's working class. It is no coincidence that the Poll Tax which occasioned the disturbances was justified in Natal's all-white legislative assembly by the Minister of Native Affairs because it would 'bring about a better state of labour'.[19]

This is not to argue that there was a conspiracy amongst colonial capitalists to wage war on Africans to force out labour in 1906; the Natal settlers and government feared an African uprising as much as they were determined to stamp it out once they were convinced that it was happening. Indeed, the severity of the response was the outcome of the apprehension of the settlers and the weakness of the colonial state rather than a measure of their confidence or its strength. And they hesitated many months before attempting to arrest Dinuzulu, despite their conviction that he was not only the reason for the continued unrest in Zululand, but the instigator of the rebellion itself. Nevertheless, like the wars waged against African communities all over South Africa in the late 1870s and 1880s,[20] the Bambatha rebellion was in a critical sense the precondition for the expansion of capitalist relations in Natal. Equally critically, it can be seen as resulting from a resistance to proletarianisation by Natal's African population, in their refusal to pay tax and in the rumours of disaffection which so alarmed the colonists.

If the causes of the Bambatha uprising are to be sought in the intolerable pressures which colonial society was exerting upon the peasant communities of Natal, our picture of the African perceptions of these realities is more fragmented and less clear. While ruling class actions, ideologies and anxieties are abundantly documented, it is extremely difficult to delineate the nature of popular consciousness in Natal at the beginning of the century. Although it is perhaps possible to establish at least some of the contours of the beliefs of the literate Chrisian Africans, or *amakholwa*, they were hardly representative of the peasants and migrant workers who constituted the mass of Natal's African population. In 1907–8, the Natal Native Affairs Commission interviewed hundreds of witnesses.[21]

Very few of them were black, and even fewer of them could be described as workers or poor peasants – let alone 'rebels', though some *amakholwa* and some of the more popular chiefs could and did on occasion represent wider grievances in a remarkably outspoken fashion.

Hence the importance of the multitude of rumours that shivered their way through the kitchens and clubrooms of colonial Natal, and which were picked up by spies and doubtless coloured by what they thought their colonial masters wanted to hear. Rumour was not simply, as I suggested in *Reluctant Rebellion*, a 'barometer' of inter-racial tension,[22] it is also an important way into popular consciousness. While some rumours may well have been 'simply "engineered" by interested parties who had ulterior motives in view',[23] in general they reflect what Antonio Gramsci has called 'non-organic' ideology: 'those less structured forms of thought that circulate among the common people, often contradictory and confused and compounded of folklore, myth and day-to-day popular experience'.[24] The vast majority of rumours reported by the intelligence officers of the state amongst Africans before, during and after the uprising revolved around the purported intentions of Dinuzulu, who was the focus of African hopes and European fears. More complex to interpret, however, are the rumours which reflected 'non-organic ideology', a fascinating meld of earlier religious beliefs and sections of the Old and New Testaments, combined with elements of the previously hegemonic ideology of the displaced Zulu ruling class, imperial liberal democratic beliefs, messages from black America and Shepstonian segregationism, interspersed with fragments of more recent political experience and bits and pieces derived from contemporary newspapers, relayed by the literate and barely literate. Refracted through existing modes of understanding, these elements are constituted into a palimpsest of popular perception.

Here, one example drawn from many similar reports in 1906 and 1907, must suffice by way of illustration:

> The main topic is . . . that a great war is imminent. They say that locusts and bees and soldiers appeared, and that Bambatha's impi appeared . . . Some say that fighting has already taken place between the English and the Basutos and that the English were exterminated by lightning . . . And magic powers in some places are said to come from Mjantji [possibly Modjadji, the Lovedu Queen in the north-eastern Transvaal, who was believed to have supernatural powers especially over locusts and disease]; others say they come from Dinuzulu, whilst others say from Bambatha. Some have great faith in as much as they believe in the serpent [a symbol of power and the ancestors in Zulu cosmology] being able to cure and that it is God Almighty. If they experience misfortune they kill a beast [i.e. make a sacrifice] believing the serpent (spirit) will assist them even in regard to the money they owe to the white people. The natives who return [from the towns] say that the white people can foresee that the country will perish . . . [The whites] want a King of their own, the [British] King has no business in this country.[25]

To some extent, behaviour may also be a clue to consciousness though it is only a clue, for in the end it cannot tell us why people act as they do, or what meaning they give their actions. By looking at the pattern of participation in the 1906 rebellion and at what is known about the leaders, it is perhaps possible to go beyond some of the cruder stereotypes of class and consciousness. Thus, contrary to some of the recent literature,[26] I would suggest that in 1906 the 'masses' – the tenants and squatters and labourers on white farms, the poor peasants on the communal lands of the reserves and the migrant workers who returned from

Durban and the Witwatersrand in response to the call of their chiefs to battle in late June – were 'restorationist' rather than 'revolutionary': they desired to 'recreate the past' rather than capture the future.[27]

As we have seen, the actual course of the disturbances suggests that none of them actually intended challenging the colonial authorities by force in 1906, though all in some measure questioned its legitimacy. Even the members of the independent Presbyterian Church of Africa under Makanda and Mjongo whose actions led to the killing of the white policeman in February were virtually prodded into action.[28] There is even less evidence that migrant workers in Durban or Johannesburg intended confronting the capitalist system in which they were embedded. There was no 'mass exodus' of migrant workers from Durban until the end of June when more than a thousand dockers, some five-hundred domestic workers, a number of rickshaw-pullers, and, to the consternation of the settlers, some 40 per cent of the African Borough Police returned to Mapumulo district in response to the call of their chiefs to repel the military invaders.[29]

They would undoubtedly have wrought far greater havoc had they taken up arms in the city at a time when it was depleted of its white adult male population. Yet the vast majority of the workers in town, for all their ability to defend the hard-won privileges of *togt* (daily-paid) labour,[30] were still deeply committed to the life of the rural chiefdoms. There was neither the class-based organisation nor the working-class consciousness to establish an urban-based opposition. Nor is there any evidence of revolt on the coal mines and sugar plantations of Natal: admittedly before 1911, the bulk of the workforce in both these areas was either Indian or drawn from the Cape and Mozambique, so that the divided nature of the workforce also militated against organised resistance. And we have no idea how many of these workers, amongst the most exploited in South Africa, headed for the Nkandhla forests to join Bambatha or for Mapumulo to join the chiefs engaged in the final phase of the disturbances, although the silence of the evidence suggests the numbers were not large.

Despite the fact that I have argued that the 1906 rebellion can be seen as resistance to proletarianisation, and that by 1906 it was estimated that some 200 000 of its African population were at work for the white man, the speed of South Africa's industrialisation, its reliance on labour migrancy rather than a fully proletarianised work force and the limited penetration of factory production, meant that the Bambatha uprising was essentially a peasant rebellion, and displayed all the classic features of a peasant revolt.

In tracing the pattern of participation in the uprising in *Reluctant Rebellion*, I was at pains to emphasise the diversity of motives which led individuals and groups to take up arms against the colonial forces: for even the appearance of the troops on the scene was not a sufficient cause for rebellion, though I argue it was a necessary one. Moreover, simply to talk of a 'peasant' revolt does not take one very far. As Eric Wolf has remarked:

> there are differences in behavior and outlook between tenants and proprietors, between poor and rich peasants, between cultivators who are also craftsmen and whose who only plow and harvest, between men who are responsible for all agricultural operations on a holding they rent or own and wage labourers . . . between peasants who live close to town markets and urban affairs and those living in more remote villages; between peasants who are beginning to send their sons and daughters to the factories and those who continue to labor within the boundaries of their parochial little worlds.[31]

Undoubtedly the vast majority of 'rebels' were the rural poor – the recently conquered 'tribesmen' of southern Zululand, fearful of further land loss like the rebels in Nqutu and Nkandhla; impoverished peasants on the reserves forced to enter the labour market as migrant workers in order to earn tax and increasingly also to find money for food, like the men who returned to Mapumulo district from Durban and the Witwatersrand at the call of their chiefs to defend their homes and cattle in June and July; tenants deprived of their independent existence by the expansion of commercial farming like the men of Richmond district where it all began, or squeezed by their landlords for more rent and labour like Bambatha's people. Significantly, 'rebellion' spread amongst all the fragmented sectors of Natal's rural population: it was not confined to tenants or wage-labourers or reserve-dwellers. And although the wealthier Christian landowners clearly stood aloof, for many poorer Christians, even those who still had a precarious independent existence, the uprising constituted a great 'crisis of commitment'. The sudden imposition of the Poll Tax in 1905 on top of the disasters already confronting rural society created a community of affliction across these lines of potential division.[32]

Yet not all the rural poor were 'rebels'. At the level of final commitment to armed rebellion, it was the minutiae of local-level politics which seemed to tip the balance. Nor should this surprise us. Again as Wolf has pointed out, 'in the village':

> transcendental ideological issues appear only in very prosaic guise ... For example, peasants may join in a national movement in order to settle scores which are age-old in their village or region ... mobilization of the peasant 'vanguard' is less an outcome of nationwide circumstances than of particular local features[33]

Eric Hobsbawm has made a similar point in relation to the peasantry in precapitalist societies:

> At the bottom of the social hierarchy ... the criteria of social definitions are either too narrow or too global for class consciousness [in precapitalist societies]. In one sense they may be entirely localized, since the village community is the only society that matters ... yet in another sense their criteria may be so general and universal as to exclude any properly social self-classification.[34]

Hobsbawm is, of course, talking of *class* consciousness, not necessarily *revolutionary* consciousness. Yet it is for these reasons that George Rudé argues that popular culture, or what he calls 'inherent ideology' may carry protesters into 'strikes, food riots, peasant rebellions (with or without success); and even into a state of awareness of the need for radical change ...; but evidently it cannot bring them all the way to revolution, even as junior partners of the bourgeoisie'.[35] Although popular protest may act as a shield against the total hegemony of ruling class ideology, to become revolutionary it needs to be supplemented by a more systematic, 'derived', and structured world view, which may be forward-looking or backward-looking.[36]

In some sense in the second phase of the uprising, during the battles in the Nkandhla forests, the leaders tried to overcome the problems of parochialism, scale and ideology through their use of Dinuzulu's name, emblem and war cry. This phase of the 'rebellion' drew most of its support from southern Zululand which had felt the brunt of civil war and colonial conquest in the recent past, and feared further land expropriation as a result of the activities of the Zululand Delimitation Commission: significantly it had not suffered the land loss yet.[37] Although chiefdoms split bitterly in this region over whether or not to join the rebels,[38] they had greater opportunities of successful opposition than in Natal proper, with its far longer history of colonisation. It was not only that the Nkandhla forests offered ideal country for guerrilla warfare, and this made resistance feasible. It was also crucial that some of the closest supporters of the Zulu kings lived in Nkandhla division and that here elements of the earlier hegemonic royal ideology were available, which could serve to unite the otherwise disparate groups of rebels. Deeply embedded in this were the kinds of millennial beliefs in a messiah figure which typify precapitalist agrarian protest.

It is not clear who made the strategic decision to associate Dinuzulu's name with the rebellion, though by all accounts the central figure was probably Bambatha's right-hand man, Cakijana, one of the most important and most talented of the military strategists in the Nkandhla forests. It is perhaps of some significance that his experiences of South Africa's burgeoning capitalist economy already went well beyond that of the peasant still intensely identified with chiefly authority in the reserves of Zululand or even the rent-paying tenantry being squeezed on the white farms of Natal: his experience was garnered in the harsher world of the New Republic at the turn of the century, and on the Witwatersrand, where, as Charles van Onselen has shown, from the 1890s, the most proletarianised Natal Africans had been carving out specific economic niches for themselves, above and under ground and in both senses of the term.[39] Yet if Cakijana's hatred of colonial rule seems undoubted, and his appreciation of the need to mobilise beyond the village level and his recognition of the utility of Dinuzulu's name for the purpose were equally clear, his revolutionary potential was far more limited. His career reveals many of the continuities between banditry, social banditry and resistance[40] and reinforces the point made by Rudé about the need for a fusion between popular and 'derived' ideology for successful peasant revolt.

On the face of it, Cakijana would seem typecast as leader of a peasant rebellion, or indeed as a social bandit. In a volume of essays touching on social banditry, it is perhaps worth dwelling on his career at some length. Born in the 1870s in what became the Vryheid district of the Transvaal and then of Natal, he was but a child in 1884 when the Boers annexed the New Republic which was inhabited by some of Dinuzulu's most loyal followers. There was little love lost between the Boer invaders and the local Zulu, as huge tracts of land were alienated to white farmers, many of them absentee owners, and the majority of Africans were converted into labour tenants. By 1907, although there were fewer than 3000 whites in the district and only 288 farms, more than half-a-million acres had been alienated to the settlers.[41] Wages were lower in Vryheid than almost anywhere else in Natal, and labour tenants were expected to render at least six months' labour, often without pay.[42]

African hatred of Boer rule and the tensions between the two communities were expressed in constant rumours in the division that Africans were about to rise to regain their land; in the 1890s this was linked with rumours of Dinuzulu's return from exile.[43] The tensions boiled over during the South African War, when Africans took the opportunity to engage in extensive cattle-rustling. Boer commandos forcibly conscripted African labourers and cattle-guards, some of them mere adolescents, shot alleged spies and commandeered horses and entire herds of cattle.[44] At Holkrantz on 6 May 1902, the Qulusi people, who were closely connected to the Zulu Royal House, fell on a Boer encampment, and killed fifty-six of the seventy men there. The 'massacre' caused a deep sense of outrage on the Boer side. General Botha called the incident 'the foullest deed of the war', and many Boers swore revenge.[45] The general atmosphere of the district was well conveyed by Lieutenant-Colonel Mills, who was appointed to enquire into 'the causes which led to ill-feeling between Boers and Zulus culminating at Holkrantz'.[46] According to Mills:

> Every Boer expressed the most bitter hatred of the Zulus. They all express a wish that the Zulus would rise now while the British troops are in the country so that they may be practically wiped out. The Boers all say that in the event of a rising, every one of them would join the British troops in order to have a chance of paying off old scores against the Zulus . . . when I first came here and asked the Boers what they thought of the advisability of keeping troops here. They all said it was most necessary, as they were afraid of the Kaffirs and it would not be safe to stay on their farms if the troops withdrew . . . Taking everything into consideration I cannot help being forced to the opinion that many Boers intend to provoke a Zulu rising if they can do so[47]

We do not know whether Cakijana took part in the attack on Holkrantz, but he apparently joined the 1500-strong regiment formed by Dinuzulu known as the 'Nkomindala'.[48] Its object was to drive into Zululand Boer cattle which might otherwise be used by the Afrikaner guerrillas, as well as to spy on the Boer forces. The Nkomindala had some access to outmoded arms and engaged in drill, and it may be here that Cakijana acquired his knowledge and expertise in warfare.[49]

Immediately after the war, according to the local magistrate, A. J. Shepstone:

> several of the Boers said that they meant 'to go for' the Natives and they were going 'to have it out of them,' and the Natives were warned by some of those Boers that they and the British were amalgamated, and before long would 'wipe the Natives off the face of the earth'. When I think of the very high rents, and, bearing in mind that naturally those very high rents will irritate the Natives, I consider that those Boers are still in hopes, by putting those men to extremes, that they will get up a row with the Natives, and so retaliate on them for the part the Natives took against them during the war.[50]

Presumably in order to recoup their war losses, farmers now demanded even more 'exorbitant' hut rents from their tenants – sometimes up to £10 a hut – while others were charged rents for the period of the war when the Africans 'were not in beneficial occupation'. Large numbers of Africans were evicted for their activities during the war, many without notice.[51] As one Vryheid chief lamented, Africans in the district had had 'a rough time during the war', and since the war 'bad seasons, locusts, and cattle diseases [meant that] many of them were in very poor circumstances; in some cases, men who had not been able to pay judgments against them, had writs of execution issued and had nothing whatever left to exist on.' Some Africans had lost all their cattle and others had very little; and a considerable number had taken refuge in the malarial thorn country in order to escape rackrenting landlords. Without exaggeration, the chief described the situation as 'an impossible one for the Natives to exist in'.[52] Not surprisingly, Shepstone maintained that since the South African war, Africans had 'gone out to labour to a very great extent. They have had to do it, otherwise they would not have lived. They go out and then send their money home for the purpose of buying food for their families . . . and for paying their rent.'[53] At the same time, some farmers were 'selling the services of their Native tenants to labour agents at 10s per head, and if the Natives will not go out they make it pretty warm for them on the farm'.[54] Shepstone's remarks are confirmed by the highly skewed male to female ratio in Vryheid in 1904: of the total number of 60 000 Africans in the division (now annexed to Natal and one of the most populous in the colony), 26 100 were male to 33 900 females, an index of the high level of labour migration.[55]

Not all the Africans in Vryheid district meekly submitted to these exactions, although the availability of work on the Rand and on the coal mines of the neighbouring Dundee district partly defused a potentially explosive situation.[56] For their part, the Africans avenged themselves on the Boers, through cattle theft and even physical assault.[57] The guns raided during the war, and the availability of the thorn country as refuge facilitated the growth of forms of social banditry in the region after the war and again after the rebellion. In 1905, Cakijana himself was in gaol on charges of cattle-maiming, an unusual crime in Natal, but quintessentially expressive of rural resistance to exploitation. It was perhaps this episode which gave him one of his many nicknames, *Gwazakanjani* (How do you stab?),[58] while his exploits in the South African war had already earned him his name, Cakijana – 'the cunning weasel who shares the honours with Unogwaja, the hare, as the trickster of Zulu folklore' – from the Boers.[59] About thirty years old and without 'that customary sign of manhood and responsibility – a headring', he was known in the district as *hlanya* – which Stuart translates as a 'firebrand', 'desperado' or 'anarchist' – terms which we might more kindly replace with 'social bandit'.[60] Yet whatever his daredevilry, and his hatred of the Boers, by the end of the 1905 or beginning of 1906 the realities of rural poverty had pushed Cakijana with his four brothers into joining the stream of migrants to the Rand.

So far this is all of a piece. Yet we should not move from this into the glib assumption that Cakijana's participation in the rebellion was either predestined or premeditated, the result of an inevitable revolutionary consciousness roused by landlord exploitation and·expropriation, or his experiences as a migrant labourer on the Rand. As Theda Skocpol has warned us: 'It is one thing to identify underlying, potential tensions rooted in objective class relations understood in a Marxist manner. It is another thing to understand how and when class members find themselves *able* to struggle effectively for their interests.'[61]

Despite the manifest class conflict between landlords and tenants and the military belief that all the chiefs in the Vryheid district were 'deeply implicated' in the rebellion and its aftermath,[62] there was no uprising in any of the districts of the 'New Republic' though some of the young men of the division undoubtedly did join the rebels as individuals. This was the more surprising, given the widespread belief that Dinuzulu was behind the uprising, and that the chiefs of Vryheid were amongst his closest and staunchest supporters. Indeed during the South African war, and contrary to the declared policy of the Natal government, Dinuzulu was called upon to exercise 'his influence upon them several times . . . and several of the headmen from there had to take refuge with him'.[63] Yet neither the group organisation nor the access to resources necessary for rebellion were available to the chiefdoms of the 'New Republic' which had been deliberately fragmented, and their chiefs ignored during the period of Boer administration.[64]

Even Cakijana's own involvement in the rebellion occurred by chance. Quite fortuitously, he was at his father's homestead in the Vryheid district having returned from the Rand with a sick brother; he had already made his plans to return to Johannesburg, and was simply waiting to see his 'girlfriend', when

Bambatha passed that way asking for a guide to Dinuzulu's homesteads at Nongoma.[65] Cakijana accompanied him and was then instructed by Dinuzulu, apparently in an attempt to get rid of the unwanted Bambatha,[66] who was already in trouble with the authorities, to accompany the Zondi chief back to Natal to find a medicine man on his behalf.

Back in Natal, Bambatha found that he had been deposed and that his uncle had been appointed regent to his place. Incensed, he determined to kill his uncle, and was only prevented from doing so by Cakijana's intervention. They then both decided to oppose the white forces sent to the area to arrest Bambatha. According to his highly unreliable evidence at his own trial on charges of rebellion, Cakijana was convinced that Bambatha had guns from Dinuzulu and that Dinuzulu backed the rebellion, and he therefore decided to assist him. Shortly thereafter, Cakijana was wounded in the calf in a skirmish with colonial forces at Macala and returned to Dinuzulu, who would have nothing to do with him. At the homestead of his father, Gezindaka, he was warned to move on as the Natal authorities were after him; at this point, Cakijana decided to throw in his lot irrevocably with Bambatha and the rebels.[67]

He now accompanied Bambatha to Sigananda's Cube chiefdom in the Nkandlha forests of southern Zululand, where he 'exercised a great influence over Bambatha's adherents'.[68] As we have seen, it was apparently Cakijana, perhaps with his wider experiences, who appreciated the need to overcome the problems of scale faced by all peasant movements, and tried to do so by alleging that Dinuzulu was behind the rebellion, and by using his name, emblem and war cry. This phase of the uprising witnesses the only concerted effort to confront the colonial forces with a military strategy.

It was brought to an end, however, by the battle of Mome Gorge on 9 June 1906, when Bambatha's followers were resoundingly defeated. Bambatha himself was killed together with large numbers of his followers; initial reports, which probably exaggerated, put the number at three thousand. In the days that followed, hundreds of rebels surrendered to the troops scouring the surrounding territory; others, including Cakijana, fled northwards. By 18 June, the Governor reported that 'there was no chance of the rebellion spreading into Natal'.[69] In fact it lasted another two months as fresh disturbances broke out in the Mapumulo division. Yet apart from the medicine man who doctored Bambatha's men for war, Cakijana and leading members of the southern Zulu chiefdoms who had participated in Bambatha's uprising did not apparently join them. Many sought refuge in the remoter districts of Zululand, some of the most prominent making their way to Dinuzulu's inaccessible homesteads at Nongoma and Babanango.

The government's failure to declare a general amnesty perpetuated the sense of insecurity and unrest, and Dinuzulu continued to be the centre of innumerable rumours of further rebellion. Despite all the police endeavours, relatively few of the rebels were apprehended. Although this was attributed at the time to the clandestine power of Dinuzulu, in fact it is clear that the sympathies of most Zulu were with the rebels and they were unlikely to betray them. When the authorities attempted to investigate the murder of the unpopular magistrate of Mhlabatini, H. M. Stainbank, who was killed in May 1906, they met with a wall of silence: it was only in 1912 that the state was able to convict Mayatana, another of Dinuzulu's Nkomindala bodyguard and Cakijana's companion in the bush.[70] A similar blank faced the investigators into the murders of two loyal chiefs and a couple of medicine men in 1907. The legitimacy of the government was clearly at stake and, in November 1907, martial law was declared once more, and troops were sent in to arrest Dinuzulu.

Finally, in March 1908, Cakijana handed himself in, perhaps persuaded by Dinuzulu's counsel to do so.[71] Harriette Colenso, daughter of the famous Bishop Colenso of Natal, and herself a champion of the Zulu, was convinced that until he was apprehended, there would be no general amnesty for rebels in Zululand. Cakijana made his way to the Colenso home at Bishopstowe, where Harriette kept him overnight, and recorded his testimony before handing him over to a somewhat astonished Chief Commissioner of Police. He was then taken to the Nkandhla district which was under martial law for six weeks, and kept away from Dinuzulu's defence team. At the end of this time he requested a government-appointed lawyer for his defence and turned state's evidence. He became the most important witness against Dinuzulu. His trial preceded that of Dinuzulu; both at these trials and subsequently he argued consistently that all his acts had been carried out under compulsion from Dinuzulu, a plea which was rejected by the Supreme Court as manifestly absurd,[72] although it was clearly fully accepted by the Natal intelligence.[73] In mid-November 1908, Cakijana was found guilty on charges of high treason, and sentenced to seven years' hard labour. He was never tried for his self-confessed murder of the medicine man, Gence, who had been accused of adultery with one of Dinuzulu's wives. In March 1910, Cakijana was discharged from prison on license, having served one year and four months of his sentence. In acknowledgment of his incomparable services as a police spy, he was not even required to report himself periodically to the police, 'nor [w]as he . . . subjected to the Police supervision ordinarily exercised in the case of prisoners on ticket-of-leave.'[74] As his defence lawyer, A. H. Hime pointed out to the Natal Minister of Justice at the time of his trial:

> You are no doubt aware that up to date Cakijana has played the game and he obtained for the government very useful information concerning the firearms which were at the Usutu and the information supplied by him led to the discovery of a considerable number of weapons which he can identify.[75]

Cakijana's subsequent career is difficult to piece together, but it is equally filled with ambiguity. In November 1910 he was fined £5 for indecently assaulting a young African girl;[76] this, together with the greater sympathy Union ministers had with the exiled king and their knowledge of Cakijana's role in Dinuzulu's downfall, may have been behind the refusal of the cabinet to sanction a request from the magistrate of Mhlabatini that he be employed as a special constable to investigate the Stainbank murder, although Harriette Colenso was convinced that he was nonetheless so employed.[77]

Although he only died in 1963,[78] I have been able to find only one further allusion to Cakijana in government records. In 1920 Dinuzulu's son Solomon informed the Chief Native Commissioner for Natal that Cakijana was staying with him, on the grounds that he 'knew of no reason for sending him away'. With some asperity, the Commissioner retorted, 'Quote me as your reason.'[79] In view of the favour in which Cakijana had been held by the former Natal administration, and his betrayal of Solomon's father, the exchange held more than a hint of irony. Under the circumstances, Henry Callaway's remarks about the nature of the Zulu trickster, Cakijana – also known as Uthlakanyana – are apposite:

> Uthlakanyana is a very cunning man; he is also very small, of the size of a weasel . . . He was also called Ukcaijana-bogconono [i.e. Cakijana] . . . a little

red animal which has a black-tipped tail. And this animal is cleverer than all others, for its cunning is great. If a trap is set for a wild cat it comes immediately to the trap, and takes away the mouse which is placed there for the cat: it takes it out first; and when the cat comes, the mouse has already been eaten by the weasel.[80]

As a dwarf-like figure, half-man, half-child, the trickster in Zulu folklore seems to have occupied the same realm of moral ambiguity as Cakijana. And it may be that the nursery tale with its acceptance of this moral ambiguity enables us to understand not only Cakijana's actions, but, more widely, elements in Zulu popular consciousness. It is, after all, not difficult to understand why Cakijana should have wished to save his own skin, especially if he felt, as many of the rebels may have done, that in the final analysis Dinuzulu had let them down by not backing the rebellion fully. More difficult to understand is the subsequent acceptance of Cakijana at the royal homestead – which was corroborated by Princess Magogo, Solomon's daughter, as being a matter of common knowledge[81] – and the contemporary regard with which Cakijana is held as a hero.[82] It suggests a remarkable popular tolerance of this 'moral ambiguity'.

Moral ambiguity, however, is not the stuff of which revolutionary consciousness is made, though it may be the way to survival. The limitations of Cakijana's vision, a limitation he shared with the peasantry he mobilised, could be transcended in the course of struggle only if it fused with a more coherent understanding of an alternative form of society. Quite crucial to peasant rebellions are the alliances they are able to forge with other class forces, and these in turn are dependent on the other class forces available.[83] What were the possibilities in 1906? Undoubtedly the key mediators in later colonial history were to be the Christian intelligentsia who formed the nationalist leaders of colonial Africa. Their forebears in Natal were the more prosperous and educated Christian, or *kholwa*, landowners of the mission reserves, 'men of improvement' who had begun to demand bourgeois democratic rights and were already forming the organisations which emerged in 1912 as the South African Native National Congress. Their numbers were minute, and their position in the tense and racist atmosphere of colonial Natal vulnerable in the extreme.[84] They were represented by men like the Rev. John Dube, churchman, educator and newspaper proprietor; Martin Luthuli, uncle of the Nobel Prize-winner, Albert Luthuli, sugar planter and the elected chief of the *kholwa* community of Groutville, and Mark Radebe, a small businessman and hotelier. Their role in 1906 was as ambiguous as their class position. There was little chance of their entering an alliance with, let alone mobilising, the dispossessed peasantry. Although the Governor of Natal characterised Dube as 'a pronounced Ethiopian who ought to be watched',[85] Dube himself was not above feeding information to the government, and strongly protested his loyalty throughout the disturbances.[86] The Secretary of the Natal Native Congress, H. C. Matiwane, also carefully distanced himself and his organisation from the uprising:

> The great bulk [of Africans in Natal] are in great sympathy with the motives that prompted the rebellion, but utterly dissociate themselves with the rebellion itself, their contention being that there are important grievances which are crying for redress . . . There are many thousands of [Kholwa] in the Colony, and to a man they have thrown in their lot with the Government, Government need have no fear of our support, believing as we do that our grievances will be considered after the rebellion is crushed.[87]

Yet characterisations of the *amakholwa* which see them as 'psychologically enslaved' and 'deculturated' are too harsh.[88] In South Africa, Archie Mafeje has argued, the *amakholwa* initially represented a 'progressive force, by introducing the art of writing and universalising metaphysical concepts in small preliterate societies which relied on simple theoretical paradigms for explanations'. Later on, he maintains (though he does not attempt a chronology) the *kholwa* community 'became reactionary, precisely by failing to come to terms with the contradiction of its own emergence in the peculiarly South African conditions'.[89] In Natal in 1906, it can be argued that they were both at the same time. At one level, the bourgeois democratic ideals they were espousing were profoundly subversive. As Eugene Genovese has pointed out, in a slave society, bourgeois democratic slogans can be revolutionary:[90] in early twentieth-century Natal, as in late twentieth-century South Africa, they question the fundamentals of white rule. Yet at another level, their prescriptions were deeply conservative: a programme of self-improvement, land purchase and the reasoned representation of 'grievances'.[91]

That the *amakholwa* landowners and intelligentsia did not rise against the whites in 1906 can hardly be held against them. The sheer slaughter tells us why. Sticks, stones and a few hundred ineffectual guns in Zululand were not much assistance against Maxim guns and dumdum bullets. At moments of crisis the petty bourgeoisie can be radicalised, as Phil Bonner has recently shown in relation to working-class action on the Witwatersrand after the First World War, and Helen Bradford has suggested in the case of the middle-ranking officials of the Industrial and Commercial Workers Union (ICU) in Natal in the 1920s, when large numbers of erstwhile landowners, clerks, teachers and skilled artisans were being pushed down into the ranks of the working class, and identified with it.[92] The situation in 1906 was different; the working class was small, and its potential for action still tied, as we have seen, to precapitalist or non-capitalist structures which the *amakholwa* saw as regressive and repressive. They could hardly share in the popular belief that if they carried out the appropriate ritual, the white man's bullets would turn into water. Under the circumstances, the turn to nationalist organisation and an attempt to create a united African front in the face of government demands suggests a valid response to their situation.

Paradoxically, the brutality with which the rebellion was crushed and the arrest of Dinuzulu may have been a potent factor in increasing the levels of interaction between this potential leadership and the more popular forces. Immediately after the uprising, when an American missionary gave a sermon based on what he had seen, it caused an uproar in his congregation and a very bitter editorial and correspondence in the columns of Dube's newspaper, *Ilange lase Natal*. Discussions with some of the leading Christian landowners revealed that the missionary 'had failed to correctly estimate the depth of feeling on the part of the people, who though not in sympathy with the rebels, could not hear a recital of its events from the lips of a white man without feeling that he was gloating over the success of his own race'.[93] In 1908, the Natal Native Affairs Commission remarked anxiously that 'the cleavage between the kraal kaffir and the Christian native' was beginning to break down.[94] And when, in 1912, Dube addressed a group of Africans gathered at Eshowe in Zululand, to introduce the new South African Native National Congress, of which he had just been elected president, and urge the need for 'unity among us black people', he was addressed by an unknown *kholwa*:

I thank Bambata. I thank Bambata very much. Would that this spirit might continue! I do not mean the Bàmbata in the bush who perished in Nkandhla, but I mean this new spirit which we have just heard explained[95]

Although they generally did not discriminate between different kinds of African Christians, it was the activities of members of the independent Christian sects which were seen as more immediately threatening by the settlers in 1906. The Natal government was convinced that 'Ethiopianism', the growth of independent churches amongst Africans, was the key to the disturbances: and one reason the colonial state reacted to the defiance of the group in Richmond in February with the declaration ofmartial law is that they were members of such a sect. They would appear to have had a vision which could have transformed the sporadic outbursts of defiance into something far more formidable. Makanda and Mjongo and their followers seem to be best described as 'middling' peasants, men who seem to have had an independent existence as tenant-farmers and woodcutters in the Enon forests near Byrnetown in the Richmond district. Both their material and their religious independence had brought them into direct collision with chiefly authority well before 1906. Like the poorer peasants, they were beginning to feel the pressures from the expanding colonial economy, with the expansion of wattle and dairy farming in the district.[96] In their case, the conjuncture of events fused with both a popular and a 'derived' ideology to make for an explosive situation.

This can be glimpsed in the evidence given by the Reverend Algernon J. Fryer, who visited the men after their court martial in order, as he put it, to bring them to realise the error of their ways, 'by God's help and by the assistance of my very able and sympathetic interpreter':

They had been taught by native preachers and evangelists that the whole of the Holy Scripture pointed to the fact that the curse on the black race which was to keep them under, was now to be removed and that it was time that blacks should take the upper hand and reduce the whites to a state of subordination. The witness [the Rev. Fryer] examined them in their knowledge of scriptures at length and he found that they were exceptionally well acquainted with the main facts, more especially of the Old Testament, but that they had ignorantly twisted . . . these facts into superstitious notions affecting their relations with Europeans.[97]

Archie Mafeje has called the very similar religious movements which he sees originating in South Africa's cities after the First World War:

the protest movements of the dispossessed, and quite distinct from the earlier 'independent' churches which were led by educated black nationalists . . . The 'primitive rebels' are a post-World War One phenomenon and, historically, they are both an affirmation and repudiation of Protestant liberalism in South Africa.[98]

Although Makanda and Mjongo were members of one of these earlier 'independent' churches, P. J. Mzimba's African Presbyterian Church, they were perhaps closest to Mafeje's and Hobsbawm's characterisation of 'primitive rebels'. Had their revolutionary mix of millennial Christianity and black nationalism been as widespread as the contemporary white paranoia about Ethiopianism feared, the 1906 'rebellion' might well have reached far greater proportions than it in fact did.

That it did not do so must in part be attributed to the continuing validity of the earlier 'moral universe' of the peasantry: in Zululand as we have seen this was still tied to the hegemonic ideology of the Zulu royal family, and this was beginning to permeate Natal. In Natal itself, the Shepstone system which bolstered chiefly authority may have acted as a buffer against the acceptance of alternative ideologies, while in some sense paving the way for a recreation even in Natal of royal ideology. As we have seen, while new elements were constantly being woven into popular consciousness, from various versions of Christianity and from humanitarian imperialism, for example, none of the alternative ideologies being presented at this stage could in any sense be described as revolutionary. The socialist ideas which were just beginning to ferment in the cities of South Africa at this time, and especially on the Witwatersrand, barely addressed agrarian issues until considerably later, if at all.[99]

It would be quite wrong, however, to attribute the crushing of the Bambatha rebellion to the limitations of ideology and popular consciousness in 1906. There can be little doubt that all Africans in Natal shared a profound sense of injustice and were aware of their exploitation. If this was to be given political expression, they needed organisation and resources from outside which were simply not available at that time. Although in retrospect one can argue that the ferocity of the response suggests the weakness rather than the strength of the colonial state in 1906, in terms of armed combat, as we have seen, the odds were very uneven indeed. The kind of space, which is necessary if peasant uprisings are to be successful, was similarly non-existent: for all the disruptions of the South African war and the weakness and lack of legitimacy of the colonial state in the eyes of the majority of its inhabitants, the colonial ruling class still had virtually total control over the means of coercion, and confidence in its own capacity to rule. As Theda Skocpol reminds us, social revolutions cannot be explained in voluntarist terms, nor without 'systematic reference to *intra*national structures and world-historical developments'.[100] Despite the differences between the Natal government and the imperial authorities over the former's handling of the 'rebellion' and the arrest of Dinuzulu, there was little doubt in ministers' minds that in the final analysis imperial forces could be called upon to bail them out. Thus, in 1906, two of the three preconditions for successful peasant revolution were lacking in Natal: a revolutionary political party 'willing and able to mobilise the peasantry' and a 'suspension of state coercive power', a weakening both in the real power of the state and in its sense of its own legitimacy.[101] Today, when there is some doubt as to whether one can talk of a peasantry any more in South Africa, those on both sides who use the experiences of 1906 in their rhetoric might ponder the transformations in these external and internal forces as well as the changes in popular consciousness.

NOTES

1 Quoted in D. Hemson, 'Are migrant workers a sub-proletariat? The dockworkers of Durban', unpublished history seminar paper, University of Warwick, 29 November 1979.

2 ibid.

3 Natal Archives, Pietermaritzburg, Chief Native Commissioner, CNC 58/7/1 N1/1/3 (32), 1932, 'The words of Nkosi Solomon ka Dinuzulu at Emakhosini, 2.8.1927', letter to *Ilanga lase Natal*, 12 August 1927, transl. by Carl Faye.

4 South African Institute of Race Relations, *Annual Survey*, vol. 30 (1976); see also, for example, his 'Message to South Africa from Black South Africa', speech to Africans at

the Jabulani Amphitheatre, Soweto, March 1976.

5   I have dealt with the causes, course and consequences of the Bambatha rebellion in detail in my *Reluctant Rebellion. The 1906–8 Disturbances in Natal* (Oxford: Clarendon Press, 1970). This outline, drawn from *Reluctant Rebellion*, is simply by way of necessary background to the present paper, which is also something of an auto-critique. As will be evident, it has been heavily influenced by the more recent literature on peasant uprisings.

6   *Times of Natal*, 28 January 1908.

7   Quoted in Marks, *Reluctant Rebellion*, p. 244.

8   For the rebellion as self-fulfilling prophecy, see Marks, *Reluctant Rebellion*, pp. 188 ff.

9   Society for the Propagation of the Gospel Archives, Letters, 1906, 24 July 1906, no. 71, quoted in Marks, *Reluctant Rebellion*, p. 242.

10 Unpublished evidence of Archdeacon Johnson before the Natal Native Affairs Commission, 1907–8, quoted in Marks, *Reluctant Rebellion*, p. 169.

11 Henry Slater, 'Land, labour and capital in Natal: the Natal Land and Colonisation Company 1860–1948', *Journal of African History*, vol. 16, no. 2 (1975), p. 275.

12 C. Bundy, *The Rise and Fall of the South African Peasantry* (London: Heinemann, 1979), p. 185.

13 Marks, *Reluctant Rebellion*, pp. 6–7.

14 Slater, 'Land, labour and capital', pp. 277–8.

15 For the pressures on land in general, see Marks, *Reluctant Rebellion*, Ch. 5, and W. Martin, 'The development of capitalism in South African agriculture: the class struggle and the capitalist state during the "phase of transition", *c*.1890–1920', unpublished paper, Binghampton, New York, 1980.

16 Marks, *Reluctant Rebellion*, p. 143; D. Hemson, 'Class consciousness and migrant workers: dockworkers of Durban' (unpublished PhD thesis, University of Warwick, 1980) has a fine analysis of worker, especially dockworker militancy during and just after the war which led to these raised wages – and to the colonial reaction.

17 Public Record Office (PRO), Kew, CO 179/222/7666, Report of Labour Conference held at Newcastle, cited in Marks, *Reluctant Rebellion*, p. 135.

18 PRO, CO 179/223/36293, Copy of confidential despatch to Adjutant-General, WO, 5 July 1902, enclosure in despatch no. 213, cited in ibid.

19 For the speech in Natal's Legislative Assembly, see Marks, *Reluctant Rebellion*, p. 132.

20 See Duncan Innes, *Anglo American and the Rise of Modern South Africa* (Heinemann, London, 1984), pp. 25–7.

21 The *Report of the Natal Native Affairs Commission* was published as a British Parliamentary Paper, Cd 3889; the evidence was not published but is to be found in the Public Record Office, Kew and the Natal Archives, Pietermaritzburg.

22 A notion I drew from Gordon Allport, *The Nature of Prejudice* (New York: Doubleday, 1958); the rumours awash in Natal-Zululand at the turn of the century are discussed in Ch. 6 of Marks, *Reluctant Rebellion*.

23 See Natal Archives, Secretary for Native Affairs, SNA 1, 1, 334, A. B. K. Farrer, Acting Magistrate of the City (Pietermaritzburg) in his unpublished report for the Annual Blue Book on Native Affairs, 1905, cited in Marks, *Reluctant Rebellion*, pp. 153–4.

24 Quoted in G. Rudé, *Ideology and Popular Protest* (London: Lawrence and Wishart, 1980), p. 9.

25 Natal Archives, SNA 1/4/17 C 51/07. Report, 'Intelligence no. 1', March 1907.

26 See, for example, Hemson, 'Are migrant workers a sub-proletariat?'.

27 Again, this should not surprise us. It conforms with what is known of peasant uprisings elsewhere: cf Ian Clegg, *Workers' Self-management in Algeria* (p. 100):

They [Algerian peasants] did have a common identity, but what bound them was the desire to recreate the past. Thus it is not true to say as some strict Marxists have, that peasants cannot participate in revolutionary struggle. Objectively the role of the Algerian peasantry in the liberation struggle was a revolutionary one. Equally, . . . it is untrue to say that because they are actively involved in the struggle against colonialism, they form a revolutionary class. Subjectively in terms of their consciousness, their role is not a revolutionary one. Quoted in J. McCulloch, *Black Soul, White Artefact* (Cambridge: Cambridge University Press, 1983), p. 157.

and Barrington Moore Jr, *Social Origins of Dictatorship and Democracy. Lord and Peasant in the Making of the Modern World* (Harmondsworth: Penguin, 1969), p. 457:

> rebellions in agrarian societies . . . take on the character of the society against which they rebel . . . The insurgents battle for the restoration of the 'old law', as in the *Bauernkrieg*, for the 'real Tsar' or the 'good Tsar' in the Russian peasant upheavals.

28 For the details, see Marks, *Reluctant Rebellion*, pp. 174–9.

29 ibid., p. 230.

30 See Hemson, 'Class consciousness and migrant workers'.

31 Eric Wolf, *Peasant Wars of the Twentieth Century* (London: Faber, 1973 edn), p. xv.

32 Marks, *Reluctant Rebellion*, chapter 12.

33 Wolf, *Peasant Wars*, p. xv.

34 E. J. Hobsbawm, 'Class consciousness in history', in István Mészáros (ed.), *Aspects of History and Class Consciousness* (London: Routledge & Kegan Paul, 1971), p. 9.

35 Rudé, *Ideology and Popular Protest*, p. 32.

36 ibid., Ch. 2, 'The ideology of popular protest'. My thinking here has been heavily influenced by Rudé's lucid discussion of the issues involved.

37 For the impact of colonial conquest and the civil wars on southern Zululand, see E. Unterhalter, 'Religion, ideology and social change in the Nquthu district of Zululand, 1879–1910' (unpublished PhD thesis, University of London 1981). Although it emphasises the neighbouring Nquthu district, rather than Nkandhla, there were many similarities.

38 This was in large measure the result of the colonial policy of introducing 'loyal' appointed chiefs into southern Zululand (Nkandhla and Nquthu) as a buffer after the Anglo-Zulu War.

39 See Charles van Onselen, *Studies in the Social and Economic History of the Witwatersrand, 1886–1914* (Harlow: Longman, 1982), esp. Vol. 2, *New Nineveh*, Chs 2 and 4.

40 See chapter by T. Ranger in this collection (Chapter 17, below).

41 *Natal Statistical Yearbook, 1907*, pp. 100–101.

42 *Natal Census*, 17 April 1904.

43 See, e.g. Natal Archival Depot, Zululand Archives (hereafter ZA) 31, Z 1196/96 Confidential. Resident Magistrate, Nqutu district, to Acting Resident Commissioner, Zululand, 24 December 1894.

44 Peter Warwick, *Black People and the South African War, 1899–1902* (Cambridge: Cambridge University Press, 1983), pp. 78–9, 87, 90–93.

45 Marks, *Reluctant Rebellion*, pp. 158–9.

46 *Report of Colonel G. A. Mills, C.B. on the causes which led to ill-feeling between Boers and Zulus culminating at Holkrantz, 6.3.1902* (Pietermaritzburg, 1902).

47 Natal Archives, Government House, G.H. 513 G. 818/02 Conf., Lieutenant G. A. Mills to Chief of Staff, Natal, 1 July 1902.

48 For the origins of the name, see Marks, *Reluctant Rebellion*, p. 112.

49 The evidence for Cajikana's membership of the Nkomindala is not conclusive, although his brother is named as a member (see Natal Archives, Colenso Collection, Statements made to R. C. Samuelson, 1907–08, pp. 61–2). The evidence of Rolela ka Fogoti (encl. in Co 179/244/3802 Dep. Secret, 11.1.08) also suggests his membership.

50 *South African Native Affairs Commission*, Vol. 3 (henceforth *SANAC*, 3), Evidence of A. J. Shepstone, RM, Vryheid, 15 April 1905, p. 266; see also Natal Archives, SNA 1/4/13 C 35/04 Memorandum of a report on Native Matters made by the magistrate, Vryheid, A. J. Shepstone to the Hon. Sec. of Native Affairs . . . . *c.* 1 June 1904.

51 *SANAC*, 3, pp. 257–8.

52 ibid., pp. 257–8, cited by Shepstone.

53 ibid., p. 258.

54 ibid., p. 264.

55 *Natal Census*, 17 April 1904.

56 Although by this time important in parts of northern Natal, the coal mining economy was only developed in the districts of the 'New Republic' after 1908, perhaps significantly after the disturbances had been crushed. In 1906 there were only fifteen Africans mining coal in Vryheid district. Coal mining eventually came to form a major sector of Vryheid's economy. (F. A. Steart, 'Coal in Natal', Paper read to the Third Triennial Empire Mining and Metallurgical Congress, South Africa, 1930; *Natal Government Gazette*, 13 March 1906, p. 293, Numbers of Natives employed on Natal Collieries in February, 1906.)

57 See, e.g. Natal Archives, SNA 1/4/19 C 260/07 Minute by R. Beachcroft, Magistrate, Vryheid, 20 December 1907, re Mabeketshiya who shot a Boer, Kemp, at the end of the war and was therefore not appointed chief of the Amafu tribe; or the case of Cajikana's companion in arms in 1906–7, the 'rebel' Mayatana, who engaged in cattle stealing after the war, and was tried for the murder of Stainbank in 1912. (Natal Archives, Colenso Collection, 171 Magis, A. W. Leslie to Dinuzulu, 10 May 1902).

58 J. Stuart, *A History of the Zulu Rebellion* (London: Macmillan and Co, 1913), p. 176, note.

59 T. Cope, *Izibongo, Zulu Praise Poems* (Oxford: Clarendon Press, 1978), p. 74, n. 5; Report of cross examination in the Special Court by W. P. Schreiner, *Natal Witness*, 15 December 1908).

60 Stuart, *History of the Zulu Rebellion*, pp. 501, 496. Cakijana was also known as *Sukabekuluma* (he who goes off whilst they are still talking); *Dakwaukwesuta* (He who becomes drunk on getting a full meal); and amongst his praises was *uSigilamikuba, ku vel'inzindaba* (the one whose pranks give rise to matters for consideration). These names may date from a slightly later period. (Stuart, *History of the Zulu Rebellion*, note on p. 176); Dinuzulu's name for him was *Ngodoyi* (mangy or wandering dog), a less flattering sobriquet. (Schreiner, *Natal Witness*, 15 December 1908.)

61 Theda Skocpol, *States and Social Revolutions. A comparative analysis of France, Russia and China* (Cambridge: Cambridge University Press, 1981 edn), p. 13.

62 PRO, CO 179/245/25140, Enclosure 5 in Despatch Secret, Gov to Sec of State, 20 June 1908, Major R. H. Wilson, Staff Officer, Intelligence to Commandant of Militia, Natal, 11 May 1908. Re. A Tour of Intelligence Duty, 1st Jan to 12th March 1908.

63 Great Britain, Parliamentary Paper, Cd 3888 (1908), *Further Correspondence relating to Native Affairs in Natal*, p. 63, CNA Eshowe to Acting PM Pietermaritzburg, 3 May 1907.

64 The policy of the Boer administration would appear to have differed from that of Natal in this respect; on the other hand, a number of the individuals involved in the establishment of the New Republic would appear to have maintained their contacts with Dinuzulu over the years, such as Coenraad Meyer and Louis Botha.

65 *The Trial of Dinuzulu on charges of High Treason at Greytown, Natal, 1908–9* (Pietermaritzburg: Government Printer, 1910), Cakijana's evidence to the Special Court summed up by Sir William Smith, p. iv.

66 C. T. Binns, *Dinuzulu. The Death of the House of Shaka* (London: Longmans, 1968), p. 194.

67 Stuart, *History of the Zulu Rebellion*, p. 242; Binns, *Dinuzulu*, pp. 196–206.

68 W. Bosman, *The Natal Rebellion of 1906* (London: Longmans, Green, 1907), p. 20.

69 Marks, *Reluctant Rebellion*, p. 225.

70 ibid., pp. 296–9.

71 ibid., p. 276.

72 *Natal Witness*, 12 November 1908.

73 Unravelling the evidence about Dinuzulu's complicity in the rebellion and especially in the unrest that followed is extraordinarily difficult. It is not made easier by the fact that it is clear that the leading witnesses against him were offered the inducement by their interrogators, members of Natal's intelligence forces, that they would lighten their own sentences by incriminating Dinuzulu. For a discussion, see Marks, *Reluctant Rebellion*, Chs 10 and 11.

74 Natal Archives, Chief Native Commissioner, CNC vol. 8, 292/1911, Actg Under Secretary of Justice, Natal to Inspector of Prisons, Natal, 28 March 1911. I am grateful to Bobby McGinn and Ruth Edgecombe for finding this for me in the Natal Archives.

75 Natal Archives, Attorney General's Office, AGO 1/7/52. Pte letter A. Hime to T. F. Carter (Minister of Justice), 14 November 1908.

76 Pretoria Archives, Governor General G.G. vol. 1591 ref. 51/801. Report by the Minister of Justice, Remission of sentence: native prisoner, Cakijana, Natal, 15 March 1911. I am grateful to Karin Schapera for finding this reference for me.

77 Natal Archives, Secretary for Native Affairs, SNA 1/4/22 C^{29}/10, Copy letter Acting Sec. Justice, Pretoria to SNA Pretoria, 23 November 1911; Colenso Collection 140, vol. III, H. E. Colenso to W. P. Schreiner, 3 May 1913.

78 Binns, *Dinuzulu*, p. 214, note.

79 Natal Archives, CNC 64/3, Brief notes of proceedings, Magistrate's Court Room, Nongoma, 31 May 1920.

80 Henry Callaway, *Nursery Tales, Traditions and Histories of the Zulus* (Westport: Negro University Press, 1970), p. 3. Callaway relates several legends about the trickster, Cakijana. The first has him speaking from his mother's womb before he is born. When he emerged from the womb he was fully formed and stood up and spoke. He was greeted immediately as a wise man. The women query whether he is a real man, as he was produced in so unnatural a way: they regard him as a prodigy who will do great things. He cheats even his own mother, and when he is captured by the cannibals, tricks them by serving them up their own mother in his stead. Many of the stories involve Cakijana outwitting and punishing cannibals or wild animals.

81 Interview with Princess Magogo, March 1969.

82 Personal communication, Mr M. B. Yengwa, August 1984.

83 This seems to be one of the few aspects of peasant uprisings on which there seems a fair degree of unanimity. See, for example, Moore, *Social Origins of Dictatorship and Democracy*, p. 479.

84 For the vulnerability of the *kholwa* at this time, see Marks, *Reluctant Rebellion*, pp. 76–80.

85 PRO, CO 179/235/22645 Gov. to Sec. of State, 30 May 1906, cited in Marks, *Reluctant Rebellion*, p. 74.

86 Marks, *Reluctant Rebellion*, pp. 333–4.

87 In the *Natal Mercury*, 1 June 1906, quoted in Marable W. Manning, 'African nationalist: the life of John Langalibalele Dube' (PhD dissertation, University of Maryland, 1976), pp. 211–12.

88 See, for example, Bernard M. Magubane, *The Political Economy of Race and Class in South Africa* (New York: Monthly Review Press, 1979), p. 57; see also Hemson, 'Are migrant workers a sub-proletariat?'.

89 Archie Mafeje, 'Religion, class and ideology in South Africa', in Michael G. Whisson and Martin West (eds), *Religion and Social Change in Southern Africa. Essays in Honour of Monica Wilson* (Cape Town: David Philip, 1975), p. 182.

90 Eugene Genovese, *From Rebellion to Revolution. Afro-American Slave Revolts of the New World* (New York: Vintage Books, 1981 edn), pp. xxi–xxii.

91 For a sympathetic but penetrating portrayal of this class elsewhere in South Africa, see Brian Willan's outstanding biography, *Sol Plaatje. South African Nationalist, 1876–1932* (London: Heinemann, 1984), especially Ch. 2. For a further discussion of the Natal *kholwa*, see S. Marks, *The Ambiguities of Dependence. State, Class and Nationalism in Early Twentieth Century Natal* (Baltimore: Johns Hopkins University Press, forthcoming), Ch. 2.

92 P. Bonner, 'The Transvaal Native Congress, 1917–20. The radicalization of the black petty bourgeoisie on the Rand', in Shula Marks and Richard Rathbone (eds), *Industrialization and Social Change in South Africa. African Class Formation, Culture and Consciousness, 1870–1930* (London: Longman, 1982), pp. 270–313, and Helen Bradford,

'Mass movements and the petty bourgeoisie: the social origins of ICU leadership, 1924–1929', *Journal of African History*, vol. 15, no. 3 (1984), pp. 295–310.

93   Marks, *Reluctant Rebellion*, p. 358.
94   *Report of the Natal Native Affairs Commission*, pp. 8–9.
95   Killie Campbell Library, Durban, Marshall Campbell papers, Bantu section, Letter from DC, Zululand, R.156/1912/32. Response of African in Zululand to an address from Dr. John Dube, explaining the formation of the South African Native National Congress, quoted in Marks, *Reluctant Rebellion*, p. 365.
96   Marks, *Reluctant Rebellion*, pp. 174–9.
97   Unpublished evidence to the 1907–8 Natal Native Affairs Commission, p. 349.
98   Mafeje, 'Religion, class and ideology in South Africa', p. 183.
99   See for example, C. Bundy, 'Land and liberation: popular rural protest and the national liberation movements in South Africa, 1920–1960', in S. Marks and S. Trapido (eds), *The Politics of Race, Class and Nationalism in Twentieth Century South Africa* (London: Longman, forthcoming).
100 Skocpol, *States and Social Revolutions*, p. 14.
101 Theda Skocpol, 'What makes peasants revolutionary?', *Comparative Politics*, vol. 14 (1982), pp. 315–75.

17 Bandits and guerrillas: the case of Zimbabwe

TERENCE RANGER

Introduction

One of the themes of this book is whether or not the concepts of banditry and social banditry can be applied to Africa. It seems relevant, then, to note that in Southern Africa the word 'bandit' is now in daily use. The governments of Angola, Zimbabwe and Mozambique describe their armed opponents as bandits. In Mozambique, for example, 1984, is 'Crush the Bandits Year'. A characteristic attack on the dissidents of Matabeleland in the *Herald* of Zimbabwe runs:

> We have armed bandits who want to overthrow a lawfully elected government and replace it with one of their own liking . . . To let them get away with murder, robbery, mutilations and rape would be to condone these heinous crimes. If bandits want to behave like beasts then they must be treated like beasts.[1]

On their side, meanwhile, the armed opponents of these regimes throw back the insulting word. On 5 March 1984, for instance, the Mozambique National Resistance (MNR) radio in Lisbon proclaimed that the Frelimo government could not save itself by means of treaties with South Africa:

> South Africa's new assistance is not coming in time to rescue the corrupt and decadent gorilla and his gang of bandits. The struggle – the peoples' struggle – is approaching Maputo . . . Who will be able to save Machel? Nobody . . . We are the people and therefore we shall win.[2]

The play on words is deliberate. To the MNR, Machel is a 'gorilla', *not* a guerrilla. For if the word 'bandit' carries with it totally negative implications, the word 'guerrilla' carries with it totally positive ones. The MNR claims that its own men are guerrillas, a title which the London *Times* at any rate is prepared to accord them.[3]

The two words are currently being applied, then, in no very analytical way. Armed men that a speaker or writer approves of are 'guerrillas'; armed men that he disapproves of are 'bandits'. It is true that on occasions finer distinctions are made. Thus in Zimbabwe the state counsel in a case against men accused of having fired on Mr Mugabe's residence 'alleged that the attack was not "a motiveless act of banditry" but part of a plan to topple the Government through a military coup and put Zapu in power.'[4] But in general the identifying criteria of banditry are not lack of coherent political motivation but rather illegitimacy and criminality. The bandit resorts to force against 'the people'; the guerrilla represents the people. Bandits behave like beasts: guerrillas behave like heroic men.

It is odd that these words of Mediterranean origin are now used in Southern Africa to describe the extremes of legitimate and illegitimate armed resistance. Are there any indigenous traditions which underlie them?

Bandits and guerrillas in colonial Zimbabwe

I shall take Zimbabwe as my case study. It is hard to argue for a *continuous* tradition of banditry or social banditry in Zimbabwe. Both are pre-eminently phenomena of peasant societies and since the first decades of colonialism in Zimbabwe saw first the emergence of an African peasantry and then its increasingly intensive exploitation, one might have expected the territory to be fertile ground for bandits. Indeed Allen Isaacman has claimed that banditry and social banditry were 'a common response to the simultaneous imposition of colonialism and incorporation of the region into the world economy around the turn of the twentieth century', and that 'it is likely that social banditry continued throughout the colonial period' in Zimbabwe.[5] However, banditry of any kind emerges not from oppressed peasant societies as such but from oppressed peasant societies with access to arms. Twentieth-century bandits without guns are hard to imagine and for most of the century Zimbabwean African society was very effectively deprived of arms. This means that we have to look for widespread banditry either at the very beginning of colonial rule, in the aftermath of the wars and uprisings of the 1890s, or at the very end of it, during and after the guerrilla war of the 1970s.

The Zimbabwean plateau in the 1890s was certainly awash with guns. Equally certainly, many guns were secreted rather than surrendered at the end of the uprisings in 1896 and 1897. The native commissioners of many districts reported the existence of armed bands, usually under the command of a young member of the chiefly family, which avoided capture and refused to join in the general surrender. Such bands raided chiefs and headmen 'loyal' to the white administration, carried off cattle, and fired on administrative agents. Some of them plainly enjoyed a good deal of popular admiration and support. None of them lasted long on the central plateau, however, and as colonial rule became more and more of a reality and disarmament was more effectively enforced, the bandit option more or less disappeared.

The belief persisted that large quantities of guns remained buried in places known only to senior elders and to spirit mediums. In his fascinating study of the interaction of mediums and guerrillas during the 1970s David Lan notes that:

> one of the guerrillas who fought the first battle of the war in 1966 at Sinoia made an attempt to contact a spirit medium named Gumbachuma who lived in the Zvimba area. His interest in this medium was sparked by the belief that the mediums knew where rifles used during the 1896 rebellion had been stored. To the weapon-starved guerrillas this information could have been a major contribution to the struggle.[6]

I came across similar ideas during my research in Makoni district in 1981 but neither there nor anywhere else have I encountered any instance in which buried guns were recovered or made use of. In the 1970s the Zimbabwean plateau was once again awash with guns but they were all brought in by the guerrillas themselves or by the 'Security Forces'.

For this reason, during most of the twentieth century the bandit 'tradition' existed only along the eastern boundaries of Zimbabwe. In these border areas a case for social banditry and for its interaction with subsequent nationalist and guerrilla resistance can be made. Isaacman's article documents the activities of a number of armed bands along the north-eastern frontier with Mozambique at the end of the nineteenth and beginning of the twentieth century. The article focuses on Kadungure Mapondera and Dambukashamba, but Isaacman notes that 'several other autonomous bands operated simultaneously within this region. In North Mazoe, for example, Kashidza and his followers repeatedly attacked British patrols and tax collectors, while across the border in Mozambique Samakangu and Capacula engaged in equally "seditious" acts'. But while some of these bands may have been merely brigands, and others were 'primary resisters' Isaacman argues that Mapondera and Dambukashamba at least are best seen as social bandits.

He sees both men as possessing the necessary 'organic relationship to the peasantry' and as committed to the necessary 'efforts to destroy the institutions that oppressed the rural populace', thereby clearly distinguishing themselves both from traditional chiefs and from those 'outlaws and brigands who indiscriminately preyed upon all sectors of society'. He shows that both men attracted 'committed followers who were drawn from a variety of oppressed groups': peasant fugitives from tax and forced labour, men who refused to acknowledge the authority of 'loyalist' chiefs, embittered labour migrants:

> The peasant outlaws repeatedly attacked the symbols of oppression that the colonial regime had imposed. Operating on both sides of the unmanned border they raided government posts, ambushed tax collectors, attacked labour recruiters and burned the rural shops which were exploiting the peasants . . . Mapondera and his followers benefitted from the material aid the peasantry provided . . . In addition to food, the rebels also received strategic assistance from the peasantry . . . Defence of the peasantry earned Dambukashamba's band popular support and material assistance. Members of the rural population smuggled its members grains, vegetables, salt and other foodstuffs . . . The rebels ambushed labour recruiters and police patrols, burned rural shops that engaged in price gauging . . . made a special effort to destroy the plantations and estates that the Companhia da Zambezia had sublet to European and Goan speculators which, in the words of a contemporary observer, were 'in the main organizations for the exploitation of the natives'.[7]

In these ways 'Mapondera and Dambukashamba created a base of popular support and a permanent bond with the rural population'.

Isaacman's later work has shown that this tradition of banditry and social banditry persisted in Zambezia well into the 1920s.[8] Moreover, he is able to show that the *memory* of the social bandits has been preserved in the region in a way which has made connection with the recent guerrilla wars. Although Dambuka-shamba was of 'humble origin', Isaacman and his student co-researchers found that 'in death, the peasant community in the region of Chioco paid Dambuka-shamba its highest honour through deification . . . To this date Dambukashamba is revered as a *mhondoro*, or national ancestor spirit, and his earthly medium regularly is propitiated and consulted in times of natural disaster and other crises'.[9] And if we can credit Michael Raeburn's imaginative reconstruction of guerrilla oral reminiscences, Kadungure Mapondera's spirit was similarly represented in a series of mediums, one of which made direct connection with the guerrilla war of the 1960s and 1970s. One of Raeburn's chapters concerns the guerrilla Goredema, who flees police pursuit to take refuge with an uncle in Mazoe district. The uncle turns out to be the medium of Mapondera's spirit; he tells

Goredema that the spirit will protect him in his struggle against the whites; and he narrates the story of the 'African hero who had fought the colonialists until his death'. Goredema comes to realise that so long as rural Africans venerated such heroic spirits, 'they could never be dejected or servile'.[10]

Another sequence of social banditry can be argued for the south-eastern corner of Zimbabwe, the so-called 'Crooks' Corner', where Zimbabwe adjoins the Transvaal and Mozambique. The Hlengwe/Shangaan people of the south-east spill over into all three territories and have had frequent occasion to wish to resist all three colonial administrations. The special form which social banditry took in the south-east was armed 'poaching' – organised defiance of the game laws of colonialism. In a recent paper Stanley Trapido has described what happened in the Transvaal part of this region:

> In 1898, to prevent Africans from hunting in the Sabi region where he was Native Commissioner, Abel Erasmus had led officials of the Boer state in rounding up people and forcing them out of the district . . . [Many] evaded Erasmus' rule by moving across the border into Portuguese-ruled territory. Settler rule was less pervasive there and an unknown number of refugees established new homesteads close to the border with the Transvaal. From there they joined forces with kinsmen and others and continued to hunt for skins and food on both sides of the Transvaal/Mozambique border . . . The next twenty-five years were to see ever-increasing raiding from the Portuguese colony as more and more poachers crossed into the South African game reserve . . . Poachers coming across the border from Mozambique – the over-whelming majority of whom were Africans – were very often armed with rifles . . . Increasing poacher activity – aggravated both by the serious drought of 1912 and by the decrease in wild-life on the Portuguese side of the border – led to a series of incidents which caused [administrators] to fear that the constant flouting of authority in this area would pose a serious challenge to colonial rule on both sides of the border.

Trapido describes how when well-known 'poachers' were arrested 'relatives and others . . . let it be known that they would henceforth shoot at unarmed policemen on sight'. And he shows how the confrontation developed into 'an increasingly desperate vendetta waged by poachers against African rangers'. Trapido's examples run up to 1927 and there is no reason to suppose that the 'poaching' bandits ended their activity then.[11]

No similar academic study has been made of the situation in the Zimbabwean south-east but there is no doubt that it was very similar. Our best insight into it, and into social banditry in the south-east, is provided by Allan Wright, who was district commissioner in the south-east for ten years from 1958. Wright describes the total public support for 'poachers' amongst the Zimbabwean Shangaan:

> Within the African community, there is absolutely no resistance to the snarer and nobody, from the Chief through the school teachers right down to the simple peasant, ostracises a man who cruelly ensnares wild animals regularly . . . The snarer is recognised as a man of standing in the community . . . The African snarer's activities, far from being discouraged, are actively applauded by every man, woman and child in the neighbourhood . . . The Shanganes of our Lowveld must rank as the most persistent poachers in all Rhodesia. One senior headman, on being taxed by me about the snaring activities of his people, replied: 'Can I stop my followers from drinking water?'

Wright tried to understand this public sympathy historically:

They, like the North American Indians, saw their hunting lands re-emerge as large European ranches or as land where they were told hunting was forbidden as it was the property of the British monarch – Crown land. The Shanganes of Rhodesia (and I suspect this applies to the other sections of the tribe in Mozambique and the Transvaal) refused to accept the validity of this ban on hunting on their old preserves . . . European ranch owners were becoming increasingly angered by the activities of poachers and snarers and the depredations of large hunting parties . . . The boldness of some of these meat-seekers was incredible – some gangs would hunt up to within a mile of a homestead situated some ten miles within the boundaries of the owner's land.

It was in this context that particular individuals achieved fame as, in effect, social bandits. Wright names two or three but describes only one in detail. This was:

a Shangane named Chitokwa and for years he was a living legend among the African people – Rhodesia's Ned Kelly, bushranger extraordinary. But Chitokwa displayed none of the brutality of that Australian outlaw – he confined his activities to theft, house-breakings, stocktheft and game poaching, and his beat was the entire lowveld area of Rhodesia and the Northern Transvaal! At the height of this activities the large European ranching region . . . was virtually in a state of emergency . . . For two long years, Chitokwa roamed the bush and defied the combined efforts of the police forces of South Africa and Rhodesia to arrest him – he became a sort of folk-hero among the tribespeople of the lowveld areas, the man the *majoni* (police) could not catch! He broke into stores, hotels, farmhouses . . . He stole a rifle and ammunition and lived off the teeming game herds. His hideouts were caves in the rough, broken koppie country . . . His principal hideout was in the Gezani area of the Sengwe T.T.L. [Tribal Trust Lands] just across the border from the European ranching area, and that all local Africans were aware of his presence in a cave on a hill but none dared, or desired, to report it. He frequently presented meat· to his temporary African neighbours.

Angry ranchers threatened to form a posse and demanded army units and helicopters: the full Riot Squad from Bulawayo moved into the area. 'The fugitive, however, used his peerless knowledge of the bush to avoid every patrol and every trap set for him and after weeks of fruitless search the Bulawayo contingent was withdrawn'. A large cash reward was offered but 'he had become a sort of modern-day Robin Hood, shooting game on European land and giving some of the meat to the protein-starved people'.

Chitokwa was not arrested until 1964. Only a year later there came a significant demonstration of how Hlengwe grievances over hunting could be mobilised by African nationalism. Joshua Nkomo and others restricted at Gonakudzingwa in Nuaanetsi district began to attract great crowds of Shangaane to their camp. Wright believed that there was a danger of insurrection. His police spies reported Nkomo as telling his Shangaane audience:

When I take the country I will look after you well. I will take my people and go back to Bulawayo and Harari and you can stay here in your own land and you will be able to hunt the wild animals as you used to do in the days before the Europeans came.[12]

In the late 1970s the Tribal Trust Lands of Nuanetsi became strongholds for the ZANLA guerrillas of ZANU/PF. It seems quite possible that the Hlengwe/Shangaan saw the guerrillas as Chitokwa writ large.

Thus in areas on the eastern border there was a long sequence of interaction between resistance, banditry and social banditry. These areas were, of course, particularly significant ones in the guerrilla struggle of the 1970s, as ZANLA began to be able to enter eastern Zimbabwe out of Mozambique. It would be fascinating to know more about the extent to which ZANLA guerrillas drew upon or were affected by or were perceived in terms of the previous sequence. It would be fascinating, too, to know how the activities of the MNR movement, which have sent many thousands of refugees fleeing into eastern Zimbabwe, are seen in terms of this background.

Elsewhere on the Zimbabwean plateau there is no evidence that these traditions of banditry existed. Nevertheless, it is plain that there were many connections waiting to be made by the guerrillas between their methods and tactics and the methods and tactics of earlier Shona warfare. One of the essential forms of knowledge, treasured and passed on by Shona elders, concerned the location of the strongholds, caves and hiding places which had been used in past wars and could be used again. In 1930 an irate Irish Jesuit, complaining to his superiors about the insufferable people of Makoni district, remarked that:

> it is hard . . . to humble a people whose age-long vice has been cowardice, when that vice is looked upon as a great virtue: to point to great rock caverns and subterranean dwellings where they crept like rabbits in quaking fear, rather than fighting like men, and to know that they smugly congratulate themselves on having sneaked away.[13]

What O'Hea was describing was a popular culture of violence highly appropriate to guerrilla war. Lan describes how spirit mediums of the *mhondoro* (royal ancestors) in the 1970s could 'reveal the location of the caves in which the rebels had concealed themselves during the first War of Independence'. As Lan writes:

> All *mhondoro* were responsible for a great deal of killing while they were alive though always in a good cause, indeed in the best of all possible causes: the establishment of the territory within which their descendants now live. And therefore all *mhondoro* know a great deal about war. Very frequently ex-guerrillas explained to me that the reason they had worked so well with the *mhondoro* mediums was that they know *zvinhu zvese svezvohondo*, everything there is to know about war. Amongst the Shona, soldiers do not form a separate section of the population. All men are potentially warriors, chiefs are the military leaders of the present, *mhondoro* were the military leaders of the past. If you think of the ZANLA guerrillas as the warriors of the past returned in new guise, their alliance with the *mhondoro* medums seems neither innovatory nor surprising.[14]

It can be seen, then, that the roles of bandit, social bandit and guerrilla each have their precedents in Zimbabwe. That is not to say, of course, that the *terms* have been used by Africans in the past, or used in what we would regard as appropriate contexts. Wright tells us, for instance, that 'short-sentence African convicts . . . are always referred to as "bandits" because their own name for themselves is *banditi*'.[15] Nor is it to deny that the conditions of the 1970s transformed everything and opened up possibilities of playing not only the guerrilla role, but also the bandit and social bandit roles on a scale not hitherto thinkable.

Guerrillas, bandits and social bandits in the 1970s

In the 1960s and especially in the 1970s guns became available on the Zimbabwean plateau from many different sources. They came in with the guerrillas from Zambia and Mozambique; they came in with the 'Security Forces' and their many auxiliary groups; white settlers also built up a heavy armament. Guerrillas captured guns in raids on white farms or in successful encounters with government forces: the administration captured guns from the guerrillas and issued them to various bands acting as pseudo-guerrillas. In this confusion many guns fell into the hands of men who were neither guerrillas nor members of one or other government unit. In any case, men who had begun as guerrillas or as government soldiers often took themselves and their guns off on private enterprises of violence. All this raised in the most acute way the questions of legitimacy and of animality: which of these men with guns represented the people and were seen to behave like heroic men; which resorted to force against the people and were seen to behave like beasts? Which, in short, were seen as guerrillas and which as bandits? And was there space for social bandits?

To begin with, of course, the government claimed legitimacy solely for its own forces and through its propaganda fixed the label of animality upon the guerrillas. Much government propaganda attacked the guerrillas as 'Communist Terrorists' and accused them of aiming to take over peasant land and peasant women. But this seems to have been much less effective than the employment of a vernacular term. The guerrillas were described by the Security Forces as *magandanga*, meaning 'wild people who live in the bush'; it was added that their animality extended to their possession of tails. The term spoke effectively to fundamental ideas about the bush as the source of danger and illegitimate violence and offered one way of situating unknown young men with guns who operated out of caves in the hills. It is clear that the disturbing force of the idea did not only affect rural peasants. In his novel *The Non-Believers Journey*, S. Nyamfukudza presents us with a sceptical urban protagonist, Sam Mapfeka. Sam cannot bring himself to believe in a 'progressive' liberation movement which works with spirit mediums. Early in the 1970s Sam visits his village home in the north-east at a time when the guerrillas have arrived in the area but have not yet begun military operations. His younger brother tells him of their presence: 'Look out, brother! They are here! "The Boys" are in the hills all around us with their bazookas and KKs, we've seen them with our own eyes! Things are moving!' Sam's reaction is revealing:

> The hills around them as they walked had assumed a startling newness and he watched them with a searching suspicion, as if every nook and cranny held secrets which would elude him if he did not ferret them out. It had been the strangest of feelings, that this rugged piece of ground over which he had run and herded cattle throughout his childhood now held a force which would wrestle with, and probably succeed in altering the course and future of the country. One could no longer look over it, much less wander through it, without coming to terms with the force it contained. He thought, at last, he had come to some understanding of what his grandmother had tried to impress on him with countless tales and legends; the forest must be treated with respect, because if you trifle with it, it will eat you up! But the idea of a truce, based on mutual respect had now assumed a new, more menacing dimension. This time the enemy was not the mysterious spirits of ancestors and of the forest who would mete out punishment for disrespect or treachery, but men just like himself who suddenly wandered into the middle of villages as if out of nowhere, with guns

strapped on to their backs, bringing the forest right into their homes as it were.[16]

Bringing the forest right into the home is, indeed, just what guerrillas do, but it is also just what bandits do. Everything turns on whether this violent power of the bush has been legitimised or whether it retains its anarchic, anti-social character. For this reason, one of the essential stories of the war, constantly repeated both by guerrillas and by peasants, is the story of the first encounter between the two, when it became clear that the guerrillas were heroic men rather than animals. All recent accounts of the war present their own variations of this story. I collected many myself in Makoni district in 1981. Isaac Tungo, for example, told me:

> I was taken to them among the rocks. We exchanged names. I can still remember some of theirs. The leader was called Soveria; then there was 'Action' and Pedzai Mabunu, finish the Boers. Those were the days when they were called *magandangas*. They said 'They call us *magandangas*. Do we have tails?' I said 'I don't see any'. 'We are human beings, you see, and we need your help'. I said I would do what I could.[17]

And one of Lan's female informants recorded in her notebook that in 1974 she:

> started hearing people talk about the *vakomana*. Some are saying they are terrorists, others freedom fighters. Some were saying they were animals. They've got tails. They can't talk. So I was curious to see them. So I travelled for a long distance . . . [The guerrillas] said: do you know us? [runs another informant's notebook]. We said no and he said: We are the terrorist you heard about, and he turned round to show us that he did not have a tail.[18]

Julie Frederikse, in her remarkable compilation, quotes from an interview with Tembo Chimedza, who met the guerrillas when he was a student in Chibi:

> I still remember the first time I saw the comrades. I was at school and when I came home in the evening I was told that the people we had heard of as *magandanga* had come . . . They were very friendly to us, so we liked them. They said, 'You have heard that *magandangas* have got tails, and they are baboons. Look, come here and we will show you that tail'. And then we saw that they were the same as us. And then they said 'You have heard that the *magandangas* are killing people. Are we killing?' Then we saw, ah, they are not killing, they are friendly to us, and they were very happy to see our mothers and fathers and sisters and brothers. So, the first time we were afraid, but when time went on, we were all friendly.[19]

Such stories carry on right to the end of the war, when for the first time the guerrillas were seen by townspeople. Thus I was told by a guerrilla that when he and his comrades were taken to the Assembly Point near Seke in early 1980, people from the township, who had been told that guerrillas had tails, came out to see if this was true.[20]

The point of all these stories, is that such encounters with guerrillas at once established their legitimacy and disproved their animality. Lan's woman diarist was encouraged when a guerrilla spoke to her politely and with respect; Frederikse's student when the guerrillas were seen to be 'not killing'. There was much more to it than that, of course. Several writers, Lan most of all, have explained how the guerrilla men of the bush were legitimated for African peasant societies. Lan has explained that through their alliance with the spirit mediums the guerrillas achieved that hero status which meant that contact with the forces of the bush gave power to represent the interests of the people rather than malevolent and destructive energy. The guerrillas, writes Lan, became autochthons instead of alien intruders, and 'autochthons of a very extreme kind. They live deep in the forest like wild semi-human creatures, so profoundly at one with nature and wild animals . . . that they are able to perceive the secret meanings contained in their behaviour.' What results is 'an extreme identification of the guerrillas with the land'. In this way the term *magandanga* becomes positive:

> Even after the human-ness of the guerrillas had been firmly established they were frequently referred to as *magandanga* . . . For the Security Forces this was a term of abuse. For the peasants it was not. It expressed a complex perception of people who lived in their territory but not in their villages, who were strangers but who claimed to be ready to risk their lives on behalf of people they hardly knew.[21]

My own recent book and Lan's work have been very properly concerned to document this process of legitimation; what Lan calls 'a remarkable feat of sustained co-operation between guerrilla and peasants'.[22] This certainly remains by far the most important thing to discuss. Yet it is not the only thing. The *magandanga* idea retained its ambiguity. Not all guerrillas achieved legitimacy and some lost it. Some armed groups never sought to achieve it.

Legitimacy in a guerrilla war, of course, derives from more than one source. It lies not only in acceptance by a peasantry but also in recognition and control by a liberation movement. If armed men are regarded as legitimate both by peasants and the party then they can be described as fully guerrillas. If they are repudiated by both peasants and the party then they are unequivocally bandits. Greater ambiguity arises when armed men are regarded as legitimate by one of the two elements but not the other. What results when armed men are accepted by the peasants but not by the party? Possibly we are here close to social banditry in the context of a guerrilla war. And what results when armed men are recognised and controlled by the party but not accepted as legitimate by the peasantry? Do we have failed guerrillas or political bandits? At any rate all four of these possibilities were realised during the 1960s and 1970s in Zimbabwe.

After the adoption of the strategy of a 'people's war' in the early 1970s most of the armed men sent into Zimbabwe achieved legitimacy in the eyes of the peasants and retained it in the eyes of the liberation movements. There was a constant emphasis within the guerrilla movement upon the rules of right conduct. ZANU/PF's *Zimbabwe News* spelt these out in 1978:

> The three main rules of discipline and nine points of attention are as follows:
> 1. Obey all orders in your actions.
> 2. Do not take a single needle or piece of thread from the masses.
> 3. Turn in everything captured.
> The nine points of attention:
> 1. Speak politely.
> 2. Pay fairly for what you buy.
> 3. Return everything you borrow.
> 4. Pay for everything you damage.
> 5. Do not hit or swear at people.
> 6. Do not damage crops.
> 7. Do not take liberty with men.
> 8. Do not take liberty with women.
> 9. Do not ill-treat captives.[23]

These same rules were embodied in *chimurenga* songs, regularly performed by the guerrillas:

> Let us refrain from expropriating the possessions of our masses,
> Let us return to them property taken from the enemy . . .
> Pay cash for the goods you've bought
> Return to the owners
> Those things you have expropriated . . .
> Educate the masses in a clear way
> And they will understand the policies of the party.[24]

Guerrilla political commissars constantly reiterated the rules, as for example in these notes for a verbatim speech to Hlengwe/Shangaan villages in south-east Zimbabwe:

> Speak politely to the masses and each other/no rebuke in public/no demand on shortage/no harassment/to teach/no strict speaking or beating/no refusal of poor food offered by the masses.[25]

Of course the guerrillas did not behave all the time with this tea-party propriety. As ZANU/PF's publicity secretary, Eddison Zvogbo, remarked:

> In every guerrilla war there is a need to demonstrate power. When you do so depends on careful calculation, because too much repulses and you lose support, while too little demonstrates weakness and you also lose support . . . Once political support has been won and the comrades move in, they are presented with lists of enemy agents by the population. It is not advisable to then arrest the person and simply kill him, because that act has no mass participation. What you need to do is to arrest the person, bring him to the people, lead evidence against him and then let the people say what should happen to him . . . He will be tried by the people, sentenced to death, if indeed his crime has lost lives, and the sentence will be carried out on the spot . . . If the purpose is to demonstrate muscle, you may order him not to be buried . . . Everybody begins to think of the price of being a traitor. Until the army comes to bury him, which further demonstrates to the people that he was really their agent.[26]

Less sympathetic chroniclers of the war phrase it rather differently but to much the same effect. Rhodesian propaganda, write Moorcraft and McLaughlin, often spoke of 'an indiscriminate reign of terror inflicted on the African population by the guerrillas'. In fact, it was a discriminate reign of terror 'against those who sympathised with or aided the Rhodesian cause. "Collaborators" or "sell-outs" were brutally murdered . . . But the targets were normally carefully selected and the local population could usually see the point'.[27] Oral testimony shows clearly that informers were detested by peasants not only because of their disloyalty to the cause but because they placed the whole community in danger of attack by the 'Security Forces', who killed very many more civilians throughout the war than the guerrillas did. Thus 'revolutionary justice' was accepted as part of the responsible behaviour of the guerrilla.

Guerrilla groups explained their conduct in these terms to African inquirers. In the archives of the Catholic Commission for Justice and Peace in Harare is a testimony by an African woman who met and questioned a group of guerrillas in Marandellas in February 1977:

> Mrs M was told that most of the Vakomana's movements, and especially their communications with villages, occur at night, both for their own safety and that of the villagers. They said they force no one to give them food asking only for what people can afford . . . Asked if there is any truth in the allegations,

particularly in the media, that the 'Terrorists' are rapists who also encourage prostitution in the areas they visit, they strongly denied them and said they would not do such things because they do not want to leave the women with babies ... They also said they neither kill, lay mines, nor rob stores indiscriminately. They would only take from a store in excess of their needs if the owner had no sympathy with their cause. In that event the extra clothes and provisions they claimed to distribute amongst those most in need of them ... Mrs M asked about a specific guerrilla execution ... The group admitted responsibility and said the man was killed because he had accused them of destroying the country by their fighting.[28]

Later on, as we shall see, reporters to the Commission were not so enthusiastic about guerrilla conduct. Yet a correspondent who wrote to complain of guerrilla behaviour in 1978 and 1979 gave the best picture of an earlier ideal guerrilla interaction with the peasantry:

In 1976 Zanla forces started to penetrate the East and West and North East of this country ... The leaders seemed to be well educated, polite and well behaved. When addressing a group they made sure that the people were well guarded and armed men encircled the crowd. At slightest sign of danger they were ready. Meetings were always held in the open ... Their first aim seemed to be to win the confidence of the people and to establish contacts. This they succeeded in doing very effectively. Right from the beginning they were determined to get rid of all 'sell outs'. These were shot but not tortured and they refused permission to bury the corpses. Rarely during those days was anybody killed who was not a genuine 'sell out' or a government agent. Young and old seemed to respect the freedom fighters who were prepared to give their lives to obtain genuine freedom for the people.[29]

By contrast with these ideal guerrillas, there were from the beginning groups of armed men who were feared by the peasants and repudiated by the parties. Some of these were 'rogue' guerrillas. One such was Silas Paul Muwira, who bore the Chimurenga name, James Bond. In 1972 Muwira:

slipped across the border with his AK-47 to settle a personal dispute with a Rhodesian village headman. The headman escaped in a hail of bullets, but Muwira was pursued into Rhodesia by Martin Rauwa, the ZANLA Field Political Commissar ... who captured him while he was at a beer-drink ... ordered him to be tied up, then force-marched him to the Chifombo ZANLA base ... for punishment.[30]

Others were more difficult for ZANLA to deal with, 'private grudge gangs' which killed old enemies or looted property under the cover of the guerrilla war.[31] By 1977 people were becoming bewildered by the number of different armed groups. 'People are disturbed by the diversity of groups', wrote Father Amstutz of the Bethlehem Fathers of Gwelo Diocese in June of that year:

freedom fighters, deserters, groups that have run amok, criminals and finally government troops, the Selous Scouts. But in the long run people are capable of distinguishing between a genuine freedom fighter and an imposter. Freedom fighters ... knock gently, do not run down doors, are very courteous and ask for specific help.[32]

In Matabeleland where both ZANU/PF and ZAPU guerrillas operated, the climax of confusion came in 1979. Lieutenant-Colonel Reid Daly, commander of the Selous Scouts, offers the following analysis:

Serious problems began being experienced by the ZIPRA (ZAPU) guerrillas in southern Matabeleland because of the threat being posed by ZANLA, who were rapidly expanding their military influence westwards across the country from Mozambique . . . ZAPU's guerrilla groups inside Rhodesia were certainly not sufficiently well organised to effectively take on ZANLA as well as the Rhodesian Security Forces, mainly due to a lot of dissension between the various guerrilla groups . . . Dissident ZIPRA guerrilla groups became commonplace, particularly in the Lupani, Que Que and Gokwe areas. Notable ones were led by Lipson Mangarirani and Sandhlana Mafuta, who openly defied orders from military leaders in Lusaka, preferring to fight the war in their own way. There were large-scale desertions too of individual terrorists, who resorted to banditry or abandoned their units . . . Needless to say, the Selous Scouts were continually in the middle, stirring things up all the time.[33]

Bandit groups who turned their weapons on the people and who stole cattle and money are remembered today with bitterness. On the other hand, there were individuals and groups who had been repudiated by their party and who had adopted a bandit style but who nevertheless retained popularity with the peasantry. These were bandits who raided European property. We can draw on Daly again for a striking example:

A renegade terrorist, Obert Dhawayo, had moved from Mount Darwin to the Chipinga area where he had become active in the lucrative field of armed robbery. The final straw for the CID in Umtali came when he hijacked a mobile bank, murdered one of the cashiers and escaped with some sixty thousand dollars. Someone, as a consequence, suggested that maybe the Selous Scouts would be able to help . . . I despatched Jeremy Strong to the Ngorima Tribal Trust Land, ostensibly to assist the CID, but in reality, to dig deeply and find out the thoughts of the tribesmen in the area and ascertain if the first stage of politicising and organising the tribesmen by ZANLA had begun . . . It was a very sobering exercise . . . They reported, even at that time [in early 1975] that the whole tribal population had given every indication of being utterly committed and sympathetic to the ZANLA cause and were eagerly awaiting . . . their arrival. The renegade terrorist, Obert Dhawayo, was not regarded as a criminal at all by those formerly very law-abiding people, but rather as a sort of Robin Hood and everyone was willing to help, feed or hide him.[34]

If Obert Dhawayo seems to have been regarded as a social bandit it is more difficult to know how to categorise other groups. These were bands of guerrillas who had entered an area bearing with them the legitimacy given by the authority of their party and who had also gained the co-operation of the peasantry, but who were later denounced as 'dissidents' as policy and leadership changed within the liberation movement. Guerrillas coming in from Mozambique came in a series of waves. The first came in the name of Ndabaningi Sithole, then still leader of ZANU. Then came groups who accepted the legitimacy of Abel Muzorewa, head of the umbrella UANC within which all the liberation movements had in theory been brought together in early 1975. There then followed a reaction against all political leaders and the emergence of the Zimbabwe People's Army, whose guerrillas recognised no civilian control. Finally, from 1977 onwards ZANLA guerrilla groups came in from Mozambique in the name of Robert Mugabe and representing ZANU/PF under his leadership. In many cases, guerrillas already in the field accepted these changes and took instructions and advice from incoming guerrilla bands. But in other cases they did not and as a result were denounced as dissidents. It was not so easy, however, to deprive them of their legitimacy and to declare them bandits rather than guerrillas. Legitimacy, after all, was accorded by peasants as well as by parties, and peasants who had learnt to trust and to work with guerrillas and to accept them as heroic men rather than as wild things did not easily abandon their old associates and accept their accusers. The veteran nationalist, Mhondiwa Remus Rungodo of Chiduku in Makoni district recalls that:

> At times the situation was aggravated by clashes between ZANLA groups operating from adjacent regions. One group accused the other as dissidents and hunted each other. As a result civilians were adversely affected. They were blamed for feeding dissidents. Some died as a result. Apart from these inconsistencies the guerrillas were in most cases good to the people. They were unlike the Security forces who harrassed the population.[35]

Overall peasant support for the guerrillas could survive cases of this sort. But the whole relationship came under intolerable strain when peasants found themselves hostile to armed men who bore their party's authority. This had often been the case during the 1960s, before the adoption of the strategy of people's war. At this period guerrilla operations usually foundered on the rock of peasant suspicion and peasant readiness to report to the administration. The nationalist parties might regard their men as guerrillas; peasants saw them as mere bandits. In the 1970s, as we have seen, all this changed. But peasant support for the guerrillas could still not be taken for granted and might be withdrawn if the rules of conduct on which it was based were no longer observed.

It was crucially important, for example, that the justice of guerrilla executions remained evident to peasant communities. But guerrilla punishment did not always follow the careful course laid down by Eddison Zvogbo; often it seemed arbitrary, terrifying and incomprehensible. I quote from an account written for me during my research in Makoni district in 1981 which describes events in Weya Tribal Trust Land in north Makoni:

> Just in the month of April 1978 comrades came by evening in the Musvosvo village. Everyone was told to assemble as usual. The comrades started singing their sorrowful songs. As soon as they stopped they called these names: Mrs Agnes Damba, Mrs Nora Damba, Mr Makora, Mr Martin, Miss Helen, the Headman and his young brother. Mrs A. Damba and Mrs N. Damba were accused of being witches, the rest being said to be sellouts. To me it was a miracle. Up till now I haven't believed that those people died. But they died in my eyes. Within seconds these people were lined up and shot on the spot. For the *Povo* (people) nothing was to say, except shedding tears of terror and fear.[36]

In assessing the impact of experiences like this, one has to bear in mind two considerations. One is that the 'Security Forces' were constantly exposing peasants to much more arbitrary and terrible sufferings. My informant's account at once goes on, for example, to describe a raid on Musvosvo village by 'these fearful creatures' of the Rhodesian army, with their blackened faces, who beat many villagers, nailed others to the floor, burnt houses and destroyed property: 'This is the time I really experienced hell. Nothing I could think except arranging ideas of death in my heart. I thought this was Doomsday.' The other was that peasants often excused guerrillas of responsibility for unjust punishments or extortionate demands, blaming these on the young boys and girls (the *mujibhas*) who acted as go-betweens. 'She is convinced,' writes Peter Chakanyuka of Mavengeni Annah, whom he interviewed in Chiduku:

that the atrocities and brutalities committed by the Security Forces alienated them from the African civilians. The guerrillas used to call at her home for food and entertainment but she has no misgivings about their deeds . . . She believes that most of the people who are said to have been killed by the guerrillas are the direct victims of the *mujibhas*. These sometimes robbed civilians, abused the populace at beer parties, and in most cases misrepresented the comrade's aims and commitments.[37]

Mhondiwa Remus Rungodo concurs:

About the prosecution of the war he thinks most people who died during the war were not the direct victims of the guerrillas but of some misguided civilians, especially the *mujibhas*. They either took steps that were not authorised to the extent of beating civilians to death, robbed their fellow Africans or misrepresented some to the guerrillas as sell-outs, or witches and wizards.[38]

In this way peasant elders directed their anger away from the young strangers with guns and towards their own unruly and over-powerful sons and daughters.[39]

The crisis of guerrilla legitimacy

It took a very great deal, then, to make peasant communities remove their support from guerrillas. Moreover, peasants did not merely respond to guerrillas but had in their own right evolved a radical politics of opposition to the regime. I have argued in my recent book that the ideology of the Zimbabwean war was not so much arrived at by means of the guerrillas transforming peasant consciousness as by means of the emergence of a joint peasant/guerrilla programme. The peasants wanted to recover their land and to repudiate government interference and coercion in peasant production; the guerrillas were committed to achieving this programme. Peasants knew that they could only get what they wanted through successful guerrilla war.[40] So even if particular guerrilla bands behaved brutally and in breach of the rules of conduct, peasants continued to back guerrilla war in principle.

This remained true right up to the elections of 1980. Nevertheless, it is important to recognise that in 1978 and 1979 there developed a much more general crisis of legitimacy than that posed by the arbitrary actions of individual guerrilla groups. Complaints became so frequent and widespread that observers both within and outside the guerrilla movements began to fear a wholesale collapse of rural support. The most perceptive reporter on the war was David Caute. In one of his articles for the *New Statesman* Caute described a conversation with an African storekeeper, through whom he made contact with a guerrilla group. 'It was in the first week of May 1979, in the seventh year of the Rhodesian war':

I drove Ellen back to her store. 'The old people, like Samuel and my brother John, don't like these Zanla "pungwes",' she explained. 'They are very dangerous. Sometimes they are held in the middle of the village, quite close to the road, and the "boys" dance to gramophone records and get very drunk on gin or brandy. It makes the village people an easy target for the security forces. There has been too little discipline among the "boys" recently. A year ago there was more fighting and less drinking. And there is one other thing you should know. All of us support the "boys", they are our only hope, but recently their behaviour with the girls has been very bad. They order the parents to bring their best blankets and then they take their daughters . . . sometimes when it rains they turn people out of their houses and take the girls inside

... This upsets the old people very much. It's against all our customs'.[41]

Caute's own conclusion was that 'during the early months of 1979 discipline within Zanla showed a notable deterioration. Local commanders were losing their grip. New units ... were attempting to discipline and sometimes disarm units which had degenerated into semi-banditry'.[42]

The major themes of the crisis of legitimacy are stated here; breakdown of discipline, drink, misconduct with girls. Caute's storekeeper added another: 'It is a pity they listen to so many false rumours, people whisper in their ears, you know, "That man is a sell-out, and that one over there . . .". The "boys" kill these people and afterwards they say sorry, they were misinformed, then they go off and kill the ones who misinformed them'. But why should all these things have come together in 1979?

To begin with, all informants agree that 1978 and in particular 1979 were the hardest years of the war. This was largely because of the deployment on the government side of the Auxiliaries. The Auxiliaries contained some ex-guerrillas and *mujibhas* who had gone over to support either Muzorewa or Ndabaningi Sithole, together with much larger numbers of urban unemployed. They were sent into the rural areas to break peasant support for the guerrillas through the use of counter-terror, and operating out of forts in the Tribal Trust Lands they rounded up hundreds of young girls and boys whom they tortured so as to extract information on guerrilla bases: 'This was the most difficult part of the war,' George Bhongoghozo, a ZANLA intelligence officer told me:

> The Auxiliaries included at least one comrade captured at Chimoio and also many *mujibhas*. They knew all our tactics and where we were. It was very easy for them to attack us. They were very brutal with the masses and in fact it was the masses who suffered by far the heaviest casualties, when the Auxiliaries attacked night-time meetings, for example.[43]

Those who gave regular aid to the guerrillas found themselves in a very dangerous position:

> Many of the Auxiliaries had been *mujibhas* or even comrades and they knew a good deal. They used to point at me and say 'We *know* you are doing it because we were in it ourselves'. Those were our toughest times. They got very cheeky. They were merciless, out to impress their masters: they knew another person killed meant another stripe on their arms.[44]

Faced with this danger, the rural population needed yet more discipline and consideration from the guerrillas they supported; it was perilous enough to meet reasonable demands, let alone unreasonable ones. But the guerrillas were also under intense pressure. One of the priests who worked with them throughout the war remarked that they 'were very far from perfect. The best were under terrific nervous strain. Many were brutal, some merely self-serving. They drank much too much, even though a command came from Mozambique against drinking.'[45] The result was that in this time of tension guerrilla demands on their supporters often became unreasonable. The late Sylvester Mudzimuremba of Rugoyi clinic in Makoni recounted that:

> The comrades never came in person to the clinic because they knew that would probably result in closure and they wanted it open. Nor did they ever call me to *povos* for the same reason . . . The dangerous year was 1979. Turned comrades

in the Auxiliaries used to tell me that they knew I was supplying drugs to the boys and threaten to kill me. 'You deny what we *know* you do'. In general the boys were too good and considerate but just before 8 August 1979 they declared that the anniversary of the Sinoia battle must be celebrated. They told me to buy beer and told the peasants to brew it. The Auxiliaries knew that this order had gone out and were watching. I made an excuse that I needed to get the pay roll and drove to Rusape. I bought 24 beers and hid them under the car seat but an informer watched me do this and I was sure I would be reported. I drove back home furiously and was not stopped at a road block. When I got back I hid the beer under cotton wool in the dispensary. I took to my house only bread I had bought in Rusape. Next day the Security men came to summon me to the fort but I refused to go. So three white and three black men came. The whites went to search my house. The blacks said they knew I had bought beer and hidden it under the car seat. They searched everywhere – except in the dispensary. Then they took my daughter and another girl to the Rugoyi fort and beat them all day but they did not know anything anyway. As soon as they had gone I got a friend with a bicycle to rush the beer into the bush. The girls were released that evening. This was the last day that was so dangerous.[46]

Mudzimuremba survived this experience and he ended his interview with me by commenting on the 1980 elections: 'The whites thought we only feared the boys. They were wrong'. Nevertheless so many other people in Makoni told me of similar dangerous and unnecessary demands made upon them to make it clear that this was a major element in peasant unease in 1979.

There were other factors at work. The increased intensity of the war meant that there were more guerrillas everywhere and this in turn meant that their basic demands put a heavy burden on peasant communities. A storekeeper, who had regularly supplied the guerrillas since their entry into his area in 1976, told me that 1978 and 1979 were very hard years: more and more guerrillas were coming in needing boots and clothes, and messages and demands were coming in from all sides. He went deeper and deeper into debt.[47] It became more difficult for peasants to provide food. At the same time the guerrillas drew in younger and younger girls and boys to assist them. 'At first when the boys were few,' a *mujibha* told me, 'they only took girls of 15 or 16: later when there were very many they took girls of 8 or 9 even to wash and cook and carry messages.'[48]

And of course many guerrillas were themselves younger than the earlier groups had been, as the training and turning-round period in Mozambique became shorter and more perfunctory. A teacher recounted that 'the first comrades were very good and behaved very well. Later they got younger and less disciplined and then they began to drink.'[49]

Finally an element of class war had entered the conflict after the Internal Settlement and the 1979 elections. ZANLA began to denounce the 'petty-bourgeois' elements who supported Muzorewa and Sithole and in many areas this led to guerrilla action against traders and master farmers and other men of relative wealth. Many of the complaints of guerrilla misconduct that began to flow into the Commission for Justice and Peace originated from these alarmed men of substance.[50] Some time in 1978 the Commission received a report on 'Terrorists' Atrocities' which began with the general assertion that 'during 1978 the guerrillas appear to have become much more ruthless and their demands for strong drinks have greatly increased'. The general body of the report listed 'attacks … frequently made on teachers and store keepers'; the harassment of clinic orderlies; denunciations of missionaries and African nuns. 'They preach against private ownership and capitalism but they themselves are real oppressors of the poor'. The report concluded:

> Do the external leaders realise that the people who were originally 100% behind the Liberation Movement have become disillusioned and would gladly return to their old way of life even if they were deprived of land and they had to get rid of their cattle? The oppressive and demanding groups of guerrillas are the greatest enemies of the Liberation Movement.[51]

It is clear that this conclusion underestimated peasant determination and that the report as a whole depended too heavily ᴏɪ the evidence of local elites.

But in the end the reality of the crisis of legitimacy is unavoidable. It is fully set out in another report to the Commission, the report I have quoted earlier in order to establish the ideal picture of legitimate guerri!ᴀ ɪnteraction with the peasantry. By contrast to this early ideal model the report went on to describe a deterioration:

> It is difficult to generalise but usually when the guerrillas were in an area for about six months there was a gradual deterioration in their behaviour. A few groups and these seem to become fewer as the war progresses are always easy to deal with. They endeavoured to protect the people and tɦeir demands were not excessive. They refrain from drinking and treat girls with respect. Unfortunately there are few such groups today. The majority are making life very difficult for all persons in the rural areas.
>
> Most commanders are frequently drunk and though their followers are not supposed to drink these too frequently drink to excess. The people are very worried about this because:
> (a) The guerrillas endanger their own lives.
> (b) They endanger the lives of the people in the area.
> (c) Their behaviour when drunk leaves much to be desired.
> (d) They worry about their daughters.
> (e) They have to provide the drink or the money to buy it when they, themselves, are often hungry . . . Beer, brandy, gin and vodka are in constant demand and it has to be provided at terrible risk to the suppliers . . .
> In most areas young girls have become pregnant . . . In general as they establish control in an area they become more cruel and sadistic. They listen to rumours and reports even from small children and without making any serious effort to find out the truth they beat innocent people to death . . . No respect is paid to elders and long established respected customs are scoffed at . . .
> Local people are asked to attend meetings at great risk to their own lives . . . Because of their desire to show their authority . . . they are defeating their causes in some areas. People are weary of the war and of suffering imposed by so many different groups.[52]

By mid-1979 complaints of this sort were becoming insistent. They reached into Salisbury Central Prison where Maurice Nyagumbo, the man who had recruited many of the guerrillas in 1975, was serving his sentence for having done so. Nyagumbo's visitors from the country brought him news of the crisis of legitimacy. He was seriously disturbed and smuggled out of prison a letter addressed to the ZANU/PF Central Committee, dated June 1979. It was a call for the re-establishment of legitimacy and a move away from a drift into banditry. Nyagumbo asked that they allow for his 'restricted circumstances' which prevented him from verifying the allegations made. 'I have thus been inevitably forced to rely excessively on third parties where the element of exaggeration can be very significant'. But he asked the Central Committee to:

investigate reports of *vakomana* who are said to take liberties with women and married life, be high-handed and arrogant towards the masses. Some are said to have resorted to pleasure-seeking pursuits – beer drinking orgies, open terror and torture of suspects prior to a thorough and impartial investigation . . . Indulgence in gossip, jealousy, grape-vine lines of communication and witch-hunting, have caused, it's reported, the loss of a good many 'innocent' lives. Such practices, if true, cripple and corrode away a principled rational approach to Chimurenga. They undermine cohesion, unity, trust and honesty (vital elements for a successful prosecution of the revolution) and plant apathy, dissension and alienate Chimurenga from the masses which by all accounts is reactionary and counter productive.

By contrast with these reports, Nyagumbo stressed once again the need for 'a sound principled internal underground organization guided and guarded by a strict Chimurenga code of conduct and discipline'. He reminded the Central Committee that:

the masses cherish well constituted married life, peace, justice, law and order, personal discipline . . . and have a potentially inexhaustible sense of fairplay. Chimurenga must of necessity adhere strictly and tenaciously to these values. *Vakomana* should acquire an inexhaustible capacity for patience to enable the masses to grasp consciously the logic of Chimurenga.

Above all, he stressed, 'the boys' operations must be distinct, unique and totally different from those of the enemy forces who are bent on mass murders'.[53]

In the event, ZANU/PF had little time to take action on Nyagumbo's recommendations before they became engrossed in the Lancaster House negotiations. I am told, however, that a plan had been drawn up in the event of these negotiations breaking down by which members of the ZANU/PF Central Committee would themselves move into Zimbabwe and take over direct control of guerrilla activities, thus ensuring a return to revolutionary legitimacy. Meanwhile the cease-fire and the disciplined appearance of the ZANLA guerrillas as they marched to the assembly points brought an end to the reservations which peasants had begun to express. The guerrillas were now given a heroes' reception once again. All my informants who told me of the breakdown of guerrilla discipline in 1978 and 1979 nevertheless went on to assure me that they realised their debt to the guerrillas for the 1980 victory. In the memory of the peasants the guerrillas now occupy a secure place as heroic men rather than as animal bandits.[54]

Banditry after 1980

The ZANU/PF government's theory of what took place in the 1980 elections is that the guerrillas had won for the peasants the chance to confer legitimacy on the state. Now that a popularly elected government rules there is no need for guerrilla or social bandits. Armed men outside the service of the state can be nothing else but bandits. Yet the conditions of the transition from Rhodesia to Zimbabwe favoured outbreaks of banditry. Moorcraft and McLaughlin write of 'a massive upsurge in banditry and crime' in January 1980 as guerrillas moved into the assembly points and the 'Security Forces' into their bases.[55] Guns were being buried by individuals and groups and whole parties. Some men were unready to give up living by the gun; some guerrillas grew disaffected as their months in the assembly areas lengthened. When I carried out field research in Makoni district in early 1981 armed bandits were reported to be holding up buses in two or three parts of the district. There were also 'dissident' ZANLA guerrillas who had taken once again to the bush and who were once again demanding clothes from stores, money from teachers and food from peasants. These men took the stance of social bandits, telling the peasants that the new government would betray its promises to redistribute land and the peasants still needed the armed assistance of fighting men to take for themselves what was due to them. In most areas of Makoni peasants wanted above all to avoid a renewal of the war, but in one particular area the dissidents *were* accepted in a social-bandit role. In this area peasants resented the reappearance of white Catholic missionaries and disliked government pressure upon them to give back to the mission goods and materials looted during the war. In this area too, there were very many members of African independent Apostolic churches and they disliked the pressure being put upon them by ZANU/PF to send their children to school, to accept innoculation, to carry the party card and to honour the symbols of the new state. In this area, finally, land shortage and land hunger was particularly acute. So for a brief period the peasants supported the dissident ZANLA guerrillas; the National Army moved in; and once again young *mujibhas* were interrogated.[56]

In Makoni the eventual redistribution of land by the new government, together with the erection of new schools and clinics, confirmed the legitimacy of the new state. Support for dissidents fell away so that today there are no guerrillas, bandits or social bandits in the district. It is notorious that this is not true for Matabeleland. But it is very much more difficult to tell what *is* true for Matabeleland. Like Maurice Nyagumbo I am 'forced to rely excessively on third parties' in dealing with Matabeleland, and indeed while he was receiving accounts direct from witnesses I am restricted to what one can learn from conversations in Harare and from reports of trials in the Zimbabwe press. I write about Matabeleland very tentatively, then, but it is necessary to advance *some* propositions about 'dissident' activity there in the context of a chapter such as this.

It is perhaps useful to apply to 'dissident' activity in Matabeleland over the last three years the criteria for legitimacy and illegitimacy I employed in discussing the relations of peasants and armed men during the guerrilla war. By these criteria it is difficult to describe the Matabeleland armed men as guerrillas since they are certainly not claimed or publicly legitimated by any party or liberation movement. Ex-ZIPRA men to whom I talked in Harare in April 1984 denied that there were any 'dissidents' in Matabeleland at all and claimed that instead there were armed criminals, government agents pretending to be dissidents, and agents infiltrated from South Africa. Joshua Nkomo in his recent autobiography refers to 'the armed bandits operating in the western province of Matabeleland' and has repudiated any link between them and ZAPU. 'The government seemed intent on creating rebellion and suggesting that I was at the head of it,' writes Nkomo:

> [It] spoke of a problem of 'dissidence'. – a problem that it did not define, and that was hard to understand. That there were gangsters at work, especially in Matabeleland, was clear. There were many unexplained deaths, robberies and beatings. A significant number of white farmers in the open ranch-land of Matabeleland were killed without explanation. By lumping all these deplorable incidents together as 'dissidence', the government implied that they were the result of an organised and politically motivated movement.

Nkomo asserts that the government has failed to prove this case and that trials of 'dissidents' have only documented armed crime.[57]

Nkomo has a clear political purpose in emphasising banditry and repudiating any idea that ZAPU guerrillas are at work. It enables him to charge that 'this was a deliberate and co-ordinated campaign to create insecurity in the Matabeleland countryside . . . A climate was being developed which would be used to justify full-scale repression'. But whatever may in the end become known about the links between the activities of armed men in Matabaleland and elements of ZAPU, the constant and widely publicised condemnations of banditry by Nkomo and other ZAPU leaders cannot but have the effect of depriving these armed men of one crucial element in guerrilla legitimacy. And while, during the war of the 1970s, the support given by Mozambique to ZANLA increased in the minds of peasants the external guarantees of guerrilla legitimacy, in the case of Matabeleland today external support for the armed men merely points up further their illegitimacy. For instance a report in the *Herald* of 7 May 1984, asserted that 'South-African trained dissidents and a band of local bandits clashed last week near Beitbridge, exchanging fire and calling each other traitors'. The *Herald* concluded that 'dissidents had been split into two hostile factions: the Zapu dissidents and South-African trained Super-Zapu dissidents'.[58]

So there are no guerrillas in Matabeleland in the sense of armed men given joint legitimacy by being publicly owned by a political movement and also accepted by the peasantry. On the other hand, it is clear that there are many bandits, in the sense of armed men repudiated both by all political movements and by the peasantry. To take one instance, the *Herald* on 29 June 1984 reported the arrest of one Nation Sengwa Dube. Dube was said to have been 'operating in his home area, Kafusi, about 70 km. south of Gwanda' and to have been 'known in the area and most of the Gwanda area as Nation'. Dube admitted that 'he had raped many women in the area and had robbed some stores. He claimed that his actions were not aimed at furthering any political object, but just to obtain some money'. Nation was 'captured by locals'.[59] Another, less extreme case, is nevertheless illuminating about the way in which the 'dissident' role can be adopted for bandit purposes. The *Herald* describes:

> a man who posed as a dissident in the Mphoengs area, Plumtree, and was apprehended by villagers . . . Robert Moyo (24) went in search of his cattle and while in the bush, he decided to go to Cde Ndeni Mpala's home. He introduced himself as a dissident and demanded food. He claimed that there were 11 other men in his company and that they were in the bush. As a proof that he was a dissident, he produced two spent cartridges from his pockets and said he had used two bullets to kill two people.

Moyo claimed to have come from Kezi district – a significant claim, as we shall see. But Mpala was not impressed. He alerted other villagers, who apprehended Moyo.[60]

In terms of my definitions, then, there are no guerrillas in Matabeleland and there are plenty of bandits. The key question, though, is whether there are any social bandits, that is to say armed men illegitimate in the eyes of the state, renounced by other political parties, but nevertheless enjoying legitimacy in the eyes of the peasantry. I am able only to draw upon official sources and upon the Zimbabwean press to answer this question. But the evidence suggests that one *can* talk about social bandits in parts at least of Matabeleland. Let us take Matobo/Kezi district. This district suffered particularly acutely from land alienation under colonialism; the rugged terrain together with a keen sense of peasant grievance made it a stronghold for ZIPRA guerrillas during the war. It is clear that peasants in Matobo/Kezi have bestowed legitimacy on the armed men operating against the government in their district in the past two or three years. For example on 20 April 1984, the *Herald* carried two reports on Independence Day celebrations in Matobo/Kezi which were attended by deputy minister of mines Sanyangare. Brigadier Edzai Chanyuka, acting commander of the Fifth Brigade, told Sanyangare that 'we have a problem in the Matopo . . . We have noticed numerous movements of dissident elements in the area. But when we approach the people they tell us they have never seen a dissident'. The local ZANU/PF chairman, Comrade Njanji, reported that 'the people in Kezi were highly politicised, and as a result he was finding it difficult to convince them to join Zapu (PF).'[61] Supported in this way by the local peasantry, 'dissidents' in Matobo/Kezi have attacked white farmers and ranchers in the district. On 8 June 1984 the Chairman of the Commercial Cattle Producers Association, Mike Huckle, announced that 'there has been a virtual abandonment of ranching in Kezi'.[62] On 14 June 1984 it was announced that the government had acquired 270 000 hectares of ranch-land which it intended to run as a state ranch. 'The scheme started as a rescue operation in a province where dissident agitation, sometimes with open support from peasants particularly in Kezi, resulted in a number of commercial farmers leaving'.[63]

It is plain that the attack on white ranching in Matobo/Kezi has enjoyed peasant support. It is also plain that peasants themselves have taken their own action against white-owned cattle. On 2 May 1984 the *Herald* reported that Security Force reports showed that 'cattle rustling, snaring and fence cutting were rife . . . Several commercial farmers . . . were complaining bitterly of the occurrences of cattle deaths trapped by wire snares set up by villagers'.[64] Whether or not the armed men have proclaimed wider political objectives, they seem to be supported as social bandits expressing the local grievances and ambitions of the peasantry. Peasants from Matobo/Kezi regularly appear before the courts charged with failing to report the presence of dissidents; or with concealing arms and ammunition on their behalf; or with succouring them when wounded.[65]

In late June 1984 Callistus Ndhlovu spoke at a mass rally at Silobini in Matobo. Ndhlovu had recently resigned from ZAPU and applied for membership of ZANU/PF. He emphasised the illegitimacy of the 'dissidents' and the legitimacy of the government:

> Liars from Zapu . . . deceive the peasant in Matabeleland and divide you on tribal lines . . . He said many Zapu members told people that the government had not yet come because Zapu had lost the election. 'These people discourage people from carrying out development projects, saying you will be co-operating with an unwanted government . . . You are your own liberators and the sooner you realise this the better. Those who continue telling you that another form of independence of Ndebele people is coming are hoodwinking you because the reality is that Mugabe is in power because he was chosen by the people'.

At the same rally the Governor of Matabeleland South, Mark Dhube, told the peasants that they were falling behind in development: 'There has been no fully completed school in this province because of petty political division. But the truth is those who really suffer as a result of this are the people in those areas and not Government as you think.'[66] It remains to be seen whether these arguments will persuade the peasants of Matobo/Kezi to withdraw legitimacy from the armed men or whether the history of social banditry in Zimbabwe has still some way to go.

NOTES

1 *The Herald*, Harare, 28 June 1984.
2 'The Voice of the Mozambique National Resistance', Lisbon, 5 March 1984, monitored in *BBC Summary of World Broadcasts*, 7 March 1984.
3 *The Times* of 20 February 1984, under the heading 'Security takes priority in Botha peace mission to Mozambique' writes of the 'guerrillas' of the MNR.
4 *The Herald*, 25 February and 12 June 1984.
5 Allen Isaacman, 'Social banditry in Zimbabwe (Rhodesia) and Mozambique, 1894–1907: an expression of early peasant protest', *Journal of Southern African Studies*, vol. 4, no. 1 (1977), pp. 3, 30.
6 David Lan, *Guns and Rain. Guerrillas and Spirit Mediums in Zimbabwe* (London: James Currey, 1985), p. 146.
7 Isaacman, 'Social banditry', pp. 12, 14, 22, 23.
8 Allen Isaacman, *The Tradition of Resistance in Mozambique. Anti-Colonial Activity in the Zambesi Valley, 1850–1921* (London: Heinemann, 1976).
9 Isaacman, 'Social banditry', p. 24.
10 Michael Raeburn, *Black Fire! Accounts of the Guerrilla War in Rhodesia* (London: Friedmann, 1978).
11 Stanley Trapido, 'Poachers, proletarians and gentry in the early twentieth century Transvaal', seminar paper, Institute of Commonwealth Studies, University of London, 18 October 1983, pp. 19–24. For background on this region see also Patrick Harries, 'A forgotten corner of the Transvaal: reconstructing the history of a relocated community through oral testimony and song', University of the Witwatersrand History Workshop, 1984.
12 Allan Wright, *Valley of the Ironwoods. A Personal Record of Ten Years served as District Commissioner in Rhodesia's Largest Administrative Area, Nuanetsi, in the South-Eastern Lowveld* (Cape Town: Bulpin, 1972), pp. 142, 143, 150, 202, 289–93, 363–72.
13 Jesuit Archives, Mount Pleasant, Harare, Box 195, Jerome O' Hea to Monsignor Robert Brown, 11 September 1930. See Terence Ranger, 'Guerrilla war and peasant violence: Makoni District, Zimbabwe', *Political Violence*, Institute of Commonwealth Studies, University of London, Collected Seminar Papers, No. 30, 1982, pp. 100–23.
14 Lan, *Guns and Rain*, p. 152.
15 Wright, *Valley of the Ironwoods*, p. 7.
16 S. Nyamfukudza, *The Non-Believers Journey* (London: Heinemann, 1980), p. 97.
17 Interview with Isaac Tsungo, Mbobo, Makoni, 14 February 1981.
18 Lan, *Guns and Rain*, p. 188.
19 Julie Frederikse, *None But Ourselves. Masses vs Media in the Making of Zimbabwe* (Johannesburg: Ravan, 1982), p. 56.
20 Interview with Lameck Madziwendira, 'Comrade Mutsawareyi', Makoni Farm, 16 February 1981.
21 Lan, *Guns and Rain*, pp. 163 and 171.
22 Terence Ranger, *Peasant Consciousness and Guerrilla War in Zimbabwe: A Comparative Study* (London: James Currey, 1985); David Lan, 'Making history. Spirit mediums and the guerrilla war in the Dande area of Zimbabwe' (unpublished PhD, London School of Economics, 1983); *idem*, *Guns and Rain*.
23 *Zimbabwe News*, September/October 1978.
24 This translation of the song is given in Frederikse, *None But Ourselves*, p. 212. A less radical translation is given in Paul Moorcraft and Peter McLaughlin, *Chimurenga! The War in Rhodesia, 1965–1980* (Marshalltown: Sygma Books, 1982), p. 130.
25 Quoted in Moorcraft and McLaughlin, *Chimurenga!*, p. 87.
26 Frederikse, *None But Ourselves*, p. 216.
27 Moorcraft and McLaughlin, *Chimurenga!*, p. 130.
28 Commission for Justice and Peace Archives, Harare (hereafter CJP), file 'Guerrilla Reports', 'The Testimony of Mrs O. M', 1 March 1977.

29   CJP, file 'Guerrilla Reports, 'Zanla Forces'.
30   Ron Reid Daly and Peter Stiff, *Selous Scouts. Top Secret War* (Alberton: Galago, 1982), p. 21.
31   Interview with Father Vernon, St Killian's, Makoni, 15 February 1981. One of Vernon's catechists was killed and his body vanished. For two years it was thought that he had been executed by guerrillas; then it was discovered that the killing had been the work of a 'private grudge' gang. Vernon approached the guerrillas who co-operated in tracing the body and ensuring a proper funeral.
32   David Caute, *Under The Skin. The Death of White Rhodesia* (London: Allen Lane, 1983), p. 127.
33   Daly and Stiff, *Selous Scouts*, pp. 406–8.
34   ibid., p. 152.
35   Interview with Mhondiwa Remus Rungodo, recorded by Peter Chakanyuka, Dewedzo, Makoni, 30 January 1981.
36   'Five scenes from Chiendambuya – East Mupururu', 1981.
37   Interview with Mavengeni Annah, recorded by Peter Chakanyuka, Dewedzo, Makoni, 8 February 1981.
38   Interview with Mhondiwa Remus Rungodo, recorded by Peter Chakanyuka, Dewedzo, Makoni, 30 January 1981.
39   On 1 February 1981 Peter Chakanyuka interviewed Mandironda Beatrice in Dewedzo. He recorded:

> She observed punishments administered to the *mujibhas* and *chimbwidos* who were staying away from home with their boyfriends, committing sexual intercourse, coming to parents asking for fowls saying they were sent by guerrillas yet not. When the guerrillas learnt about this all the youths involved and the parents of the villages supporting the guerrillas were gathered together one night. The case was made known to the parents/villagers. The guerrillas told parents plainly that they treat people who misbehave in a war situation by beating. It was a sad moment for most parents but they felt pleased about this correction. Beatrice's daughter was among the *chimbwidos* who were punished this day by beating.

Strictly speaking *mujibhas* are male assistants to the guerrillas, and *chimbwidos* are female and this is the sense in which the words are used in this quotation. Very often, however, the word *mujibha* is used irrespective of sex and this is my practice in the text of this chapter.

40   Ranger, *Peasant Consciousness and Guerrilla War*.
41   Caute, *Under The Skin*, p. 18. *Pungwe* means a night-time political education meeting.
42   ibid., p. 314.
43   Interview with George Bhongoghozo, Rusape, 16 February 1981.
44   Interview with Father Vernon, St Killian's, Makoni, 15 February 1981.
45   Interview with Father Michael Kenny, St Barbara's, 26 February 1981.
46   Interview with Sylvester Zwidzidzayi Mudzimuremba, Rugoyi, Makoni, 22 March 1981.
47   Interview with Isaac Tsungo, Mbobo, Makoni, 14 February 1981.
48   Interview with Happiness Chiwandamira, Harare, 19 August 1982.
49   Interview with Malachias Chapurendima, Toriro, Makoni, 24 March 1981.
50   For discussion of this 'class war' element in 1978/9 and quotations of desperate protests by storekeepers and native purchase farmers see Ranger, *Peasant Consciousness and Guerrilla War*, Ch. 6.
51   CJP, file 'Guerrilla Reports', 'Terrorists' Atrocities'.
52   CJP, file 'Guerrilla Reports', 'Zanla Forces'.
53   Maurice Nyagumbo to Robert Mugabe and the Central Committee, June 1979, copy in author's possession.
54   Early in 1984 I met in London a Zimbabwean exile who had at varying times been a member of both ZIPRA and ZANLA but who had become convinced of the inadequacies of the ideology of both parties. In his view no changes for the fundamental benefit of the people have taken place since independence. He told me: 'Despite

everything that the people suffered in the war to support the guerrillas they have gained nothing. This means that we must regard even the best of the guerrillas as having been nothing but bandits'. This is a highly idiosyncratic view however.

55 Moorcraft and McLaughlin, *Chimurenga!*, p. 231.

56 The area to which I refer was the central belt of Chiduku communal land around St Theresa's mission. My account of the reason for peasant support of the dissidents is derived particularly from an interview with an ex-guerrilla, now teaching in the area, Comrade Believe.

57 Joshua Nkomo, *The Story of My Life* (London: Methuen, 1984), pp. 1 and 230.

58 *Herald*, 7 May 1984, 'Rival bandit gangs shoot it out near Beit-bridge'.

59 *Herald*, 29 June 1984, 'Dissident admits rape and robbery'.

60 *Herald*, 17 May 1984, 'Jail for bogus bandit caught by villagers'.

61 *Herald*, 20 April 1984.

62 *Herald*, 8 June 1984.

63 *Herald*, 14 June 1984, 'Ranch shot in arm for Matabele folk'.

64 *Herald*, 2 May 1984.

65 *Herald*, 23 May 1984.

66 *Herald*, 28 June 1984, 'Peace is in your hands'.

INDEX